Til Kjell,

Med vennlig hilsen fra

Amljit

System Reliability Theory

WILEY SERIES IN PROBABILITY AND MATHEMATICAL STATISTICS

Established by WALTER A. SHEWHART and SAMUEL S. WILKS

Editors: *Vic Barnett, Ralph A. Bradley, Nicholas I. Fisher, J. Stuart Hunter, J. B. Kadane, David G. Kendall, David W. Scott, Adrian F. M. Smith, Stephen M. Stigler, Jozef L. Teugels, Geoffrey S. Watson*

A complete list of the titles in this series appears at the end of this volume

System Reliability Theory

Models and Statistical Methods

ARNLJOT HØYLAND and MARVIN RAUSAND
The Norwegian Institute of Technology

A Wiley-Interscience Publication
JOHN WILEY & SONS, INC.
New York • Chichester • Brisbane • Toronto • Singapore

This text is printed on acid-free paper.

Copyright © 1994 by John Wiley & Sons, Inc.

All rights reserved. Published simultaneously in Canada.

Reproduction or translation of any part of this work beyond that permitted by Section 107 or 108 of the 1976 United States Copyright Act without the permission of the copyright owner is unlawful. Requests for permission or further information should be addressed to the Permissions Department, John Wiley & Sons, Inc., 605 Third Avenue, New York, NY 10158-0012.

Library of Congress Cataloging in Publication Data:
Høyland, Arnljot, 1924–
    System reliability theory: models and statistical methods/ Arnljot Høyland and Marvin Rausand.
        p.   cm — (Wiley series in probability and mathematical statistics. Applied probability and statistics)
    "A Wiley-Interscience publication."
    Includes bibliographical references and index.
    ISBN 0-471-59397-4
    1. Reliability (Engineering)—Statistical methods.   I. Rausand, Marvin.   II. Title.   III. Series.
TA169.H68   1994
620′.00452—dc20   94-10362

Printed in the United States of America

10 9 8 7 6 5 4 3 2 1

Contents

Preface   ix

1. Introduction   1

    1.1   A Brief History, 1
    1.2   Scope of the Text, 2
    1.3   Basic Concepts, 3
    1.4   Reliability Management, 6
    1.5   Application Areas, 7
    1.6   Failures and Failure Classification, 10
    1.7   Model Construction, 12
    1.8   Standards and Directives, 16

2. Failure Models   18

    2.1   Time to Failure, 18
    2.2   Reliability Function, 20
    2.3   Failure Rate, 20
    2.4   Mean Time to Failure, 24
    2.5   The Poisson Process, 26
    2.6   The Exponential Distribution, 31
    2.7   The Gamma Distribution, 33
    2.8   The Pareto Distribution, 35
    2.9   The Weibull Distribution, 37
    2.10   The Normal Distribution, 40
    2.11   The Lognormal Distribution, 42
    2.12   The Birnbaum-Saunders Distribution, 46
    2.13   The Inverse Gaussian Distribution, 48
    2.14   The Extreme Value Distributions, 53

2.15 Stressor-Dependent Modeling, 59
2.16 Some Families of Distributions, 60
2.17 Summary of Failure Models, 65
2.18 Problems, 65

3. Qualitative System Analysis — 73

3.1 Introduction, 73
3.2 FMEA/FMECA, 73
3.3 Fault Tree Analysis, 81
3.4 Reliability Block Diagrams, 93
3.5 System Structure Analysis, 101
3.6 Problems, 118

4. Systems of Independent Components — 127

4.1 Introduction, 127
4.2 System Reliability, 128
4.3 Nonrepairable Systems, 132
4.4 Quantitative Fault Tree Analysis, 138
4.5 Exact System Reliability, 145
4.6 Redundancy, 153
4.7 Repairable Systems, 159
4.8 Problems, 185

5. Component Importance — 195

5.1 Birnbaum's Measure, 195
5.2 Criticality Importance, 200
5.3 Vesely-Fussell's Measure, 203
5.4 Improvement Potential, 208
5.5 A Brief Comparison, 210
5.6 Problems, 211

6. Markov Models — 214

6.1 Basic Concepts, 214
6.2 Markov Processes, 216
6.3 State Equations, 218
6.4 Time-Dependent Solution, 220
6.5 Asymptotic Solution, 224
6.6 Mean Time to Failure, 239

CONTENTS vii

    6.7 Standby Systems, 251
    6.8 Problems, 259

7. Counting Processes   263

    7.1 Introduction, 263
    7.2 Homogeneous Poisson Processes, 271
    7.3 Renewal Processes, 275
    7.4 Nonhomogeneous Poisson Processes, 314
    7.5 Problems, 322

8. Dependent Failures   325

    8.1 Introduction, 325
    8.2 How to Obtain Reliable Systems, 328
    8.3 Modeling of Dependent Failures, 330
    8.4 Special Models, 330
    8.5 Associated Variables, 340
    8.6 Problems, 352

9. Life Data Analysis   355

    9.1 Introduction, 355
    9.2 Complete Data Sets, 356
    9.3 Censored Data Sets, 385
    9.4 Other Applications, 411
    9.5 Problems, 414

10. Accelerated Life Testing   419

    10.1 Introduction, 419
    10.2 Experimental Designs for ALT, 420
    10.3 Parametric Models Used in SALT, 422
    10.4 Nonparametric Models Used in ALT, 431
    10.5 Problems, 434

11. Bayesian Reliability Analysis   435

    11.1 Introduction, 435
    11.2 Basic Concepts, 436
    11.3 Bayesian Point Estimation, 440
    11.4 Credibility Intervals, 443
    11.5 Choice of Prior Distribution, 443
    11.6 Bayesian Life Test Sampling Plans, 450

11.7 Interpretation of the Prior Distribution, 454
11.8 The Predictive Density, 455
11.9 Problems, 457

12. Reliability Data Sources — 460

12.1 Data Books and Data Banks, 460
12.2 Data Quality, 469

APPENDIXES — 471

Appendix A. The Gamma and Beta Functions — 471

Appendix B. Distribution Theorems — 474

Appendix C. Maximum Likelihood Estimation — 477

Appendix D. Laplace Transforms — 482

Appendix E. Basic Concepts — 485

Appendix F. Statistical Tables — 492

References — 495

Author Index — 507

Subject Index — 511

Preface

The main purpose of this book is to present a comprehensive introduction to system reliability theory. We have structured our presentation such that the book may be used as a text in introductory as well as graduate level courses. For this reason we treat simple situations first. Then we proceed to more complicated situations requiring advanced analytical tools.

At the same time the book has been developed as a reference and handbook for industrial statisticians and reliability engineers.

The reader ought to have some knowledge of calculus and of elementary probability theory and statistics.

In the first five chapters we confine ourselves to situations where the state variables of components and systems are binary and independent. Failure models, qualitative and quantitative system analysis and reliability importance are discussed. These chapters constitute a fairly elementary, though comprehensive, introduction to system reliability theory. They may be covered in a one-semester course with three weekly lectures over fourteen weeks.

The remaining part of the book is more advanced and can be used as a text for a graduate course. In Chapter 6 situations where the components and systems may be in two or more states are discussed. This situation is modeled by Markov processes. Renewal theory is treated in Chapter 7, and dependent failures in Chapter 8. A rather broad introduction to life data analysis is given in Chapter 9, accelerated life testing in Chapter 10, and Bayesian reliability analysis in Chapter 11. The book concludes with information about reliability data sources in Chapter 12.

The book contains a large number of worked examples, and each chapter ends with a selection of problems, providing exercises and additional applications.

A forerunner of this book, written in Norwegian by professor Arne T. Holen and the present authors, appeared in 1983 as an elementary introduction to reliability analysis. It was published by TAPIR and reprinted in 1988. However, we have rewritten all the chapters of the earlier book and added new material as well as several new chapters. The present book contains approximately twice as many pages as its forerunner and can be considered as a completely new book.

We have already tried much of the material in the present book in courses on reliability and risk analysis at the university level in Norway and Sweden, including continuing education courses for engineers working in industry. The feedback from participants in these courses has significantly improved the quality of the book.

We are grateful to Bjarne Stolpnessæter for drawing many of the figures and to Anne Kajander for typing a first draft of the manuscript. We are further grateful for economic support by Conoco Norway. Permission from various publishers to reproduce tables and figures is also appreciated.

ARNLJOT HØYLAND
MARVIN RAUSAND

University of Trondheim
The Norwegian Institute of Technology
1993

CHAPTER 1

Introduction

1.1 A BRIEF HISTORY

Reliability, as a human attribute, has been praised for a very long time. For technical systems, however, the reliability concept has not been applied for more than some 50 years. It emerged with a technological meaning just after World War I; it was then used in connection with comparing operational safety of one-, two-, and four-engine airplanes. The reliability was measured as the number of accidents per hour of flight time.

At the beginning of the 1930s, Walter Shewhart, Harold F. Dodge, and Harry G. Romig laid down the theoretical basis for utilizing statistical methods in quality control of industrial products. Such methods were, however, not brought into use to any great extent until the beginning of World War II. Products that were composed of a large number of parts, often did not function, despite the fact that they were made up of individual high-quality components.

During World War II a group in Germany was working under Wernher von Braun developing the V-1 missile. After the war it was reported that the first ten V-1 missiles were all fiascos. Despite attempts to provide high-quality parts and careful attention to details, all of the first missiles either exploded on the launching pad, or landed "too soon" (in the English Channel). Robert Lusser, a mathematician, was called in as a consultant. His task was to analyze the missile system, and he quickly derived the *product probability law of series components*. This theorem concerns systems functioning only if all the components are functioning and is valid under special assumptions. It says that the reliability of such a system is equal to the product of the reliabilities of the individual components that make up the system. If the system comprises a large number of components, the system reliability may be rather low, even though the individual components have high reliabilities.

In the United States attempts were made to compensate a low system reliability by improving the quality of the individual components. Better raw materials and better designs for the products were demanded. A higher system reliability was obtained, but extensive systematic analysis of the problem was probably not carried out at that time.

After World War II the development continued throughout the world as increasingly more complicated products were produced, composed of an ever-increasing number of components (television sets, electronic computers, etc.). With automation, the need for complicated control and safety systems also became steadily more pressing.

Toward the end of the 1950s and the beginning of the 1960s, interest in the United States was concentrated on intercontinental ballistic missiles and space research, especially connected to the Mercury and Gemini programs. In the race with the Russians to be the first nation to put a man on the moon, it was very important that the launching of a manned spacecraft was a success. An association for engineers working with reliability questions was soon established. The first journal on the subject, *IEEE—Transactions on Reliability*, came out in 1963, and a number of textbooks on the subject were published in the 1960s.

In the 1970s interest increased, in the United States as well as in other parts of the world, in risk and safety aspects connected to building and operation of nuclear power plants. In the United States a large research commission, led by Norman Rasmussen was set up to analyze the problem. The multimillion dollar project resulted in the so-called Rasmussen report, WASH-1400. Despite its weaknesses this report represents the first serious safety analysis of so complicated a system as a nuclear power plant.

Similar work has also been carried out in Europe and Asia. In the majority of industries a lot of effort is presently put on the analysis of risk and reliability problems. The same is true in Norway, particularly within the offshore oil industry. The offshore oil and gas development in the North Sea is presently progressing into deeper and more hostile waters, and an increasing number of remotely operated subsea production systems are put into operation. The importance of the reliability of subsea systems is in many respects parallel to the reliability of spacecrafts. A low reliability cannot be compensated by extensive maintenance.

A more detailed history of reliability technology is presented, for example, by Knight (1991) and Villemeur (1992).

1.2 SCOPE OF THE TEXT

We can distinguish between four main branches of reliability:

- Component and system reliability
- Structural reliability
- Human reliability
- Software reliability

The present textbook is concerned with the first of these branches: component and system reliability. The main objectives of this book are:

1. To present and discuss the terminology and the main models used in reliability studies.
2. To present the analytical methods that are fundamental within reliability engineering and analysis of reliability data.

The methods described in the book are applicable during any phase of a system's lifetime. They have, however, their greatest value during the design phase. During this phase reliability engineering can have the greatest effect for enhancing the system's safety, quality, and operations regularity.

Some of the methods described in the book may also be applied during the operational phase of the system. During this phase the methods will aid in the evaluation of the system and in improving the maintenance and the operating procedures.

Management aspects of reliability are very briefly discussed in Section 1.4. The book does not specifically deal with *how* to build a reliable system. The main topics of the book are connected to how to evaluate, measure, and predict the reliability of a system.

1.3 BASIC CONCEPTS

The main concept of this book is *reliability*. During the preceding sections the concept of reliability has been used without a precise definition. It is, however, very important that all main concepts are defined in an unambiguous way. We fully agree with Kaplan (1990) who states: "When the words are used sloppily, concepts become fuzzy, thinking is muddled, communication is ambiguous, and decisions and actions are suboptimal, to say the least."

A precise definition of reliability, and some associated concepts like quality, availability, safety, security, and dependability, is given below. All of these concepts are more or less interconnected, and there is a considerable controversy concerning which is the broadest and most general concept.

Until the 1960s reliability was defined as the probability that an item will perform a required function under stated conditions for a stated period of time. Some authors still prefer this definition, for example, Smith (1988) and Lakner and Anderson (1985). We will, however, in this book use the more general definition of reliability given in standards like ISO 8402 and British Standard BS 4778:

Reliability
The ability of an item to perform a required function, under given environmental and operational conditions and for a stated period of time.

- The term "item" is used here to denote any component, subsystem, or system that can be considered as an entity.
- A "required function" may be a single function or a combination of functions that is necessary to provide a specified service.
- All technical items (components, subsystems, systems) are designed to perform one or more (required) functions. Some of these functions are active and some functions are passive. Containment of fluid in a pipeline is an example of a passive function. Complex systems (e.g., an automobile) usually have a wide range of required functions. To assess the reliability (e.g., of an automobile), we must first specify the required function(s) that we are considering.
- For a hardware item to be reliable, it must do more than meet an initial factory performance or quality specification—it must operate satisfactorily for a specified period of time in the actual application for which it is intended.

Quality
The totality of features and characteristics of a product or service that bear on its ability to satisfy stated or implied needs (ISO 8402).

- "Quality" is also sometimes defined as conformance to specifications (e.g., see Smith 1988).
- The quality of a product is characterized not only by its conformity to specifications at the time it is supplied to the user, but also by its ability to meet these specifications over its entire lifetime.

However, according to common usage, quality denotes the conformity of the product to its specification as manufactured, while reliability denotes its ability to continue to comply with its specification over its useful life. *Reliability is therefore an extension of quality into the time domain.*

Availability
The ability of an item (under combined aspects of its reliability, maintainability, and maintenance support) to perform its required function at a stated instant of time or over a stated period of time (BS 4778).

- We may distinguish between the availability $A(t)$ at time t and the average availability A_{av}. The availability at time t is

    $$A(t) = P(\text{item is functioning at time } t)$$

    where $P(A)$ denotes the probability of event A. The term "functioning" means here that the item is either in active operation or is able to operate if required.

BASIC CONCEPTS

The average availability A_{av} denotes the mean proportion of time the item is functioning. If we have an item that is repaired to an "as good as new" condition every time it fails, the average availability is

$$A_{av} = \frac{\text{MTTF}}{\text{MTTF} + \text{MTTR}}$$

where MTTF (mean time to failure) denotes the mean functioning time of the item and MTTR (mean time to repair) denotes the mean downtime or repair time after a failure.

- When considering a production system, the average availability of the production (i.e., the mean proportion of time the system is producing), is sometimes called the *production regularity*.

Maintainability
The ability of an item, under stated conditions of use, to be retained in, or restored to, a state in which it can perform its required functions, when maintenance is performed under stated conditions and using prescribed procedures and resources (BS 4778).

- "Maintainability" is a main factor determining the availability of the item.
- RAM is often used as an acronym for reliability, availability, and maintainability. We also use the notions RAM studies and RAM engineering.

Safety
Freedom from those conditions that can cause death, injury, occupational illness, or damage to or loss of equipment or property (MIL-STD-882).

- This definition has caused considerable controversy. A number of alternative definitions have therefore been proposed. The main controversy concerns the use of the term "freedom from." Most activities involve some sort of risk and are never totally *free* from risk. In most of the alternative definitions safety is defined as an *acceptable level of risk*.

Security
Dependability with respect to the prevention of unauthorized access and/or handling of information (Laprie 1992).

Dependability
The collective term used to describe the availability performance and its influencing factors: reliability performance, maintainability performance, and maintenance support performance (IEC 300-1).

- A slightly different definition is given by Laprie (1992). He defines dependability as: "Trustworthiness of a system such that reliance can justifiably be placed on the service it delivers." In his comments on this definition, Laprie (1992) claims that dependability is a global concept that subsumes the attributes of reliability, availability, safety, and security. This is also in accordance with the definition used by Villemeur (1992).
- If safety and security are included in the definition of dependability as influencing factors, dependability will be identical to the RAMS concept (RAMS is an acronym for reliability, availability, maintainability, and safety).
- According to Laprie (1992) the definition of dependability is synonymous to the definition of reliability. Some authors, however, prefer to use the concept of dependability instead of reliability. This is also reflected in the important series of standards IEC 300 "Dependability Management."

In this book we will, however, use reliability as a global, or general, concept with the same main attributes as listed under the definition of dependability.

The reliability of an item may be measured in different ways depending on the circumstances. For example,

1. Mean time to failure (MTTF)
2. Number of failures per time unit (*failure rate*)
3. The probability that the item does not fail in a time interval $(0, t]$ (*survival probability*)
4. The probability that the item is able to function at time t (*availability at time t*)

If the item is not repaired after failure, measures 3 and 4 coincide. All of these measures are given a mathematically precise definition in Chapter 2 with concepts from probability theory.

1.4 RELIABILITY MANAGEMENT

Procedures for a total and integrated reliability program approach in the design and development phase, the manufacturing and installation phase, the operation and maintenance phase, and the disposal phase of a system's life cycle are presented, for example, by Lakner and Anderson (1985), Guthrie et al. (1990), Strandberg (1992), Ireson and Coombs (1988), O'Connors (1991), Smith (1988), and more specifically in IEC 300, MIL-STD 785 and British Standard BS 5760.

The following definitions are adapted from IEC 300-2 (where "dependability" is replaced with "reliability"):

Reliability Program
The organizational structure, responsibilities, procedures, processes and resources used for managing reliability.

Reliability Program Tasks
A set of activities addressing reliability aspects of an item or its support in producing output of a certain kind.

Reliability Program Elements
A reliability program task, or set of tasks, usually performed by some kind of subject area experts or personnel category.

The various reliability program elements and tasks are described in detail in the references listed in the beginning of this section. Figure 1.1 presents a diagram illustrating in which phases of the item's life cycle the various tasks should be carried out. The figure is copied from BS 5760.

1.5 APPLICATION AREAS

The main objective of a reliability study should always be to provide information as a basis for decisions. Before a reliability study is initiated, the decision maker should clarify the decision problem and then specify the objectives, boundary conditions, and limitations for the study such that relevant information is at hand.

Reliability technology has a potentially wide range of application areas. Some of these areas are listed below to illustrate the wide scope of application of reliability technology.

1. *Safety/risk analyses.* Reliability analysis is a well-established part of most risk and safety studies. The causal part of a risk analysis is normally accomplished by reliability techniques like failure mode and effects analysis (FMEA) and fault tree analysis.

    Reliability techniques are often applied in risk analysis to evaluate the availability and applicability of safety systems, ranging from single component safety barriers (e.g., valves) up to complex computer-based process safety systems.

2. *Environmental protection.* Reliability studies may be used to improve the design and operational regularity of antipollution systems like gas/water cleaning systems.

    Many industries have realized that the majority of the pollution from their plants is caused by production irregularities and that consequently the production regularity of the plant is the most important factor in reducing pollution. Reliability and regularity studies are among the most important tools used to optimize production regularity.

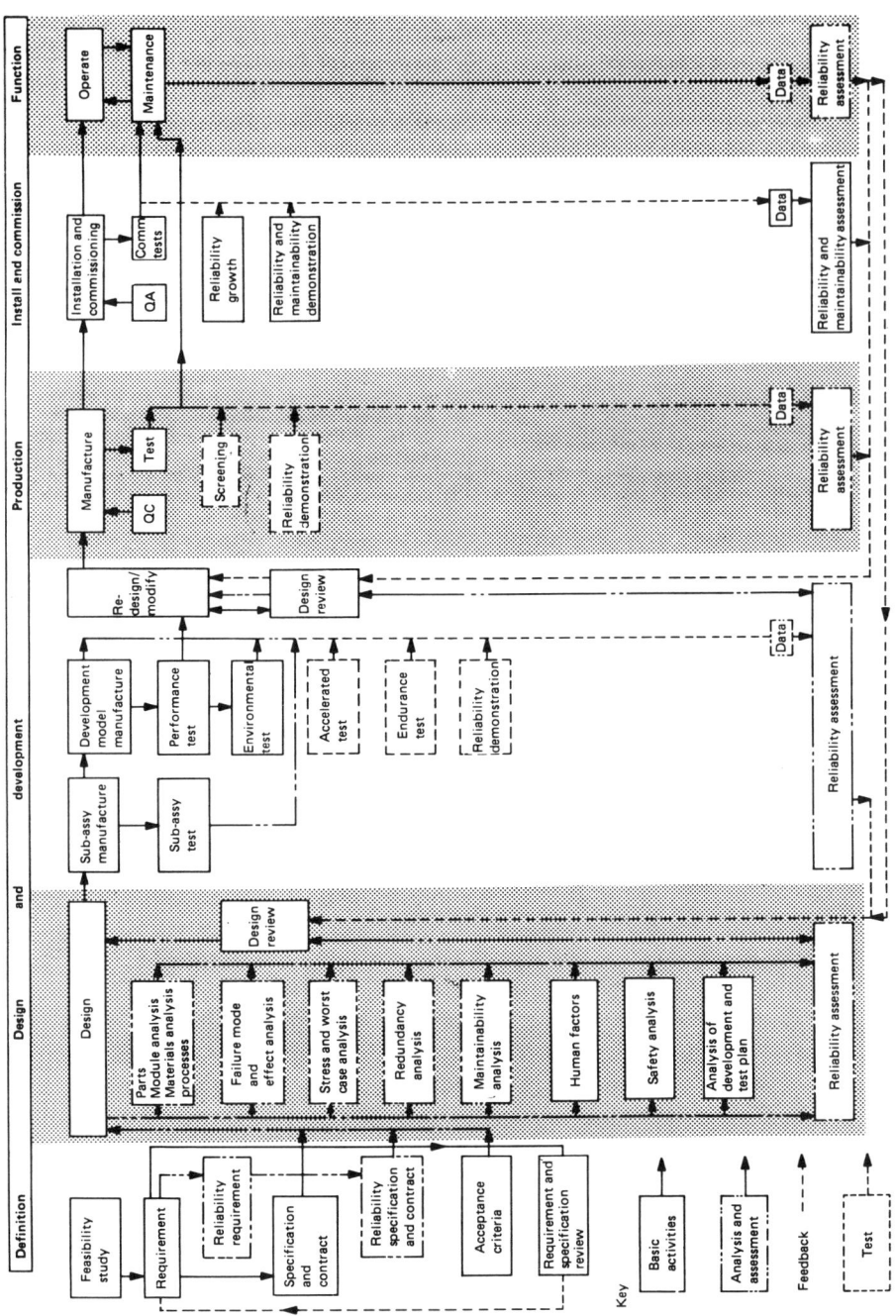

Figure 1.1 Reliability program (Extracts from British Standard BS 5760: Part 1: 1985 are reproduced with the permission of BSI. Complete copies can be obtained through national standards bodies.)

3. *Quality*. Quality management and assurance is increasingly focused, stimulated by the almost compulsory application of the ISO 9000 series of standards.

The concepts of quality and reliability are closely connected. Reliability may in some respects be considered to be a quality characteristic (perhaps its most important characteristic). Complementary systems are therefore being developed and implemented for reliability management and assurance as part of a total quality management (TQM) system. Note the relation between the ISO 9000 and the IEC 300 series of standards as discussed by Strandberg (1992).

4. *Optimization of maintenance and operation*. Maintenance is carried out to prevent system failures, as well as to restore the system function when a failure has occurred. The prime objective of maintenance is thus to maintain or improve the system reliability and production/operation regularity.

Many industries (e.g., the nuclear power, aviation, defense, and the offshore and shipping industry) have fully realized the important connection between maintenance and reliability and have implemented the reliability centered maintenance (RCM) methodology. The RCM methodology aims to improve the cost-effectiveness and control of maintenance in all types of industries, and hence to improve availability and safety. Reliability assessment is also an important element of the following applications: Life Cycle Cost (LCC), Life Cycle Profit (LCP), Logistic support, spare part allocation, and manning level analysis.

5. *Engineering design*. Reliability is considered to be one of the most important quality characteristics of technical products. Reliability assurance should therefore be one of the most important topics during the engineering design process.

Many industries have integrated a reliability program in the design process. This is especially the case within the nuclear power, the aviation, the aerospace, the automobile, and the offshore industries. Such integration may be accomplished through the concept of concurrent engineering (CE) which focuses on the total product perspective from inception through product delivery; see, for example, Keene (1992).

6. *Verification of quality/reliability*. A number of authorities require that the producer and/or the user of technical systems are able to verify that their equipment satisfies specified requirements. Such requirements usually have a basis in safety and/or environmental protection. Some industries also meet strict requirements with respect to production regularity. This is especially the case within the power generation and petroleum industries.

As part of the formation of the European Union a number of new EU directives have been issued. Among these are the machine safety directive, the product safety directive, and the product liability directive. The producers of equipment must, according to these directives, verify that their equipment comply with the requirements. Reliability analyses and reliability demonstration testing will be necessary tools in the verification process.

During the last few years it has become more common for buyers of technical equipment to require a quantitative assessment of the quality and reliability as part of the total system documentation. The documentation required varies a lot, from filled-in FMEA forms to detailed results from life testing of the equipment (e.g., accelerated testing). Documented quality/reliability has been required by some industries for many years (e.g., aircraft, aerospace, automobile, nuclear, and defense).

1.6 FAILURES AND FAILURE CLASSIFICATION

The main concern of a reliability engineer is to prevent failures. According to British Standard BS 4778 failure of an item is defined as follows:

Failure
The termination of an item's ability to perform a required function.

Failures may be classified in failure modes:

Failure Mode
The effect by which a failure is observed on the failed item (EuReDatA, 1983).

- All technical items are designed to fulfill one or more functions. A failure mode is thus defined as nonfulfillment of one of these functions.

Most items will show a number of different failure modes. Failure modes may generally be subdivided into two classes (EuReDatA 1983):

1. Demanded change of state is not achieved
2. Change of conditions (states)

An automatic valve may, for example, show the following failure modes:

- Fail to open on command (FTO)
- Fail to close on command (FTC)
- Leakage (through the valve) in closed position (LCP)
- Leakage to the environment (LTE)

The failure modes FTO and FTC belong to class 1, demanded change of state is not achieved, while the failure modes LCP and LTE are of class 2, change of conditions (states).

FAILURES AND FAILURE CLASSIFICATION

It is important to realize that a failure mode is a manifestation of the failure as seen from the *outside* (i.e., the termination of one or more functions). Leakage (through the valve) in closed position is thus a failure mode of a valve, since the valve looses its required function to close in the fluid. Erosion of the valve seal, however, represents a cause of failure and is hence not a failure mode of the valve.

Failures may be classified in a number of different ways:

1. *Sudden versus gradual failures.* A failure may represent a sudden loss of function, or be a gradual "drifting out" of the specified range of performance values. Sudden failures are easy to recognize; at any instant within the operating time, the item is either functioning well or not at all. The recognition of drift failures requires the comparison of actual device performance with a performance specification, and this may often be a difficult task.

2. *Hidden versus evident failures.* Some failure modes are called *evident* failures. Evident failures are detected instantly when they occur. The failure mode "spurious stop" of a pump is an example of an evident failure. Another type of failure is called *hidden* failures. A hidden failure is normally detected only during testing of the item. The failure mode "fail to start" of a pump is an example of a hidden failure.

3. *Classification of failures according to effects or severity.* Failures may be classified by their effects in many different ways. The following classification is used in OREDA (1992):

   - *Critical failure.* A failure that is sudden and causes cessation of one or more fundamental functions. (Note: This failure requires immediate corrective action in order to return the item to a satisfactory condition.)

   - *Degraded failure.* A failure that is gradual, partial, or both. (Note: Such a failure does not cease the fundamental functions but compromises one or several functions. The function may be compromised by any combination of reduced, increased, or erratic outputs. In time such a failure may develop into a critical failure.)

   - *Incipient failure.* An imperfection in the state or condition of an item so that a degraded or critical failure can be expected to result if corrective action is not taken.

   In US MIL-STD 882 "System Safety Program Requirements" the following classification is used with respect to severity:

   - *Catastrophic.* Any failure that results in death or system loss.
   - *Critical.* Any failure that results in severe injury, severe occupational illness, or major system damage.
   - *Marginal.* Any failure that results in minor injury, minor occupational illness, or minor system damage.

- *Negligible.* Any failure that results in less than minor injury, occupational illness, or system damage.

4. *Primary failures, secondary failures, and command faults.* Failures of an item may be classified as primary failures, secondary failures, or command faults (e.g., see Henley and Kumamoto 1981; Villemeur 1992).

    A *primary failure* is a failure caused by natural aging of the item. The primary failure occurs under conditions within the design envelope of the item. A repair action is necessary to return the item to a functioning state.

    A *secondary failure* is a failure caused by excessive stresses outside the design envelope of the item. Such stresses may be shocks from thermal, mechanical, electrical, chemical, magnetic, or radioactive energy sources. The stresses may be caused by neighboring components, the environment, or by system operators/plant personnel. A repair action is necessary to return the item to a functioning state.

    A *command fault* is a failure caused by an improper control signal or noise. A repair action is usually not required to return the item to a functioning state. Command faults are often referred to as *transient* failures.

The primary failures, secondary failures and command faults, and their causes are illustrated in Figure 1.2.

It is important to realize the fundamental difference between the two concepts: failure mode and failure mechanism. By "failure mechanism" we mean the physical, chemical, or other processes that are deteriorating the item, leading to a failure. The cause of a failure may in some cases occur at a time different from the time when the actual failure mode is manifested. Consider, for example, the case for the failure mode FTC for a valve that is normally open. Foreign particles and corrosion may make it impossible to close the valve. The FTC failure is, however, not manifested until the operation of the valve has been tried.

When starting on a reliability study of a system, it is important to define all relevant failure modes in a clear and unambiguous way. A fundamental technique for identifying and classifying failure modes is the failure mode and effect analysis (FMEA). FMEA is further discussed in Chapter 3.

1.7 MODEL CONSTRUCTION

In practical situations the analyst derives stochastic models of the system at hand or at least chooses from several possible models. To be "realistic" the model must describe the essential features of the system; the details do not necessarily have to be exact. One of the pioneers in mathematical statistics, Jerzy Neyman (1954), expresses this in the following way:

> Every attempt to use mathematics to study some real phenomena must begin with building a mathematical model of these phenomena. Of necessity, the model simplifies the matters to a greater or lesser extent and a number of details are ignored. The

MODEL CONSTRUCTION

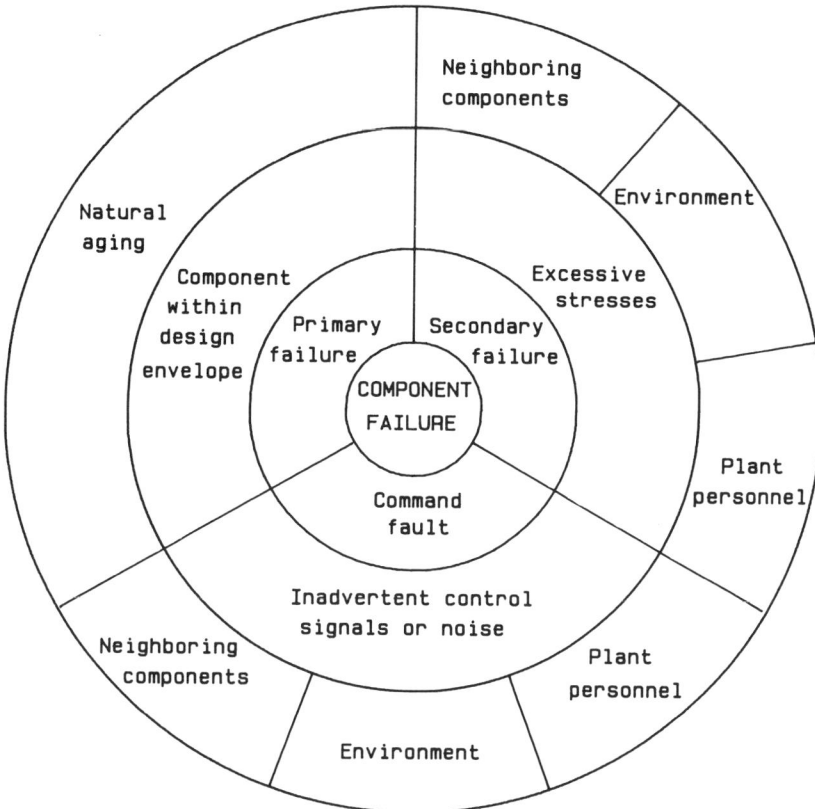

Figure 1.2 Component failure characteristics (adapted from Henley and Kumamoto 1981, p. 60, © 1992 IEEE)

success depends on whether or not the details ignored are really unimportant in the development of the phenomena studied. The solution of the mathematical problem may be correct and you may be in violent conflict with realities, simply because the original assumptions of the mathematical model diverge essentially from the conditions of the practical problem considered. Beforehand, it is impossible to predict with certainty whether or not a given mathematical model is adequate. To find this out, it is necessary to deduce a number of consequences of the model and to compare them with observation.

Another pioneer in statistics, George E. P. Box, repeatedly points out that "no model is absolutely correct. In particular situations, however, some models are more useful than others."

In reliability and safety studies of technical systems, one always has to work with models of the systems. These models may be graphical (networks of different types) or mathematical. A mathematical model is necessary to enable one to bring in data and use mathematical and statistical methods to estimate reliability, safety, or risk parameters. For such models two conflicting interests always apply:

- The model should be sufficiently simple to be handled by available mathematical and statistical methods.
- The model should be sufficiently "realistic" such that the deducted results are of practical relevance.

We should, however, always bear in mind that we are working with an idealized, simplified model of the system. Furthermore the results we derive are, strictly speaking, valid only for the model and are accordingly only "correct" to the extent that the model is realistic.

Now and then there is the contention that use of "probabilistic reliability analysis" has only bounded validity and is of little practical use. When something goes wrong, it is usually attributed to human error. That is to say, someone has failed to do what should have been done in a certain situation or has done something that should *not* have been done. In principle, however, there is nothing to prevent key persons from counting as "components" of a system in the way that technical components do. It would obviously be difficult to derive numerical estimates for the probability of human errors in a given situation, but that is another issue.

From what has been said, we understand that many subject areas are involved in a reliability analysis of technical systems:

- Detailed knowledge is needed of the technical aspects of the system and of the physical mechanisms that may lead to failure.
- Knowledge of mathematical/statistical concepts and statistical methods is a requisite (but far from sufficient) to be able to carry out such analyses.
- If humans are treated as components in the system, medical, psychological, and sociological insight into their behavior patterns is needed, including information on how these humans react under stress.
- Data must be available for estimation of parameters and checking of models.
- Analysis of complicated systems must be accompanied by appropriate computer programs.

The above list is not complete, but it illustrates that a reliability analysis involves many different areas of knowledge and has to be considered a multidiscipline task.

Boundary Conditions for the Analysis

A reliability analysis of a system is based on a wide range of assumptions and boundary conditions. Here we briefly mention a few such considerations:

MODEL CONSTRUCTION

- Precisely which parts of the system are going to be included in the analysis and which parts are not?
- Precisely what are the objectives of the analysis? Different objectives may necessitate different approaches.
- What system interfaces will be used? Operator and software interfaces have to be identified and defined.
- What level of detail is required?
- Which operational phases are to be included in the analysis (e.g., start-up, steady state, maintenance, disposal)?
- What are the environmental conditions for the system?
- Which external stresses should be considered (e.g., sabotage, earthquakes, lightning strikes)?

Binary Representation

Whether we are dealing with a single-component or a complex system, we will in the main part of this book content ourselves to classifying a system as being in one of two possible states, either in a functioning state (according to more closely defined criteria) or in a failed state. For the sake of brevity we say in the first case that the unit is *functioning*, and in the second that it is *failed*. When the unit changes from functioning state, to failed state we say that it fails. Failure is therefore an event that occurs at a certain moment in time.

For some units, it is easy to decide the exact moment of failure. A light bulb may be functioning perfectly one moment but fail the next moment. The time of failure is easily observed. Other units deteriorate continuously, and it may be difficult to decide the exact moment of failure. For example, a leakage rate may increase gradually, and it may be difficult to observe the exact moment when it exceeds a certain level.

We will assume that the time for the component (system) failure can be noted exactly. The state of a given unit at time t can be expressed by a binary variable $x(t)$ where

$$x(t) = \begin{cases} 1 & \text{if the unit is functioning at time } t \\ 0 & \text{if the unit is failed at time } t \end{cases} \quad (1.1)$$

The assumption of binary state variables is rather restrictive. It especially creates problems for the analysis of units with a gradually decreasing functionability. In the present theory one has to define an acceptable limit for functionability and classify all states with a lower level of performance as failure. If we look at a fixed choke unit in a pipeline system, the choke effect will deteriorate continuously due to erosion. In this case it is very difficult to define the exact time when the "failure" occurs. During the last few years a research activity has been initiated to develop a more general multistate theory (e.g., see Natvig, 1985a).

1.8 STANDARDS AND DIRECTIVES

A wide range of standards and directives containing requirements with respect to reliability and safety have been issued. Application of such standards is often mandatory as a basis for reliability analyses. Any reliability engineer should be familiar with the standards and directives that are applicable within his or her subject areas. Some relevant standards and directives are listed below. A more detailed listing, especially of MIL-STD's is presented, for example, by Lakner and Anderson (1985, app. B).

1. British Standard BS 5760 "Reliability of systems, equipments and components." This standard is divided in five parts:

    Part 0 "Introductory guide to reliability"

    Part 1 "Guide to reliability and maintainability programme management"

    Part 2 "Guide to the assessment of reliability"

    Part 3 "Guide to reliability practices: Examples"

    Part 4 "Guide to specification clauses relating to the achievement and development of reliability in new and existing items"

2. MIL-STD 785 "Reliability Program for Systems and Equipment—Development and Production"
3. MIL-STD 882 "System Safety Program Requirements"
4. MIL-STD 756 "Reliability Prediction"
5. IEEE Std. 352 "General Principles of Reliability Analysis of Nuclear Power Generating Station Protection Systems"
6. ISO series 9000 Quality Systems Standards (which is identical to the European CEN/CENELEC EN29000 series, and the American National Standards ANSI/ASQC Q90 series). These contractual standards do not, as they are now formatted, separate out specific requirements for reliability and dependability areas.
7. IEC 300 "Dependability Management." The IEC 300 series of standards extends beyond the scope of the ISO 9000 series by addressing standardization of reliability assurance technology and dependability technology. IEC 300-1 "Dependability Management Part 1: Dependability Assurance of Products" provides contract requirements for product reliability assurance throughout the life cycle of a product. IEC 300-2 "Dependability Management Part 2: Dependability Programme Elements and Tasks" outlines reliability program elements and tasks in a prescriptive manner but permits tailoring. IEC 300-3 is a series of application guides addressing such topics as

    Life cycle costing

    Maintainability and logistics

Risk analysis
Reliability/availability performance testing
Design review
Human performance testing
Software
Data collection and failure prevention
System dependability analysis
8. The European Union (EU) has issued a wide range of directives and associated standards. The most relevant directives are
The EU machine safety directive
The EU product safety directive
The EU product liability directive
The EU major hazards directive (the seveso directive)

Further EU directives and standards are given in Neal and Wright (1992).

CHAPTER 2

Failure Models

We now introduce several quantitative measures for the reliability of a unit that is not repaired. This unit can be anything from a small component to a large system. The three most important measures for the reliability of an unrepaired unit are

- The reliability (survivor) function $R(t)$
- The failure rate $z(t)$
- The mean time to failure (MTTF)

2.1 TIME TO FAILURE

By the *time to failure* of a unit we mean the time elapsing from when the unit is put into operation until it fails for the first time. We set $t = 0$ as the starting point. At least to some extent the time to failure is subject to chance variations. It is therefore natural to interpret the time to failure as a random variable T.

The state of the unit at time t may be described by the state variable $X(t)$, which is also a random variable:

$$X(t) = \begin{cases} 1 & \text{if the unit is functioning at time } t \\ 0 & \text{if the unit is in a failed state at time } t \end{cases}$$

The connection between the state variable $X(t)$ and the time to failure T is illustrated in Figure 2.1. Note that the time to failure T is not always measured in calendar time. It may also be measured by more indirect time concepts, such as

the number of times a switch is operated
the number of kilometers driven by a car
the number of rotations of a bearing
the number of cycles for a periodically working unit

TIME TO FAILURE

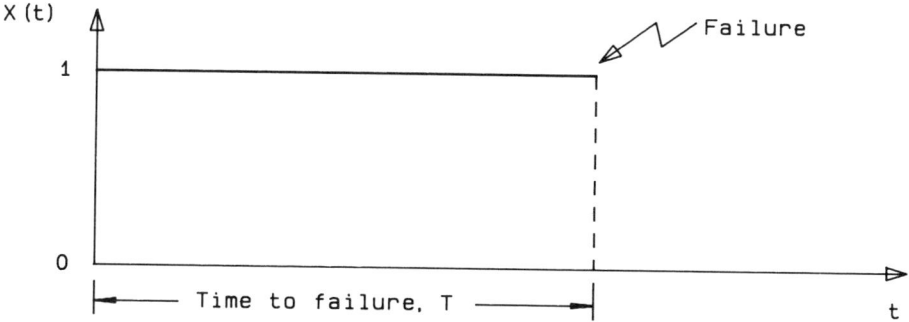

Figure 2.1 The state variable and the time to failure of a unit

From these examples we notice that time to failure may often be a discrete variable. A discrete variable can, however, be approximated by a continuous variable. Here, unless stated otherwise, we will assume that the time to failure T is continuously distributed with probability density $f(t)$ and distribution function

$$F(t) = P(T \leq t) = \int_0^t f(u)\,du \qquad \text{for } t > 0 \qquad (2.1)$$

$F(t)$ thus denotes the probability that the unit fails within the time interval $(0, t]$. The probability density $f(t)$ is defined as

$$f(t) = \frac{d}{dt} F(t) = \lim_{\Delta t \to 0} \frac{F(t+\Delta t) - F(t)}{\Delta t} = \lim_{\Delta t \to 0} \frac{P(t < T \leq t + \Delta t)}{\Delta t}$$

This implies that when Δt is small,

$$P(t < T \leq t + \Delta t) \approx f(t) \cdot \Delta t \qquad (2.2)$$

The distribution function $F(t)$ and the probability density $f(t)$ are illustrated in Figure 2.2.

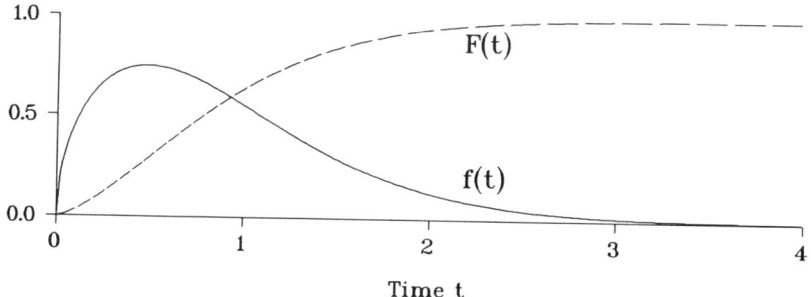

Figure 2.2 Distribution function $F(t)$ and probability density $f(t)$

2.2 RELIABILITY FUNCTION

The reliability function of a unit is defined by

$$R(t) = 1 - F(t) = P(T > t) \qquad \text{for } t > 0 \qquad (2.3)$$

Hence $R(t)$ is the probability that the unit does not fail in the time interval $(0, t]$ or, equivalently, the probability that the unit survives the time interval $(0, t]$ and is still functioning at time t. The reliability function $R(t)$ is also called the *survivor function*, and is illustrated in Figure 2.3.

2.3 FAILURE RATE

The probability that a unit will fail in the time interval $(t, t + \Delta t]$, when we know that the unit is functioning at time t, is

$$P(t < T \leq t + \Delta t | T > t) = \frac{P(t < T \leq t + \Delta t)}{P(T > t)} = \frac{F(t + \Delta t) - F(t)}{R(t)}$$

By dividing this probability by the length of the time interval Δt and letting $\Delta t \to 0$, we get the failure rate $z(t)$ of the unit:

$$\begin{aligned} z(t) &= \lim_{\Delta t \to 0} \frac{P(t < T \leq t + \Delta t | T > t)}{\Delta t} \\ &= \lim_{\Delta t \to 0} \frac{F(t + \Delta t) - F(t)}{\Delta t} \frac{1}{R(t)} = \frac{f(t)}{R(t)} \end{aligned} \qquad (2.4)$$

This implies that when Δt is small,

$$P(t < T \leq t + \Delta t | T > t) \approx z(t) \cdot \Delta t \qquad (2.5)$$

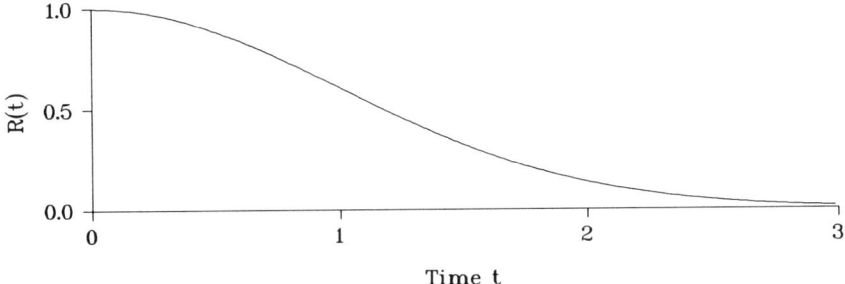

Figure 2.3 The reliability (survivor) function $R(t)$

FAILURE RATE

Note the difference between the probability density $f(t)$ and the failure rate $z(t)$. Say that we start out with a new unit at time $t = 0$ and at time $t = 0$ ask, "What is the probability that this unit will fail in the interval $(t, t + \Delta t]$?" According to (2.2) this probability is approximately equal to the probability density $f(t)$ at time t multiplied by the length of the interval Δt. Next consider a unit that has survived until time t, and ask, "What is the probability that this unit will fail in the next interval $(t, t + \Delta t]$?" This (conditional) probability is according to (2.5) approximately equal to the failure rate $z(t)$ at time t multiplied by the length of the interval Δt.

If we put a large number of identical units into operation at time $t = 0$, then $z(t) \cdot \Delta t$ will roughly represent the relative proportion of the units still functioning at time t, failing in $(t, t + \Delta t]$. Since

$$f(t) = \frac{d}{dt} F(t) = \frac{d}{dt} (1 - R(t)) = -R'(t)$$

then

$$z(t) = -\frac{R'(t)}{R(t)} = -\frac{d}{dt} \ln R(t) \qquad (2.6)$$

Since $R(0) = 1$, then

$$\int_0^t z(t) \, dt = -\ln R(t) \qquad (2.7)$$

and

$$R(t) = e^{-\int_0^t z(u) \, du} \qquad (2.8)$$

The reliability (survivor) function $R(t)$ and the distribution function $F(t) = 1 - R(t)$ are therefore uniquely determined by the failure rate $z(t)$. From (2.4) and (2.8) we see that the probability density $f(t)$ can be expressed by

$$f(t) = z(t) e^{-\int_0^t z(u) \, du} \qquad \text{for } t > 0 \qquad (2.9)$$

In actuarial statistics the failure rate is called the *force of mortality* (FOM). This term has also been adopted by several authors of reliability textbooks to avoid the confusion between the failure rate and the *rate of occurrence of failures* (ROCOF) of a repairable unit. The failure rate (FOM) is a function of the life distribution of a single unit and an indication of the "proneness to failure" of the unit after time t has elapsed, while ROCOF is the occurrence rate of failures for a stochastic process; see Chapter 7. A thorough discussion of these concepts is given by Ascher and Feingold (1984). Some authors (e.g. Thompson, 1988) prefer the term *hazard rate* instead of failure rate. The term "failure rate" is, however, now well established in applied reliability. We have therefore decided to use this term instead of FOM in this textbook, although we realize that the use of this term may lead to some confusion.

The relationships between the functions $F(t), f(t), R(t)$, and $z(t)$ are presented in Table 2.1.

From (2.8) we see that the reliability (survivor) function $R(t)$ is uniquely determined by the failure rate $z(t)$. To determine the form of $z(t)$ for a given type of units, the following experiment may be carried out:

Split the time interval $(0, t)$ into disjoint intervals of equal length Δt. Then put n identical units into operation at time $t = 0$. When a unit fails, note the time and leave that unit out. For each interval record:

- The number of units $n(i)$ that fail in interval i.

- The functioning times for the individual units $(T_{1i}, T_{2i}, \ldots, T_{ni})$ in interval i. Hence T_{ji} is the time unit j has been functioning in time interval i. T_{ji} is therefore equal to 0 if unit j has failed before interval i, where $j = 1, 2, \ldots, n$.

Thus $\sum_{j=1}^{n} T_{ji}$ is the total functioning time for the units in interval i. Now

Table 2.1 Relationships between the functions $F(t), f(t), R(t)$, and $z(t)$

Expressed by	$F(t)$	$f(t)$	$R(t)$	$z(t)$
$F(t) =$	—	$\int_0^t f(u)\,du$	$1 - R(t)$	$1 - e^{-\int_0^t z(u)du}$
$f(t) =$	$\dfrac{d}{dt} F(t)$	—	$-\dfrac{d}{dt} R(t)$	$z(t)\, e^{-\int_0^t z(u)du}$
$R(t) =$	$1 - F(t)$	$\int_t^\infty f(u)\,du$	—	$e^{-\int_0^t z(u)du}$
$z(t) =$	$\dfrac{dF(t)/dt}{1 - F(t)}$	$\dfrac{f(t)}{\int_t^\infty f(u)\,du}$	$-\dfrac{d}{dt} \ln R(t)$	—

FAILURE RATE

$$z(i) = \frac{n(i)}{\sum_{j=1}^{n} T_{ji}}$$

which shows the number of failures per unit functioning time in interval i, is a natural estimate of the "failure rate" in interval i for the units that are functioning at the start of this interval.

Let $m(i)$ denote the number of units which are functioning at the start of interval i:

$$z(i) \approx \frac{n(i)}{m(i)\Delta t}$$

and hence

$$z(i)\Delta t \approx \frac{n(i)}{m(i)}$$

A histogram depicting $z(i)$ as a function of i typically is of the form given in Figure 2.4. If n is very large, we may use very small time intervals. If we let $\Delta t \to 0$, it is expected that the step function $z(i)$ will tend toward a "smooth" curve, as illustrated in Figure 2.5, which may be interpreted as an estimate for the failure rate $z(t)$.

The curve in Figure 2.5 is usually called a *bathtub curve* after its characteristic shape. The failure rate is often high in the initial phase. This can be explained by the fact that there may be undiscovered defects (known as "infant mortality") in the units; these soon show up when the units are activated. When the unit has survived the infant mortality period, the failure rate often stabilizes at a level where it remains for a certain amount of time until it starts to increase

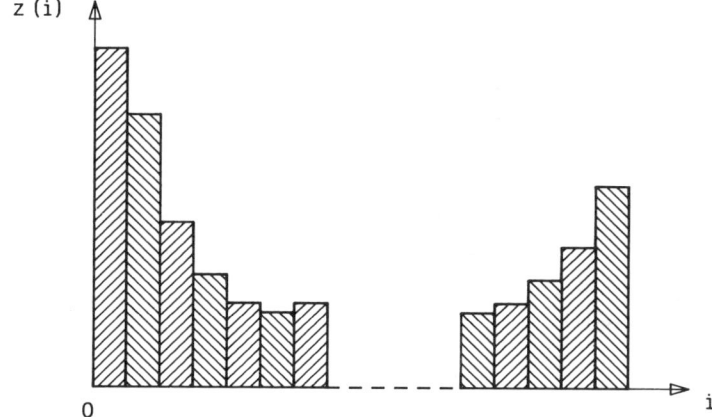

Figure 2.4 Empirical bathtub curve

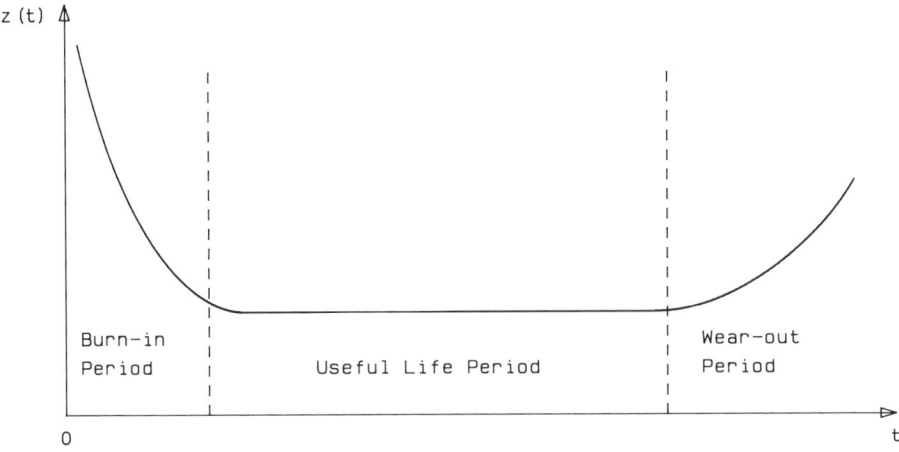

Figure 2.5 Bathtub curve

as the unit begins to wear out. From the shape of the bathtub curve, the lifetime of a unit may be divided into three typical intervals: the burn-in period, the useful life period, and the wear-out period. The useful life period is also called the *chance failure period*. Often the units are tested at the factory before they are distributed to the users, and thus much of the infant mortality will be removed before the units are delivered for use. For the majority of mechanical units the failure rate will usually show a slightly increasing tendency in the useful life period.

2.4 MEAN TIME TO FAILURE

The mean time to failure (MTTF) of a unit is defined by

$$\text{MTTF} = E(T) = \int_0^\infty t f(t)\, dt \qquad (2.10)$$

When the time required to repair or replace a failed unit is very short compared to MTTF, MTTF also represents the mean time between failures (MTBF). If the repair time cannot be neglected, MTBF also includes the mean time to repair (MTTR). This is illustrated in Figure 2.6.

Since $f(t) = -R'(t)$,

$$\text{MTTF} = -\int_0^\infty t R'(t)\, dt$$

MEAN TIME TO FAILURE

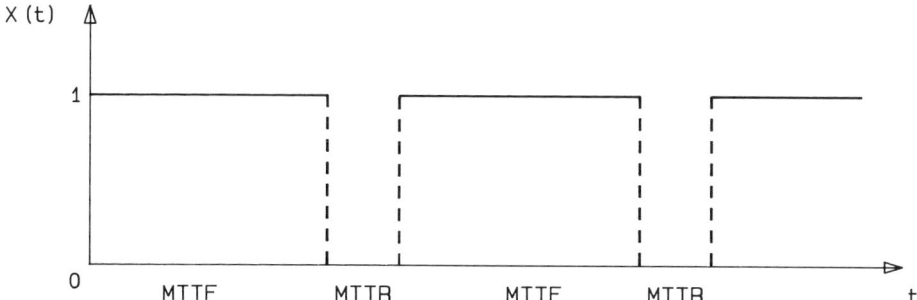

Figure 2.6 "Average behavior" of a unit

By partial integration

$$\text{MTTF} = -[tR(t)]_0^\infty + \int_0^\infty R(t)\,dt$$

If MTTF $< \infty$, it can be shown that $[tR(t)]_0^\infty = 0$. In that case

$$\text{MTTF} = \int_0^\infty R(t)\,dt \qquad (2.11)$$

It is often easier to determine MTTF by (2.11) than by (2.10).

The mean time to failure of a unit may also be derived by the following approach: The Laplace transform of the survivor function $R(t)$ is (see Appendix D)

$$R^*(s) = \int_0^\infty R(t)\,e^{-st}\,dt \qquad (2.12)$$

When $s = 0$, we get

$$R^*(0) = \int_0^\infty R(t)\,dt = \text{MTTF} \qquad (2.13)$$

The MTTF may thus be derived from the Laplace transform $R^*(s)$ of the survivor function $R(t)$ by setting $s = 0$.

2.5 THE POISSON PROCESS

Suppose that we are studying the occurrence of a certain event $\mathcal{A}$ in the course of a given time interval. $\mathcal{A}$ may, for example, be the emission of a particle from a radioactive source, or a telephone call at a switchboard. If we consider components where repair time is negligible, then $\mathcal{A}$ may be the event that the component fails.

A homogeneous Poisson process (HPP) requires the following assumptions:

1. $\mathcal{A}$ can occur at any time in the interval, and the probability of $\mathcal{A}$ occurring in the interval $(t, t + \Delta t]$ is independent of t and may be written as $\lambda \cdot \Delta t + o(\Delta t)$,[1] where λ is a positive constant.
2. The probability of more that one event $\mathcal{A}$ in the interval $(t, t + \Delta t]$ is $o(\Delta t)$.
3. Let $(t_{11}, t_{12}], (t_{21}, t_{22}], \ldots$ be any sequence of disjoint intervals in the time period in question. Then the events "$\mathcal{A}$ occurs in $(t_{j1}, t_{j2}]$," $j = 1, 2, \ldots$, are independent.

The process is said to have the intensity λ. Without loss of generality we let $t = 0$ be the starting point of the process.

Let $N(t)$ denote the number of times the event $\mathcal{A}$ occurs during the interval $(0, t]$, and let $p(n, t) = P(N(t) = n)$ denote the probability that $\mathcal{A}$ occurs exactly n times in the time interval $(0, t]$. Then

$$p(0, t + \Delta t) = p(0, t)(1 - \lambda \Delta t - o(\Delta t))$$
$$= p(0, t) - \lambda p(0, t)\Delta t - p(0, t)o(\Delta t) \quad (2.14)$$

Hence

$$\frac{p(0, t + \Delta t) - p(0, t)}{\Delta t} = -\lambda p(0, t) - \frac{o(\Delta t) \cdot p(0, t)}{\Delta t}$$

We now let $\Delta t \to 0$, and assuming $p(0, t)$ to be differentiable with respect to t, we get

$$\frac{d}{dt} p(0, t) = -\lambda p(0, t)$$

Hence

$$p(0, t) = Ce^{-\lambda t} \quad \text{for } t > 0$$

where C is a constant.

---
[1] $o(\Delta t)$ denotes a function of Δt with the property that $\lim_{\Delta t \to 0} [o(\Delta t)/\Delta t] = 0$.

THE POISSON PROCESS

In this situation it seems natural to put $p(0,0) = 1$. This implies that $C = 1$, and hence that

$$p(0, t) = e^{-\lambda t} \quad \text{for } t > 0$$

Before we proceed determining $p(n, t)$ for n values greater than zero, we will take a look at the time intervals between occurrences of the event $\mathcal{A}$.

Let T_1 denote the time point when $\mathcal{A}$ occurs for the first time. T_1 is then a random variable and

$$F_{T_1}(t) = 0 \quad \text{for } t < 0$$

while

$$\begin{aligned} F_{T_1}(t) &= P(T_1 \leq t) = 1 - P(T_1 > t) \\ &= 1 - p(0, t) = 1 - e^{-\lambda t} \quad \text{for } t > 0 \end{aligned}$$

The time T_1 to the first $\mathcal{A}$ is said to be *exponentially distributed* with parameter λ.

The corresponding probability density is

$$f_{T_1}(t) = \begin{cases} \lambda e^{-\lambda t} & \text{for } t > 0, \lambda > 0 \\ 0 & \text{otherwise} \end{cases}$$

and the mean value of T_1 is

$$E(T_1) = \int_0^\infty t f_{T_1}(t)\, dt = \lambda \int_0^\infty t e^{-\lambda t}\, dt = \frac{1}{\lambda} \qquad (2.15)$$

The exponential distribution is further discussed in Section 2.6. When the event $\mathcal{A}$ has occurred, the process starts over again, and thus the waiting time T between consecutive occurrences in a homogeneous Poisson process (HPP) is exponentially distributed. A rigorous proof is given in Cox and Miller (1965, p. 148). If λ is large, the mean waiting time is short, and if λ is small, the mean waiting time is long. This is why λ is called the *intensity of the process*.

Next let us determine $p(n, t)$ for $n = 1, 2, \ldots$. Using the same type of argument as in the case where $n = 0$, we get

$$p(n, t + \Delta t) = p(n, t) + \lambda[p(n-1, t) - p(n, t)]\Delta t + o(\Delta t)$$

Hence

$$\frac{p(n, t + \Delta t) - p(n, t)}{\Delta t} = \lambda [p(n - 1, t) - p(n, t)] + \frac{o(\Delta t)}{\Delta t}$$

By letting $\Delta t \to 0$, and assuming $p(n, t)$ to be differentiable with respect to t, we get

$$\frac{d}{dt} p(n, t) = \lambda [p(n - 1, t) - p(n, t)] \qquad (2.16)$$

Hence $p(1, t)$ has to satisfy the differential equation

$$\frac{d}{dt} p(1, t) = \lambda e^{-\lambda t} - \lambda p(1, t)$$

Since $p(n, 0) = 0$ for $n \geq 1$,

$$p(1, t) = \lambda t e^{-\lambda t}$$

By recursive solution of (2.16) and use of $p(n, 0) = 0$ for $n \geq 1$,

$$p(n, t) = \frac{(\lambda t)^n}{n!} e^{-\lambda t} \qquad \text{for } n = 0, 1, 2, \ldots$$

Accordingly, if $N(t)$ denotes the number of times $\mathcal{A}$ occurs in the time interval $(0, t]$,

$$P(N(t) = n) = \frac{(\lambda t)^n}{n!} e^{-\lambda t} \qquad \text{for } n = 0, 1, 2, \ldots \qquad (2.17)$$

This distribution is called the *Poisson distribution* with parameter λt.

$$E(N(t)) = \sum_{n=0}^{\infty} n \cdot \frac{(\lambda t)^n}{n!} e^{-\lambda t} = \lambda t \qquad (2.18)$$

Since the expected number of occurrences of $\mathcal{A}$ per unit time ($t = 1$) is λ, λ expresses the intensity of the process.

Let us consider a homogeneous Poisson process (HPP) with intensity λ, and assume that we are interested in determining the distribution of the time point T_k where $\mathcal{A}$ occurs for the kth time (k is accordingly an integer). We let t be an arbitrarily chosen time point on the positive real axis. The event $(T_k > t)$ is then obviously synonymous with the event that $\mathcal{A}$ is occurring at most $(k - 1)$

times in the time interval $(0, t]$:

$$1 - F_{T_k}(t) = \sum_{j=0}^{k-1} \frac{(\lambda t)^j}{j!} e^{-\lambda t}$$

Hence

$$F_{T_k}(t) = 1 - \sum_{j=0}^{k-1} \frac{(\lambda t)^j}{j!} e^{-\lambda t} \qquad (2.19)$$

The probability density $f_{T_k}(t)$ is obtained from (2.19) by differentiating with respect to t:

$$\begin{aligned}
f_{T_k}(t) &= -\sum_{j=1}^{k-1} \frac{j\lambda(\lambda t)^{j-1}}{j!} e^{-\lambda t} + \lambda \sum_{j=0}^{k-1} \frac{(\lambda t)^j}{j!} e^{-\lambda t} \\
&= \lambda e^{-\lambda t} \left(\sum_{j=0}^{k-1} \frac{(\lambda t)^j}{j!} - \sum_{j=1}^{k-1} \frac{(\lambda t)^{j-1}}{(j-1)!} \right) \\
&= \lambda e^{-\lambda t} \left(\sum_{j=0}^{k-1} \frac{(\lambda t)^j}{j!} - \sum_{j=0}^{k-2} \frac{(\lambda t)^j}{j!} \right) \\
&= \lambda e^{-\lambda t} \frac{(\lambda t)^{k-1}}{(k-1)!}
\end{aligned}$$

Accordingly

$$f_{T_k}(t) = \frac{\lambda}{(k-1)!} (\lambda t)^{k-1} e^{-\lambda t} \qquad \text{for } t > 0, \lambda > 0. \qquad (2.20)$$

where k is a positive integer. This distribution is called the *gamma* distribution with parameters k and λ. The gamma distribution is further discussed in Section 2.7.

We can therefore conclude that the waiting time until the kth occurrence of $\mathcal{A}$ in a homogeneous Poisson process with intensity λ is gamma distributed (k, λ). Note that the gamma distribution $(1, \lambda)$ is an exponential distribution with parameter λ. The homogeneous Poisson process (HPP) is further discussed in Section 7.2. See also Cox and Miller (1965), Ross (1983), Taylor and Karlin (1984), and Thompson (1988).

Example 2.1

Suppose that exactly one event (failure) of a homogeneous Poisson process (HPP) with intensity λ is known to have occurred some time in the interval $(0, t_0]$. We want to determine the distribution of the time T_1 at which the event occurred.

$$\begin{aligned} P(T_1 \leq t | N(t_0) = 1) &= \frac{P(T_1 \leq t \cap N(t_0) = 1)}{P(N(t_0) = 1)} \\ &= \frac{P(1 \text{ event in } (0, t] \cap 0 \text{ events in } (t, t_0])}{P(N(t_0) = 1)} \\ &= \frac{P(1 \text{ event in } (0, t]) \cdot P(0 \text{ events in } (t, t_0])}{P(N(t_0) = 1)} \\ &= \frac{\lambda t e^{-\lambda t} e^{-\lambda(t_0 - t)}}{\lambda t_0 e^{-\lambda t_0}} \\ &= \frac{t}{t_0} \qquad \text{for } 0 < t \leq t_0 \end{aligned}$$

When we know that exactly one failure has taken place in the time interval $(0, t_0]$; the time at which the first failure occurs is uniformly distributed over $(0, t_0]$. Thus each interval of equal length in $(0, t_0]$ has the same probability of containing the failure. The expected time at which the event occurs is

$$E(T_1 | N(t_0) = 1) = \frac{t_0}{2} \qquad \square$$

Example 2.2

Suppose that failures of a system are occurring in accordance with a homogeneous Poisson process (HPP) with intensity λ. Some failures develop into a consequence C, and others do not. The failures developing into a consequence C is denoted a C failure. The probability that a failure develops into consequence C is denoted p and is assumed constant for each failure. The failure consequences are further assumed to be independent of each other. Let $N(t)$ denote the number of failures in the time interval $(0, t]$, and let $Y_C(t)$ denote the number of C failures in the same interval. When $N(t)$ is equal to n, $Y_C(t)$ will have a binomial distribution

$$P(Y_C(t) = y | N(t) = n) = \binom{n}{y} p^y (1-p)^{n-y} \qquad \text{for } y = 0, 1, 2, \ldots, n$$

The marginal distribution of $Y_C(t)$ is

THE EXPONENTIAL DISTRIBUTION

$$P(Y_C(t) = y) = \sum_{n=y}^{\infty} \binom{n}{y} p^y (1-p)^{n-y} \frac{(\lambda t)^n}{n!} e^{-\lambda t}$$

$$= \frac{p^y e^{-\lambda t}}{y!} (\lambda t)^y \sum_{n=y}^{\infty} \frac{[\lambda t (1-p)]^{n-y}}{(n-y)!}$$

$$= \frac{(p\lambda t)^y e^{-\lambda t}}{y!} \sum_{x=0}^{\infty} \frac{[\lambda t(1-p)]^x}{x!}$$

$$= \frac{(p\lambda t)^y e^{-\lambda t}}{y!} e^{\lambda t(1-p)}$$

$$= \frac{(p\lambda t)^y}{y!} e^{-p\lambda t}$$

$Y_C(t)$ is thus a homogeneous Poisson process with intensity $p\lambda$, and the mean number of C consequences in the time interval $(0, t]$ is

$$E(Y_C(t)) = p\lambda t \qquad \square$$

2.6 THE EXPONENTIAL DISTRIBUTION

Let us assume that the time to failure T of a unit is exponentially distributed with parameter λ. The probability density is then given by

$$f(t) = \begin{cases} \lambda e^{-\lambda t} & \text{for } t > 0, \lambda > 0 \\ 0 & \text{otherwise} \end{cases}$$

The reliability (survivor) function becomes

$$R(t) = P(T > t) = \int_t^{\infty} f(u)\,du = e^{-\lambda t} \qquad \text{for } t > 0 \qquad (2.21)$$

The mean time to failure is

$$\text{MTTF} = \int_0^{\infty} R(t)\,dt = \int_0^{\infty} e^{-\lambda t}\,dt = \frac{1}{\lambda} \qquad (2.22)$$

and the failure rate is

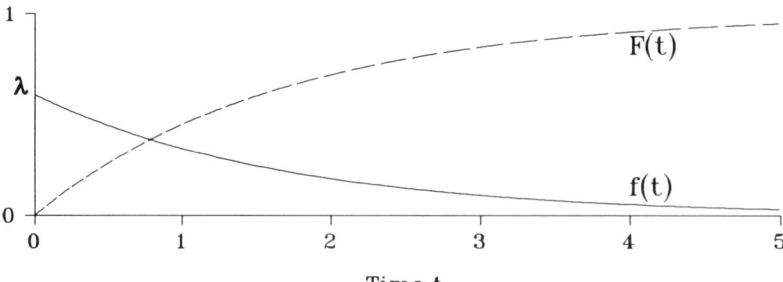

Figure 2.7 Exponential distribution.

$$z(t) = \frac{f(t)}{R(t)} = \frac{\lambda e^{-\lambda t}}{e^{-\lambda t}} = \lambda \tag{2.23}$$

The variance of T is

$$\text{var}(T) = \frac{1}{\lambda^2}$$

Accordingly the failure rate of a unit with exponential life distribution is constant (i.e., independent of time). By comparing with Figure 2.5, we see that the exponential distribution may be a realistic life distribution for a unit during its useful life period, at least for certain types of units.

The results (2.22) and (2.23) compare well with the use of the concepts in everyday language. If a unit on the average has $\lambda = 4$ failures/year, the mean time to failure MTTF of the unit is 1/4 year.

Now suppose that a unit has exponential time to failure T. For such a unit

$$P(T > t + x \mid T > t) = \frac{P(T > t + x)}{P(T > t)} = \frac{e^{-\lambda(t+x)}}{e^{-\lambda t}} = e^{-\lambda x} = P(T > x)$$

This implies that the probability that a unit will be functioning at time $t + x$, given that it is functioning at time t, is equal to the probability that a new unit has a time to failure longer than x. Hence the remaining lifetime of a unit, functioning at time t, is independent of t. The exponential distribution has no "memory."

Therefore an assumption of exponentially distributed lifetime implies that

- A used unit is stochastically *as good as new*, so there is no reason to replace a functioning unit.
- For the estimation of the reliability function, the mean time to failure, and

THE GAMMA DISTRIBUTION 33

so on, it is sufficient to collect data on the number of hours of observed time in operation and the number of failures. The age of the units is of no interest in this connection.

The exponential distribution is the most commonly used life distribution in applied reliability analysis. The reason for this is its mathematical simplicity and that it leads to realistic lifetime models for certain types of units.

Example 2.3
Consider a system of two independent components with failure rates λ_1 and λ_2, respectively. We want to determine the probability that component 1 fails before component 2. This probability is

$$P(T_2 > T_1) = \int_0^\infty P(T_2 > t | T_1 = t) f_{T_1}(t) \, dt$$

$$= \int_0^\infty e^{-\lambda_2 t} \lambda_1 e^{-\lambda_1 t} \, dt$$

$$= \lambda_1 \int_0^\infty e^{-(\lambda_1 + \lambda_2) t} \, dt = \frac{\lambda_1}{\lambda_1 + \lambda_2}$$

This result can easily be generalized to a system of n independent components with failure rates $\lambda_1, \ldots \lambda_n$. The probability that component j is the first component to fail is

$$P(\text{component } j \text{ fails first}) = \frac{\lambda_j}{\sum_{i=1}^n \lambda_i} \qquad \square$$

2.7 THE GAMMA DISTRIBUTION

Consider a unit that is exposed to a series of shocks which occur according to a homogeneous Poisson process with intensity λ. The time intervals $T_1, T_2, T_3, \ldots$, between consecutive shocks are then independent and exponentially distributed with parameter λ (see Section 2.5). Assume that the unit fails exactly at shock k, and not earlier. The time to failure of the unit is then

$$T = T_1 + T_2 + \cdots + T_k$$

and according to (2.20) T is gamma distributed (k, λ) with probability density

$$f(t) = \frac{\lambda}{\Gamma(k)} (\lambda t)^{k-1} e^{-\lambda t} \qquad (2.24)$$

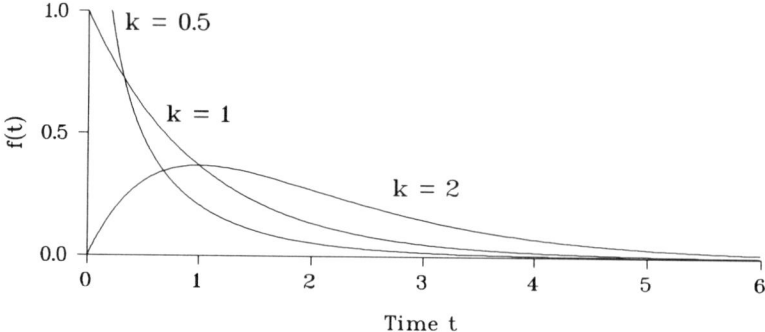

Figure 2.8 The gamma probability density, $\lambda = 1.0$

where $\Gamma(\cdot)$ denotes the gamma function (see Appendix A), $t > 0$, $\lambda > 0$, and k is a positive integer. $f(t)$ is sketched in Figure 2.8 for selected values of k.

The parameter λ denotes the shock intensity and is an external parameter for the unit. k may be interpreted as a measure of the ability to resist the shocks and will from now on generally not be restricted to integer values but be a positive constant. Equation (2.24) will still be a probability density function.

From (2.24) we find that

$$\text{MTTF} = \frac{k}{\lambda}$$
$$\text{var}(T) = \frac{k}{\lambda^2}$$

When k is an integer, the gamma distribution is also known as the *Erlangian* distribution. For integer values of k the reliability function (see 2.19 and Figure 2.9) is given by

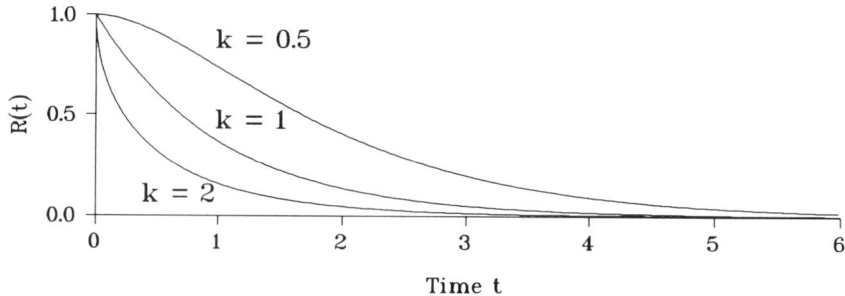

Figure 2.9 Reliability function for the gamma distribution, $\lambda = 1.0$

THE PARETO DISTRIBUTION

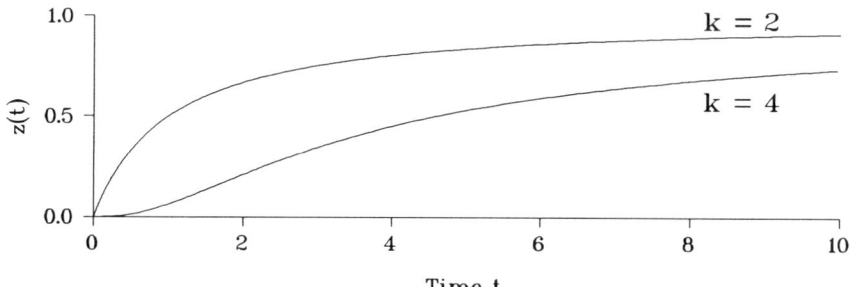

Figure 2.10 Failure rate of the gamma distribution, $\lambda = 1$

$$R(t) = 1 - F(t) = \sum_{x=0}^{k-1} \frac{(\lambda t)^x}{x!} e^{-\lambda t} \qquad (2.25)$$

The corresponding failure rate is

$$z(t) = \frac{f(t)}{R(t)} = \frac{\lambda(\lambda t)^{k-1} e^{-\lambda t}/\Gamma(k)}{\sum_{x=0}^{k-1}(\lambda t)^x e^{-\lambda t}/x!} \qquad (2.26)$$

The failure rate (2.26) presupposing that k is an integer, is always nondecreasing as a function of t, as illustrated in Figure 2.10. This will be discussed in Section 2.16.

2.8 THE PARETO DISTRIBUTION

A random variable X is said to have a *Pareto* distribution $P(c,d)$ if its distribution function is given by

$$F_X(x) = 1 - \left(\frac{d}{x}\right)^c \qquad \text{for } x > d,\ c > 0 \qquad (2.27)$$

A certain set of assumptions made in connection with the exponential life distribution leads to the Pareto distribution $P(k, 1)$.

Suppose that we have a population of components with exponentially distributed times to failure T and failure rate λ. λ is, however, allowed to vary across the individual components. The probability density of T, given λ, is thus

$$f(t|\lambda) = \lambda e^{-\lambda t} \qquad \text{for } t > 0$$

The variation in λ is often modeled by a gamma distribution with parameters α and k.

$$\pi(\lambda) = \frac{\alpha}{\Gamma(k)} (\alpha\lambda)^{k-1} e^{-\alpha\lambda} \qquad \text{for } \lambda > 0, \; \alpha > 0, \; k > 0$$

The marginal density of T is thus

$$f(t) = \int_0^\infty f(t|\lambda)\pi(\lambda)\, d\lambda = \frac{k\alpha^k}{(\alpha+t)^{k+1}} \qquad (2.28)$$

and the corresponding distribution function becomes

$$F(t) = 1 - \frac{\alpha^k}{(\alpha+t)^k} = 1 - \left(1 + \frac{t}{\alpha}\right)^{-k}$$

Now introduce

$$Y = \left(\frac{T}{\alpha} + 1\right)$$

Then

$$F_Y(y) = P(Y \le y) = P\left(\frac{T}{\alpha} + 1 \le y\right) = P(T \le \alpha(y-1))$$
$$= 1 - \left(1 + \frac{\alpha(y-1)}{\alpha}\right)^{-k} = 1 - \left(\frac{1}{y}\right)^k$$
$$\text{for } y > 1, \; k > 0 \qquad (2.29)$$

Hence $Y = T/\alpha + 1$ has a Pareto distribution $P(1, k)$.

The reliability (survivor) function is in this case

$$R(t) = P(T > t) = \int_t^\infty f(u)\, du = \frac{\alpha^k}{(\alpha+t)^k} = \left(1 + \frac{t}{\alpha}\right)^{-k} \qquad (2.30)$$

The mean time to failure is

$$\text{MTTF} = \int_0^\infty R(t)\,dt = \frac{\alpha}{k-1} \qquad \text{for } k > 1 \tag{2.31}$$

Note that the MTTF does not exist for $0 < k \leq 1$. The failure rate is

$$z(t) = \frac{f(t)}{R(t)} = \frac{k}{\alpha + t} \tag{2.32}$$

and hence is monotonically *decreasing* as a function of t.

2.9 THE WEIBULL DISTRIBUTION

As stated in Section 2.3, the failure rate $z(t)$ will often decrease in the burn-in period and increase in the wear-out period. For the majority of mechanical units $z(t)$ will also be slightly increasing in the useful life period. In such situations the exponential distribution will not be a realistic model.

A distribution often used when $z(t)$ is monotonic is the *Weibull distribution*. Its applicability to a wide range of failure situations was discussed by Weibull (1951). The time to failure T of a unit is said to be Weibull distributed with parameters $\alpha(> 0)$ and $\lambda(> 0)$ if the distribution function is given by

$$F(t) = P(T \leq t) = \begin{cases} 1 - e^{-(\lambda t)^\alpha} & \text{for } t > 0 \\ 0 & \text{otherwise} \end{cases} \tag{2.33}$$

The corresponding probability density is

$$f(t) = \frac{d}{dt} F(t) = \begin{cases} \alpha \lambda (\lambda t)^{\alpha-1} e^{-(\lambda t)^\alpha} & \text{for } t > 0 \\ 0 & \text{otherwise} \end{cases} \tag{2.34}$$

where λ is a *scale* parameter, and α is a *shape* parameter.

The reliability (survivor) function is

$$R(t) = P(T > 0) = e^{-(\lambda t)^\alpha} \qquad \text{for } t > 0 \tag{2.35}$$

and the failure rate is

$$z(t) = \frac{f(t)}{R(t)} = (\alpha \lambda)(\lambda t)^{\alpha-1} \qquad \text{for } t > 0 \tag{2.36}$$

When $\alpha = 1$, the failure rate is constant, and the Weibull distribution reduces

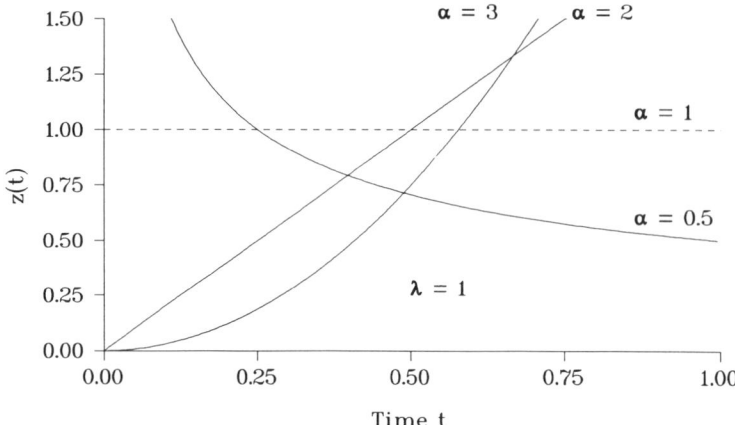

Figure 2.11 Failure rate of the Weibull distribution, $\lambda = 1$

to an exponential distribution as illustrated in Figure 2.11. When $\alpha > 1$, the failure rate is increasing, and when $0 < \alpha < 1$, $z(t)$ is decreasing. When $\alpha = 2$, the resulting distribution is known as the *Rayleigh distribution*.

The Weibull distribution is flexible and may be used to model life distributions, where the failure rate is decreasing, constant, or increasing.

From (2.35) it follows that

$$R\left(\frac{1}{\lambda}\right) = \frac{1}{e} \approx 0.3679 \qquad \text{for all } \alpha > 0$$

Hence $P(T > 1/\lambda) = 1/e$, independent of α. The quantity $1/\lambda$ is sometimes called the *characteristic* lifetime.

The mean time to failure is

$$\text{MTTF} = \int_0^\infty R(t)\,dt = \frac{1}{\lambda} \cdot \Gamma\left(\frac{1}{\alpha} + 1\right) \qquad (2.37)$$

The variance of T is

$$\text{var}(T) = \frac{1}{\lambda^2}\left(\Gamma\left(\frac{2}{\alpha} + 1\right) - \Gamma^2\left(\frac{1}{\alpha} + 1\right)\right) \qquad (2.38)$$

Note that $\text{MTTF}/\sqrt{\text{var}(T)}$ is independent of λ.

The Weibull distribution also arises as a limit distribution for the smallest of a large number of independent, identically distributed, nonnegative random variables. The Weibull distribution is therefore often called the *weakest link distribution*. This is further discussed in Section 2.14.

The Weibull distribution has been widely used in reliability analysis of semiconductors, ball bearings, engines, spot weldings, and biological organisms, for example.

Example 2.4

Consider a series system of n components. The corresponding lifetimes $T_1, T_2, \ldots, T_n$ are assumed to be independent and Weibull distributed:

$$T_i \sim \text{Weibull}(\alpha, \lambda_i) \quad \text{for } i = 1, 2, \ldots, n$$

A series system fails as soon as the first component fails. The time to failure of the system, T_s is thus

$$T_s = \min\{T_1, T_2, \ldots, T_n\}$$

The survivor function of this system becomes

$$R_s(t) = P(T > t) = P(\min_{1 \le i \le n} T_i > t) = \prod_{i=1}^{n} P(T_i > t)$$

$$= \prod_{i=1}^{n} e^{-(\lambda_i t)^\alpha} = e^{-(\sum_{i=1}^{n} \lambda_i t)^\alpha} = e^{-(\sum_{i=1}^{n} \lambda_i^\alpha) \cdot t^\alpha}$$

Hence a series system of independent components with Weibull life distribution and the same shape parameter α again has a Weibull distribution, with the shape parameter being unchanged. □

The Weibull distribution we have discussed so far, is a two-parameter distribution with shape parameter $\alpha > 0$ and scale parameter $\lambda > 0$. A natural extension of this distribution is the three-parameter Weibull distribution (α, λ, ξ) with distribution function

$$F(t) = P(T \le t) = \begin{cases} 1 - e^{-[\lambda(t-\xi)]^\alpha} & \text{for } t > \xi \\ 0 & \text{otherwise} \end{cases}$$

The corresponding density is

$$f(t) = \frac{d}{dt} F(t) = \alpha \lambda [\lambda(t - \xi)]^{\alpha-1} e^{-[\lambda(t-\xi)]^\alpha} \quad \text{for } t > \xi$$

The third parameter ξ is sometimes called the *guarantee* or *threshold* parameter, since a failure occurs before time ξ with probability 0 (e.g., see Mann, Schafer, and Singpurwalla 1974, p. 185).

Since $(T - \xi)$ obviously has a two-parameter Weibull distribution (α, λ), the mean and variance of the three-parameter Weibull distribution (α, λ, ξ) follows from (2.37) and (2.38):

$$\text{MTTF} = \xi + \frac{1}{\lambda} \Gamma\left(\frac{1}{\alpha} + 1\right)$$

$$\text{var}(T) = \frac{1}{\lambda^2}\left(\Gamma\left(\frac{2}{\alpha} + 1\right) - \Gamma^2\left(\frac{1}{\alpha} + 1\right)\right)$$

In statistical literature reference to the Weibull distribution usually means the two-parameter family, unless otherwise specified.

2.10 THE NORMAL DISTRIBUTION

The most commonly used distribution in statistics is the normal (Gaussian) distribution. A random variable T is said to be normally distributed with mean ν and variance τ^2, $T \sim \mathcal{N}(\nu, \tau^2)$, when the probability density of T is

$$f(t) = \frac{1}{\sqrt{2\pi} \cdot \tau} e^{-(t-\nu)^2/2\tau^2} \qquad \text{for } -\infty < t < \infty \qquad (2.39)$$

The $\mathcal{N}(0, 1)$ distribution is called the *standard normal distribution*. The distribution function of the standard normal distribution is usually denoted by $\Phi(\cdot)$. The probability density of the standard normal distribution is

$$\phi(t) = \frac{1}{\sqrt{2\pi}} e^{-t^2/2} \qquad (2.40)$$

The distribution function of $T \sim \mathcal{N}(\nu, \tau^2)$ may be written as

$$F(t) = P(T \leq t) = \Phi\left(\frac{t - \nu}{\tau}\right) \qquad (2.41)$$

The normal distribution is sometimes used as a lifetime distribution, even though it allows negative values with positive probability.

The survivor function is

THE NORMAL DISTRIBUTION

$$R(t) = 1 - \Phi\left(\frac{t-\nu}{\tau}\right) \quad (2.42)$$

The failure rate of the normal distribution is

$$z(t) = -\frac{R'(t)}{R(t)} = \frac{1}{\tau} \cdot \frac{\phi((t-\nu)/\tau)}{1 - \Phi((t-\nu)/\tau)} \quad (2.43)$$

If $z_\Phi(t)$ denotes the failure rate of the standard normal distribution, the failure rate of $\mathcal{N}(\nu, \tau^2)$ is seen to be

$$z(t) = \frac{1}{\tau} \cdot z_\Phi\left(\frac{t-\nu}{\tau}\right)$$

The failure rate of the standard normal distribution, $\mathcal{N}(0, 1)$, is illustrated in Figure 2.12. The failure rate is *increasing* for all t and approaches $z(t) = t$ when $t \to \infty$.

When a random variable has a normal distribution but with an upper bound and/or a lower bound for the values of the random variable, the resulting distribution is called a *truncated normal distribution*. When there is only a lower bound, the distribution is said to be left truncated. When there is only an upper bound, the distribution is said to be right truncated. Should there be an upper as well as a lower bound, it is said to be *doubly truncated*.

A normal distribution, left truncated at 0, is sometimes used as a *life distribution*. This left-truncated normal distribution has survivor function

$$R(t) = P(T > t | T > 0) = \frac{\Phi((\nu - t)/\tau)}{\Phi(\nu/\tau)} \quad \text{for } t \geq 0 \quad (2.44)$$

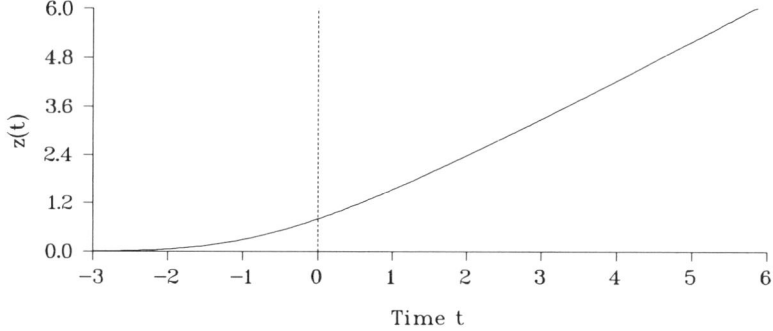

Figure 2.12 Failure rate of the standard normal distribution $\mathcal{N}(0, 1)$.

The corresponding failure rate becomes

$$z(t) = \frac{-R'(t)}{R(t)} = \frac{1}{\tau} \cdot \frac{\phi((t-\nu)/\tau)}{1 - \Phi((t-\nu)/\tau)} \quad \text{for } t \geq 0$$

Note that the failure rate of the left truncated normal distribution is identical to the failure rate of the (untruncated) normal distribution when $t \geq 0$.

2.11 THE LOGNORMAL DISTRIBUTION

The time to failure T of a unit is said to be lognormally distributed if $Y = \ln T$ is normally (Gaussian) distributed. If ν and τ are, respectively, the mean and variance of Y [i.e., $Y \sim \mathcal{N}(\nu, \tau^2)$], then the corresponding probability density of T is

$$f(t) = \begin{cases} \frac{1}{\sqrt{2\pi}} \frac{1}{\tau} \frac{1}{t} e^{-(\ln t - \nu)^2/2\tau^2} & \text{for } t > 0 \\ 0 & \text{otherwise} \end{cases} \quad (2.45)$$

The lognormal probability density is sketched in Figure 2.13 for selected values of ν and τ.

The mean time to failure is

$$\text{MTTF} = e^{\nu + \tau^2/2} \quad (2.46)$$

The variance of T is

$$\text{var}(T) = e^{2\nu}(e^{2\tau^2} - e^{\tau^2}) \quad (2.47)$$

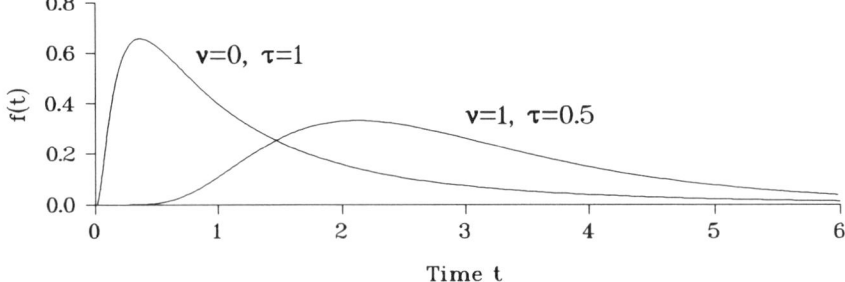

Figure 2.13 Probability density of the lognormal distribution

The reliability (survivor) function becomes

$$R(t) = P(T > t) = P(\ln T > \ln t)$$
$$= P\left(\frac{\ln T - \nu}{\tau} > \frac{\ln t - \nu}{\tau}\right) = \Phi\left(\frac{\nu - \ln t}{\tau}\right) \quad (2.48)$$

where $\Phi(\cdot)$ is the distribution function of the standard normal distribution. The median t_M, satisfying $R(t_M) = 0.5$, is

$$t_M = e^{\nu} \quad (2.49)$$

The failure rate of the lognormal distribution is

$$z(t) = -\frac{d}{dt}\left(\ln \Phi\left(\frac{\nu - \ln t}{\tau}\right)\right) = \frac{\phi((\nu - \ln t)/\tau)/\tau t}{\Phi((\nu - \ln t)/\tau)} \quad (2.50)$$

where $\phi(t)$ denotes the probability density of the standard normal distribution.

The shape of $z(t)$, which is illustrated in Figure 2.14, is discussed in detail by Sweet (1990) who describes an iterative procedure to compute the time t for which the failure rate attains its maximum value. He also proves that $z(t) \to 0$ when $t \to \infty$.

Repair Time Distribution

The lognormal distribution is commonly used as a distribution for repair time. The *repair rate* is defined analogous to the failure rate. When modeling the repair time, it is natural to assume that the repair rate is increasing, at least in a first phase. This means that the probability of completing the repair action within a short interval increases with the elapsed repair time. When the repair

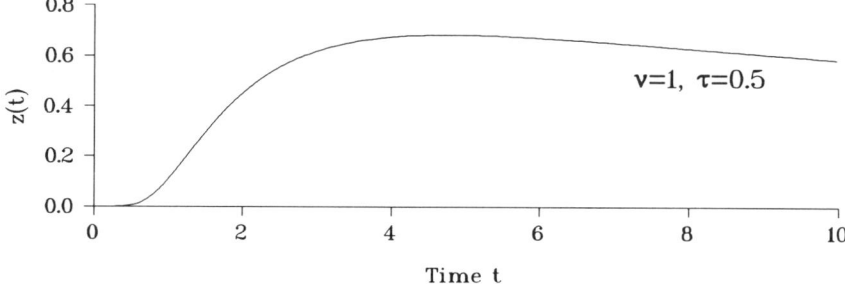

Figure 2.14 Failure rate of the lognormal distribution

has been going on for a rather long time, this indicates serious problems, for example, that there are no spare parts available on the site. It is therefore natural to believe that the repair rate is decreasing after a certain period of time, namely, that the repair rate has the same shape as the failure rate of the lognormal distribution illustrated in Figure 2.14.

Median and Error Factor

In some cases we may be interested to find an interval (t_L, t_U) such that $P(t_L < T \leq t_U) = 1 - 2\alpha$, for example. If the interval is symmetric in the sense that $P(T \leq t_L) = \alpha$ and $P(T > t_U) = \alpha$, it is easy to verify that $t_L = e^{\nu - u_\alpha \tau}$ and $t_U = e^{\nu + u_\alpha \tau}$, where u_α is the upper $\alpha\%$ percentile of the standard normal distribution [i.e., $\Phi(u_\alpha) = 1 - \alpha$]. By introducing the median $t_M = e^\nu$ and $k = e^{u_\alpha \tau}$, the lower limit t_L and the upper limit t_U may be written

$$t_L = \frac{t_M}{k} \quad \text{and} \quad t_U = k \cdot t_M \qquad (2.51)$$

The factor k is often called the $(1 - 2\alpha)$ *error factor*. α is usually chosen to be 0.05.

Example 2.5
In many situations the failure rate λ may vary from one unit to another. In the Reactor Safety Study, WASH-1400 (1975), the variation (uncertainty) in λ was modeled by a lognormal distribution; that is, the failure rate λ is regarded as a random variable with a lognormal distribution. The uncertainty in λ was expressed by an interval (λ_L, λ_U) where $P(\lambda_L < \lambda \leq \lambda_U) = 0.90$. Since a lognormal distribution is assumed, this interval can according to (2.51) be presented as a median failure rate λ_M and a 90% error factor k such that

$$P\left(\frac{\lambda_M}{k} < \lambda \leq k \cdot \lambda_M\right) = 0.90$$

If we, as an example, choose the median to be $\lambda_M = 6.0 \cdot 10^{-5}$ failures per hour and an error factor $k = 3$, then a 90% interval is equal to $(2.0 \cdot 10^{-5}, 1.8 \cdot 10^{-4})$. The parameters ν and τ of the lognormal distribution can now be determined from (2.49) and (2.51):

$$\nu = \ln(\lambda_M) = \ln 6.0 \cdot 10^{-5} \approx -9.721$$
$$\tau = \frac{1}{1.645} \ln k = \frac{1}{1.645} \ln 3 \approx 0.668$$

With these parameter values the MTTF is equal to

$$\text{MTTF} = e^{\nu + \tau^2/2} \approx 1.47 \cdot 10^{-4} \qquad \square$$

Example 2.6: Fatigue Analysis
The lognormal distribution is also commonly used in the analysis of fatigue failures. Consider the following simple situation: A smooth, polished test rod of steel is exposed to sinusoidal stress cycles with a given stress range (double amplitude) s. We want to estimate the time to failure of the test rod (i.e., the number of stress cycles N) until fracture occurs. In this situation it is usually assumed that N is lognormally distributed. The justification for this is partly physical and partly mathematical convenience. A fatigue crack will always start in an area with local yield, normally caused by an impurity in the material. It seems reasonable that in the beginning the failure rate increases with the number of stress cycles. If the test rod survives a large number of stress cycles, this indicates that there are very few impurities in the material. It is therefore to be expected that the failure rate will decrease when the possibility for impurities in the material is reduced.

It is known that within a limited area of the stress range s, the number N of cycles to failure will roughly satisfy the equation

$$Ns^b = c \qquad (2.52)$$

where b and c are constants depending on the material and the geometry of the test rod. They may also depend on the surface treatment and the environment in which the rod is used.

By taking the logarithms of both sides of (2.52), we get

$$\ln N = \ln c - b \ln s \qquad (2.53)$$

If we introduce $Y = \ln N$, $\alpha = \ln c$, $\beta = -b$, and $x = \ln s$, it follows from (2.53) that Y roughly can be expressed by the relation

$$Y = \alpha + \beta x + \text{random error}$$

If N is assumed to be lognormally distributed, then $Y = \ln N$ will be normally distributed, and the usual theory for linear regression models (e.g., see Dudewicz and Mishra 1988) applies when estimating the expected number of cycles to failure for a given stress range s. Equation (2.52) represents the so-called Wöhler or s-N diagram for the test rod. Such a diagram is illustrated in Figure 2.15.

46 FAILURE MODELS

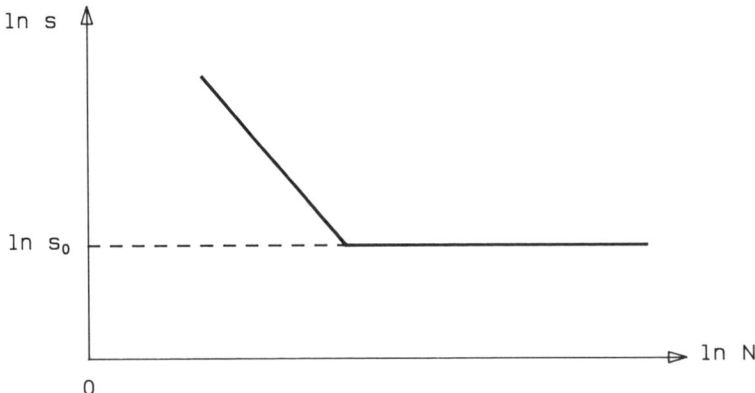

Figure 2.15 Wöhler or s-N diagram

When the stress range is below a certain value s_0, the test rod will not fracture, irrespective of how many stress cycles it is exposed to. Formula (2.53) will therefore be valid only for stress values above s_0. The stress range s_0 is called the *fatigue limit*. For certain materials such as aluminium the Wöhler curve has no horizontal asymptote. Such materials therefore have no fatigue limit. In a corrosive environment such as salt water, neither does steel have any fatigue limit. □

2.12 THE BIRNBAUM-SAUNDERS DISTRIBUTION

If the fatigue cannot be attributed to sinusoidal stress cycles, formula (2.52) cannot be applied directly. The oldest, simplest, and most used procedure in this situation is Miner's rule (Miner 1945): The various applied stress ranges are divided into a certain number m of discrete stress ranges $s_1, s_2, \ldots, s_m$. Let n_j denote the numbers of stress cycles occurring under stress range s_j, $j = 1, 2, \ldots, m$. Then, according to Miner's rule, fatigue fracture will occur when

$$\frac{n_1}{N_1} + \frac{n_2}{N_2} + \cdots + \frac{n_m}{N_m} = 1 \qquad (2.54)$$

where N_j is the hypothetic number of stress cycles until fracture if the unit were exposed to pure sinusoidal stress cycles with stress range s_j, $j = 1, 2, \ldots, m$. Experience shows that by using Miner's rule the average time to failure will often be overestimated.

Birnbaum and Saunders (1969) introduced a life distribution based on a stochastic interpretation of Miner's rule. They argue as follows: Consider a unit

THE BIRNBAUM-SAUNDERS DISTRIBUTION

that is exposed to a sequence of work cycles each of which is composed of m individual stresses. Each of the work cycles is the source of a partial damage that is stochastic and may depend on factors such as the material and the number of earlier stresses. Assume that the increase Z_j in partial damage in work cycle j is a random variable with mean value μ and variance σ^2 for all $j = 1, 2, \ldots$. Furthermore assume that increases in partial damage due to different work cycles are independent of each other.

Let $W_n = Z_1 + Z_2 + \cdots + Z_n$ denote the total partial damage after n work cycles. Further let N be the smallest number of work cycles for which W_n exceeds the critical value that causes a fatigue fracture to occur. Then we have

$$P(N \leq n) \approx P(W_n > \omega) = 1 - P\left(\sum_{i=1}^{n} Z_i \leq \omega\right)$$

$$= 1 - P\left(\sum_{i=1}^{n} \frac{Z_i - \mu}{\sigma\sqrt{n}} \leq \frac{\omega - n\mu}{\sigma\sqrt{n}}\right) \quad (2.55)$$

Since the Z_i's are assumed to be independent and identically distributed, W_n for large n is approximately normally distributed $\mathcal{N}(n\mu, n\sigma^2)$ by the central limit theorem (e.g., see Dudewicz and Mishra 1988, p. 315). Thus

$$P(N \leq n) \approx \Phi\left(\frac{\omega - n\mu}{\sigma\sqrt{n}}\right) \quad (2.56)$$

This discrete model is now extended to a continuous model by replacing "life length" N by T, where T is a continuous variable.

The life distribution for the unit is then

$$F(t) = P(T \leq t) \approx 1 - \Phi\left(\frac{\omega - t\mu}{\sigma\sqrt{t}}\right) \quad (2.57)$$

By introducing $\alpha = \sigma/\sqrt{\mu\omega}$ and $\lambda = \mu/\omega$, we obtain

$$F(t) \approx \Phi\left(\frac{1}{\alpha}\left(\sqrt{\lambda t} - \frac{1}{\sqrt{\lambda t}}\right)\right) \quad (2.58)$$

The distribution on the right-hand side is called the *Birnbaum-Saunders* distribution with shape parameter α and scale parameter λ.

The probability density of the Birnbaum-Saunders distribution is

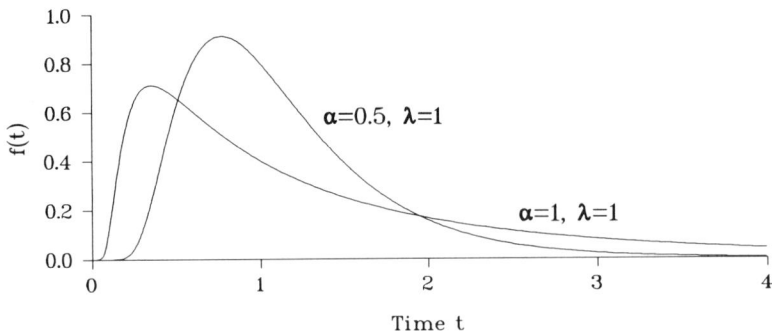

Figure 2.16 Probability density of the Birnbaum-Saunders distribution

$$f(t) = \frac{\sqrt{\lambda t} + 1/\sqrt{\lambda t}}{2\alpha t} \frac{1}{\sqrt{2\pi}} e^{-(\sqrt{\lambda t} - (1/\sqrt{\lambda t}))^2/\alpha^2} \quad \text{for } t > 0 \quad (2.59)$$

This probability density is plotted for selected values of α in Figure 2.16. It can furthermore be shown that

$$U = \frac{1}{\alpha}\left(\sqrt{\lambda T} - \frac{1}{\sqrt{\lambda T}}\right)$$

has a standard normal distribution and that

$$\text{MTTF} = \frac{1}{\lambda}\left(1 + \frac{\alpha^2}{2}\right) \quad (2.60)$$

$$\text{var}(T) = \frac{\alpha^2}{\lambda^2}\left(1 + \frac{5\alpha^2}{4}\right) \quad (2.61)$$

2.13 THE INVERSE GAUSSIAN DISTRIBUTION

In some situations the failure rate $z(t)$ appears to be an increasing function of time t from $t = 0$ to $t = t_0$ (unknown) and from t_0 on monotonically decreasing. Among the life distributions having the mentioned property is the lognormal distribution. This distribution is in such situations often "shown to be the appropriate one" by use of some goodness of fit criterion, and then used without

THE INVERSE GAUSSIAN DISTRIBUTION

further consideration. However, if one studies the lognormal distribution a little more closely, it can be shown (Sweet 1990) that the corresponding failure rate, after having reached its maximum, decreases monotonically to *zero* as $t \to \infty$. A natural question is then: Is this a reasonable assumption? In many practical situations this will *not* be the case, and hence the lognormal life distribution should *not* be used as model, even if the goodness-of-fit test does not reject it.

We now turn to another life distribution, namely the *inverse Gaussian* distribution. The inverse Gaussian, like the lognormal distribution, has a failure rate that increases with time t, from $t = 0$ to $t = t_0$ (unknown), and thereafter monotonically decreases *not toward zero* but toward a limit, depending on the parameters of the distribution function. (As we will see, the name "inverse Gaussian" is rather misleading.)

For the reason mentioned above, we claim that in many situations where the lognormal model has been used, it would have been more appropriate to apply the inverse Gaussian distribution. The probability density of the inverse Gaussian distribution can be expressed in many different ways, depending on how the distribution is parameterized. We are going to use the following form:

$$f_T(t; \mu, \lambda) = \sqrt{\frac{\lambda}{2\pi t^3}}\, e^{-(\lambda/2\mu^2)[(t-\mu)^2/t]} \quad \text{for } t > 0,\ \mu > 0,\ \lambda > 0$$
$$= 0 \quad \text{otherwise} \tag{2.62}$$

This distribution will be referred to as the inverse Gaussian distribution with parameters μ and λ; $I_G(\mu, \lambda)$. The shape of the probability density can be varied considerably by varying μ and λ (see Figures 2.17 and 2.18). Figure 2.19 shows graphs of the corresponding failure rate for $\mu = 1$ and selected values of λ. In Chhikara and Folks (1977, p. 155) the failure rate is shown to converge to $\lambda/2\mu^2$ as $t \to \infty$.

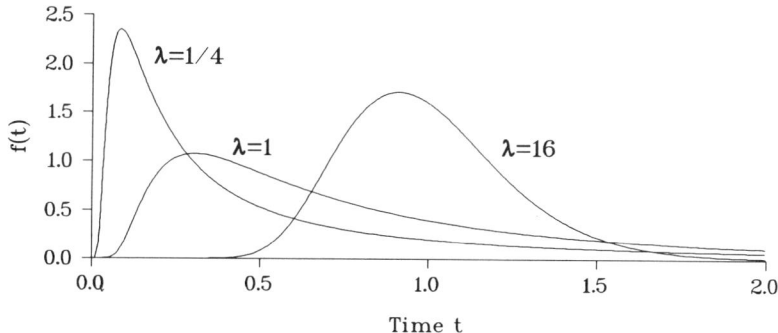

Figure 2.17 The probability density of the inverse Gaussian distribution for the parameters $\mu = 1$, and $\lambda = \frac{1}{4}, 1, 16$

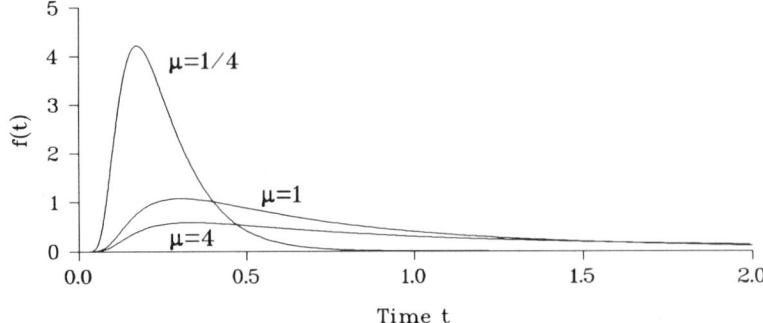

Figure 2.18 The probability density of the inverse Gaussian distribution for the parameters $\mu = \frac{1}{4}, 1, 4,$ and $\lambda = 1$

In Section 2.12 we presented a heuristic argument leading to the Birnbaum-Saunders distribution. Their derivation has been criticized by Bhattacharayya and Fries (1982), who instead propose that the accumulated fatigue in the time interval $(0, t]$, $W(t)$ should be governed by a Wiener process $\{W(t); 0 < t < \infty\}$ with positive drift η and diffusion constant δ^2; that is, for any $(0 \leq s < t)$ the distribution of $(W(t) - W(s))$ is $N(\eta(t-s), \delta^2(t-s))$ (e.g., see Cox and Miller 1965). Furthermore the unit in question is supposed to fail when $W(t)$ for the first time exceeds ω (the critical level of failure). Accordingly the time T to failure is defined as

$$T = \inf_{t} \{t; W(t) > \omega\}$$

Under these assumptions it follows (Cox and Miller 1965, p.221) that the distribution of T is given by

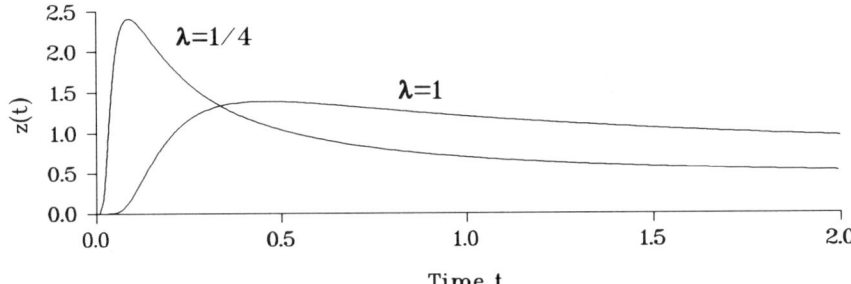

Figure 2.19 Failure rate of the inverse Gaussian distribution ($\mu = 1$ and $\lambda = 0.25, 1$)

THE INVERSE GAUSSIAN DISTRIBUTION

$$F_T(t) = \Phi\left(\frac{\eta}{\delta}\sqrt{t} - \frac{\omega}{\delta}\frac{1}{\sqrt{t}}\right)$$
$$+ \Phi\left(-\frac{\eta}{\delta}\sqrt{t} - \frac{\omega}{\delta}\frac{1}{\sqrt{t}}\right) \cdot e^{2\eta\omega/\delta^2} \quad \text{for } t > 0 \quad (2.63)$$

where $\Phi(\cdot)$ denotes the distribution function of the standard normal distribution. By introducing new parameters

$$\mu = \frac{\omega}{\eta}, \quad \lambda = \frac{\omega^2}{\delta^2} \quad \text{for } \mu > 0, \lambda > 0 \quad (2.64)$$

$F_T(t)$ can be written

$$F_T(t) = \Phi\left(\frac{\sqrt{\lambda}}{\mu}\sqrt{t} - \sqrt{\lambda}\frac{1}{\sqrt{t}}\right)$$
$$+ \Phi\left(-\frac{\sqrt{\lambda}}{\mu}\sqrt{t} - \sqrt{\lambda}\frac{1}{\sqrt{t}}\right) \cdot e^{2\lambda/\mu}$$
$$\text{for } t > 0, \mu > 0, \lambda > 0. \quad (2.65)$$

The corresponding probability density becomes

$$f_T(t) = \sqrt{\frac{\lambda}{2\pi t^3}} \cdot e^{-(\lambda/2\mu^2)[(t-\mu)^2/t]} \quad \text{for } t > 0, \mu > 0, \lambda > 0 \quad (2.66)$$

which we recognize as the probability density of $IG(\mu, \lambda)$. Without loss of generality, ω can be chosen to be 1. Then μ corresponds to $1/\eta$ and λ to $1/\delta^2$.

This distribution was first derived by Schrödinger (1915) in connection with his studies of Brownian motion. In an attempt to extend Schrödingers results, Tweedie (1957) noticed the inverse relationship between the cumulant-generating function of the time to cover a unit distance and the cumulant-generating function of the distance covered in unit time. In 1956 he used for the first time "inverse Gaussian" for the first passage time of the Brownian motion. Thereby the distribution got the rather misleading name "inverse Gaussian."

Wald (1947) derived the same distribution as a limiting distribution for the sample size in connection with a certain sequential probability ratio test. A heuristic derivation is given by Withmore and Seshadre (1987).

The moment-generating function (see Dudewicz and Mishra 1988, p. 255) of the $IG(\mu, \lambda)$ turns out to be

$$M_T(t;\mu,\lambda) = e^{\lambda(1-(1-2\mu^2 t/\lambda)^{1/2}/\mu}$$ (2.67)

and the corresponding cumulant-generating function, defined by

$$L_T(t;\mu,\lambda) = \ln M_T(t;\mu,\lambda)$$ (2.68)

becomes

$$L_T(t;\mu,\lambda) = \frac{\lambda}{\mu}\left(1 - \left(1 - \frac{2\mu^2 t}{\lambda}\right)^{1/2}\right)$$ (2.69)

The mean and variance of $IG(\mu,\lambda)$ are now easily determined

$$\text{MTTF} = \kappa_1 = \frac{d}{dt}L_T(t;\mu,\lambda)|_{t=0} = \mu$$ (2.70)

$$\text{var}(T) = \kappa_2 = \frac{d^2}{dt^2}L_T(t;\mu,\lambda)|_{t=0} = \frac{\mu^3}{\lambda}$$ (2.71)

Notice that μ enters in MTTF $= \mu$, as well as into var(T). Hence (μ, λ) are *not* location/scale parameters in the usual sense. Also notice that

$$\lambda = \frac{\kappa_1^3}{\kappa_2}$$ (2.72)

From (2.67) it follows that moments of arbitrarily high (positive) order exist for $IG(\mu,\lambda)$. From the result stated in Problem 26 in Section 2.18, it follows that moments of arbitrarily high "negative" order also exist.

The failure rate of $IG(\mu,\lambda)$ is

$$z(t) = z(t;\mu,\lambda) = \frac{f_T(t;\mu,\lambda)}{1 - F_T(t;\mu,\lambda)} \quad \text{for } t > 0, \mu > 0, \lambda > 0$$

$$= \frac{\sqrt{\frac{\lambda}{2\pi t^3}}\, e^{-(\lambda/2\mu^2)[(t-\mu^2)/t]}}{\Phi(\sqrt{\lambda}/\sqrt{t} - \sqrt{\lambda}\sqrt{t}/\mu) - \Phi(-\sqrt{\lambda}/\sqrt{t} - \sqrt{\lambda}\sqrt{t}/\mu)\cdot e^{2\lambda/\mu}}$$ (2.73)

for $t > 0$, $\mu > 0$, and $\lambda > 0$. In Figure 2.19 graphs of the failure rate (2.73) are given for $\mu = 1$ and selected values of λ.

Chhikara and Folks (1989, sec. 1.3) point out a surprising analogy between sampling results for the inverse Gaussian distribution $IG(\mu,\lambda)$ and sampling results known for the normal distribution $\mathcal{N}(\mu,\tau^2)$. If one considers a random sample $T_1, T_2, \ldots, T_n$ from the inverse Gaussian distribution $IG(\mu,\lambda)$,

1. $\overline{T} = 1/n \sum_{j=1}^{n} T_j$ is inverse Gaussian $IG(\mu, n\lambda)$
2. $n\lambda \sum_{j=1}^{n}(1/T_j - 1/\overline{T})$ is χ^2 distributed with $(n-1)$ degrees of freedom
3. $\overline{T}$ and $\sum_{j=1}^{n}(1/T_j - 1/\overline{T})$ are independently distributed
4. The term in the exponent of the probability density function of $IG(\mu,\lambda)$ is $-\frac{1}{2}$ times a χ^2 distributed variable

The explanation of these and other analogies with the normal distribution has yet not been revealed.

2.14 THE EXTREME VALUE DISTRIBUTIONS

Extreme value distributions play an important role in reliability analysis. They arise in a natural way, for example, in the analysis of engineering systems made up of n identical units with a series structure, and also in the study of corrosion of metals, of material strength, and of breakdown of dielectrics.

Let $T_1, \ldots, T_n$ be independent, identically distributed random variables (not necessarily life lengths) with a continuous distribution function $F_T(t)$, which, for the sake of simplicity, is assumed to be strictly increasing for $F_T^{-1}(0) < t < F_T^{-1}(1)$. Then

$$T_{(1)} = \min\{T_1, \ldots, T_n\} = U_n \qquad (2.74)$$
$$T_{(n)} = \max\{T_1, \ldots, T_n\} = V_n \qquad (2.75)$$

are called the *extreme values*.

The distribution functions of U_n and V_n are easily expressed by $F_T(\cdot)$ in the following way (e.g., see Cramér 1946, p. 370; Mann, Schafer, and Singpurwalla 1974, p. 102):

$$F_{U_n}(u) = 1 - (1 - F_T(u))^n = L_n(u) \qquad (2.76)$$

and

$$F_{V_n}(v) = F_T(v)^n = H_n(v) \qquad (2.77)$$

Despite the simplicity of (2.76) and (2.77), these formulas are usually not easy to work with. If $F_T(t)$, say, represents a normal distribution, one is led to work with powers of $F_T(t)$, which may be cumbersome.

However, in many practical applications to reliability problems, n is very large. Hence one is led to look for asymptotic techniques, which under general conditions on $F_T(t)$ may lead to simple representations of $F_{U_n}(u)$ and $F_{V_n}(v)$.

Cramér (1946) suggested the following approach: Introduce

$$Y_n = nF_T(U_n) \qquad (2.78)$$

where U_n is defined as in (2.74). Then for $y \geq 0$,

$$\begin{aligned} P(Y_n \leq y) &= P\left(F_T(U_n) \leq \frac{y}{n}\right) \\ &= P\left[U_n \leq F_T^{-1}\left(\frac{y}{n}\right)\right] \\ &= F_{U_n}\left[F_T^{-1}\left(\frac{y}{n}\right)\right] \\ &= 1 - \left[1 - F_T\left(F_T^{-1}\left(\frac{y}{n}\right)\right)\right] \\ &= 1 - \left(1 - \frac{y}{n}\right)^n \end{aligned} \qquad (2.79)$$

As $n \to 0$

$$P(Y_n \leq y) \to 1 - e^{-y} \qquad \text{for } y > 0 \qquad (2.80)$$

Since the right-hand side of (2.80) expresses a distribution function,[2] continuous for $y > 0$, this implies that Y_n converges in distribution to a random variable Y, with distribution function

$$F_Y(y) = 1 - e^{-y} \qquad \text{for } y > 0 \qquad (2.81)$$

Hence it follows from (2.76) that the sequence of random variables U_n converges in distribution to the random variable $F_T^{-1}(Y/n)$.

$$U_n \xrightarrow{\mathcal{L}} F_T^{-1}\left(\frac{Y}{n}\right) \qquad (2.82)$$

Similarly, let

$$Z_n = n(1 - F_T(V_n)) \qquad (2.83)$$

where V_n is defined in (2.75). By an analogous argument it can be shown that

---

[2] This is the exponential distribution with parameter $\lambda = 1$.

for $z > 0$:

$$P(Z_n \leq z) = 1 - \left(1 - \frac{z}{n}\right)^n \tag{2.84}$$

which implies that

$$V_n \xrightarrow{\mathscr{L}} F_T^{-1}\left(1 - \frac{Z}{n}\right) \tag{2.85}$$

where Z has distribution function

$$P(Z \leq z) = 1 - e^{-z} \quad \text{for } z > 0 \tag{2.86}$$

It is to be expected that the limit distribution of U_n and V_n will depend on the type of distribution $F_T(\cdot)$. However, it turns out that there are only three possible types of limiting distributions for the minimum extreme U_n and only three possible types of limiting distributions for the maximum extreme V_n.

For a comprehensive discussion of the application of extreme value theory to reliability analysis, the reader is referred to Mann, Schafer, and Singpurwalla (1974, p. 106), Nelson (1982, p. 39), Lawless (1982, p. 169), and Johnson and Kotz (1970, vol. 1, sec. 2.1). Here we will content ourselves with mentioning three of the possible types of limiting distributions and indicate areas where they are applied.

The Gumbel Distribution of the Smallest Extreme

If the probability density $f_T(t)$ of the T_i's approaches zero exponentially as $t \to \infty$, then the limiting distribution of $U_n = T_{(1)} = \min\{T_1, \ldots, T_n\}$ is of the form

$$F_{T_{(1)}}(t) = 1 - e^{-e^{(t-\vartheta)/\alpha}} \quad -\infty < t < \infty \tag{2.87}$$

where $\alpha > 0$ and ϑ are constants. α is the mode and ϑ is a scale parameter.

The corresponding "survivor" function is

$$R_{T_{(1)}}(t) = 1 - F_{T_{(1)}}(t) = e^{-e^{(t-\vartheta)/\alpha}} \quad -\infty < t < \infty \tag{2.88}$$

Gumbel (1958) denotes this distribution the type I asymptotic distribution of the smallest extreme. It is now called the *Gumbel distribution of the smallest extreme*. If standardized variables

$$Y = \frac{T - \vartheta}{\alpha} \tag{2.89}$$

are introduced, the distribution function takes the form

$$F_{Y_{(1)}}(y) = 1 - e^{-e^y} \quad \text{for } -\infty < y < \infty$$

with probability density

$$f_{Y_{(1)}}(y) = e^y \cdot e^{-e^y} \quad \text{for } -\infty < y < \infty \tag{2.90}$$

The corresponding "failure rate" is

$$z_{Y_{(1)}}(y) = \frac{f_{Y_{(1)}}(y)}{1 - F_{Y_{(1)}}(y)} = e^y \quad \text{for } -\infty < y < \infty \tag{2.91}$$

The mean value of $T_{(1)}$ is (see Lawless 1974, p. 19)

$$E(T_{(1)}) = \vartheta - \alpha\gamma$$

where $\gamma = 0.5772\ldots$ is known as *Euler's constant*.

Since $T_{(1)}$ can take negative values, (2.87) is not a valid life distribution. A valid life distribution is, however, obtained by left truncating (2.87) at $t = 0$. In this way we get the *truncated* Gumbel distribution of the smallest extreme, which is given by the survivor function

$$R^0_{T_{(1)}}(t) = P(T_{(1)} > t \mid T_{(1)} > 0) = \frac{P(T_{(1)} > t)}{P(T_{(1)} > 0)}$$

$$= \frac{e^{-e^{(t-\vartheta)/\alpha}}}{e^{-e^{-\vartheta/\alpha}}} = e^{-e^{-(\vartheta/\alpha)(e^{t/\alpha}-1)}} \quad \text{for } t > 0 \tag{2.92}$$

By introducing new parameters $\beta = e^{-\vartheta/\alpha}$ and $\rho = 1/\alpha$, the truncated Gumbel distribution of the smallest extreme is given by the survivor function

$$R^0_{T_{(1)}}(t) = e^{-\beta(e^{\rho t}-1)} \quad \text{for } t > 0 \tag{2.93}$$

The failure rate of the truncated distribution is

$$z^0_{T_{(1)}}(t) = -\frac{d}{dt} \ln R^0_{T_{(1)}}(t) = \frac{d}{dt} \beta(e^{\rho t} - 1) = \beta\rho e^{\rho t} \quad \text{for } t \geq 0 \tag{2.94}$$

The Gumbel Distribution of the Largest Extreme

If the probability density $f_T(t)$ approaches zero exponentially as $t \to \infty$, then the limiting distribution of $V_n = T_{(n)} = \max\{T_1, \ldots, T_n\}$ is of the form

$$F_{T_{(n)}}(t) = e^{-e^{-(t-\vartheta)/\alpha}} \qquad \text{for } -\infty < t < \infty$$

where $\alpha > 0$ and ϑ are constants. Gumbel (1958) denotes this distribution the type I asymptotic distribution of the largest extreme. It is now called the *Gumbel distribution of the largest extreme.*

If standardized variables are introduced (see 2.89), the distribution takes the form

$$F_{Y_{(n)}}(y) = e^{-e^{-y}} \qquad \text{for } -\infty < y < \infty \tag{2.95}$$

with probability density

$$f_{Y_{(n)}}(y) = e^{-y} \cdot e^{-e^{-y}} \qquad \text{for } -\infty < y < \infty \tag{2.96}$$

The Weibull Distribution of the Smallest Extreme

Another type of such limiting distributions for the smallest extreme is the Weibull distribution:

$$F_{T_{(1)}}(t) = 1 - e^{((t-\vartheta)/\eta)^\beta} \qquad \text{for } t \geq \vartheta \tag{2.97}$$

where $\beta > 0$, $\eta > 0$, and $\vartheta > 0$ are constants.

Introducing standardized variables (see 2.89),

$$F_{Y_{(1)}}(y) = 1 - e^{-y^\beta} \qquad \text{for } y > 0, \beta > 0 \tag{2.98}$$

This distribution is also denoted the type III asymptotic distribution of the smallest extreme.

Example 2.7: Pitting Corrosion

Consider a steel pipe with wall thickness θ which is exposed to corrosion. Initially the surface has a certain number n of microscopic pits. Pit i has a depth D_i, for $i = 1, 2, \ldots, n$. Due to corrosion the depth of each pit will increase with time. Failure occurs when the first pit penetrates the surface, that is, when $\max\{D_1, \ldots, D_n\} = \theta$.

Let T_i denote the time pit i will need to penetrate the surface, for $i = 1, \ldots, n$. Then the time to failure T of the unit is

$$T = \min\{T_1, \ldots, T_n\}$$

We will assume that the time to penetration T_i is proportional to the remaining wall thickness, that is, $T_i = k \cdot (\theta - D_i)$. We will further assume that k is independent of time, which implies that the corrosion rate is constant.

Assume next that the random initial depths of the pits $D_1, \ldots, D_n$ are independent and identically distributed with a right-truncated exponential distribution. Then the distribution function of D_i is

$$F_{D_i}(d) = P(D_i \leq d | D_i \leq \theta) = \frac{P(D_i \leq d)}{P(D_i \leq \theta)}$$

$$= \frac{1 - e^{-\eta d}}{1 - e^{-\eta \theta}} \qquad \text{for } 0 \leq d \leq \theta.$$

The distribution function of the time to penetration T_i is thus

$$F_{T_i}(t) = P(T_i \leq t) = P(k(\theta - D_i) \leq t) = P\left(D_i \geq \theta - \frac{t}{k}\right)$$

$$= 1 - F_{D_i}\left(\theta - \frac{t}{k}\right) = \frac{e^{\eta t/k} - 1}{e^{\eta \theta} - 1} \qquad \text{for } 0 \leq t \leq k\theta \quad (2.99)$$

and the survivor function $R(t)$ of the unit becomes

$$R(t) = P(T > t) = [1 - F_{T_i}(t)]^n \qquad \text{for } t \geq 0$$

If we assume that the number n of pits is very large, then as $n \to \infty$ we get

$$R(t) = [1 - F_{T_i}(t)]^n \approx e^{-nF_{T_i}(t)} \qquad \text{for } t \geq 0$$

By using (2.99)

$$R(t) \approx e^{-n\frac{e^{\eta t/k} - 1}{e^{\eta \vartheta} - 1}} \qquad \text{for } t \geq 0$$

By introducing new parameters $\beta = n/(e^{\eta \vartheta} - 1)$ and $\rho = \eta/k$, we get

$$R(t) \approx e^{-\beta(e^{\rho t} - 1)} \qquad \text{for } t \geq 0$$

STRESSOR-DEPENDENT MODELING

which is equal to (2.93), namely the time to failure caused by pitting corrosion has approximately a truncated Gumbel distribution of the smallest extreme. This example is also discussed by Lloyd and Lipow (1962, p. 140), Mann, Schafer, and Singpurwalla (1974, p. 131), and Kapur and Lamberson (1977, p. 44).

□

2.15 STRESSOR-DEPENDENT MODELING

So far we have mainly considered parametric families of time to failure distributions, tacitly assuming that the units tested are exposed to constant stress (normal stress). If the units tested are exposed to planned variations in operational and environmental stresses (temperature, pressure, etc.) as in accelerated testing (see Chapter 10), a stressor dependent model of the life distribution is needed.

A typical approach that, if successful, leads to a parametric stressor-dependent family of time-to-failure distributions is the following:

1. Establish an appropriate parametric family of time-to-failure distributions valid under normal stress as well as under overstress.
2. Establish the functional relationship between the stressors and the parameters of the time-to-failure distribution in question.
3. Combine the result of these two steps into a stressor-dependent family of time-to-failure distributions.

Example 2.8
Let there be only one stressor s, say, temperature. Suppose that the time-to-failure distribution under normal stress as well as under overstress is a Weibull distribution (see Section 2.9). Furthermore suppose that the functional relationship between the stressor s and the parameters α and λ can be expressed as

$\lambda(s) = cs^b$, where c and b are unknown constants (power rule model)

$\alpha(s) = \alpha$, independent of s

When these two assumptions are combined, we obtain the following parametric stressor dependent time to failure distribution:

$$P(T \leq t; s) = 1 - e^{-(cs^b t)^\alpha}$$

Such models are discussed in more detail in Chapter 10. (In Chapter 10 semiparametric and nonparametric stress-dependent models are also briefly discussed.)

□

2.16 SOME FAMILIES OF DISTRIBUTIONS

IFR and DFR Distributions

In Section 2.2 we showed the following one-to-one correspondence between the distribution function $F(t)$ of a continuous life distribution and the corresponding failure rate $z(t)$:

$$F(t) = 1 - e^{-\int_0^t z(u)\,du} \qquad \text{for } t > 0 \qquad (2.100)$$

A special family of such distributions are life distributions that have increasing[3] failure rate. These are called *IFR distributions*. Similarly life distributions that have decreasing failure rate are called *DFR distributions*. A definition of IFR and DFR distributions which is not restricted to continuous distributions is given below:

Definition 2.1. A life distribution F is said to be IFR if $-\ln(1 - F(t))$ is convex for $0 < t < F^{-1}(1)$. A life distribution F is said to be DFR if $-\ln(1-F(t))$ is concave for $0 < t < F^{-1}(1)$. □

Remember that when the life distribution is continuous, the failure rate $z(t)$ can be written

$$z(t) = \frac{f(t)}{1 - F(t)} = \frac{d}{dt}(-\ln(1 - F(t))) \qquad \text{for } t > 0 \qquad (2.101)$$

Since a differentiable convex function always has an increasing derivative (for continuous life distributions), the definition given above of the IFR property corresponds to an increasing $z(t)$ (for continuous life distributions). Since a differentiable concave function always has a derivative that is decreasing, by analogy the definition of the DFR property given above corresponds to a decreasing $z(t)$ (for continuous life distributions). Let us consider some commonly used life distributions and see whether they are IFR, DFR, or neither of these.

Example 2.9: The Uniform Distribution over $(0, a]$
Let T be uniformly distributed over $(0, a]$. Then

$$F(t) = \frac{t}{a} \qquad \text{for } 0 < t \leq a$$
$$f(t) = \frac{1}{a} \qquad \text{for } 0 < t \leq a$$

---
[3] In this section the concepts "increasing" and "decreasing" are used in place of "nondecreasing" and "nonincreasing," respectively.

SOME FAMILIES OF DISTRIBUTIONS 61

Hence

$$z(t) = \frac{1/a}{1 - (t/a)} = \frac{1}{a - t} \qquad \text{for } 0 < t \leq a \qquad (2.102)$$

is strictly increasing in t for $0 < t \leq a$. The uniform distribution is accordingly IFR.

The same conclusion follows by considering $-\ln(1 - F(t))$ which in this case becomes $-\ln(1 - (t/a))$ and hence is convex for $0 < t \leq a$. □

Example 2.10: The Exponential Distribution
Let T be exponentially distributed with probability density

$$f(t) = \lambda e^{-\lambda t} \qquad \text{for } t > 0$$

Then

$$z(t) = \lambda \qquad \text{for } t > 0$$

$z(t)$ is thus constant, that is, both nonincreasing and nondecreasing.

The exponential distribution therefore belongs to the IFR family as well as the DFR family. Alternatively, one could argue that $-\ln(1 - F(t)) = \lambda t$, which is convex and concave as well. □

Hence the families of IFR distributions and DRF distributions are not disjoint. The exponential distribution can be shown to be the only continuous distribution which belongs to both families (see Barlow and Proschan, 1975, p. 73).

Example 2.11: The Weibull Distribution
The distribution function of the Weibull distribution with parameters α (>0) and λ (>0) is given by

$$F(t) = 1 - e^{-(\lambda t)^\alpha} \qquad \text{for } t \geq 0$$

It follows that

$$-\ln(1 - F(t)) = -\ln(e^{-(\lambda t)^\alpha}) = (\lambda t)^\alpha \qquad (2.103)$$

Since $(\lambda t)^\alpha$ is convex in t when $\alpha \geq 1$ and concave in t when $\alpha \leq 1$, the Weibull distribution is IFR for $\alpha \geq 1$ and DFR for $\alpha \leq 1$. For $\alpha = 1$ the distribution is "reduced" to an exponential distribution with parameter λ, and hence is IFR as well as DFR. □

Example 2.12: The Gamma Distribution

The gamma distribution is defined by the probability density

$$f(t) = \frac{\lambda}{\Gamma(\alpha)} (\lambda t)^{\alpha-1} e^{-\lambda t} \quad \text{for } t > 0$$

where $\alpha > 0$ and $\lambda > 0$. To determine whether the gamma distribution (α, λ) is IFR, DFR, or neither of these, we consider the failure rate

$$z(t) = \frac{[\lambda(\lambda t)^{\alpha-1} e^{-\lambda t}]/\Gamma(\alpha)}{\int_t^\infty [\lambda(\lambda u)^{\alpha-1} e^{-\lambda u}]/\Gamma(\alpha)\, du}$$

By dividing the denominator by the numerator we get

$$z(t)^{-1} = \int_t^\infty \left(\frac{u}{t}\right)^{\alpha-1} e^{-\lambda(u-t)}\, du$$

By introducing $v = (u - t)$ as a new variable of integration, we get

$$z(t)^{-1} = \int_0^\infty \left(1 + \frac{v}{t}\right)^{\alpha-1} e^{-\lambda v}\, dv \qquad (2.104)$$

□

First, suppose that $\alpha \geq 1$. Then $(1 + (v/t))^{\alpha-1}$ is nonincreasing in t. Accordingly the integrand is a decreasing function of t. Thus $z(t)^{-1}$ is decreasing in t. When $\alpha \geq 1$, then $z(t)$ is, in other words, increasing in t, and the gamma distribution (α, λ) is IFR. This will in particular be the case when α is an integer (the Erlangian distribution).

Next suppose that $\alpha \leq 1$. Then by an analogous argument $z(t)$ will be decreasing in t, which means that the gamma distribution (α, λ) is DRF. For $\alpha = 1$, the gamma distribution (α, λ) is reduced to an exponential distribution with parameter λ:

ooo OOO ooo

The plot of the failure rate given in Figure 2.14 for a lognormal distribution indicates that this distribution is neither IFR nor DFR. The following result may be useful when deciding whether a continuous distribution is DFR or not.

SOME FAMILIES OF DISTRIBUTIONS

Theorem 2.1. If a continuous life distribution is to be DFR, its probability density $f(t)$ must be nonincreasing.

Proof
If the time to failure distribution is DFR and continuous, $z(t) = f(t)/(1 - F(t))$ must be decreasing. Knowing that $1 - F(t)$ is decreasing in t, then $f(t)$ must decrease by at least as much as $1 - F(t)$ in order for $z(t)$ to be decreasing. □

IFRA and DFRA Distributions

Consider a life distribution $F(t)$ with failure rate $z(t)$. The cumulative failure rate is, according to Section 2.3,

$$-\int_0^t z(u)\,du = \ln R(t) = \ln(1 - F(t))$$

Definition 2.2. A distribution $F(t)$ with failure rate function $z(t)$ is said to have an increasing failure rate average (F is IFRA) if

$$-\frac{1}{t}\ln(1 - F(t)) \quad \text{increases with } t \geq 0$$

A distribution $F(t)$ with failure rate function $z(t)$ is said to have a decreasing failure rate average (F is DFRA) if

$$-\frac{1}{t}\ln(1 - F(t)) \quad \text{decreases with } t \geq 0 \qquad \square$$

The IFRA (DFRA) property demands less of the failure rate $z(t)$ than the IFR (DFR) property does. It is straightforward to verify that if F is IFR (DFR), then it is also IFRA (DFRA).

NBU and NWU Distributions

$R(t|x)$ is called the conditional survivor function of a unit of age x. The probability that a unit of age x survives an additional time interval of length t is

$$R(t|x) = P(T > t + x | T > x) = \frac{P(T > t + x)}{P(T > x)} = \frac{R(t + x)}{R(x)}$$

Definition 2.3. A distribution $F(t)$ is said to be "new better than used" (F is NBU) if

$$R(t|x) \leq R(t) \quad \text{for } t \geq 0, \quad x \geq 0$$

A distribution $F(t)$ is said to be "new worse than used" (F is NWU) if

$$R(t|x) \geq R(t) \quad \text{for } t \geq 0, \quad x \geq 0 \qquad \square$$

A distribution $F(t)$ is thus NBU (NWU) if the conditional survivor function $R(t|x)$ of a unit of age x is less (greater) than the corresponding survivor function $R(t)$ of a new unit.

NBUE and NWUE Distributions

The mean remaining lifetime of a unit at age x is

$$\text{MTTF}_x = \int_x^\infty R(t|x)\,dt$$

Definition 2.4. A life distribution $F(t)$ is said to be "new better than used in expectation" (F is NBUE) if

1. F has a finite mean μ
2. $\text{MTTF}_x \leq \mu \quad$ for $t \geq 0$

A life distribution $F(t)$ is said to be "new worse than used in expectation" (F is NWUE) if

1. F has a finite mean μ
2. $\text{MTTF}_x \geq \mu \quad$ for $t \geq 0 \qquad \square$

A Brief Comparison

The families of life distributions presented above are further discussed for example by Barlow and Proschan (1975) and Gertsbakh (1989) who proves the following chain of implications:

$$\begin{array}{ccccccc}
\text{IFR} & \Rightarrow & \text{IFRA} & \Rightarrow & \text{NBU} & \Rightarrow & \text{NBUE} \\
\text{DFR} & \Rightarrow & \text{DFRA} & \Rightarrow & \text{NWU} & \Rightarrow & \text{NWUE}
\end{array}$$

2.17 SUMMARY OF FAILURE MODELS

In the previous sections of this chapter we have discussed a number of different failure models or life distributions. Table 2.2 provides a brief reference of the main characteristics of these models.

2.18 PROBLEMS

1. A component with time to failure T has constant failure rate

$$z(t) = \lambda = 2.5 \cdot 10^{-5} \text{ (hours)}^{-1}$$

   (a) Determine the probability that the component survives a period of 2 months without failure.
   (b) Find the mean time to failure (MTTF) of the component.
   (c) Find the probability that the component survives its MTTF.

2. A machine with constant failure rate λ will survive a period of 100 hours without failure, with probability 0.50.
   (a) Determine the failure rate λ.
   (b) Find the probability that the machine will survive 500 hours without failure.
   (c) Determine the probability that the machine will fail within 1000 hours, when you know that the machine was functioning at 500 hours.

3. A component with time to failure T has failure rate

$$z(t) = kt \quad \text{for } t > 0 \text{ and } k > 0$$

   (a) Determine the probability that the component survives 200 hours, when $k = 2.0 \cdot 10^{-6}$ (hours)$^{-2}$.
   (b) Determine the mean time to failure, MTTF, of the component when $k = 2.0 \cdot 10^{-6}$ (hours)$^{-2}$.
   (c) Determine the probability that a component, which is functioning after 200 hours, is still functioning after 400 hours, when $k = 2.0 \cdot 10^{-6}$(hours)$^{-1}$.
   (d) Does this distribution belong to any of the distribution classes described in Chapter 2?

Table 2.2 Summary of Life Distributions and Parameters

Distribution	Probability Density $f(t)$	Survivor Function $R(t)$	Failure Rate $z(t)$	MTTF
Exponential	$\lambda e^{-\lambda t}$	$e^{-\lambda t}$	λ	$\dfrac{1}{\lambda}$
Gamma	$\dfrac{\lambda}{\Gamma(k)}(\lambda t)^{k-1}e^{-\lambda t}$	$\sum_{x=0}^{k-1}\dfrac{(\lambda t)^x}{x!}e^{-\lambda t}$	$\dfrac{f(t)}{R(t)}$	$\dfrac{k}{\lambda}$
Pareto	$\dfrac{k\alpha^k}{(\alpha+t)^{k+1}}$	$\left(1+\dfrac{t}{\alpha}\right)^{-k}$	$\dfrac{k}{\alpha+t}$	$\dfrac{\alpha}{k-1},\ k>1$
Weibull	$\alpha\lambda(\lambda t)^{\alpha-1}e^{-(\lambda t)^\alpha}$	$e^{-(\lambda t)^\alpha}$	$\alpha\lambda(\lambda t)^{\alpha-1}$	$\dfrac{1}{\lambda}\Gamma\left(\dfrac{1}{\alpha}+1\right)$
Lognormal	$\dfrac{1}{\sqrt{2\pi}}\dfrac{1}{\tau}\dfrac{1}{t}e^{-(\ln t-\nu)^2/2\tau^2}$	$\Phi\left(\dfrac{\nu-\ln t}{\tau}\right)$	$\dfrac{f(t)}{R(t)}$	$e^{\nu+\tau^2/2}$
Birnbaum–Saunders	$\dfrac{\sqrt{\lambda t}+\dfrac{1}{\sqrt{\lambda t}}}{2\alpha t}\dfrac{1}{\sqrt{2\pi}}e^{-(\sqrt{\lambda t}-1/\sqrt{\lambda t})^2/\alpha^2}$	$\Phi\left(\dfrac{1}{\alpha}\left(\dfrac{1}{\sqrt{\lambda t}}-\sqrt{\lambda t}\right)\right)$	$\dfrac{f(t)}{R(t)}$	$\dfrac{1}{\lambda}\left(1+\dfrac{\alpha^2}{2}\right)$
Gumbel— smallest extreme	$e^t e^{-e^t}$	e^{-e^t}	e^t	—
Inverse-Gaussian	$\sqrt{\dfrac{\lambda}{2\pi t^3}}e^{-(\lambda/2\mu^2)[(t-\mu)^2/t]}$	$\Phi\left(\sqrt{\lambda}\dfrac{1}{\sqrt{t}}-\dfrac{\sqrt{\lambda}}{\mu}\sqrt{t}\right)$ $-\Phi\left(-\sqrt{\lambda}\dfrac{1}{\sqrt{t}}-\dfrac{\sqrt{\lambda}}{\mu}\sqrt{t}\right)e^{2\lambda/\mu}$	$\dfrac{f(t)}{R(t)}$	μ

PROBLEMS

4. A component with time to failure T has failure rate

$$z(t) = \frac{t}{1+t} \quad \text{for } t > 0$$

   (a) Make a sketch of the failure rate function.
   (b) Determine the corresponding probability density function $f(t)$.
   (c) Determine the mean time to failure (MTTF) of the component.
   (d) Does this distribution belong to any of the distribution classes described in Chapter 2?

5. Let N_1 and N_2 be independent Poisson random variables with $E(N_1) = \lambda_1$ and $E(N_2) = \lambda_2$.
   (a) Determine the distribution of $N_1 + N_2$.
   (b) Determine the conditional distribution of N_1 given that $N_1 + N_2 = n$.

6. The time to failure T of a unit is assumed to have a Weibull distribution with scale parameter $\lambda = 5.0 \cdot 10^{-5}$ (hours)$^{-1}$ and shape parameter $\alpha = 1.5$. Compute MTTF and var(T).

7. The time to failure T of a unit is assumed to have a Weibull distribution with scale parameter λ and shape parameter α. Show that the rth moment of T is

$$E(T^r) = \frac{1}{\lambda^r} \Gamma\left(\frac{r}{\alpha} + 1\right)$$

8. Let T have a three-parameter Weibull distribution (α, λ, ξ) with probability density

$$f(t) = \frac{d}{dt} F(t) = \alpha\lambda[\lambda(t - \xi)]^{\alpha-1} e^{-[\lambda(t-\xi)]^\alpha} \quad \text{for } t > \xi$$

   (a) Show that the density is unimodal if $\alpha > 1$. Also show that the density decreases monotonically with t if $\alpha < 1$.
   (b) Show that the failure rate is $\alpha\lambda[\lambda(t - \xi)]^{\alpha-1}$ for $t > \xi$, and hence is increasing, constant, and decreasing with t, respectively, as $\alpha > 1$, $\alpha = 1$, and $\alpha < 1$.

9. The failure rate of a unit is $z(t) = t^{-1/2}$. Derive
   (a) The probability density function $f(t)$.

(b) The survivor function $R(t)$.

(c) The mean time to failure (MTTF).

(d) The variance of the time to failure T, $\text{var}(T)$.

10. The time to failure T has survivor function $R(t)$. Show that if $E(T^r) < \infty$, then

$$E(T^r) = \int_0^\infty rt^{r-1} R(t)\, dt \qquad \text{for } r = 1, 2, \ldots$$

11. Consider a unit with time to failure T and failure rate function $z(t)$. Show that

$$P(T > t_2 | T > t_1) = e^{-\int_{t_1}^{t_2} z(u)\, du} \qquad \text{for } t_2 > t_1$$

12. Assume the time to failure T to be lognormally distributed such that $Y = \ln T$ is $\mathcal{N}(\nu, \tau^2)$. Show that

$$E(T) = e^{\nu + \frac{1}{2}\tau^2}$$
$$\text{var}(T) = e^{2\nu}(e^{2\tau^2} - e^{\tau^2})$$

13. Let $z(t)$ denote the failure rate function of the lognormal distribution. Show that $z(0) = 0$, that $z(t)$ increases to a maximum, and then decreases with $z(t) \to 0$ as $t \to \infty$.

14. The time to failure T of a component is assumed to be uniformly distributed over $(a, b]$. The probability density is thus

$$f(t) = \frac{1}{b-a} \qquad \text{for } a < t \le b$$

Derive the corresponding survivor function $R(t)$ and failure rate $z(t)$. Draw a sketch of $z(t)$.

15. The time to failure T of a component has probability density $f(t)$ as shown in Figure 2.20.

(a) Determine c such that $f(t)$ is a valid probability density.

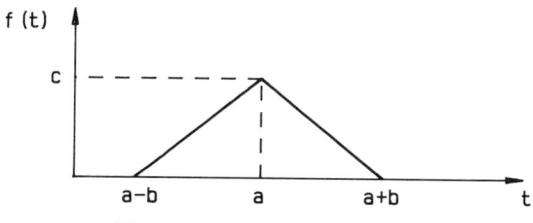

Figure 2.20 Probability density

(b) Derive the corresponding survivor function $R(t)$.

(c) Derive the corresponding failure rate $z(t)$, and make a sketch of $z(t)$.

16. Consider a system of n independent components with failure rates $\lambda_1, \lambda_2, \ldots, \lambda_n$, respectively. Show that the probability that component i fails first is

$$\frac{\lambda_i}{\sum_{j=1}^{n} \lambda_j}$$

17. A component may fail due to two different causes, excessive stresses and aging. A large number of this type of component have been tested. It has been shown that the time to failure T_1 caused by excessive stresses is exponentially distributed with density function

$$f_1(t) = \lambda_1 \cdot e^{-\lambda_1 t} \qquad \text{for } t \geq 0$$

while the time to failure T_2 caused by aging has density function

$$f_2(t) = \frac{1}{\Gamma(k)} \lambda_2 (\lambda_2 t)^{k-1} e^{-\lambda_2 t} \qquad \text{for } t \geq 0$$

(a) Describe the rationale behind using

$$f(t) = p \cdot f_1(t) + (1 - p) \cdot f_2(t) \qquad \text{for } t \geq 0$$

as the probability density function for the time to failure T of the component.

(b) Explain the meaning of p in this model.

(c) Let $p = 0.1$, $\lambda_1 = \lambda_2$, and $k = 5$, and determine the failure rate $z(t)$ corresponding to T. Calculate $z(t)$ for some selected values of t, for example,

$$t = 0, \tfrac{1}{2}, 1, 2, \ldots,$$

and make a sketch of $z(t)$.

18. A component may fail due to two different causes, A and B. It has been shown that the time to failure T_A caused by A is exponentially distributed with density function

$$f_A(t) = \lambda_A \cdot e^{-\lambda_A t} \qquad \text{for } t \geq 0$$

while the time to failure T_B caused by B has density function

$$f_B(t) = \lambda_B \cdot e^{-\lambda_B t} \qquad \text{for } t \geq 0$$

(a) Describe the rationale behind using

$$f(t) = p \cdot f_A(t) + (1 - p) \cdot f_B(t) \qquad \text{for } t \geq 0$$

as the probability density function for the time to failure T of the component.

(b) Explain the meaning of p in this model.

(c) Show that a component with probability density $f(t)$ has a decreasing failure rate (DFR).

19. Let $F(t)$ denote the distribution of the time to failure T. Assume $F(t)$ to be strictly increasing. Show that

(a) $F(T)$ is uniformly distributed over $[0, 1]$.

(b) If U is a uniform $[0, 1]$ random variable, then $F^{-1}(U)$ has distribution F, where $F^{-1}(y)$ is that value of x such that $F(x) = y$.

20. Let T_1 and T_2 be independent lifetimes with failure rates $z_1(t)$ and $z_2(t)$, respectively. Show that

$$P(T_1 < T_2 | \min\{T_1, T_2\} = t) = \frac{z_1(t)}{z_1(t) + z_2(t)}$$

PROBLEMS

21. Derive the Laplace transform of the survivor function $R(t)$ of the exponential distribution with failure rate λ, and use the Laplace transform to determine the MTTF of this distribution.

22. Let T be Weibull distributed with parameter λ and α. Show that $Y = \ln T$ has a type I asymptotic distribution of the smallest extreme (2.87). Find the mode and the scale parameter of this distribution.

23. Show that the median m of the lognormal distribution is e^{ν}. (The median m is defined by $R(m) = \frac{1}{2}$.) Compute k such that $P(m/k \leq T \leq km) = 0.90$.

24. Show that the failure rate of the Birnbaum-Saunders distribution is not monotonic.

25. Show that the inverse Gaussian distribution $IG(\mu, \lambda)$ is unimodal with mode

$$t_M = -\frac{3\mu^2}{2\lambda} + \mu\sqrt{1 + \frac{9\mu^2}{4\lambda^2}}$$

26. Show that when T has an inverse Gaussian distribution $IG(\mu, \lambda)$, then

$$E(T^{-r}; \mu, \lambda) = \frac{1}{\mu^{2r+1}} E(T^{r+1}; \mu, \lambda)$$

Show that this implies that

$$E(T^{-r}; 1, \lambda) = E(T^{r+1}; 1, \lambda)$$

and furthermore that

$$E(T^{-1}; 1, \lambda) = E(T^2; 1, \lambda)$$

27. Prove that

$$\int_0^{t_0} z(t)\, dt \to \infty \quad \text{when} \quad t_0 \to \infty$$

28. Let X and Y be a pair of random variables with cumulant generating function (cgf) $L_X(t)$ and $L_Y(t)$, respectively. According to Tweedie's definition

(Tweedie 1946) the distributions of X and Y are said to constitute a pair of "inverse distributions" if there exists a function $L(t)$ such that

$$L_X(t) = \alpha L(t)$$
$$L_Y(t) = \beta L^{-1}(t) \qquad [\text{i.e., } L_Y(t) = \gamma L_X^{-1}(t)]$$

for all values of t belonging to the domain of both cfg's, where $L(L^{-1}(t)) = t$, and α and β are appropriate constants.

Show that with Tweedie's definition the following distributions constitute pairs of "inverse distributions":
(a) The binomial and the negative binomial distributions.
(b) The Poisson and the gamma distributions.
(c) The normal and the inverse Gaussian distributions.

CHAPTER 3

Qualitative System Analysis

3.1 INTRODUCTION

A technical system will normally comprise a number of functional blocks that are interconnected in such a way that the system is able to perform a set of required functions. The term "functional block" will, in this context, not be given a precise definition. The blocks may range from single components up to rather large subsystems, depending on the system type and the boundary conditions laid down for the study. The functional blocks will, for the sake of simplicity, in the following mainly be referred to as *components*.

The structural relations between a system and its components may be described in a number of different ways. All these approaches try to "illustrate" how a specified system failure may, or may not, occur. The best approach to use will depend on the following:

- The objective(s) of the study
- The system's characteristics and the system's layout
- The relevant system failure mode(s)
- The system's operational and maintenance procedures

3.2 FMEA/FMECA

Failure Mode and Effects Analysis

Failure mode and effects analysis (FMEA) was one of the first systematic techniques for failure analysis. It was developed by reliability engineers in the late 1950s to study problems that might arise from malfunctions of military systems.

A failure mode and effects analysis is often the first step of a systems reliability study. It involves reviewing as many components, assemblies, and subsystems as possible to identify failure modes, causes, and effects of such failures. For each component the failure modes and their effects on the rest of the system are written on a specific FMEA worksheet. There are a variety of such worksheets. An example of an FMEA worksheet is shown in Figure 3.1.

An FMEA becomes a failure mode, effects, and criticality analysis (FMECA) if criticalities or priorities are assigned to the failure mode effects. More detailed information on how to conduct a failure mode and effects analysis (and an FMECA) may be found in MIL-STD-1629A, IEC 812, SAE ARP 926, and IEEE Std. 352.

Objectives of FMEA

According to IEEE Std. 352 the objectives of a failure mode and effects analysis are as follows:

1. Assist in selecting design alternatives with high reliability and high safety potential during early design phase.
2. Ensure that all conceivable failure modes and their effects on operational success of the system have been considered.
3. List potential failures, and identify the magnitude of their effects.
4. Develop early criteria for test planning and the design of the test and checkout systems.
5. Provide a basis for quantitative reliability and availability analyses.
6. Provide historical documentation for future reference to aid in analysis of field failures and consideration of design changes.
7. Provide input data for trade-off studies.
8. Provide basis for establishing corrective action priorities.
9. Assist in the objective evaluation of design requirements related to redundancy, failure detection systems, fail-safe characteristics, and automatic and manual override.

A failure mode and effects analysis is mainly a qualitative analysis, and it should be carried out by the designers during the design stage of a system. The purpose is to identify design areas where improvements are needed to meet reliability requirements. An updated FMEA is an important basis for design reviews and inspections.

"Bottom-up versus Top-down Approach"

The failure mode and effect analysis can be carried out either by starting at the component level and expanding upward (the bottom-up approach), or from the system level downward (the top-down approach). The two approaches are

FMECA

System:
Ref. drawing no.:

Performed by:
Date:

Page: of:

Description of unit			Description of failure			Effect of failure		Failure rate	Severity ranking	Risk reducing measures	Comments
Ref. no.	Function	Operational mode	Failure mode	Failure mechanisms	Detection of failure	On components in the subsystem	On the system function				
(1)	(2)	(3)	(4)	(5)	(6)	(7)	(8)	(9)	(10)	(11)	(12)

Figure 3.1 Example of an FMEA worksheet.

discussed in detail in SAE ARP 926 where the bottom-up approach is called the *hardware approach*, while the top-down approach is called the *functional approach*.

To decide to which component level the analysis should be conducted is a difficult task. It is often necessary to make compromises, since the workload could be overwhelming even for a system of moderate size. It is, however, a general rule to expand the analysis down to a level at which failure rate estimates are available or can be obtained.

Most failure mode and effects analyses are carried out according to the bottom-up approach. One may, however, for some systems save considerable effort by adopting the top-down approach. By this approach the analysis is carried out in two or more stages. The first stage is to split the system into a number of subsystems, and to identify possible failure modes and failure effects of each subsystem based on knowledge of the subsystem's required functions or on experience with similar equipment. One then proceeds to the next stage, where the components within each subsystem are analyzed. If a subsystem has no failure modes that are critical, then no further analysis of that subsystem needs to be performed. By this screening, it is possible to save time and effort. A weakness of this top-down approach is that it does not ensure that all failure modes of a subsystem have been identified.

Procedure

A failure mode and effects analysis is simple to conduct. It does not require any advanced analytical skills of the personnel performing the analysis. It is, however, necessary to know and understand the purpose of the system and the constraints under which it has to operate. The basic questions to be answered by an FMEA are according to IEEE Std. 352:

1. How can each part conceivably fail?
2. What mechanisms might produce these modes of failure?
3. What could the effects be if the failures did occur?
4. Is the failure in the safe or unsafe direction?
5. How is the failure detected?
6. What inherent provisions are provided in the design to compensate for the failure?

The analysis may be performed according to the following scheme:

1. Definition and delimitation of the system (which components are within the boundaries of the system and which are outside).
2. Definition of the main functions (missions) of the system.
3. Description of the operational modes of the system.

4. System breakdown into subsystems that can be handled effectively.
5. Review of system functional diagrams and drawings to determine interrelationships between the various subsystems. These interrelations may be illustrated by drawing functional block diagrams where each block corresponds to a subsystem.
6. Preparation of a complete component list for each subsystem.
7. Description of the operational and environmental stresses that may affect the system and its operation. These are reviewed to determine the adverse effects that they could generate on the system and its components.

The various entries in the FMEA worksheet are best illustrated by going through a specific worksheet column by column. We will use the FMEA worksheet in Figure 3.1 as an example.

Reference (column 1). The name of the unit or a drawing, for example, is given in the first column.

Function (column 2). The function(s) of the unit is (are) described in this column.

Operational mode (column 3). The unit may have various operational modes, for example, running or standby. Operational modes for an airplane include, for example, taxi, takeoff, climb, cruise, descent, approach, flare-out, and roll.

Failure mode (column 4). For each component's function and operational mode all the failure modes are identified and recorded. Note the definition of a *failure mode*. A failure mode is the manner by which a failure is observed. All units are designed to fulfil one or more functions. A failure mode is thus defined as non-fulfilment of one of these functions.

Failure mechanisms (column 5). The possible failure mechanisms (corrosion, erosion, fatigue, etc.) that may produce the identified failure modes are recorded in this column.

Detection of failure (column 6). The various possibilities for detection of the identified failure modes are then recorded. These may involve different alarms, testing, human perception, and so on. Some failure modes are called *evident failures*. Evident failures are detected instantly when they occur. The failure mode "spurious stop" of a pump with operational mode "running" is an example of an evident failure. Another type of failures is the called *hidden failure*. A hidden failure is normally detected only during testing of the unit. The failure mode "fail to start" of a pump with operational mode "standby" is an example of a hidden failure.

Effects on other components in the same subsystem (column 7). All the main effects of the identified failure modes on other components in the subsystem are recorded.

Effects on the primary function of the system (column 8). All the main effects of the identified failure mode on the primary function of the system are then recorded. The resulting operational status of the system after the failure may also be recorded, that is, whether the system is functioning or is switched over to another operational mode.

Failure rate (column 9). Failure rates for each failure mode are then recorded. In many cases it is more suitable to classify the failure rate in rather broad classes. An example of such a classification is

Very unlikely	Once per 1.000 years or more seldom
Remote	Once per 100 years
Occasional	Once per 10 years
Probable	Once per year
Frequent	Once per month or more often

Note that the failure rate with respect to a failure mode might be different for the various operational modes. The failure mode "Leakage to the environment" for a valve may, as an example, be more likely when the valve is closed and pressurized than when the valve is open.

Severity (column 10). By the severity of a failure mode we mean the worst potential consequence of the failure, determined by the degree of injury, property damage, or system damage that could ultimately occur. The following ranking categories (See Hammer, 1972, p. 152) are often adopted:

Catastrophic	Any failure that can result in deaths or injuries or prevent performance of the intended mission
Critical	Any failure that will degrade the system beyond acceptable limits and create a safety hazard (cause death or injury if corrective action is not immediately taken)
Major	Any failure that will degrade the system beyond acceptable limits but can be adequately counteracted or controlled by alternate means
Minor	Any failure that does not degrade the overall performance beyond acceptable limits—one of the nuisance variety

Slightly different categories are adopted in MIL-STD-882 ("System Safety Program Requirements").

Risk-reducing measures (column 11). Possible actions to correct the failure and restore the function or prevent serious consequences are then recorded. Actions that are likely to reduce the frequency of the failure modes may also be recorded.

Comments (column 12). This column may be used to record pertinent information not included in the other columns.

Table 3.1 Frequency/Consequence Diagram of the Different Failure Modes

Failure Rate	Severity Group			
	Minor	Major	Critical	Catastrophic
Frequent				
Probable				
Occational	(×)			
Remote		(×)		
Very unlikely	(×)		(×)	

By combining the failure rate (column 9) and the severity (column 10), we can obtain a ranking of the criticality of the different failure modes. This ranking can be illustrated as in Table 3.1 by a so-called criticality matrix. In this example we have classified the failure rate in five classes as described under column 9. The severity is in the same way classified in four classes as described under column 10. The most critical failure modes will be represented by an (×) in the upper right corner of the matrix, while the least critical failure modes will have an (×) in the lower left corner of the matrix.

Applications

Many industries require that an FMECA be integrated in the design process of technical systems and that FMECA worksheets be part of the system documentation. This is, for example, a common practice for suppliers to the defense, the aerospace, and the car industry. The same type of requirements are also becoming more usual within the offshore industry.

Subcontractors to the car industry are usually met with requirements for both hardware and process FMECA. A hardware FMECA is a traditional FMECA of the technical units that are supplied to the car manufacturer. A process FMECA is an analysis of the producer's in-house production system.

The FMECA is usually carried out during the design phase of a system. The main objective of the analysis is to reveal weaknesses and potential failures at an early stage, to enable the designer to incorporate corrections and barriers in the design. The results from the FMECA may also be useful during modifications of the system and for maintenance planning.

Many industries are introducing a reliability centered maintenance (RCM) program for maintenance planning. The RCM concept was introduced by the aviation industry and has formed the basis for the scheduled maintenance planning of a number of new airplane systems. The RCM concept is today applied in a wide range of industries, especially in nuclear power plants and within the offshore industry. FMECA is one of the basic analytical tools of the RCM concept. The RCM concept is further discussed by Nowlan and Heap (1978), Anderson and Neri (1990), and Sandtorv and Rausand (1991). Since all failure modes, failure mechanisms and symptoms are documented in the FMECA, this also provides valuable information as a basis for failure diagnostic procedures and for a repairperson's checklists.

A failure mode and effects analysis may be very effective when applied to a system where system failures most likely are the results of single component failures. During the analysis each failure is considered individually as an independent occurrence with no relation to other failures in the system. Thus a failure mode and effect analysis is not suitable for analysis of systems with a fair degree of redundancy. For such systems a fault tree analysis would be a much better alternative. An introduction to fault tree analysis is given in Section 3.3. In addition the failure mode and effect analysis is not well suited for analyzing systems where common cause failures are considered to be a significant problem. Common cause failures are discussed in Chapter 8.

A second limitation of the failure mode and effects analysis is the inadequate attention generally given to human errors. This is mainly due to the concentration on hardware failures. Perhaps the worst drawback is that all component failures are examined and documented, including those that do not have any significant consequences. For large systems, especially systems with a high degree of redundancy, the amount of unnecessary documentation work is a major disadvantage.

Computer Programs

A number of FMEA/FMECA programs for personal computers has been developed. Among these are

FMECA	This PC program is part of the CARA program, which also contains modules for fault tree analysis, cause consequence analysis, and life data analysis. CARA is developed by SINTEF Safety and Reliability. N-7034 Trondheim, Norway.
FME	FMECA program developed by Powertronic Systems Inc. P.O. Box 29109 New Orleans, LA 70189
Relex FMECA	FMECA program developed by Innovative Software Designs, Inc. Two English Elm Court, Baltimore, MD 21228
FMECA	FMECA program for Windows™ available from Systems Effectiveness Associates, Inc. 20 Vernon Street, Norwood, MA 02062
FailMode	FMECA program developed by OMI Logistics Limited, Item Software, Wellow House, 14 Little Park Farm Road, Fareham, Hampshire, PO15 5TD, England
Reliability Toolbox	Includes an FMECA program. Available from Innovative Timely Solutions, 6401 Lakerest Court, Raleigh, NC 27612

3.3 FAULT TREE ANALYSIS

The fault tree technique was introduced in 1962 at Bell Telephone Laboratories, in connection with a safety evaluation of the launching system for the intercontinental *Minuteman* missile. The Boeing Company improved the technique and introduced computer programs for both qualitative and quantitative fault tree analysis. Today fault tree analysis is by far the most commonly used technique for risk and reliability studies. In particular, fault tree analysis has been used with success to analyse safety systems in nuclear power stations, such as the Reactor Safety Study, WASH-1400 (1975).

A fault tree is a logic diagram that displays the interrelationships between a potential critical event (accident) in a system and the reasons for this event. The reasons may be environmental conditions, human errors, normal events (events which are expected to occur during the life span of the system), and specific component failures.

A fault tree analysis may be qualitative, quantitative, or both, depending on the objectives of the analysis. Possible results from the analysis may, for example, be

- A listing of the possible combinations of environmental factors, human errors, normal events and component failures that may result in a critical event in the system.
- The probability that the critical event will occur during a specified time interval.

Only qualitative fault tree analysis is covered in this chapter. Quantitative fault tree analysis is discussed in Chapter 4. Fault tree analysis is thoroughly described in the literature; see, for example, Henley and Kumamoto (1981) and Vesely et al. (1981).

Fault Tree Construction

A fault tree illustrates the states of the system's components (basic events) and the connections between these basic events and the system's state (TOP event). The graphical symbols used to represent these connections are called *logic gates*. The output from a logic gate is determined by the input events.

The graphical layout of the fault tree symbols are dependent on what standard we choose to follow. Table 3.2 shows the most commonly used fault tree symbols, together with a brief description of their interpretation. A number of more advanced fault tree symbols are available but will not be covered in this book. A thorough description may be found, for example, in Vesely et al. (1981).

Table 3.2 Fault Tree Symbols

	Symbol	Description
Logic gates	OR-gate	The OR-gate indicates that the output event A occurs if any of the input events E_i occur.
	AND-gate	The AND-gate indicates that the output event A occurs only when all the input events E_i occur simultaneously.
Input events	Basic event	The Basic event represents a basic equipment failure that requires no further development of failure causes.
	Undeveloped event	The Undeveloped event represents an event that is not examined further because information is unavailable or because its consequence is insignificant.
Description of state	Comment rectangle	The Comment rectangle is for supplementary information.
Transfer symbols	Transfer-out Transfer-in	The Transfer-our symbol indicates that the fault tree is developed further at the occurrence of the corresponding Transfer-in symbol.

A fault tree analysis is normally carried out in five steps[1]:

1. Definition of the problem and the boundary conditions
2. Construction of the fault tree
3. Identification of minimal cut and/or path sets
4. Qualitative analysis of the fault tree
5. Quantitative analysis of the fault tree

[1]The procedure described below is derived from AIChE (1985).

Definition of the Problem and the Boundary Conditions
The first activity of fault tree analysis clearly consists of two substeps:

- Definition of the critical event (the accident) to be analyzed
- Definition of the boundary conditions for the analysis

The critical event (accident) to be analyzed is normally called the TOP event. It is very important that the TOP event be given a clear and unambiguous definition. If not, the analysis will often be of limited value. As an example, the event description "Fire in the plant" is far too general and vague. The description of the TOP event should always give answer to the questions *what*, *where*, and *when*:

What. Describes what type of critical event (accident) is occurring (e.g., fire)

Where. Describes where the critical event occurs (e.g., in the process oxidation reactor)

When. Describes when the critical event occurs (e.g., during normal operation)

A more precise TOP event description is thus: "Fire in the process oxidation reactor during normal operation."

To get a consistent analysis, it is important that the boundary conditions for the analysis are carefully defined. By boundary conditions we understand the following:

- The physical boundaries of the system. Which parts of the system are to be included in the analysis, and which parts are not?
- The initial conditions. What is the operational state of the system when the TOP event is occurring? Is the system running on full/reduced capacity? Which valves are open/closed, which pumps are functioning and so on?
- Boundary conditions with respect to external stresses. What type of external stresses should be included in the analysis? By external stresses we here mean stresses from war, sabotage, earthquake, lightning, and so on.
- The level of resolution. How far down in detail should we go to identify potential reasons for a failed state? Should we, for example, be satisfied when we have identified the reason to be a "valve failure," or should we break it further down to failures in the valve housing, valve stem, actuator, and so forth. When determining the preferenced level of resolution, we should remember that the detailedness in the fault tree should be comparable to the detailedness of the information available.

Construction of the Fault Tree

The fault tree construction always starts with the TOP event. We must thereafter carefully try to identify all fault events that are the immediate, necessary, and sufficient causes of the TOP event. These causes are connected to the TOP event by a logic gate. It is important that the first level of causes under the TOP event be put up in a structured way. This first level is often referred to as the *TOP structure* of the fault tree. The TOP structure causes are often taken to be failures in the prime modules of the system, or in the prime functions of the system. We then proceed, level by level, until all fault events have been developed to the prescribed level of resolution. The analysis is, in other words, deductive and is carried out by repeated asking "What are the reasons for this event?"

Rules for Fault Tree Construction

1. *Describe the fault events.* Each of the basic events must be carefully described (what, where, when) in a "rectangle."

2. *Evaluate the fault events.* As explained in Section 1.6, component failures may be divided in three groups: *primary failures*, *secondary failures*, and *command faults*.

   The "normal" basic events in a fault tree are primary failures identifying the equipment that is responsible for the failure. Secondary failures and command faults are intermediate events that require a further investigation to identify the prime reasons.

   When evaluating a fault event, we ask the question, "Can this fault be a primary failure?" If the answer is yes, we classify the fault event as a "normal" basic event. If the answer is no, we classify the fault event either as an intermediate event, which has to be further developed, or as a "secondary" basic event. The "secondary" basic event is often called an *undeveloped* event and represents a fault event that is not examined further because information is unavailable or because its consequence is insignificant.

3. *Complete the gates.* All inputs to a specific gate should be completely defined and described before proceeding to the next gate. The fault tree should be completed in levels, and each level should be completed before beginning the next level.

Example 3.1: Fire Detector System

Let us consider a simplified version of a fire detector system located in a production room. (Observe that this system is not a fully realistic fire detector system.)

The fire detector system is divided into two parts, heat detection and smoke detection. In addition there is an alarm button that can be operated manually.

FAULT TREE ANALYSIS

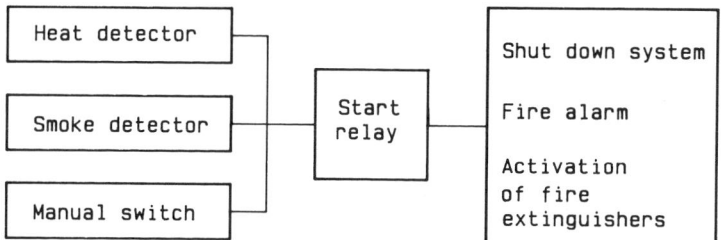

Figure 3.2 System overview of fire detector system

The fire detector system can be described schematically, as shown in Figures 3.2 and 3.3.

Heat Detection
In the production room there is a closed, pneumatic pipe circuit with four identical fuse plugs, *FP*1, *FP*2, *FP*3, and *FP*4. These plugs let air out of the circuit if they are exposed to temperatures higher than 72°C. The pneumatic system

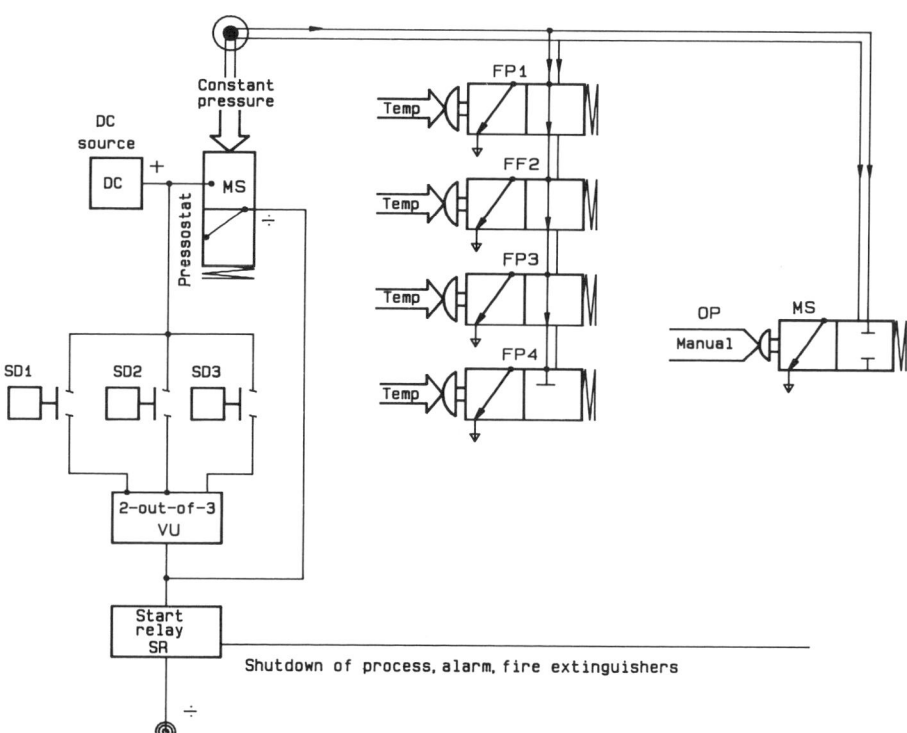

Figure 3.3 Schematic layout of the fire detector system

has a pressure of 3 bars and is connected to a pressure switch (pressostat) *PS*. If one or more of the plugs are activated, the switch will be activated and give an electrical signal to the start relay for the alarm and shutdown system. To transmit an electrical signal, the dc source, *DC*, must be intact.

Smoke Detection
The smoke detection system consists of three optical smoke detectors, *SD*1, *SD*2, and *SD*3; all are independent and have their own batteries. These detectors are very sensitive and can give warning of fire at an early stage. To avoid false alarms, the three smoke detectors are connected by a logical 2-out-of-3 voting unit *VU*. This means that at least two detectors must give fire signal before the fire alarm is activated. If at least two of the three detectors are activated, the 2-out-of-3 voting unit will give an electric signal to the start relay *SR* for the alarm and shutdown system. Again the dc source, *DC*, must be intact to obtain an electrical signal.

Manual Activation
Together with the pneumatic pipe circuit with the four fuse plugs, there is also a manual switch *MS* that can be turned to relieve the pressure in the pipe circuit. If the operator *OP*, who should be continually present, notices a fire, he can activate this switch. When the switch is activated, the pressure in the pipe circuit is relieved and the pressure switch *PS* is activated; this gives an electric signal to the start relay *SR*. Again the dc source must be intact.

The Start Relay
When the start relay *SR* receives an electrical signal from the detection systems, it is activated and gives a signal to

- shut down the process
- activate the alarm and the fire extinguishers

Assume now that a fire starts. The fire detector system should detect and give warning about the fire. Let the TOP event be "*No signals from the start relay SR when a fire condition is present.*" (Remember *what*, *where*, and *when*.) A possible fault tree for this TOP event is presented in Figure 3.4. The fault tree is constructed by CAFTAN, the fault tree module of the CARA (ref.) program. □

Identification of Minimal Cut and Path Sets

A fault tree provides valuable information about possible combinations of fault events that result in a critical failure (accident) in the system. Such a combination of fault events is called a *cut set*. In the fault tree terminology, a cut set is defined as follows:

FAULT TREE ANALYSIS

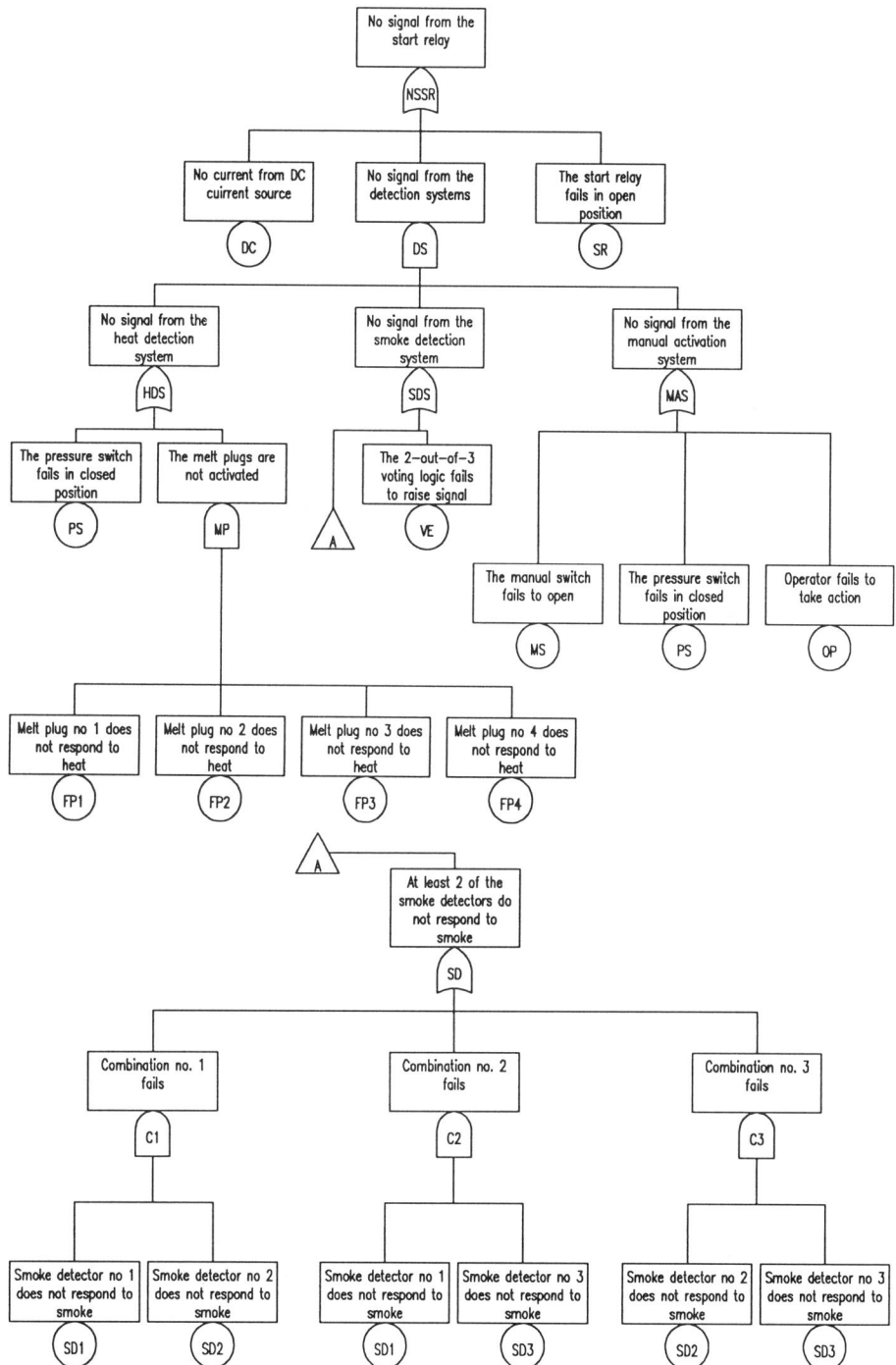

Figure 3.4 Fault tree for the fire detector system (printout from the CARA program)

88   QUALITATIVE SYSTEM ANALYSIS

Definition 3.1. A cut set in a fault tree is a set of basic events whose (simultaneous) occurrence ensures that the TOP event occurs. A cut set is said to be minimal if the set cannot be reduced without loosing its status as a cut set.   □

The number of different basic events in a minimal cut set is called the *order* of the cut set. For small and simple fault trees it is feasible to identify the minimal sets by inspection without any formal procedure/algorithm. For large or complex fault trees we need an efficient algorithm.

MOCUS

MOCUS (method for obtaining cut sets) is an algorithm that can be used to find the minimal cut sets in a fault tree. The algorithm is best explained by an example. Consider the fault tree in Figure 3.5, where the gates are numbered from $G0$ to $G6$. (The example fault tree is copied from Barlow and Lambert 1975.)

The algorithm starts at the $G0$ gate directly under the TOP event. If this is an "OR" gate, each input to the gate is written in separate rows. (The inputs may

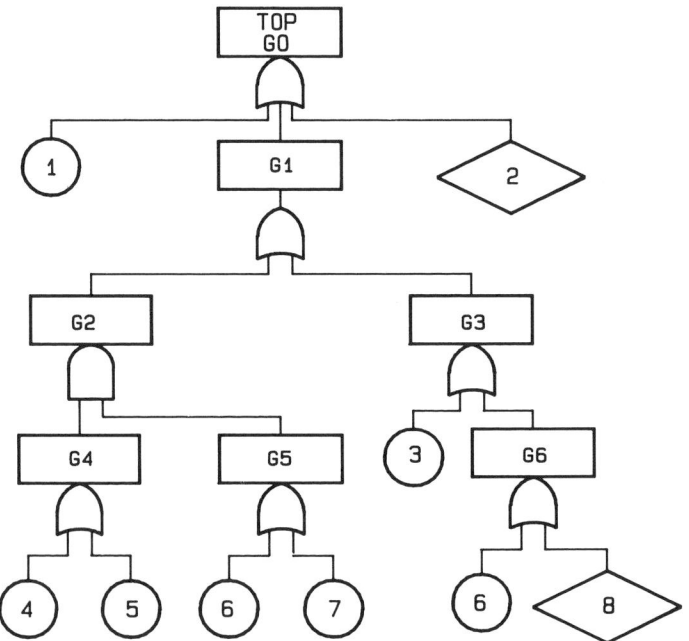

Figure 3.5 Example of a fault tree (from Barlow and Lambert 1975)

FAULT TREE ANALYSIS

be new gates.) Similarly, if the $G0$ gate is an "AND" gate, the inputs to the gate are written in separate columns.

In our example, $G0$ is an OR gate; hence we start with

$$\begin{array}{c} 1 \\ G1 \\ 2 \end{array}$$

Since each of the three inputs, 1, $G1$ and 2 will cause the TOP event to occur, each of them will constitute a cut set.

The idea is to successively replace each gate with its inputs (basic events and new gates) until one has gone through the whole fault tree and is left with just the basic events. When this procedure is completed, the rows in the established matrix represent the cut sets in the fault tree.

Since $G1$ is an OR gate: Since $G2$ is an AND gate:
$$\begin{array}{cc} 1 & 1 \\ G2 & G4, G5 \\ G3 & G3 \\ 2 & 2 \end{array}$$

Since $G3$ is an OR gate: Since $G4$ is an OR gate:
$$\begin{array}{cc} 1 & 1 \\ G4, G5 & 4, G5 \\ 3 & 5, G5 \\ G6 & 3 \\ 2 & G6 \\ & 2 \end{array}$$

Since $G5$ is an OR gate: Since $G6$ is an OR gate:
$$\begin{array}{cc} 1 & 1 \\ 4, 6 & 4, 6 \\ 4, 7 & 4, 7 \\ 5, 6 & 5, 6 \\ 5, 7 & 5, 7 \\ 3 & 3 \\ G6 & 6 \\ 2 & 8 \\ & 2 \end{array}$$

The fault tree that we are left with has the following 9 cut sets:

{1}   {4, 6}
{2}   {4, 7}
{3}   {5, 6}
{6}   {5, 7}
{8}

Since {6} is a cut set, {4, 6} and {5, 6} are not minimal. If we leave these out, we are left with the following list of minimal cut sets:

{1}, {2}, {3}, {6}, {8}, {4, 7}, {5, 7}

In other words, five minimal cut sets are of order 1 and two minimal cut sets of order 2. The reason that the algorithm in this case leads to nonminimal cut sets is that basic event 6 occurs several places in the fault tree.

In some situations it may also be of interest to identify the possible combinations of components which by functioning secures that the system is functioning. Such a combination of components (basic events) is called a *path set*. In the fault tree terminology a path set is defined as follows:

Definition 3.2. A path set in a fault tree is a set of basic events whose nonoccurrence (simultaneously) ensures that the TOP event does not occur. A path set is said to be minimal if the set cannot be reduced without loosing its status as a path set. □

The number of different basic events in a minimal path set is called the *order* of the path set. To find the minimal path sets in the fault tree, we may start with the so-called dual fault tree. This can be obtained by replacing all the AND gates in the original fault tree with OR gates, and vice versa. In addition we let the events in the dual fault tree be complements of the corresponding events in the original fault tree. The same procedure as was described above is applied to the dual fault tree, and it now yields the minimal path sets.

For relatively "simple" fault trees one can apply the MOCUS algorithm by hand. More complicated fault trees require the use of a computer. A number of computer programs for minimal cut (path) set identification are available. Some of these are based on MOCUS. The algorithm developed for the CARA (ref.) program is described by Vatn (1992).

Example 3.2: Offshore Separator
Consider the process on an offshore gas production installation. Gas from the various wells is collected in a wellhead manifold and led into two identical

FAULT TREE ANALYSIS

process trains. The gas from the process trains is then collected in a compressor manifold and led to the gas export pipeline via compressors.

Figure 3.6 shows a (simplified) section of one of the process trains. Gas from the wellhead manifold is led into the first stage separator. On the inlet pipeline, there are installed two process shutdown (PSD) valves, PSD_1 and PSD_2 in series. The valves are fail-safe close and are held open by hydraulic pressure. When the hydraulic pressure is bled off, the valves will close by the force of a precharged actuator. Another PSD valve, PSD_3 of the same type is installed on the gas outlet from the separator.

Due to the fail-safe design of the PSD valves, these valves may close spuriously, that is, without the presence of a hazardious situation. The two most important failure modes of the PSD valves are thus "fail to close on command" (FTC) and "premature closure" (PC).

A pressure safety valve (PSV) is installed to relieve the pressure in the separator in case the pressure increases beyond a specified high pressure p. The PSV is equipped with a spring-loaded actuator, which may be adjusted to the preset pressure p. The most critical failure mode for the PSV is "fail to open" (FTO); that is, the valve does not open when the pressure in the separator increases beyond the preset pressure p.

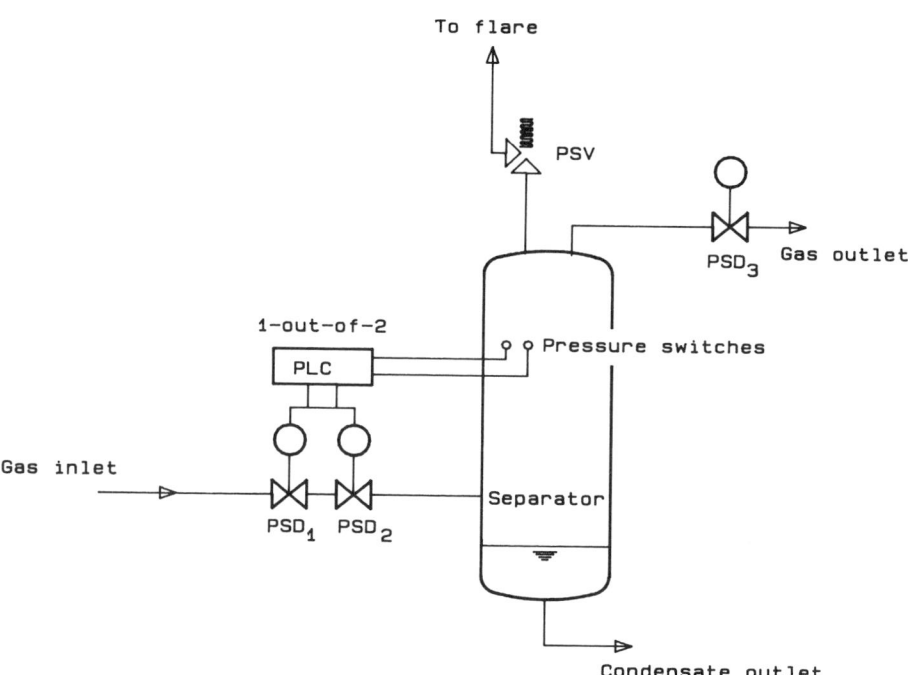

Figure 3.6 Sketch of a first-stage gas separator

A hazardous situation occurs if PSD_3 closes spuriously, because this will lead to a rapid pressure build-up in the separator. Two identical pressure switches, PS_1 and PS_2, are installed in the separator. The pressure switches are on–off devices that are preset to be activated at a pressure p_1 which is less than the activation pressure p for the pressure safety valve. When activated, the pressure switches will provide signal via the Programmable Logic Controller (PLC) to close the process shutdown valves PSD_1 and PSD_2. The PLC employs a 1-out-of-2 voting logic.

Assume now that the outlet valve PSD_3 closes spuriously during normal production and that the pressure in the separator increases rapidly. This situation may be studied by a fault tree analysis with respect to the TOP event *"overpressure in the first-stage separator."* A possible fault tree for this TOP event is presented in Figure 3.7. The fault tree is constructed by CAFTAN, the fault tree module of the CARA (ref.) program. □

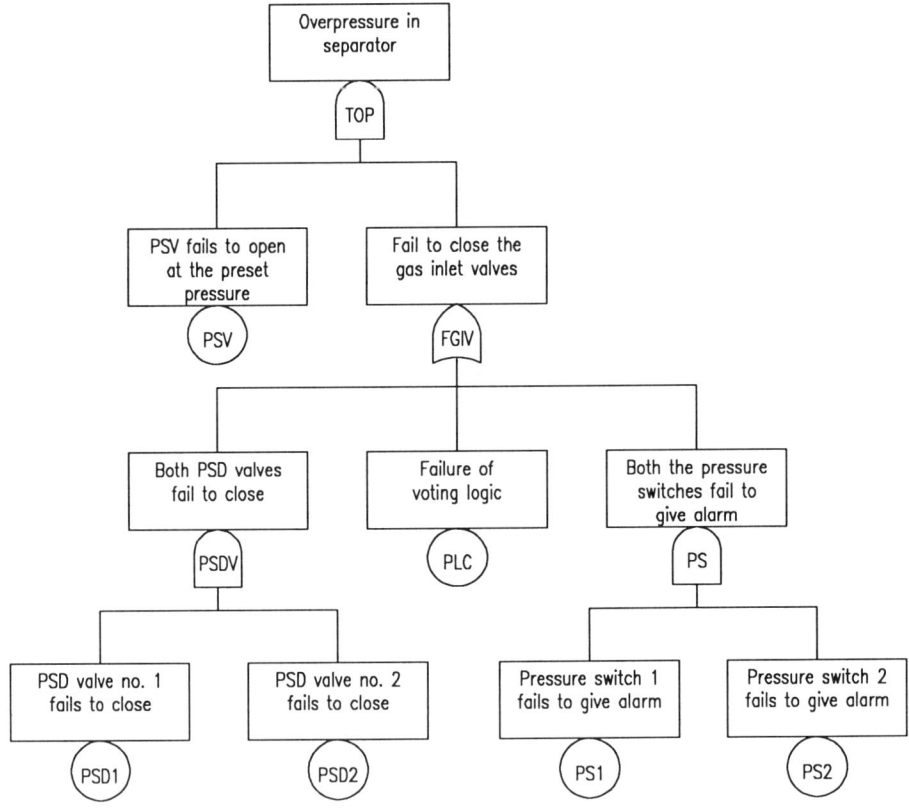

Figure 3.7 Fault tree for the first-stage separator in Example 3.3 (printout from the CARA program)

Qualitative Evaluation of the Fault Tree

A qualitative evaluation[2] of the fault tree may be carried out on the basis of the minimal cut sets. The criticality of a cut set depends obviously on the number of basic events in the cut set (i.e., the order of the cut set). A cut set of order one is usually more critical than a cut set of order two, or more. When we have a cut set of order one, the TOP event will occur as soon as the corresponding basic event occurs. When a cut set has two basic events, both of these have to occur simultaneously to cause the TOP event to occur.

Another important factor is the type of basic events of a minimal cut set. We may rank the criticality of the various cut sets according to the following ranking of basic events:

1. Human error
2. Active equipment failure
3. Passive equipment failure

This ranking is based on the assumption that human errors occur more frequently than active equipment failures and that active equipment is more prone to failure than passive equipment (e.g., an active or running pump is more exposed to failures than a passive standby pump). Based on this ranking, we get the ranking in Table 3.3 of the criticality of minimal cut sets of order two. (Rank 1 is the most critical one.)

3.4 RELIABILITY BLOCK DIAGRAMS

In this section we will illustrate the structure of a system by what is known as a *reliability block diagram*. A reliability block diagram is a success-oriented network describing the *function* of the system. If the system has more than one function, each function is considered individually, and a separate reliability block diagram is established for each system function.

[2]This section is based on AIChE (1985).

Table 3.3 Criticality Ranking of Minimal Cut Sets of Order Two

Rank	Basic event 1 (type)	Basic event 2 (type)
1	Human error	Human error
2	Human error	Active equipment failure
3	Human error	Passive equipment failure
4	Active equipment failure	Active equipment failure
5	Active equipment failure	Passive equipment failure
6	Passive equipment failure	Passive equipment failure

94 QUALITATIVE SYSTEM ANALYSIS

Figure 3.8 Component i illustrated by a block

Consider a system with n different components. Each of the n components is illustrated by a block as shown is Figure 3.8. When there is connection between the end points (a) and (b) in Figure 3.8, we say that component i is functioning. This does not necessarily mean that component i functions in all respects. It only means that one, or a specified set of functions, is achieved (i.e., that some specified failure mode(s) do not occur). What is meant by functioning must be specified in each case and will depend on the objectives of the study.

The way the n components are interconnected to fulfill a specified system function may be illustrated by a reliability block diagram, as shown in Figure 3.9. When we have connection between the end points (a) and (b) in Figure 3.9, we say that the specified system function is achieved, which means that some specified system failure mode(s) do(es) not occur.

Series Structure

A system that is functioning if and only if all of its n components are functioning is called a *series structure*. The corresponding reliability block diagram is shown in Figure 3.10. We have "connection" between the end points (a) and (b) (i.e., the system is functioning) if and only if we have "connection" through all the n blocks representing the components.

Parallel Structure

A system that is functioning if at least one of its n components is functioning is called a *parallel structure*. The corresponding reliability block diagram is shown in Figure 3.11. In this case we have "connection" between the end points (a) and (b) (i.e., the system is functioning) if we have "connection" through at least one of the blocks representing the components.

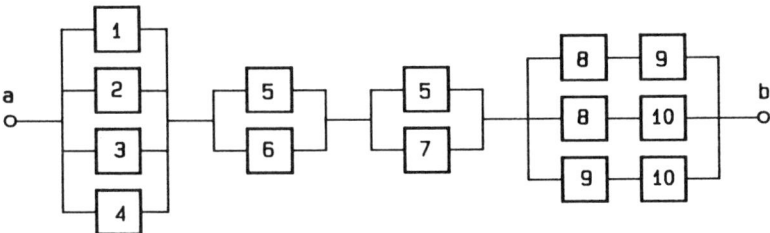

Figure 3.9 System function illustrated by a reliability block diagram

RELIABILITY BLOCK DIAGRAMS

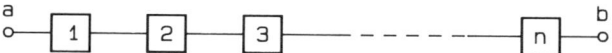

Figure 3.10 Reliability block diagram of a series structure

Example 3.3

Consider a pipeline with two independent safety valves V_1 and V_2 that are physically installed in series. The valves are supplied with a spring-loaded, failsafe, close hydraulic actuator. The valves are opened and held open by hydraulic control pressure and is closed automatically by spring force whenever the control pressure is removed or lost. In normal operation both valves are held open. The main function of the valves is to act as a safety barrier, that is, to close and stop the flow in the pipeline in case of an emergency.

Since it is sufficient that one of the valves closes in order to stop the flow, the valves will form a *parallel* system with respect to the safety barrier function, as shown in Figure 3.12*b*. The valves may close spuriously, that is, without a control signal, and stop the flow in the pipeline. In this case it is sufficient that only one of the valves fails in order to stop the flow. The valves will thus form a *series* system with respect to spurious closures, as shown in Figure 3.12*c*. □

Reliability Block Diagrams versus Fault Trees

In practical applications, we may choose whether to model the system structure by a fault tree or by a reliability block diagram. When the fault tree is limited to only OR-gates and AND-gates, both methods yield the same result, and we may convert the fault tree to a reliability block diagram, and vice versa.

In a reliability block diagram, "connection" through a block means that the component represented by the block is functioning. This again means that one failure mode, or a specified set of failure modes of the component, is not occurring. In a fault tree we may let a basic event be the occurrence of the same failure mode, or the same specified set of failure modes, for the component. When the TOP event in the fault tree represents "system failure" and the basic

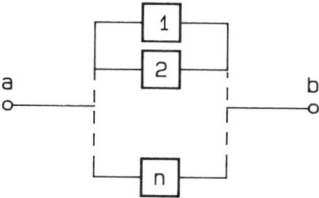

Figure 3.11 Reliability block diagram of a parallel structure

96 QUALITATIVE SYSTEM ANALYSIS

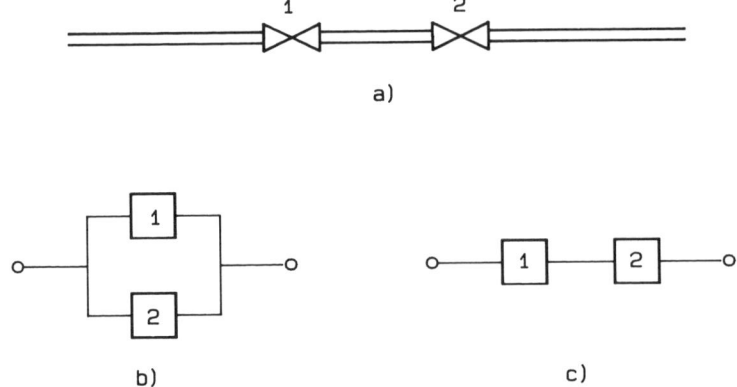

Figure 3.12 Two safety valves in a pipeline. (*a*) Physical layout, (*b*) reliability block diagram with respect to the safety barrier function, and (*c*) reliability block diagram with respect to spurious closure

events are defined as above, it is easy to see, for instance, that a series structure is equivalent to a fault tree where all the basic events are connected through an OR-gate. The TOP event occurs if either component 1 or component 2 or component 3 or ... component n fails.

In the same way a parallel structure may be represented as a fault tree where all the basic events are connected through an AND gate. The TOP event occurs (i.e., the parallel structure fails), if component 1 and component 2 and component 3 and ... component n fail. The relationship between some simple reliability block diagrams and fault trees is illustrated in Figure 3.13.

Example 3.1 (*cont.*)
It is usually an easy task to convert a fault tree to a reliability block diagram. The reliability block diagram corresponding to the fault tree for the fire detector system in Figure 3.4 is shown in Figure 3.14. In this conversion we start from the TOP event and replace the gates successively. OR-gates are replaced by series structures of the "components" directly beneath the gate, and AND-gates are replaced by a parallel structure of the "components" directly beneath the gate. □

From Figure 3.14 we observe that some of the components are represented in two different locations in the diagram. It should be emphasized that a reliability block diagram is not a physical layout diagram for the system. It is a logic diagram, illustrating the function of the system.

Most people prefer to use a fault tree to establish the system's structure. They claim this to be a more natural approach, at least for systems that do not have a physical layout similar to a network. For further evaluations, however, it is

RELIABILITY BLOCK DIAGRAMS 97

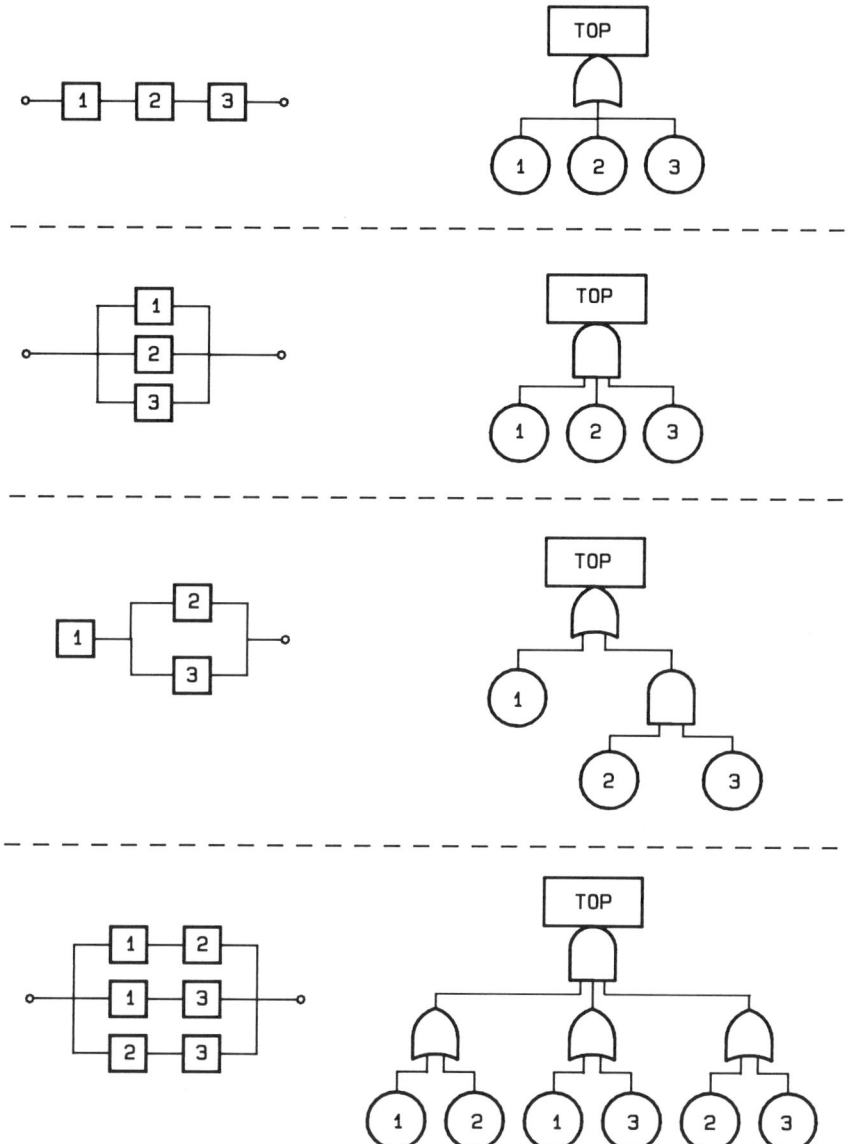

Figure 3.13 Relationship between some simple reliability block diagrams and fault trees

often more natural to base these on a reliability block diagram. A fault tree will therefore often be converted to a reliability block diagram for qualitative and quantitative analyses. This is the main reason why we have chosen to focus on reliability block diagrams in the rest of this book and only look upon fault trees as an alternative approach.

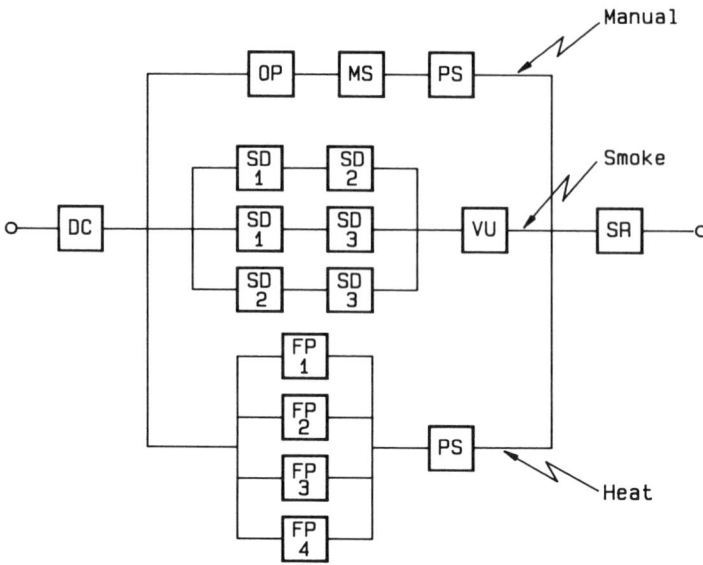

Figure 3.14 Reliability block diagram for the fire detector system

Structure Function

A system that is composed of n components will be denoted in this discussion as a system of *order n*. The components are assumed to be numbered consecutively from 1 to n.[3]

In this chapter we will confine ourselves to situations where it suffices to distinguish between only two states, a functioning state and a failed state. This applies to each component as well as to the system itself. The state of component i, $i = 1, 2, \ldots, n$, can then be described by a binary[4] variable x_i, where

$$x_i = \begin{cases} 1 & \text{if component } i \text{ is functioning} \\ 0 & \text{if component } i \text{ is in a failed state} \end{cases}$$

$\mathbf{x} = (x_1, x_2, \ldots, x_n)$ is called the state vector. Furthermore we assume that by knowing the states of all the n components, we also know whether the system is functioning or not.

Similarly the state of the system can then be described by a binary function

$$\phi(\mathbf{x}) = \phi(x_1, x_2, \ldots, x_n)$$

where

---

[3]The remaining sections of this chapter are strongly influenced by Barlow and Proschan (1975).
[4]In this context a binary variable (function) is a variable (function) that can take only the two values, 0 or 1.

RELIABILITY BLOCK DIAGRAMS

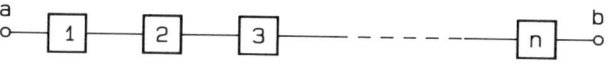

Figure 3.15 Series structure

$$\phi(\mathbf{x}) = \begin{cases} 1 & \text{if the system is functioning} \\ 0 & \text{if the system is in a failed state} \end{cases}$$

$\phi(\mathbf{x})$ is called the *structure function of the system*, or just the *structure*. In the following sections we will often talk about structures instead of systems. Examples of simple structures are given there.

Series Structure

A system that is functioning if and only if *all* of its n components are functioning is called *a series structure*. The structure function is

$$\phi(\mathbf{x}) = x_1 \cdot x_2 \cdots x_n = \prod_{i=1}^{n} x_i \qquad (3.1)$$

A series structure of order n can be illustrated as in Figure 3.15. Connection between (a) and (b) is interpreted as "the structure (system) is functioning."

Parallel Structure

A system that is functioning if at least one of its n components is functioning is called *a parallel structure*. A parallel structure of order n is illustrated in Figure 3.16. In this case the structure function can be written

$$\phi(\mathbf{x}) = 1 - (1 - x_1)(1 - x_2) \cdots (1 - x_n) = 1 - \prod_{i=1}^{n} (1 - x_i) \qquad (3.2)$$

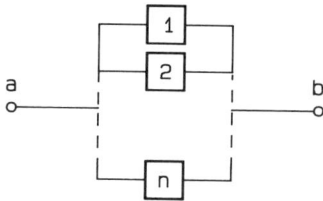

Figure 3.16 Parallel structure

The expression on the right-hand side of (3.2) is often written as $\coprod_{i=1}^{n} x_i$ where $\coprod$ is read "ip."

Hence a parallel structure of order 2 has structure function

$$\phi(x_1, x_2) = 1 - (1 - x_1)(1 - x_2) = \coprod_{i=1}^{2} x_i$$

The right-hand side may also be written: $x_1 \amalg x_2$. Note that

$$\phi(x_1, x_2) = x_1 + x_2 - x_1 x_2$$

Since x_1 and x_2 are binary variables, $x_1 \amalg x_2$ will be equal to the maximum of the x_i's. Similarily

$$\coprod_{i=1}^{n} x_i = \max_{i=1,2,\ldots,n} x_i$$

k-out-of-n Structure

A system that is functioning if and only if at least k of the n components are functioning, is called a *k-out-of-n structure*. A series structure is therefore an n-out-of-n structure, and a parallel structure is a 1-out-of-n structure.

The structure function of a k-out-of-n structure can be written

$$\phi(\mathbf{x}) = \begin{cases} 1 & \text{if } \sum_{i=1}^{n} x_i \geq k \\ 0 & \text{if } \sum_{i=1}^{n} x_i < k \end{cases} \quad (3.3)$$

As an example consider a 2-out-of-3 structure, which is illustrated in Figure 3.17. A three-engined airplane which can stay in the air if and only if at least two of its three engines are functioning, is an example of a 2-out-of-3 structure.

SYSTEM STRUCTURE ANALYSIS

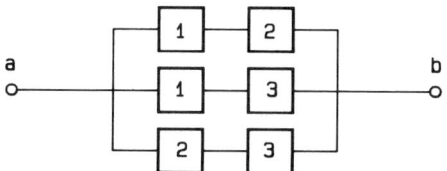

Figure 3.17 2-out-of-3 structure

The structure function of the 2-out-of-3 structure in Figure 3.17 may also be written

$$\begin{aligned}
\phi(\mathbf{x}) &= x_1x_2 \amalg x_1x_3 \amalg x_2x_3 \\
&= 1 - (1 - x_1x_2)(1 - x_1x_3)(1 - x_2x_3) \\
&= x_1x_2 + x_1x_3 + x_2x_3 - x_1^2x_2x_3 - x_1x_2^2x_3 - x_1x_2x_3^2 + x_1^2x_2^2x_3^2 \\
&= x_1x_2 + x_1x_3 + x_2x_3 - 2x_1x_2x_3
\end{aligned}$$

Note that since x_i is a binary variable, $x_i^k = x_i$ for all i and k.

3.5 SYSTEM STRUCTURE ANALYSIS

Coherent Structures

When establishing the structure of a system, it seems reasonable first to leave out all components that do not play any *direct* role for the functioning ability of the system. The components we are left with are called *relevant*. The components that are not relevant are called *irrelevant*.

If component i is irrelevant, then

$$\phi(1_i, \mathbf{x}) = \phi(0_i, \mathbf{x}) \qquad \text{for all } (\cdot_i, \mathbf{x}) \tag{3.4}$$

where $(1_i, \mathbf{x})$ represents a state vector where the state of the ith component = 1, $(0_i, \mathbf{x})$ represents a state vector where the state of the ith component = 0, and $(\cdot_i, \mathbf{x})$ represents a state vector where the state of the ith component = 0 or 1. Figure 3.18 illustrates a system of order 2, where component 2 is irrelevant.

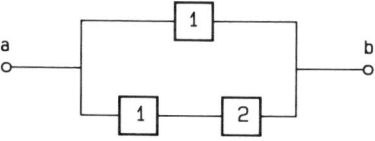

Figure 3.18 Component 2 is irrelevant

Remark
The notation "relevant/irrelevant" is sometimes misleading, as it is easy to find examples of components of great importance for a system without being relevant in the above sense. Devices that are installed to protect a system against extreme loads are usually not included in the reliability block diagram and are consequently irrelevant according to our definition. Nevertheless, the protective devices may be of great importance for the system's ability to function.

Now we will assume that the system will not run worse than before if we replace a component in a failed state with one that is functioning. This is obviously the same as requiring that the structure function shall be nondecreasing in each of its arguments.

ooo OOO ooo

Let us now define what is meant by a *coherent system:*

Definition 3.3. A system of components is said to be coherent if all its components are relevant and the structure function is nondecreasing in each argument. □

All the systems that we have considered so far (except the one in Figure 3.18) are coherent. One might get the impression that all systems of interest must be coherent, but this is not the case. It is, for example, easy to find systems where the failure of one component prevents another component from failing. This complication will be discussed later.

General Characteristics of Coherent Systems

Theorem 3.1. Let $\phi(\mathbf{x})$ be the structure function of a coherent system. Then

$$\phi(\mathbf{0}) = 0 \quad \text{and} \quad \phi(\mathbf{1}) = 1$$

In other words, Theorem 3.1 merely says that

- If all the components in a coherent system are functioning, then the system is functioning
- If all the components in a coherent system are in a failed state, then the system is in a failed state

Proof
The argument uses the fact that $\phi(\mathbf{x})$ is binary, that is, that it can only assume the values 0 and 1.

If $\phi(\mathbf{0}) = 1$, then we must have $\phi(\mathbf{0}) = \phi(\mathbf{1}) = 1$, since $\phi(\mathbf{x})$ is assumed to be nondecreasing in each argument. This implies that all the components in

SYSTEM STRUCTURE ANALYSIS

the system are irrelevant, which contradicts the assumption that the system is coherent. Hence $\phi(\mathbf{0}) = 0$.

Similarly $\phi(\mathbf{1}) = 0$ implies that $\phi(\mathbf{0}) = 0$, that is, that all the components are irrelevant. This contradicts the assumption of coherence. Hence $\phi(\mathbf{1}) = 1$. □

Theorem 3.2. Let $\phi(\mathbf{x})$ be the structure function of a coherent system of order n. Then

$$\prod_{i=1}^{n} x_i \leq \phi(\mathbf{x}) \leq \coprod_{i=1}^{n} x_i \qquad (3.5)$$

Theorem 3.2 states that any coherent system is functioning at least as well as a corresponding system where all the n components are connected in series and at most as well as a system where all the n components are connected in parallel.

Proof
First note that $\prod_{i=1}^{n} x_i$ and $\coprod_{i=1}^{n} x_i$ are both binary. Assume that $\prod_{i=1}^{n} x_i = 0$. Since we already know that $\phi(\mathbf{x}) \geq 0$, the left-hand side of (3.5) is satisfied.

Assume that $\prod_{i=1}^{n} x_i = 1$, that is, $\mathbf{x} = \mathbf{1}$. According to Theorem 3.1, then $\phi(\mathbf{x}) = 1$. Hence the left-hand side of (3.5) is always satisfied.

Further assume that $\coprod_{i=1}^{n} x_i = 0$, that is, $\mathbf{x} = \mathbf{0}$. Then according to Theorem 3.1, $\phi(\mathbf{x}) = 0$, and the right-hand side of (3.5) is satisfied. Finally, assume that $\coprod_{i=1}^{n} x_i = 1$. Since we already know that $\phi(\mathbf{x}) \leq 1$, the right-hand side in (3.5) is automatically satisfied. □

Let $\mathbf{x} = (x_1, x_2, \ldots, x_n)$ and $\mathbf{y} = (y_1, y_2, \ldots, y_n)$ be state vectors, and let $\mathbf{x} \cdot \mathbf{y}$ and $\mathbf{x} \amalg \mathbf{y}$ be defined as follows:

$$\mathbf{x} \cdot \mathbf{y} = (x_1 y_1, x_2 y_2, \ldots, x_n y_n)$$
$$\mathbf{x} \amalg \mathbf{y} = (x_1 \amalg y_1, x_2 \amalg y_2, \ldots, x_n \amalg y_n)$$

We will now prove the following important result:

Theorem 3.3. Let ϕ be a coherent structure. Then

$$\phi(\mathbf{x} \amalg \mathbf{y}) \geq \phi(\mathbf{x}) \amalg \phi(\mathbf{y}) \qquad (3.6)$$

$$\phi(\mathbf{x} \cdot \mathbf{y}) \leq \phi(\mathbf{x}) \cdot \phi(\mathbf{y}) \qquad (3.7)$$

Proof
For (3.6) we know that $x_i \amalg y_i \geq x_i$ for all i. Since ϕ is coherent, ϕ is nondecreasing in each argument, and therefore

$$\phi(\mathbf{x} \amalg \mathbf{y}) \geq \phi(\mathbf{x})$$

For symmetrical reasons

$$\phi(\mathbf{x} \amalg \mathbf{y}) \geq \phi(\mathbf{y})$$

Furthermore $\phi(\mathbf{x})$ and $\phi(\mathbf{y})$ are both binary. Therefore

$$\phi(\mathbf{x} \amalg \mathbf{y}) \geq \phi(\mathbf{x}) \amalg \phi(\mathbf{y})$$

For (3.7) we know that $x_i \cdot y_i \leq x_i$ for all i. Since ϕ is coherent, then

$$\phi(\mathbf{x} \cdot \mathbf{y}) \leq \phi(\mathbf{x})$$

Similarly

$$\phi(\mathbf{x} \cdot \mathbf{y}) \leq \phi(\mathbf{y})$$

Since $\phi(\mathbf{x})$ and $\phi(\mathbf{y})$ are binary, then

$$\phi(\mathbf{x} \cdot \mathbf{y}) \leq \phi(\mathbf{x}) \cdot \phi(\mathbf{y})$$

Let us interpret (3.6) in common language. Consider the structure in Figure 3.19 with structure function $\phi(\mathbf{x})$. Assume that we also have an identical structure $\phi(\mathbf{y})$ with state vector $\mathbf{y}$. Figure 3.20 illustrates a structure with "redundancy at system level." The structure function for this system is $\phi(\mathbf{x}) \amalg \phi(\mathbf{y})$.

Next consider the system we get from Figure 3.19 when we connect each pair x_i, y_i in parallel; see Figure 3.21. This figure illustrates a structure with "redundancy at component level." The structure function is $\phi(\mathbf{x} \amalg \mathbf{y})$.

According to Theorem 3.3, $\phi(\mathbf{x} \amalg \mathbf{y}) \geq \phi(\mathbf{x}) \amalg \phi(\mathbf{y})$. This means that

We obtain a "better" system by introducing redundancy at component level than by introducing redundancy at system level.

This principle, which is well known to designers, is further discussed by

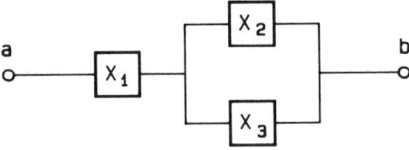

Figure 3.19 Example system

SYSTEM STRUCTURE ANALYSIS

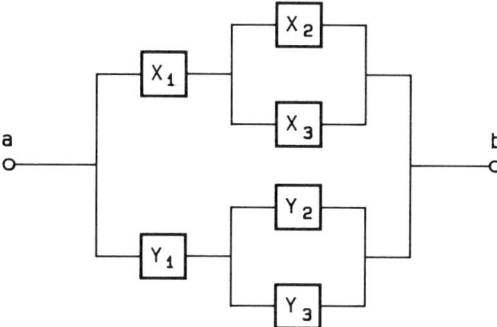

Figure 3.20 Redundancy at system level

Shooman (1968, pp. 281–289). The principle is, however, not obvious when the system has two ore more failure modes, for example, "fail to function" and "false alarm" of a fire detection system. The concept of redundancy is further discussed in Section 4.5.

Structures Represented by Paths and Cuts

A structure of order n consists of n components numbered from 1 to n. The set of components is denoted by

$$C = \{1, 2, 3, \ldots, n\}$$

Definition 3.4: Path sets, minimal path sets
A path set P is a set of components in C which by functioning ensures that the system is functioning. A path set is said to be minimal if it cannot be reduced without loosing its status as a path set. □

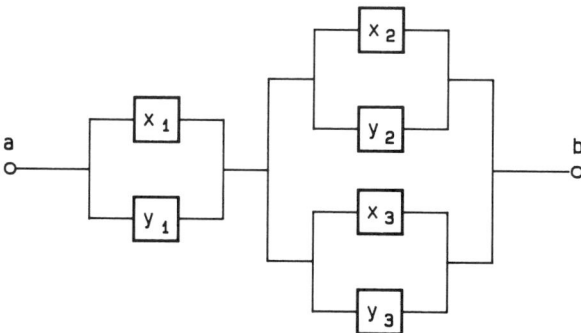

Figure 3.21 Redundancy at component level

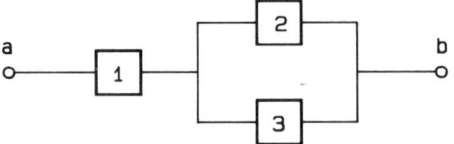

Figure 3.22 Reliability block diagram—Example 3.4

Definition 3.5: Cut sets, minimal cut sets
A cut set K is a set of components in C which by failing causes the system to fail. A cut set is said to be minimal if it cannot be reduced without loosing its status as a cut set. □

Example 3.4
Consider the reliability block diagram in Figure 3.22. The component set is $C = \{1,2,3\}$

$$\begin{array}{ll} \text{Path sets} & \text{Cut sets} \\ \{1,2\}* & \{1\}* \\ \{1,3\}* & \{2,3\}* \\ \{1,2,3\} & \{1,2\} \\ & \{1,3\} \\ & \{1,2,3\} \end{array}$$

The minimal cut sets and path sets are marked with an ∗.
In this case the minimal path sets are

$$P_1 = \{1,2\} \quad \text{and} \quad P_2 = \{1,3\}$$

while the minimal cut sets are

$$K_1 = \{1\} \quad \text{and} \quad K_2 = \{2,3\} \qquad \square$$

Example 3.5: Bridge Structure
Consider a bridge structure such as that given by the physical network in Figure 3.23. The minimal path sets are

$$P_1 = \{1,4\}, \quad P_2 = \{2,5\}, \quad P_3 = \{1,3,5\} \quad \text{and} \quad P_4 = \{2,3,4\}$$

The minimal cut sets are

$$K_1 = \{1,2\}, \quad K_2 = \{4,5\}, \quad K_3 = \{1,3,5\} \quad \text{and} \quad K_4 = \{2,3,4\}$$

□

SYSTEM STRUCTURE ANALYSIS

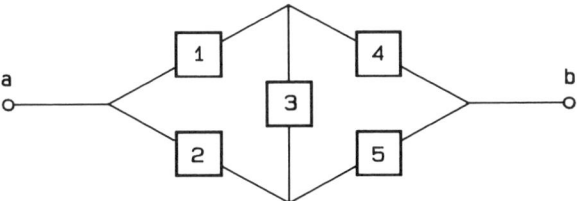

Figure 3.23 Bridge structure

In these examples the number of minimal cut sets coincides with the number of minimal path sets. This will usually not be the case.

The Designer's Point of View
Consider a designer who wants to ensure that a system is functioning with the least possible design effort. What the designer needs is a list of the minimal path sets from which one will be chosen for the design.

The Saboteur's Point of View
Next consider a saboteur who wants to bring the system into a failed state, again with the least possible effort on his or her part. What the saboteur needs is a list of the minimal cut sets from which to choose one for the sabotage plan.

ooo OOO ooo

Consider an arbitrary structure with minimal path sets $P_1, P_2, \ldots, P_p$ and minimal cut sets $K_1, K_2, \ldots, K_k$. To the minimal path set P_j, we associate the binary function

$$\rho_j(\mathbf{x}) = \prod_{i \in P_j} x_i; \quad j = 1, 2, \ldots, s \tag{3.8}$$

Note that $\rho_j(\mathbf{x})$ represents the structure function of a series structure composed of the components in P_j. $\rho_j(\mathbf{x})$ is therefore called the jth minimal *path series structure*.

Since we know that the structure is functioning if and only if at least one of the minimal path series structures is functioning,

$$\phi(\mathbf{x}) = \coprod_{j=1}^{p} \rho_j(\mathbf{x}) = 1 - \prod_{j=1}^{p} (1 - \rho_j(\mathbf{x})) \tag{3.9}$$

Hence our structure may be interpreted as a parallel structure of the minimal path series structures. From (3.8) and (3.9) we get

$$\phi(\mathbf{x}) = \coprod_{j=1}^{p} \prod_{i \in P_j} x_i \qquad (3.10)$$

Example 3.5 (cont.)
In the bridge structure in Figure 3.23, the minimal path sets were $P_1 = \{1,4\}$, $P_2 = \{2,5\}$, $P_3 = \{1,3,5\}$, and $P_4 = \{2,3,4\}$. The corresponding minimal path series structures are

$$\rho_1(\mathbf{x}) = x_1 \cdot x_4$$
$$\rho_2(\mathbf{x}) = x_2 \cdot x_5$$
$$\rho_3(\mathbf{x}) = x_1 \cdot x_3 \cdot x_5$$
$$\rho_4(\mathbf{x}) = x_2 \cdot x_3 \cdot x_4$$

Accordingly the structure function may be written

$$\begin{aligned}
\phi(\mathbf{x}) &= \coprod_{j=1}^{4} \rho_j(\mathbf{x}) = 1 - \prod_{j=1}^{4} (1 - \rho_j(\mathbf{x})) \\
&= 1 - (1 - \rho_1(\mathbf{x}))(1 - \rho_2(\mathbf{x}))(1 - \rho_3(\mathbf{x}))(1 - \rho_4(\mathbf{x})) \\
&= 1 - (1 - x_1 x_4)(1 - x_2 x_5)(1 - x_1 x_3 x_5)(1 - x_2 x_3 x_4) \\
&= x_1 x_4 + x_2 x_5 + x_1 x_3 x_5 + x_2 x_3 x_4 - x_1 x_3 x_4 x_5 - x_1 x_2 x_3 x_5 \\
&\quad - x_1 x_2 x_3 x_4 - x_2 x_3 x_4 x_5 - x_1 x_2 x_4 x_5 + 2 x_1 x_2 x_3 x_4 x_5
\end{aligned}$$

(Note that since x_i is a binary variable, $x_i^k = x_i$ for all i and k.)

Hence the bridge structure can be represented by the reliability block diagram in Figure 3.24. □

Similarly we can associate the following binary function to the minimal cut set K_j:

$$\kappa_j(\mathbf{x}) = \coprod_{i \in K_j} x_i = 1 - \prod_{i \in K_j} (1 - x_i), \qquad j = 1, 2, \ldots, k \qquad (3.11)$$

We see that $\kappa_j(\mathbf{x})$ represents the structure function of a parallel structure composed of the components in K_j. $\kappa_j(x)$ is therefore called the *j*th minimal *cut parallel structure*.

Since we know that the structure is failed if and only if at least one of the

SYSTEM STRUCTURE ANALYSIS

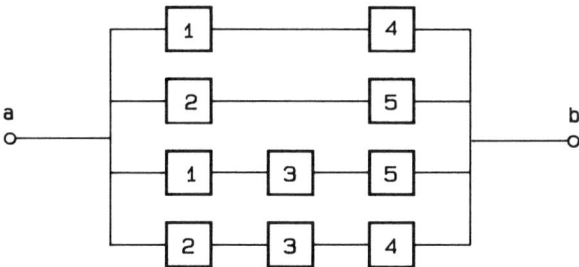

Figure 3.24 The bridge structure as a parallel structure of minimal path series structures

minimal cut parallel structures is failed, then

$$\phi(\mathbf{x}) = \prod_{j=1}^{k} \kappa_j(\mathbf{x}) \tag{3.12}$$

Hence we can regard this structure as a series structure of the minimal cut parallel structures. By combining (3.11) and (3.12), we get

$$\phi(\mathbf{x}) = \prod_{j=1}^{k} \coprod_{i \in K_j} x_i \tag{3.13}$$

Example 3.5 (cont.)
In the bridge structure the minimal cut sets were $K_1 = \{1,2\}$, $K_2 = \{4,5\}$, $K_3 = \{1,3,5\}$, and $K_4 = \{2,3,4\}$. The corresponding minimal cut parallel structures become

$$\kappa_1(\mathbf{x}) = x_1 \amalg x_2 = 1 - (1 - x_1)(1 - x_2)$$
$$\kappa_2(\mathbf{x}) = x_4 \amalg x_5 = 1 - (1 - x_4)(1 - x_5)$$
$$\kappa_3(\mathbf{x}) = x_1 \amalg x_3 \amalg x_5 = 1 - (1 - x_1)(1 - x_3)(1 - x_5)$$
$$\kappa_4(\mathbf{x}) = x_2 \amalg x_3 \amalg x_4 = 1 - (1 - x_2)(1 - x_3)(1 - x_4)$$

and we may find the structure function of the bridge structure by inserting these expressions into (3.12). The bridge structure may therefore be represented by the reliability block diagram in Figure 3.25. ☐

Structural Importance of Components

Some components in a system may obviously be more important than others in determining whether the system is functioning or not. A component in series

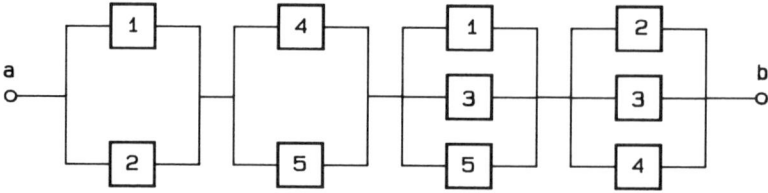

Figure 3.25 The bridge structure as a series structure of minimal cut parallel structures

with the rest of the system will, for example, be at least as important as any other component in the system. It would be useful to have a quantitative measure of the importance of the individual components in the system. Before we can establish such a measure, we need to define some new concepts.

Definition 3.6
A critical path vector for component i is a state vector $(1_i, \mathbf{x})$ such that

$$\phi(1_i, \mathbf{x}) = 1 \quad \text{while} \quad \phi(0_i, \mathbf{x}) = 0 \qquad \square$$

This is equivalent to requiring that

$$\phi(1_i, \mathbf{x}) - \phi(0_i, \mathbf{x}) = 1 \qquad (3.14)$$

In other words, given the states of the other components $(\cdot_i, \mathbf{x})$, the system is functioning if and only if component i is functioning. It is therefore natural to call $(1_i, \mathbf{x})$ a *critical* path vector for component i.

Definition 3.7
A critical path set $C(1_i, \mathbf{x})$ corresponding to the critical path vector $(1_i, \mathbf{x})$ for component i is defined by

$$C(1_i, \mathbf{x}) = \{i\} \cup \{j; x_j = 1, \ j \neq i\} \qquad (3.15)$$

$$\square$$

The total number of critical path sets (path vectors) for component i is

$$\eta_\phi(i) = \sum_{(\cdot_i, \mathbf{x})} [\phi(1_i, \mathbf{x}) - \phi(0_i, \mathbf{x})] \qquad (3.16)$$

Since the x_j's are binary variables and thus can take only two possible values, 0 and 1, the total number of state vectors $(\cdot, \mathbf{x}) = (x_1, \ldots, x_{i-1}, \cdot, x_{i+1}, \ldots, x_n)$ is 2^{n-1}.

SYSTEM STRUCTURE ANALYSIS

Birnbaum (1969) proposed the following measure for the structural importance of component i:

$$B_\phi(i) = \frac{\eta_\phi(i)}{2^{n-1}} \qquad (3.17)$$

$B_\phi(i)$ expresses the relative proportion of the 2^{n-1} possible state vectors $(\cdot_i, \mathbf{x})$ which are critical path vectors for component i. The components in the system can now be (partially) ranked according to the size of $B_\phi(i)$.

Example 3.6
Consider the 2-out-of-3 structure in Figure 3.26. For component 1, we have

$(\cdot, x_2, x_3)$	$\phi(1, x_2, x_3) - \phi(0, x_2, x_3)$	$C(1, x_2, x_3)$
$(\cdot 00)$	0	
$(\cdot 01)$	1	$\{1, 3\}$
$(\cdot 10)$	1	$\{1, 2\}$
$(\cdot 11)$	0	

In this case the total number of critical path vectors for component 1 is 2:

$$\eta_\phi(1) = 2$$

while the total number of state vectors $(\cdot_1, x_2, x_3)$ is $2^{3-1} = 4$. Hence

$$B_\phi(1) = \tfrac{2}{4} = \tfrac{1}{2}$$

For symmetrical reasons

$$B_\phi(1) = B_\phi(2) = B_\phi(3) = \tfrac{1}{2}$$

Hence the components 1, 2, and 3 are of equal structural importance. □

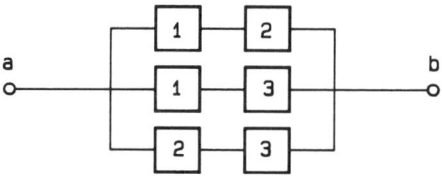

Figure 3.26 A 2-out-of-3 structure

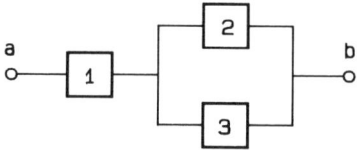

Figure 3.27 Reliability block diagram—Example 3.7

Example 3.7
Consider the structure in Figure 3.27. Here the structure function is

$$\phi(x_1, x_2, x_3) = x_1 \cdot (x_2 \amalg x_3) = x_1(1 - (1 - x_2)(1 - x_3))$$
$$= x_1 x_2 + x_1 x_3 - x_1 x_2 x_3$$

Consider component 1:

$(\cdot, x_2, x_3)$	$\phi(1, x_2, x_3) - \phi(0, x_2, x_3)$	$C(1, x_2, x_3)$
$(\cdot 00)$	0	
$(\cdot 01)$	1	$\{1, 3\}$
$(\cdot 10)$	1	$\{1, 2\}$
$(\cdot 11)$	1	$\{1, 2, 3\}$

The structural importance of component 1 is therefore

$$B_\phi(1) = \frac{\eta_\phi(1)}{2^{3-1}} = \frac{3}{4}$$

Now consider component 2.

$(x_1, \cdot, x_3)$	$\phi(x_1, 1, x_3) - \phi(x_1, 0, x_3)$	$C(x_1, 1, x_3)$
$(\cdot 00)$	0	
$(\cdot 01)$	0	
$(\cdot 10)$	1	$\{1, 2\}$
$(\cdot 11)$	0	

The structural importance of component 2 is therefore

$$B_\phi(2) = \frac{\eta_\phi(2)}{2^{3-1}} = \frac{1}{4}$$

For symmetrical reasons

SYSTEM STRUCTURE ANALYSIS

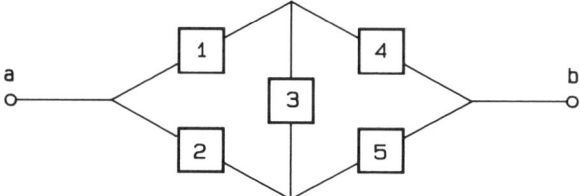

Figure 3.28 Bridge structure

$$B_\phi(2) = B_\phi(3) = \frac{\eta_\phi(2)}{2^{3-1}} = \frac{1}{4}$$

Component 1 is accordingly of greater structural importance than components 2 and 3. □

Pivotal Decomposition

The following identity holds for every structure function $\phi(\mathbf{x})$:

$$\phi(\mathbf{x}) \equiv x_i \phi(1_i, \mathbf{x}) + (1 - x_i)\phi(0_i, \mathbf{x}) \qquad \text{for all } \mathbf{x} \qquad (3.18)$$

We can easily see that this identity is correct from the fact that

$$x_i = 1 \Rightarrow \phi(\mathbf{x}) = 1 \cdot \phi(1_i, \mathbf{x}) \quad \text{and} \quad x_i = 0 \Rightarrow \phi(\mathbf{x}) = 1 \cdot \phi(0_i, \mathbf{x}).$$

By repeated use of (3.18) we arrive at $\phi(\mathbf{x}) = \sum_{\mathbf{y}} \prod_j x_j^{y_j}(1-x_j)^{1-y_j} \phi(\mathbf{y})$, where the summation is taken over all n-dimensional binary vectors.

Example 3.8: Bridge Structure
Consider the bridge structure in Figure 3.28. The structure function $\phi(\mathbf{x})$ of this system can be determined by pivotal decomposition with respect to component 3:

$$\phi(\mathbf{x}) = x_3 \phi(1_3, \mathbf{x}) + (1 - x_3)\phi(0_3, \mathbf{x})$$

Here, $\phi(1_3, \mathbf{x})$ is the structure function of the system in Figure 3.29,

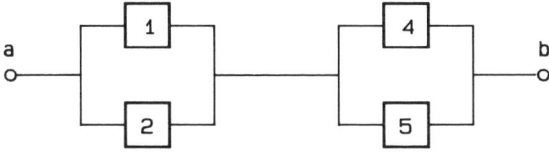

Figure 3.29 The structure $\phi(1_3, \mathbf{x})$ of the bridge structure

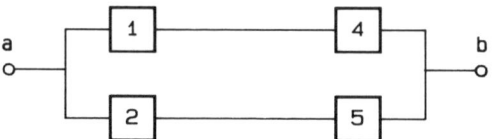

Figure 3.30 The structure $\phi(0_3, \mathbf{x})$ of the bridge structure

$$\phi(1_3, \mathbf{x}) = (x_1 \amalg x_2)(x_4 \amalg x_5) = (x_1 + x_2 - x_1x_2)(x_4 + x_5 - x_4x_5)$$

while $\phi(0_3, \mathbf{x})$ is the structure function of the system in Figure 3.30,

$$\phi(0_3, \mathbf{x}) = x_1x_4 \amalg x_2x_5 = x_1x_4 + x_2x_5 - x_1x_2x_4x_5$$

Hence the structure function of the bridge system becomes

$$\phi(\mathbf{x}) = x_3(x_1 + x_2 - x_1x_2)(x_4 + x_5 - x_4x_5)$$
$$+ (1 - x_3)(x_1x_4 + x_2x_5 - x_1x_2x_4x_5) \qquad \square$$

Modules of Coherent Structures

Let us assume that we want to find out under which conditions the structure in Figure 3.31 is functioning. It seems natural, first of all, to interpret this as a simple structure of "composed components," such as in Figure 3.32, where $\boxed{\text{I}}$, $\boxed{\text{II}}$, and $\boxed{\text{III}}$ are as defined in Figure 3.33. Then the subsystems $\boxed{\text{I}}$, $\boxed{\text{II}}$, and $\boxed{\text{III}}$ are analyzed individually. Finally, the results are put together logically. Regarding this logical connection, it is important that the partitioning into subsystems is done in such a way that *each single component never appears within more than one of the subsystems.*

When this partitioning is carried out in a specific way, described later, the procedure is called a *modular decomposition of the system*. In the following discussion, we will denote a system with (C, ϕ) where C is the set of components and ϕ the structure function. Let A represent a subset of C,

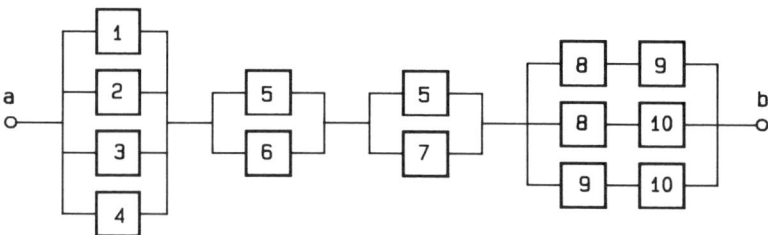

Figure 3.31 Reliability block diagram

SYSTEM STRUCTURE ANALYSIS

Figure 3.32 Structure of "composed components"

$$A \subseteq C$$

and A^c denote the complement of A with respect to C,

$$A^c = C - A$$

We denote the elements in A by $i_1, i_2, \ldots, i_\nu$, where $i_1 < i_2 < \cdots < i_\nu$. Let $\mathbf{x}^A$ be the state vector corresponding to the elements in A:

$$\mathbf{x}^A = (x_{i_1}, x_{i_2}, \ldots, x_{i_\nu})$$

and let

$$\chi(\mathbf{x}^A) = \chi(x_{i_1}, x_{i_2}, \ldots, x_{i_\nu})$$

be a binary function of $\mathbf{x}^A$. Obviously (A, χ) can be interpreted as a system (a structure).

In our example, $C = \{1, 2, \ldots, 10\}$. Let us choose $A = \{5, 6, 7\}$ and $\chi(\mathbf{x}^A) = (x_5 \amalg x_6)(x_5 \amalg x_7)$. (A, χ) then represents the substructure $\boxed{\text{II}}$. With this

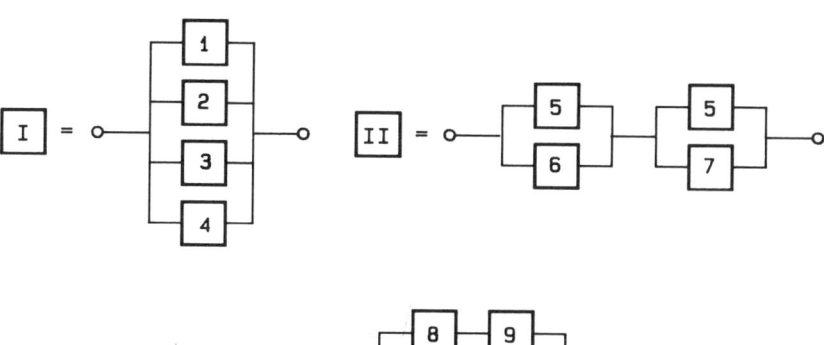

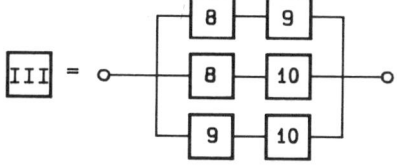

Figure 3.33 The three substructures

notation, a precise definition of the concept of a coherent module can be given as follows:

Definition 3.8: Coherent Modules
Let the coherent structure (C, ϕ) be given, and let $A \subseteq C$. Then (A, χ) is said to be a coherent module of (C, ϕ), if $\phi(\mathbf{x})$ can be written as a function of $\chi(\mathbf{x}^A)$ and $\mathbf{x}^{A^c}$, $\psi(\chi(\mathbf{x}^A), \mathbf{x}^{A^c})$ where ψ is the structure function of a coherent system. □

A is called a modular set of (C, ϕ), and if in particular $A \subset C$, (A, χ) is said to be a *proper* module of (C, ϕ).

What we actually do here is consider all the components with the index belonging to A as one "component" with state variable $\chi(\mathbf{x}^A)$. When we interpret the system in this way, the structure function will be

$$\psi(\chi(\mathbf{x}^A), \mathbf{x}^{A^c}).$$

In our example we chose $A = \{5, 6, 7\}$. Then

$$\psi(\chi(\mathbf{x}^A), \mathbf{x}^{A^c}) = \chi(x_5, x_6, x_7)\left(\coprod_{i=1}^{4} x_i\right)(x_8 x_9 \amalg x_8 x_{10} \amalg x_9 x_{10})$$

Since $A \subset C$, (A, χ) will be a proper module of (C, ϕ). Now let us define the concept of modular decomposition.

Definition 3.9: Modular Decomposition
A modular decomposition of a coherent structure (C, ϕ) is a set of disjoint modules (A_i, χ_i), $i = 1, \ldots, r$, together with an organizing structure ω, such that

1. $C = \bigcup_{i=1}^{r} A_i$ where $A_i \cap A_j = \emptyset$ for $i \neq j$.
2. $\phi(\mathbf{x}) = \omega[\chi_1(\mathbf{x}^{A_1}), \chi_2(\mathbf{x}^{A_2}), \ldots, \chi_r(\mathbf{x}^{A_r})]$ □

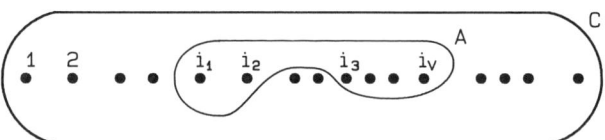

Figure 3.34 Modular decomposition

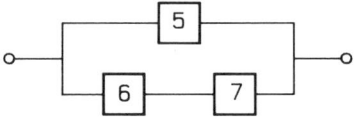

Figure 3.35 Module II

Remark
Is "disjoint" superfluous here? Elements in a module can certainly not appear outside the module. But even so, it does not prevent the possibility for *two* modules to be defined in such a way that they overlap each other. Thus "disjoint" is necessary.

The "finest" partitioning into modules that one can have, is obviously to let each individual component constitute one module. The "coarsest" partitioning into modules is to let the whole system constitute one module. To be of practical use, a module decomposition should, if possible, be something between these two extremes. A module that cannot be partitioned into smaller modules without letting each components represent a module, is called a *prime module* (an analogy to prime number).

In our example III represents a prime module. But II is not a prime module, since it may be described as in Figure 3.35, and hence can be partitioned into two modules *IIa* and *IIb* as in Figure 3.36. This gives no guidance on how to determine individual prime modules in a system. However, algorithms have been developed, for example, by Chatterjee (1975), that can be used to find all the prime modules in a fault tree or in a reliability block diagram.

In Chapter 4 we will justify the fact that it is natural to interpret the state vector as stochastic. In accordance with what we did in probability theory, we will, from now on, denote the state variables with *capital* letters from the end of the alphabet, for example, $X_1, \ldots, X_n$.

Occasionally two or more of the state variables can be stochastically dependent. In such situations it is advisable to try to "collect" the variables in modules in such a way that dependency occurs only *within the modules*. If one succeeds in this, the individual modules can be considered independent. This will make the further analysis simpler.

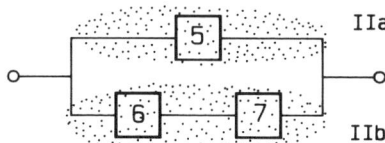

Figure 3.36 Two prime modules

3.6 PROBLEMS

1. Consider the pressure safety valve (PSV) in Example 3.2. The main purpose of the PSV, which is normally closed, is to relieve the pressure in the separator when the pressure increases beyond a specified high pressure p. The PSV is equipped with a spring-loaded actuator that may be adjusted to the preset pressure p. A sketch of the pressure safety valve (PSV) in closed position is shown in Figure 3.37.

   Carry out an FMECA analysis of the PSV according to the procedure described in Section 3.2.

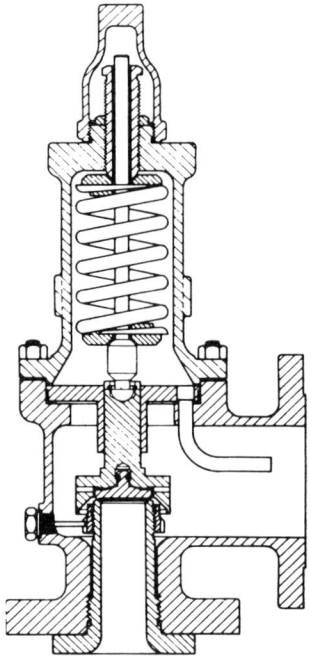

Figure 3.37 Pressure safety valve in closed position

2. Consider the process shutdown (PSD) valve A_1 in Example 3.2. The valve, which is illustrated in Figure 3.38, is a spring-loaded, fail-safe close gate valve that is held open by hydraulic pressure. The gate is a solid block with a cylindrical hole with the same diameter as the pipeline. To open the valve, hydraulic pressure is applied on the upper side of the piston. The pressure forces the piston, the piston rod, and the gate downward until the hole in the gate is in line with the pipeline. When the pressure is bled off,

PROBLEMS   **119**

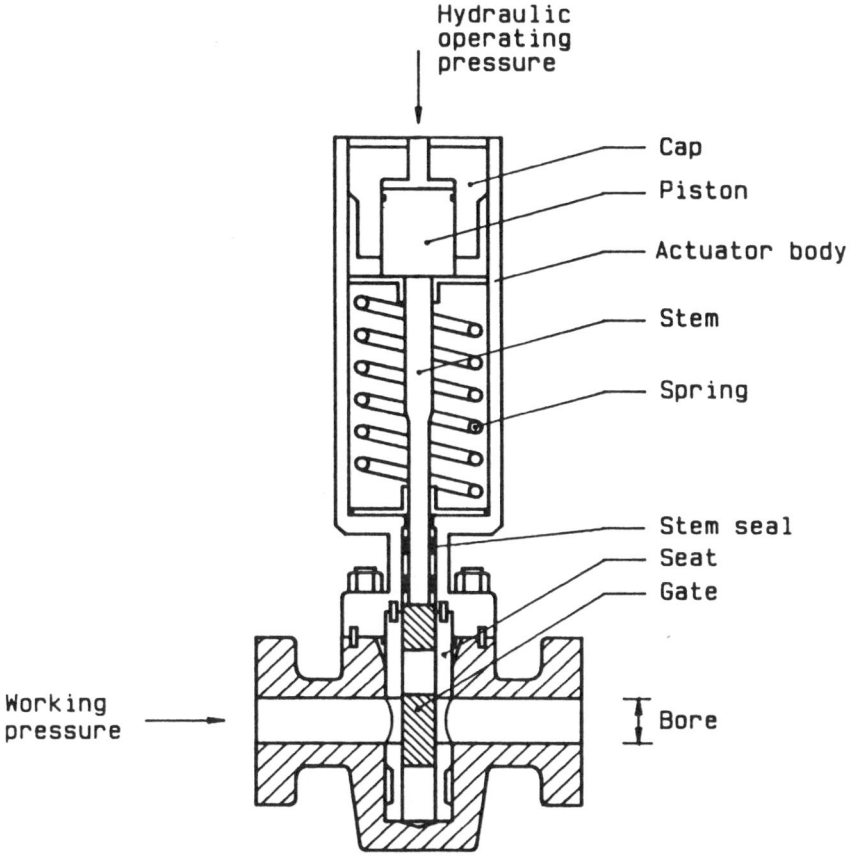

Figure 3.38 Hydraulically operated gate valve

the spring forces the piston upward until the hole in the gate is no longer in contact with the pipeline conduct. The solid part of the gate is now pressed against the seat seal and the valve is closed.

Carry out an FMECA analysis of the PSD valve according to the procedure described in Section 3.2. (*Remark*: In many process systems fail-safe close ball valves are used as PSD valves instead of gate valves.)

3. Figure 3.39 shows a sketch of a steam boiler system that supplies steam to a process system at a specified pressure. Water is led to the boiler through a pipeline with a regulator valve LICV. Fuel (oil) is led to the burner chamber through a pipeline with a regulator valve PCV. The valve PCV is installed in parallel with a bypass valve V-1 together with two isolation valves to facilitate inspection and maintenance of PCV during normal operation.

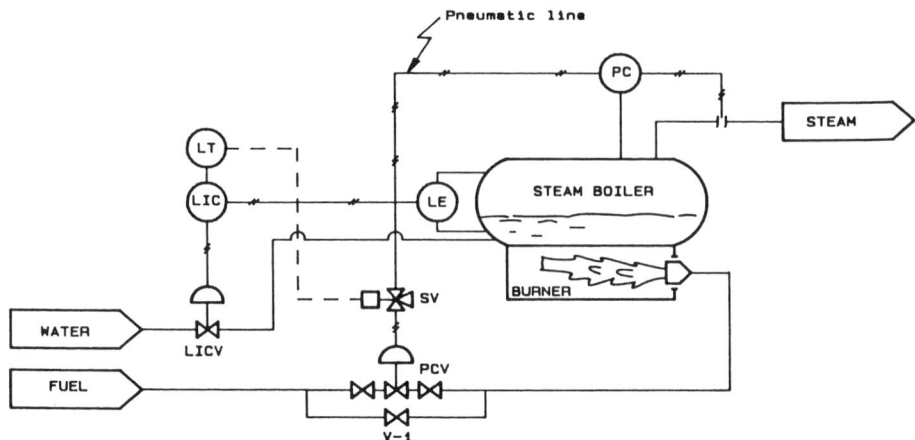

Figure 3.39 Steam boiler system

The level of the water in the boiler is surveyed by a level emitter (LE). The water level is maintained in an interval between a specified *low* level and a specified *high* level by a pneumatic control circuit connected to the water regulator valve LICV. The level indicator controller (LIC) translates the pneumatic "signal" from LE to a pneumatic "signal" controlling the valve LICV.

It is very important that the water level not falls below the specified *low* level. When the water level approaches the *low* level, a pneumatic "signal" is passed from LIC to the level transmitter (LT). LT translates the pneumatic "signal" to an electrical "signal" which is sent to the solenoid valve (SV). The solenoid valve again controls the valve PCV on the fuel inlet pipeline. This circuit is thus installed to cut off the fuel supply in the event that the water level falls below the specified *low* level.

The pressure in the boiler and in the steam outlet pipeline is surveyed by a pressure controller (PC) which is connected to the valve SV, and thereby to the valve PCV on the fuel inlet pipeline. This circuit is thus installed to cut off the fuel supply in the event that the pressure in the boiler increases above a specified *high* pressure. Again, the symbols that concern us are as follows:

V-1: bypass valve
LE: level emitter
LIC: level indicator controller
LICV: level indicator controller valve
LT: level transmitter
PC: pressure controller

PROBLEMS   **121**

PCV: pressure controller valve
SV: solenoid valve

A *critical* situation occurs if the boiler is boiled dry. Then the pressure in the vessel will increase very rapidly and the vessel may explode.

(a) Construct a fault tree where the TOP event is the *critical* situation mentioned above. Secondary failure causes are not to be included. Write down assumptions and limitations you have to make during the fault tree construction.

(b) Establish a reliability block diagram corresponding to the fault tree in (a), and determine the structure function $\phi(\mathbf{x})$.

(c) Determine all minimal cut sets in the fault tree (reliability block diagram).

4. Figure 3.40 shows a sketch of the lubrication system on a ship engine. The separator separates water from the oil lubricant. The separator will only function satisfactory when the oil is heated to a specified temperature. When the water content in the oil is too high, the quality of the lubrication will be too low, and this may lead to damage or breakdown of the engine. The engine will generally require

- Sufficient throughput of oil/lubricant
- Sufficient quality of the oil/lubricant

The oil throughput is sufficient when at least one cooler is functioning, at least one filter is open (i.e., not clogged), and the pump is func-

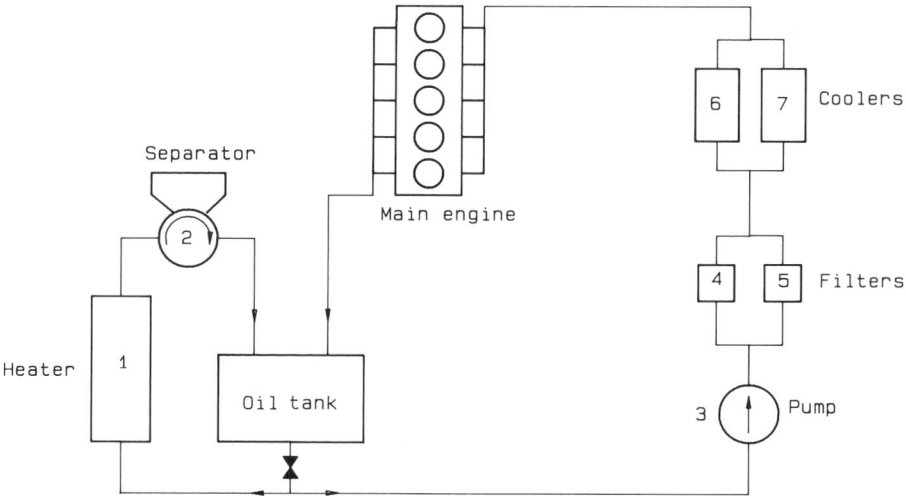

Figure 3.40 Lubrication system on a ship engine

tioning. In addition all necessary pipelines must be open, no valves must be unintentionally closed, the lubrication channels in the engine must be open (not clogged), and the lubrication system must not have significant leakages to the environment. We will here assume that the probabilities of these "additional" events are very low and that these events therefore may be neglected.

The quality of the oil is sufficient when

- Both coolers are functioning (with full throughput) such that the temperature of the oil to the engine is sufficiently low
- None of the filters is clogged, and there are no holes in the filters
- The separator system is functioning

   (a) Construct a fault tree with respect to the TOP event, "Too low throughput of oil/lubricant."

   (b) Construct a fault tree with respect to the TOP event, "Too low quality of the oil/lubricant."

5. Use MOCUS to identify all the minimal cut sets of the fault tree in Figure 3.4.

6. Consider the system in Figure 3.41.

   (a) Derive the corresponding structure function by a direct approach.

   (b) Determine the path sets and the cut sets, and then the minimal path sets and the minimal cut sets.

   (c) Derive the structure function of the system by using the property that the structure may be considered as a parallel structure of the minimal path series structures.

   (d) Derive the structure function of the system by using the property that the structure may be considered as a series structure of the minimal cut parallel structures.

   (e) Compare the results obtained in (a) and (b).

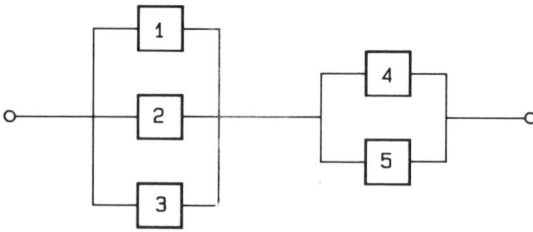

Figure 3.41 Reliability block diagram—Problem 6

PROBLEMS

7. Reduce the reliability block diagram in Figure 3.42 to the simplest possible form and determine the structure function.

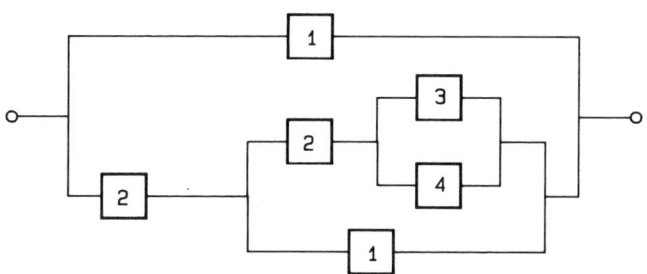

Figure 3.42 Reliability block diagram—Problem 7

8. Show that
   (a) If ϕ represents a parallel structure, then:

   $$\phi(\mathbf{x} \amalg \mathbf{y}) = \phi(\mathbf{x}) \amalg \phi(\mathbf{y})$$

   (b) If ϕ represents a series structure, then

   $$\phi(\mathbf{x} \cdot \mathbf{y}) = \phi(\mathbf{x}) \cdot \phi(\mathbf{y})$$

9. The dual structure $\phi^D(\mathbf{x})$ to a given structure $\phi(\mathbf{x})$ is defined by

   $$\phi^D(\mathbf{x}) = 1 - \phi(\mathbf{1} - \mathbf{x})$$

   where $(\mathbf{1} - \mathbf{x}) = (1 - x_1, 1 - x_2, \ldots, 1 - x_n)$.
   (a) Show that the dual structure of a k-out-of-n structure is a $(n - k + 1)$-out-of-n structure.
   (b) Show that the minimal cut sets for ϕ are minimal path sets for ϕ^D, and vice versa.

10. Determine the structural importance of the different components of the structure studied in Problem 6.

11. Determine the structure function of the system in Figure 3.43 by applying an appropriate modular decomposition.

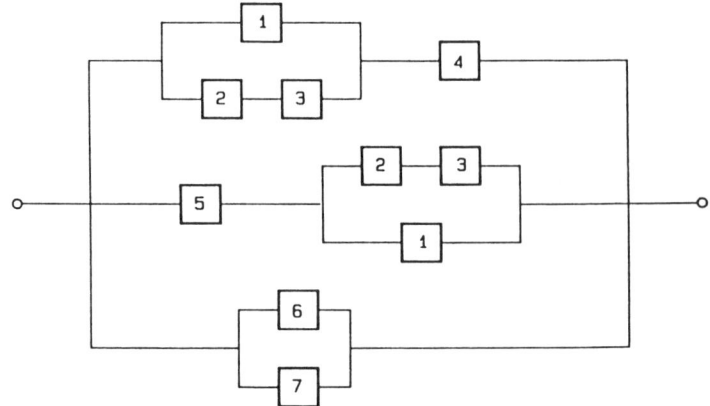

Figure 3.43 Reliability block diagram—Problem 11

12. Consider the fault tree in Figure 3.44.

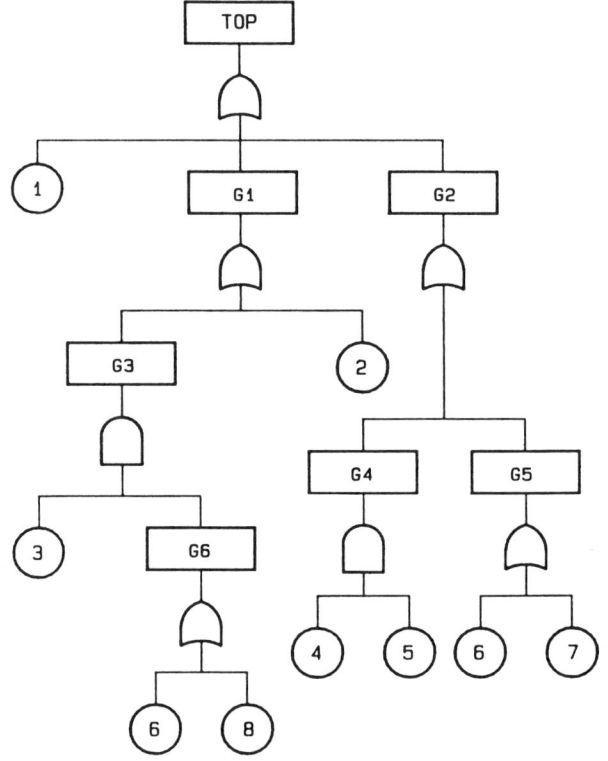

Figure 3.44 Example fault tree

(a) Use MOCUS to identify all the minimal path sets of the fault tree.

(b) Show that the system may be represented by the reliability block diagram in Figure 3.45.

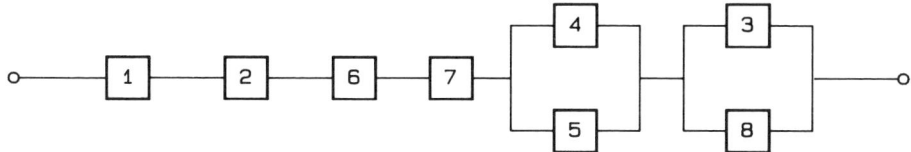

Figure 3.45 Reliability block diagram—Problem 12

13. Use appropriate pivotal decompositions to determine the structure function of the system in Figure 3.46.

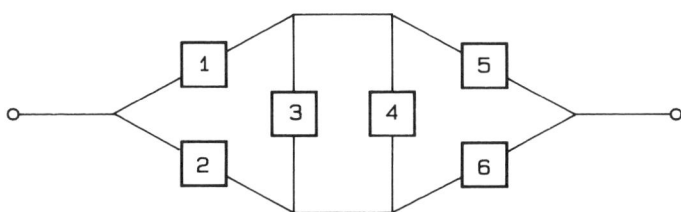

Figure 3.46 Reliability block diagram—Problem 13

14. Determine the structure function of the structure in Figure 3.47 by use of appropriate pivotal decompositions.

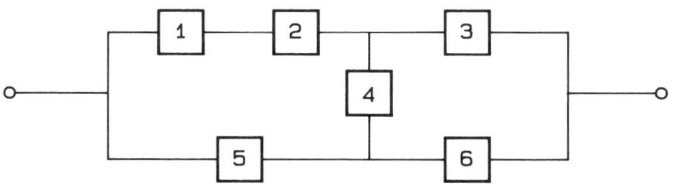

Figure 3.47 Reliability block diagram—Problem 14

15. Determine the structure function of the structure in Figure 3.48.

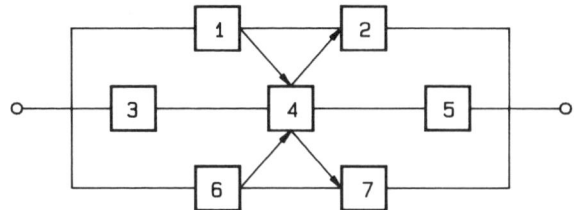

Figure 3.48 Reliability block diagram—Problem 15

16. Show that the structure function of the system in Figure 3.49 may be written

$$\phi(\mathbf{x}) = [x_1(x_2 + x_3 - x_2x_3)x_4x_5(x_6 + x_7 - x_6x_7)x_8 \\ + x_9x_{10}x_{11} - x_1(x_2 + x_3 - x_2x_3)x_4x_5 \\ \cdot (x_6 + x_7 - x_6x_7)x_8x_9x_{10}x_{11}]x_{12}$$

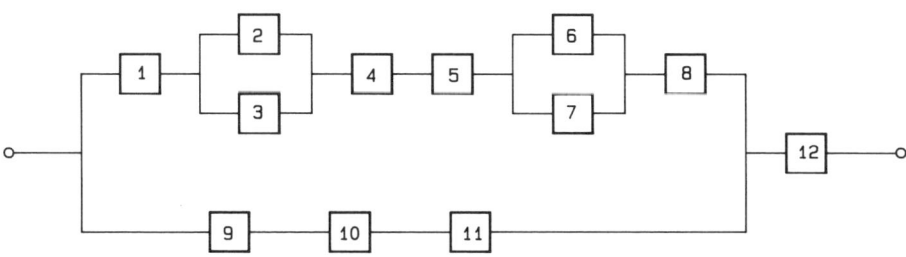

Figure 3.49 Reliability block diagram—Problem 16

17. Consider the structure in Figure 3.50.
    (a) Find the minimal path sets $P_1, \ldots, P_p$ and the minimal cut sets $K_1, \ldots, K_k$ of the structure.
    (b) Derive the structure function of this system.

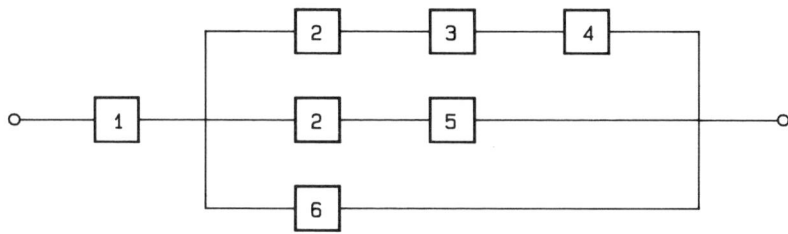

Figure 3.50 Reliability block diagram—Problem 17

CHAPTER 4

Systems of Independent Components

4.1 INTRODUCTION

In Chapter 3 we discussed the structural relationship between a system and its components and showed how a *deterministic* model of the structure can be established by using a reliability block diagram or a fault tree. Whether or not a given component will be in a failed state after t time units, cannot usually be predicted with certainty, however. Rather, when studying the occurrence of such failures, one looks for statistical regularity. Hence it seems reasonable to interpret the state variables of the n components at time t as random variables.

We denote the state variables by

$$X_1(t), X_2(t), \ldots, X_n(t)$$

Correspondingly the state *vector* and the structure function are denoted, respectively, by

$$X(t) = (X_1(t), \ldots, X_n(t)) \quad \text{and} \quad \phi(X(t))$$

The following probabilities are of interest:

$$P(X_i(t) = 1) = p_i(t) \quad \text{for } i = 1, 2, \ldots, n \quad (4.1)$$
$$P(\phi(X(t)) = 1) = p_S(t) \quad (4.2)$$

Throughout Chapter 4 we will restrict ourselves to studying systems where failures of the individual components can be interpreted as *independent* events. This implies that the state variables at time t, $X_1(t), \ldots, X_n(t)$ can be considered as being stochastically independent. Unfortunately, independence is often assumed just to "simplify" the analysis, even though it is unrealistic. (This is discussed in more detail in Chapter 8.)

In the main part of this chapter we consider *nonrepairable* components and systems, which are discarded the first time they fail. In that case (4.1) and (4.2) correspond to what in Chapter 2 is called the *survivor function* of component i and of the system, respectively.

Components and systems that are *replaced* or *repaired* after failure are (in this book) called *repairable*. Then (4.1) and (4.2) correspond with what we in Section 1.6 called the *availability* at time t of component i and of the system, respectively. Repairable components and systems that are considered only until the first failure are treated as nonrepairable. A thorough discussion of the reliability of repairable systems is given by Ascher and Feingold (1984).

During most of this chapter, $p_i(t)$ will, for brevity, be called the *reliability* of component i at time t for $i = 1, 2, \ldots, n$. In the same way, $p_S(t)$ will be called the *system reliability* at time t. In Sections 4.1 and 4.2 we will first study the general characteristics of system reliability. Nonrepairable and repairable systems will be discussed in Sections 4.3 and 4.7, respectively.

4.2 SYSTEM RELIABILITY

Since the state variables $X_i(t)$ for $i = 1, \ldots, n$ are binary, then

$$E[X_i(t)] = 0 \cdot P(X_i(t) = 0) + 1 \cdot P(X_i(t) = 1)$$
$$= p_i(t) \quad \text{for } i = 1, 2, \ldots, n \quad (4.3)$$

Similarly the system reliability (at time t) is

$$p_S(t) = E(\phi(X(t))) \quad (4.4)$$

It can be shown (see Problem 4.1) that when the components are independent, the system reliability, $p_S(t)$ will be a function only of the $p_i(t)$'s. Hence $p_S(t)$ may be written

$$p_S(t) = h(p_1(t), p_2(t), \ldots, p_n(t)) = h(\boldsymbol{p}(t)) \quad (4.5)$$

Unless we state otherwise, we will use the letter h to express system reliability in situations *where the components are independent*. Now let us determine the reliability of some simple structures.

Reliability of Series Structures

In Section 3.4 (p. 99) we found that a series structure of order n has the structure function

# SYSTEM RELIABILITY

$$\phi(X(t)) = \prod_{i=1}^{n} X_i(t)$$

Since $X_1(t), \ldots, X_n(t)$ are assumed to be independent, the system reliability is

$$h(p(t)) = E(\phi(X(t))) = E\left(\prod_{i=1}^{n} X_i(t)\right)$$
$$= \prod_{i=1}^{n} E(X_i(t)) = \prod_{i=1}^{n} p_i(t) \qquad (4.6)$$

Note that

$$h(p(t)) \leq \min_i (p_i(t))$$

In other words, a series structure is *at most* as reliable as the *least* reliable component.

In particular, if all the components have the same reliability $p(t)$, then the system reliability of a series structure of order n will be

$$p_S(t) = p(t)^n \qquad (4.7)$$

If, for example, $n = 10$ and $p(t) = 0.95$, then

$$p_S(t) = 0.95^{10} \approx 0.60$$

Hence the system reliability of a series structure is low already when $n = 10$, even if the component reliability is relatively high ($= 0.95$).

Reliability of Parallel Structures

In Section 3.4 (p. 99) we found that a parallel structure of order n has the structure function

$$\phi(X(t)) = \coprod_{i=1}^{n} X_i(t) = 1 - \prod_{i=1}^{n} (1 - X_i(t))$$

Hence

$$h(p(t)) = E(\phi(X(t))) = 1 - \prod_{i=1}^{n}(1 - E(X_i(t)))$$

$$= 1 - \prod_{i=1}^{n}(1 - p_i(t)) = \coprod_{i=1}^{n} p_i(t) \qquad (4.8)$$

In particular, if all the components have the same reliability $p(t)$, then the system reliability of a parallel structure of order n is

$$p_S(t) = 1 - (1 - p(t))^n \qquad (4.9)$$

Table 4.1 illustrates the relationship between $p_S(t)$ and n for selected values of $p(t)$ and n.

Reliability of k-out-of-n Structures

In Section 3.4 (p. 100) we found that a k-out-of-n structure has the structure function

$$\phi(X(t)) = \begin{cases} 1 & \text{if } \sum_{i=1}^{n} X_i(t) \geq k \\ 0 & \text{if } \sum_{i=1}^{n} X_i(t) < k \end{cases}$$

Table 4.1 Reliability of a Parallel Structure With n Identical Components for Selected Values of $p(t)$

$p(t)$	1	2	3	4	6	8	10
0.10	0.100	0.190	0.271	0.344	0.469	0.570	0.651
0.20	0.200	0.360	0.488	0.590	0.738	0.832	0.893
0.30	0.300	0.510	0.657	0.760	0.882	0.942	0.972
0.40	0.400	0.640	0.784	0.870	0.953	0.983	0.994
0.50	0.500	0.750	0.875	0.938	0.984	0.996	
0.60	0.600	0.840	0.936	0.974	0.996		
0.70	0.700	0.910	0.973	0.992			
0.80	0.800	0.960	0.992				
0.90	0.900	0.990					

SYSTEM RELIABILITY

Let us for simplicity consider a k-out-of-n structure where all the n components have identical reliabilities $p_i(t) = p(t)$ for $i = 1, 2, \ldots, n$.

Since we have assumed that failures of individual components are independent events, then at a given time t, $Y(t) = \sum_{i=1}^{n} X_i(t)$ will be binomially distributed $(n, p(t))$:

$$P(Y(t) = y) = \binom{n}{y} p(t)^y (1 - p(t))^{n-y} \quad \text{for} \quad y = 0, 1, \ldots, n$$

Hence the reliability of a k-out-of-n structure of components with identical reliabilities is

$$p_S(t) = P(Y(t) \geq k) = \sum_{y=k}^{n} \binom{n}{y} p(t)^y (1 - p(t))^{n-y} \quad (4.10)$$

Finally, let us see how the system reliability of a more complex structure can be determined.

Example 4.1

Figure 4.1 shows the reliability block diagram of a simplified automatic alarm system for gas leakage. In the case of gas leakage, "connection" is established between (a) and (b) so that at least one of the alarm bells (7 and 8) will start ringing. The system has three independent gas detectors (1, 2 and 3) which are connected to a 2-out-of-3 voting unit (4); that is, at least two detectors must indicate gas leakage before an alarm is triggered off. Component 5 is a power supply unit, and component 6 is a relay.

We will consider the system at a given time t_0 and introduce the state variables

$$X_i = \begin{cases} 1 & \text{if component } i \text{ is functioning at time } t_0 \\ 0 & \text{otherwise}, \ i = 1, 2, \ldots, 8 \end{cases}$$

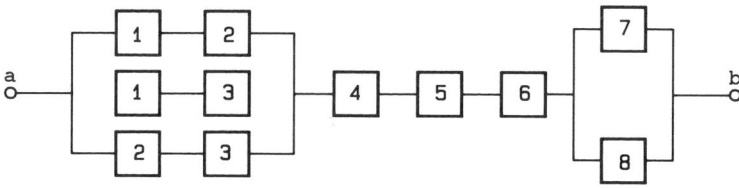

Figure 4.1 Reliability block diagram of a simplified automatic alarm system for gas leakage

Figure 4.2 Modularized reliability block diagram of a simplified automatic alarm system for gas leakage

First, let us find the minimal cut and path sets:

Minimal path sets	Minimal cut sets
$P_1 = \{1,2,4,5,6,7\}$	$K_1 = \{1,2\}$
$P_2 = \{1,2,4,5,6,8\}$	$K_2 = \{1,3\}$
$P_3 = \{1,3,4,5,6,7\}$	$K_3 = \{2,3\}$
$P_4 = \{1,3,4,5,6,8\}$	$K_4 = \{4\}$
$P_5 = \{2,3,4,5,6,7\}$	$K_5 = \{5\}$
$P_6 = \{2,3,4,5,6,8\}$	$K_6 = \{6\}$
	$K_7 = \{7,8\}$

The system may now be represented by the series structure in Figure 4.2 of the three modules, $\boxed{I}$, $\boxed{II}$, and $\boxed{III}$. The organizing structure of the modules is

$$\omega(\chi_I, \chi_{II}, \chi_{III}) = \chi_I \cdot \chi_{II} \cdot \chi_{III} \tag{4.11}$$

where χ_i represents the structure function of module i, $i = I, II, III$. Hence the structure function of the system is

$$\phi(X) = (X_1 X_2 + X_1 X_3 + X_2 X_3 - 2 X_1 X_2 X_3)$$
$$\cdot (X_4 X_5 X_6)(X_7 + X_8 - X_7 X_8)$$

If the component reliability at time t_0 of component i is denoted by p_i, $i = 1, \ldots, 8$, and $X_1, \ldots, X_8$ are independent, then the system reliability at time t_0 is

$$p_S = (p_1 p_2 + p_1 p_3 + p_2 p_3 - 2 p_1 p_2 p_3) \cdot p_4 p_5 p_6 \cdot (p_7 + p_8 - p_7 p_8) \tag{4.12}$$

□

4.3 NONREPAIRABLE SYSTEMS

As explained in Section 4.1 the component reliability and the survivor function will coincide for nonrepairable components:

$$p_i(t) = R_i(t) \quad \text{for} \quad i = 1, 2, \ldots, n$$

Nonrepairable Series Structures

According to (4.6) the survivor function of a nonrepairable series structure consisting of independent components is

$$R_S(t) = \prod_{i=1}^{n} R_i(t) \tag{4.13}$$

Furthermore, according to (2.8),

$$R_i(t) = e^{-\int_0^t z_i(u)\,du} \tag{4.14}$$

where $z_i(t)$ denotes the failure rate of component i at time t.
By inserting (4.14) into (4.13), we get

$$R_S(t) = \prod_{i=1}^{n} e^{-\int_0^t z_i(u)\,du} = e^{-\int_0^t \sum_{i=1}^{n} z_i(u)\,du} \tag{4.15}$$

Hence the failure rate $z_S(t)$ of a series structure (of independent components) is equal to the sum of the failure rates of the individual components:

$$z_S(t) = \sum_{i=1}^{n} z_i(t) \tag{4.16}$$

The mean time to failure of this series structure is

$$\text{MTTF} = \int_0^\infty R_S(t)\,dt = \int_0^\infty e^{-\int_0^t \sum_{i=1}^{n} z_i(u)\,du}\,dt$$

In the particular case where all the components have constant failure rates, $z_i(t) = \lambda_i$, $i = 1, 2, \ldots, n$, then

$$\text{MTTF} = \int_0^\infty e^{-\sum_{i=1}^{n} \lambda_i t}\,dt = \frac{1}{\sum_{i=1}^{n} \lambda_i} \tag{4.17}$$

Nonrepairable Parallel Structures

According to (4.8) the survivor function of a nonrepairable parallel structure of independent components is

$$R_S(t) = 1 - \prod_{i=1}^{n}(1 - R_i(t)) \tag{4.18}$$

In the particular case where all the components have constant failure rates $z_i(t) = \lambda_i$ for $i = 1, 2, \ldots, n$, then

$$R_S(t) = 1 - \prod_{i=1}^{n}(1 - e^{-\lambda_i t}) \tag{4.19}$$

The survivor function of a parallel structure of two independent components with constant failure rates λ_1 and λ_2, respectively, is

$$R_S(t) = 1 - (1 - e^{-\lambda_1 t})(1 - e^{\lambda_2 t}) = e^{-\lambda_1 t} + e^{-\lambda_2 t} - e^{-(\lambda_1+\lambda_2)t} \tag{4.20}$$

Note that the time to failure T of this parallel structure is *not* exponentially distributed, even if both components have exponentially distributed times to failure.

The mean time to failure of this system is

$$\text{MTTF} = \int_0^\infty R_S(t)\,dt = \frac{1}{\lambda_1} + \frac{1}{\lambda_2} - \frac{1}{\lambda_1 + \lambda_2} \tag{4.21}$$

Example 4.2
Consider a nonrepairable parallel structure of two components with lifetimes T_1 and T_2, which are assumed to be independent and exponentially distributed with failure rates λ_1 and λ_2, respectively.

The survivor function of this system is

$$R_S(t) = e^{-\lambda_1 t} + e^{-\lambda_2 t} - e^{-(\lambda_1+\lambda_2)t} \tag{4.22}$$

The corresponding failure rate is given by

$$z(t) = -\frac{R'_S(t)}{R_S(t)} \tag{4.23}$$

Hence

$$z(t) = \frac{\lambda_1 e^{-\lambda_1 t} + \lambda_2 e^{-\lambda_2 t} - (\lambda_1 + \lambda_2)e^{-(\lambda_1+\lambda_2)t}}{e^{-\lambda_1 t} + e^{-\lambda_2 t} - e^{-(\lambda_1+\lambda_2)t}} \tag{4.24}$$

NONREPAIRABLE SYSTEMS

It is now easy find a time t_0 so that $z(t)$ is increasing in the interval $(0, t_0)$ while $z(t)$ is decreasing in the interval (t_0, ∞). This t_0 will depend on λ_1 and λ_2. In Figure 4.3, $z(t)$ is sketched for selected combinations of λ_1 and λ_2, such that $\lambda_1 + \lambda_2 = 1$. □

This example illustrates that even if the individual components of a system have *constant* failure rates, the system itself may not have a constant failure rate.

Nonrepairable 2-out-of-3 Structures

According to Section 3.4 (p. 101) the structure function of a 2-out-of-3 structure is

$$\phi(X(t)) = X_1(t)X_2(t) + X_1(t)X_3(t) + X_2(t)X_3(t) - 2X_1(t)X_2(t)X_3(t)$$

Thus the survivor function of the 2-out-of-3 structure is

$$R_S(t) = R_1(t)R_2(t) + R_1(t)R_3(t) + R_2(t)R_3(t) - 2R_1(t)R_2(t)R_3(t)$$

In the special case where all the three components have the common constant failure rate λ, then

$$R_S(t) = 3e^{-2\lambda t} - 2e^{-3\lambda t} \tag{4.25}$$

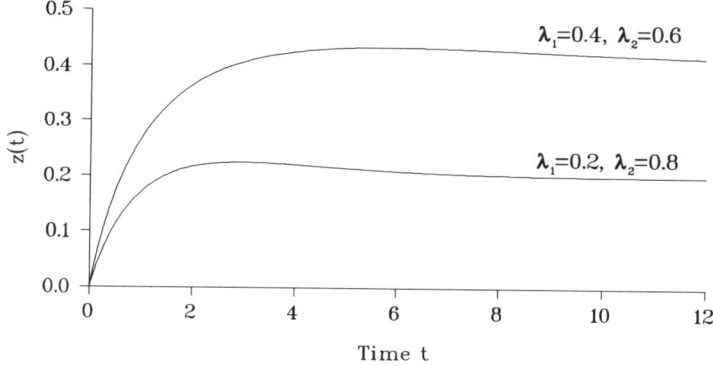

Figure 4.3 Failure rate for a parallel structure of two independent components for selected values of λ_1 and λ_2 ($\lambda_1 + \lambda_2 = 1$)

The mean time to failure of this 2-out-of-3 structure is

$$\text{MTTF} = \int_0^\infty R_S(t)\,dt = \frac{3}{2\lambda} - \frac{2}{3\lambda} = \frac{5}{6}\frac{1}{\lambda} \qquad (4.26)$$

A Brief Comparison

Let us now compare the three simple systems:

1. A single component
2. A parallel structure of two identical components
3. A 2-out-of-3 system with identical components

All the components are assumed to be independent with a common constant failure rate λ. A brief comparison of the three systems is presented in Table 4.2. Note that a single component has a higher MTTF than the 2-out-of-3 structure. The survivor functions of the three simple systems are also compared in Figure 4.4

The introduction of a 2-out-of-3 structure instead of a single component hence reduces the MTTF by about 16%. The 2-out-of-3 structure has, however, a significantly higher reliability (probability of functioning) in the interval $(0, t]$ for $t < \ln 2/\lambda$.

Nonrepairable k-out-of-n Structures

Assume that we have a k-out-of-n structure of n identical and independent components with constant failure rate λ. The survivor function of the k-out-of-n structure is, from (4.10),

$$R(t) = \sum_{x=k}^{n} \binom{n}{x} e^{-\lambda t x}(1 - e^{-\lambda t})^{n-x} \qquad (4.27)$$

According to (2.11) the mean time to failure is

$$\text{MTTF} = \sum_{x=k}^{n} \binom{n}{x} \int_0^\infty e^{-\lambda t x}(1 - e^{-\lambda t})^{n-x}\,dt \qquad (4.28)$$

By introducing $v = e^{-\lambda t}$, we obtain (see Appendix A)

NONREPAIRABLE SYSTEMS

Table 4.2 A Brief Comparison of the Systems (1), (2), and (3)

System	Survivor Function $R_S(t)$	Mean Time to Failure MTTF
1oo1	$e^{-\lambda t}$	$\dfrac{1}{\lambda}$
1oo2	$2e^{-\lambda t} - e^{-2\lambda t}$	$\dfrac{3}{2}\dfrac{1}{\lambda}$
2oo3	$3e^{-2\lambda t} - 2e^{-3\lambda t}$	$\dfrac{5}{6}\dfrac{1}{\lambda}$

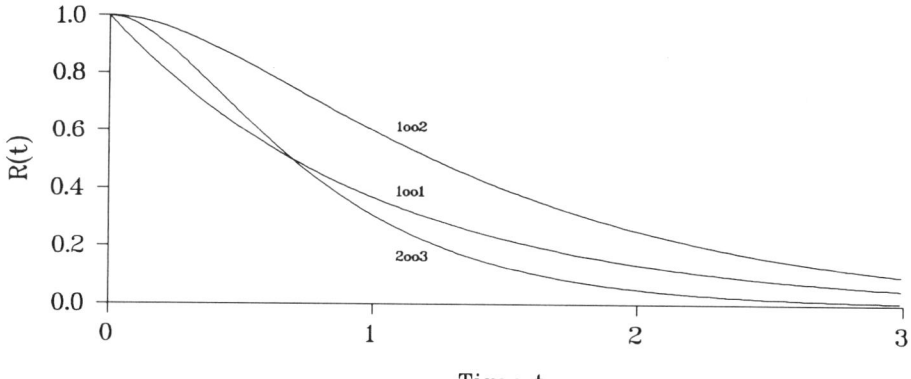

Figure 4.4 The survivor functions of three systems

Table 4.3 MTTF of Some k-Out-of-n Systems of Identical and Independent Components With Constant Failure Rate λ

$k\backslash n$	1	2	3	4	5
1	$\dfrac{1}{\lambda}$	$\dfrac{3}{2\lambda}$	$\dfrac{11}{6\lambda}$	$\dfrac{25}{12\lambda}$	$\dfrac{137}{60\lambda}$
2	—	$\dfrac{1}{2\lambda}$	$\dfrac{5}{6\lambda}$	$\dfrac{13}{12\lambda}$	$\dfrac{77}{60\lambda}$
3	—	—	$\dfrac{1}{3\lambda}$	$\dfrac{7}{12\lambda}$	$\dfrac{47}{60\lambda}$
4	—	—	—	$\dfrac{1}{4\lambda}$	$\dfrac{9}{20\lambda}$
5	—	—	—	—	$\dfrac{1}{5\lambda}$

$$\begin{aligned} \text{MTTF} &= \sum_{x=k}^{n} \binom{n}{x} \frac{1}{\lambda} \int_0^1 v^{x-1}(1-v)^{n-x}\, dv \\ &= \sum_{x=k}^{n} \binom{n}{x} \frac{1}{\lambda} \frac{\Gamma(x)\cdot\Gamma(n-x+1)}{\Gamma(n+1)} \\ &= \frac{1}{\lambda} \sum_{x=k}^{n} \binom{n}{x} \frac{(x-1)!(n-x)!}{n!} = \frac{1}{\lambda} \sum_{x=k}^{n} \frac{1}{x} \end{aligned} \qquad (4.29)$$

The MTTF of some simple k-out-of-n systems computed by (4.29) are listed in Table 4.3. Note that a 1-out-of-n system is a parallel system, while a n-out-of-n system is a series system.

4.4 QUANTITATIVE FAULT TREE ANALYSIS

We will now see how we can analyze a structure that is modeled as a fault tree. Let n denote the number of different basic events in the fault tree. In analogy with our notation for reliability block diagrams, the fault tree is said to be of order n. The n basic events are numbered, and the following state variables are introduced:

$$Y_i(t) = \begin{cases} 1 & \text{if the } i\text{th basic event occurs at time } t \\ 0 & \text{otherwise,} \quad i = 1, 2, \ldots, n \end{cases}$$

$Y(t) = (Y_1(t), \ldots, Y_n(t))$ denotes the state vector for the structure at time t.

QUANTITATIVE FAULT TREE ANALYSIS

The purpose of a "quantitative analysis" of a fault tree usually is to determine the probability of the TOP event (system failure). The procedure is completely analogous to the one for reliability block diagrams.

The state of the TOP event at time t can be described by the binary variable $\psi(Y(t))$, where

$$\psi(Y(t)) = \begin{cases} 1 & \text{if the TOP event occurs at time } t \\ 0 & \text{otherwise} \end{cases}$$

It is assumed that if we know the states of all the n basic events, we also know whether or not the TOP event occurs:

$$\psi(Y(t)) = \psi(Y_1(t), Y_2(t), \ldots, Y_n(t)) \tag{4.30}$$

The function $\psi(Y(t))$ is called the *structure function* of the fault tree.

Let $q_i(t)$ denote the probability that basic event i occurs at time t for $i = 1, 2, \ldots, n$. Then

$$P(Y_i(t) = 1) = E(Y_i(t)) = q_i(t) \quad \text{for} \quad i = 1, 2, \ldots, n$$

Let $Q_0(t)$ denote the probability that the TOP event (system failure) occurs at time t. Then

$$Q_0(t) = P(\psi(Y(t)) = 1) = E(\psi(Y(t))) \tag{4.31}$$

If the basic event i means that component i in the system is in a failed state for $i = 1, 2, \ldots, n$, then

$$P(Y_i(t) = 1) = q_i(t) = 1 - p_i(t) \quad \text{for} \quad i = 1, 2, \ldots, n$$

where $p_i(t)$ is the probability that component i is in a functioning state at time t. $q_i(t)$ is called the *unreliability* of *component i* at time t, while $Q_0(t)$ denotes the *unreliability* of the *system* at the same point of time.

In this case

$$Q_0(t) = 1 - h(p(t)) = 1 - h[1 - q_1(t), 1 - q_2(t), \ldots, 1 - q_n(t)]$$

where $p_i(t)$ for $i = 1, 2, \ldots, n$ and $h(p(t))$ are as defined in (4.3) and (4.5). Note that $Q_0(t)$ in this case is a function of the $q_i(t)$'s only. In general it can be shown that if the basic events are independent, then $Q_0(t)$ will be a function of $q_i(t)$ only for $i = 1, 2, \ldots, n$. Hence $Q_0(t)$ may be written

$$Q_0(t) = g(q_1(t), q_2(t), \ldots, q_n(t)) = g(q(t)) \tag{4.32}$$

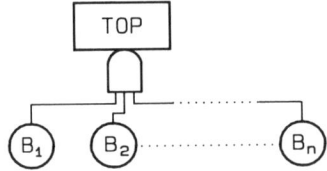

Figure 4.5 Fault tree with a single "AND"-gate

Fault Trees with a Single AND-Gate

Consider the fault tree in Figure 4.5. Here the TOP event occurs if and only if all the basic events $B_1, B_2, \ldots, B_n$ occur simultaneously. The structure function of this fault tree is

$$\psi(Y(t)) = Y_1(t) \cdot Y_2(t) \cdot \,\cdots\, \cdot Y_n(t) = \prod_{i=1}^{n} Y_i(t)$$

Since the basic events are assumed to be independent, then

$$\begin{aligned} Q_0(t) &= E(\psi(Y(t))) = E(Y_1(t) \cdot Y_2(t) \cdots Y_n(t)) \\ &= E(Y_1(t)) \cdot E(Y_2(t)) \cdots E(Y_n(t)) \\ &= q_1(t) \cdot q_2(t) \cdots q_n(t) = \prod_{i=1}^{n} q_i(t) \end{aligned} \qquad (4.33)$$

The probability $Q_0(t)$ of the TOP event may also be determined directly by the following argument: Let $B_i(t)$ denote that basic event B_i occurs at time t; $i = 1, 2, \ldots, n$. Then

$$\begin{aligned} Q_0(t) &= P(B_1(t) \cap B_2(t) \cap \cdots \cap B_n(t)) \\ &= P(B_1(t)) \cdot P(B_2(t)) \cdots P(B_n(t)) \\ &= q_1(t) \cdot q_2(t) \cdots q_n(t) = \prod_{i=1}^{n} q_i(t) \end{aligned}$$

Fault Tree with a Single OR-Gate

Consider the fault tree in Figure 4.6. The TOP event occurs if at least one of the basic events $B_1, B_2, \ldots, B_n$ occurs. The structure function of this fault tree is

QUANTITATIVE FAULT TREE ANALYSIS

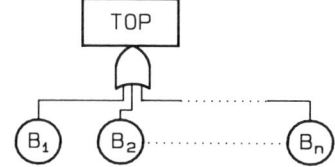

Figure 4.6 Fault tree with a single OR-gate

$$\psi(Y(t)) = Y_1(t) \amalg Y_2(t) \amalg \cdots \amalg Y_n(t)$$
$$= 1 - \prod_{i=1}^{n}(1 - Y_i(t))$$

Since the basic events are assumed to be independent, then

$$Q_0(t) = E(\psi(Y(t))) = 1 - \prod_{i=1}^{n} E(1 - Y_i(t))$$
$$= 1 - \prod_{i=1}^{n}(1 - E(Y_i(t))) = 1 - \prod_{i=1}^{n}(1 - q_i(t)) \quad (4.34)$$

$Q_0(t)$ can also be determined directly in the following way: Let $B_i^c(t)$ denote that basic event B_i does *not* occur at time t. Then

$$P(B_i^c(t)) = 1 - P(B_i(t)) = 1 - q_i(t) \quad \text{for} \quad i = 1, 2, \ldots, n$$

$$Q_0(t) = P(B_1(t) \cup B_2(t) \cup \cdots \cup B_n(t))$$
$$= 1 - P(B_1^c(t)) \cap B_2^c(t) \cap \cdots \cap B_n^c(t)$$
$$= 1 - P(B_1^c(t)) \cdot P(B_2^c(t)) \cdots P(B_n^c(t))$$
$$= 1 - \prod_{i=1}^{n}(1 - q_i(t))$$

Approximation Formula for $Q_0(t)$

As illustrated in Example 4.1, calculation of the TOP event's probability by the structure function can be both time-consuming and cumbersome. Hence there is a need for approximation formulas.

Consider a system (fault tree) with k minimal cut sets $K_1, K_2, \ldots, K_k$. This system may be represented by a series structure of the k minimal cut parallel structures. Hence it can be depicted by the reliability block diagram, as shown in Figure 4.7.

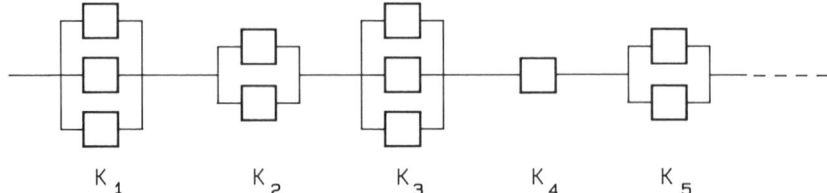

Figure 4.7 A series structure of the minimal cut parallel structures

The TOP event occurs if at least one of the k minimal cut parallel structures fails. A minimal cut parallel structure fails if each and all the basic events in the minimal cut set occur simultaneously. Note that the same input event may enter in many different cut sets.

Let $\check{Q}_j(t)$ denote the probability that minimal cut parallel structure j fails at time t. If the basic events are assumed to be independent, then

$$\check{Q}_j(t) = \prod_{i \in K_j} q_i(t) \tag{4.35}$$

Let $Q_0(t)$ denote the probability that the TOP event (system failure) occurs at time t. If all the k minimal cut parallel structure were independent, then

$$Q_0(t) = \coprod_{j=1}^{k} \check{Q}_j(t) = 1 - \prod_{j=1}^{k} (1 - \check{Q}_j(t)) \tag{4.36}$$

But, since the same basic event can occur in several minimal cut sets, the minimal cut parallel structures are obviously positively dependent (or associated; see Chapter 8). In Chapter 8 it is shown that

$$Q_0(t) \leq 1 - \prod_{j=1}^{k} (1 - \check{Q}_j(t)) \tag{4.37}$$

Hence the right-hand side of (4.37) may be used as an upper bound (conservative) for the probability of system failure.

When all the $q_i(t)$'s are very small, then it can be shown that with good approximation

$$Q_0(t) \approx 1 - \prod_{j=1}^{k} (1 - \check{Q}_j(t)) \tag{4.38}$$

QUANTITATIVE FAULT TREE ANALYSIS

This approximation is called the *upper bound approximation*, and it is used in a number of computer programs for fault tree analysis, for example, in CAFTAN, the fault tree module of the CARA program. Equation (4.38), however, has to be used with care when at least one of the $q_i(t)$'s is of order 10^{-2} or larger.

Assume now that all the $\check{Q}_j(t)$'s are so small that we can disregard their products. In this case (4.38) may be approximated by

$$Q_0(t) \approx 1 - \prod_{j=1}^{k}(1 - \check{Q}_j(t)) \approx \sum_{j=1}^{k} \check{Q}_i(t) \qquad (4.39)$$

It is straightforward to verify that the last approximation is more conservative than the first one:

$$Q_0(t) \leq 1 - \prod_{j=1}^{k}(1 - \check{Q}_j(t)) \leq \sum_{j=1}^{k} \check{Q}_i(t)$$

Example 4.3:* TOP *Event Frequency
In risk analyses we are often more interested in the *frequency* of the TOP event than in the probability of the TOP event in a specific situation. Consider a system that is exposed to a set of threats or hazards $H_1, H_2, \ldots, H_m$. The threats may be extreme loads (caused by fire, explosion, earthquake, lightning, etc.), component failures, or operational or maintenance errors. Some of these threats (hazards) have been identified during system design, and barriers and/or protective systems may have been established to withstand the threats.

To have a system failure (accident), one of the threats must manifest itself *and* the protective system must fail. As an example, assume that the threat H_1 denotes a fire in a specified system module. The expected frequency of such fires has been estimated to be λ_{H_1}.

The fire protection and extinguisher systems that have been installed to withstand the fire may be studied by a fault tree analysis with the TOP event "failure of the fire protection or extinguisher system." Assume that we find the TOP event probability $Q_{H_1}(t)$. The expected frequency of system failures (accidents) caused by this type of fires is, according to Example 2.2, equal to $\lambda_{H_1} Q_{H_1}(t)$. This situation is illustrated by the fault tree in Figure 4.8.

The total expected frequency of system failures may now be determined by combining all the identified threats $H_1, H_2, \ldots, H_m$. Protection systems may be available for only a limited number of the threats. Some of the threats may, on the other hand, have a number of redundant protection systems. For some of the threats it may be impossible to establish protection systems or barriers. The total expected frequency of system failures (accidents) is illustrated in Figure 4.9.

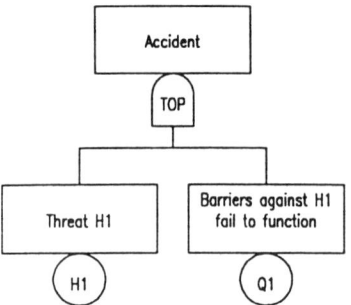

Figure 4.8 System failure caused by fire (threat H_1)

Let $Q_{H_j}(t)$ denote the probability that all the protection systems against the threat H_j fail for $j = 1, 2, \ldots, m$. If the threat H_j has n_j redundant and independent protection systems, then $Q_{H_j}(t) = \prod_{i=1}^{n_j} Q_{H_{j,i}}$, where $Q_{H_{j,i}}(t)$ denotes the probability that the ith protection system against the threat H_j fails at time t. When no protection system is available for the threat H_j, then $Q_{H_j}(t) = 1$.

The total expected frequency of system failures (accidents) λ_S may now be approximated by

$$\lambda_S \approx \sum_{j=1}^{m} \lambda_{H_j} Q_{H_j}(t)$$

Note that for this approach to be valid, the fault tree in Figure 4.9 must have one and only one threat in each minimal cut set. This application of fault tree analysis is further discussed in AIChE (1989). □

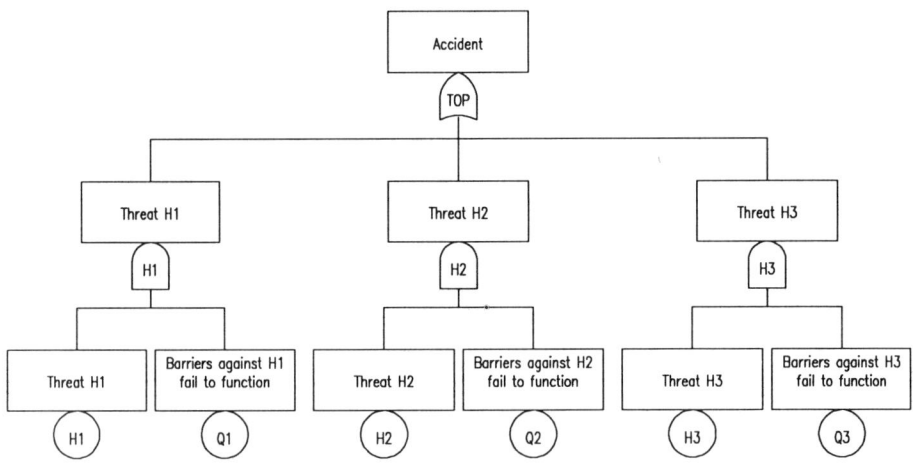

Figure 4.9 System failure caused by the various threats $H_1, H_2, \ldots, H_m$

Computer Programs for Fault Tree Analysis

A number of computer programs for Fault Tree construction and analysis are available. Among these are

CAFTAN	Part of the CARA program with modules for FMECA, cause consequence analysis, and life data analysis. CARA is developed by SINTEF Safety and Reliability, N-7034 Trondheim, Norway
SALP-PC	Developed by the Joint Research Centre of the Commission of the European Communities, 21020 Ispra (Varese), Italy
IRRAS	Developed by Idaho National Engineering Laboratory, EG&G Idaho, Inc. Idaho Falls, Idaho 83415
CAFTA	Developed by Science Applications International Corporation, 5150 El Camino Real, Suite C-31, Los Altos, CA 94022
FRANTIC ABC	Developed by Applied Biomathematics, 100 North Country Road, Setauket, NY 11733
Reliability Toolbox	Includes a fault tree analysis program module. Available from Innovative Timely Solutions, 6401 Lakerest Court, Raleigh, NC 27612
Risk Spectrum Fault Tree	Developed by Relcon Teknik AB, Box 1288, S-17206 Sundbyberg, Sweden, also available from Innovative Software Designs, Inc. Two English Elm Court, Baltimore, MD 21228
FaultTree	Developed by OMI Logistics Limited, Item Software, Wellow House, 14 Little Park Farm Road, Fareham, Hampshire, PO15 5TD, England
FaultrEASE	Developed by Arthur D. Little, Inc. Acorn Park, Cambridge, Massachusetts 02140-2390 (available for PC/Windows and Macintosh)

4.5 EXACT SYSTEM RELIABILITY

In this section we will describe some different methods for calculation of exact system reliability at a given time t_0 when the n components are *independent*. To simplify the notation, we replace the component reliabilities $p_j(t_0)$ by p_j for $j = 1, 2, \ldots, n$.

Computation Based on the Structure Function

The most straightforward method for computation of the exact system reliability is illustrated in Example 4.1. The structure function of the system is established, and powers of the X_i's are deleted ($i = 1, 2, \ldots, n$). The exact system reliability is then obtained by replacing the X_i's by the corresponding p_i's.

Computation Based on Pivotal Decomposition

By repeated pivotal decomposition (see p. 113) a structure function can always be written as

$$\phi(X) = \sum_y \prod_{j=1}^n X_j^{y_j}(1 - X_j)^{1-y_j} \phi(y) \qquad (4.40)$$

where the summation over y is a summation over all n-dimensional binary vectors y, and $0^0 \equiv 1$. If $X_1, X_2, \ldots, X_n$ are assumed to be independent, $X_j^{y_j}(1 - X_j)^{1-y_j}$ are also independent for all j. Since y_j can only take the values 0 and 1, then

$$E[X_j^{y_j}(1 - X_j)^{1-y_j}] = p_j^{y_j}(1 - p_j)^{1-y_j} \qquad \text{for } j = 1, 2, \ldots, n$$

Therefore the system reliability may be written

$$p_S = E[\phi(X)] = \sum_y \prod_{j=1}^n p_j^{y_j}(1 - p_j)^{1-y_j} \phi(y) \qquad (4.41)$$

The approach is the following:

1. First, determine by experiment the value of $\phi(y)$ for all 2^n possible y-vectors.
2. Next, put the obtained values of $\phi(y)$ for the different vectors y into (4.41). [$\phi(y)$ is set equal to 1 if the system is in a functioning state, and equal to 0 otherwise.]

Thereby an expression is obtained for the desired system reliability.

Computation Based on Minimal Cut (Path) Sets

When all the minimal cut sets $K_1, K_2, \ldots, K_k$ and/or all the minimal path sets $P_1, P_2, \ldots, P_P$ are determined, the structure function may be written

EXACT SYSTEM RELIABILITY

$$\phi(X) = \prod_{j=1}^{k} \coprod_{i \in K_j} X_i \qquad (4.42)$$

alternatively

$$\phi(X) = \coprod_{j=1}^{p} \prod_{i \in P_j} X_i \qquad (4.43)$$

The structure function is thus written on a multilinear form. (Remember that since X_i, $i = 1, 2, \ldots, n$, are binary, all exponents can be omitted.) Since $X_1, X_2, \ldots, X_n$ are assumed to be independent, the system reliability is obtained by replacing all the X_i's in the structure function by the corresponding p_i's.

Example 4.4
The structure in Figure 4.10 has the following three minimal path sets:

$$P_1 = \{1, 2, 3\}, \quad P_2 = \{1, 2, 4\}, \quad P_3 = \{1, 3, 4\}$$

Hence

$$\begin{aligned}
\phi(X) &= \coprod_{j=1}^{3} \prod_{i \in P_j} X_i = X_1 X_2 X_3 \amalg X_1 X_2 X_4 \amalg X_1 X_3 X_4 \\
&= 1 - (1 - X_1 X_2 X_3)(1 - X_1 X_2 X_4)(1 - X_1 X_3 X_4) \\
&= X_1 X_2 X_3 + X_1 X_2 X_4 + X_1 X_3 X_4 - X_1^2 X_2^2 X_3 X_4 \\
&\quad - X_1^2 X_2 X_3^2 X_4 - X_1^2 X_2 X_3 X_4^2 + X_1^3 X_2^2 X_3^2 X_4^2 \\
&= X_1 X_2 X_3 + X_1 X_2 X_4 + X_1 X_3 X_4 - 2 X_1 X_2 X_3 X_4
\end{aligned}$$

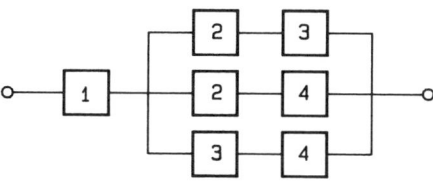

Figure 4.10 Reliability block diagram for Example 4.4

Since the components are independent, the system reliability is

$$h(p) = p_1 p_2 p_3 + p_1 p_2 p_4 + p_1 p_3 p_4 - 2 p_1 p_2 p_3 p_4$$

Time t_0 is reintroduced by replacing p_j by $p_j(t_0)$ for $j = 1, 2, \ldots, n$ in the above expression. □

The Inclusion-Exclusion Principle

In this section we will study how the so-called inclusion-exclusion principle can be applied to determine the *unreliability* of a system. The same approach can also be used to determine the system *reliability*. This is shown at the end of this section.

A system of n independent components has the minimal cut sets $K_1, \ldots, K_k$. Let E_j denote the event that the components of the minimal cut set K_j are all in a failed state, that is, that the jth minimal cut parallel structure is failed at time t. According to (4.35),

$$P(E_j) = \check{Q}_j = \prod_{i \in K_j} q_i$$

where q_i denotes the unreliability of component i at time t_0 for $i = 1, 2, \ldots, n$.

Since the system fails as soon as one of its minimal cut parallel structures fails, the system reliability may be expressed by

$$Q_0 = P\left(\bigcup_{j=1}^{k} E_j\right) \qquad (4.44)$$

In general, the individual events E_j, $j = 1, 2, \ldots, k$ are not disjoint. Hence the probability $P(\bigcup_{j=1}^{k} E_j)$ is determined by using the general addition theorem in probability theory (e.g., see Dudewicz and Mishra 1988, p. 45):

$$Q_0 = \sum_{j=1}^{k} P(E_j) - \sum_{i<j} P(E_i \cap E_j) + \cdots$$
$$+ (-1)^{k+1} P(E_1 \cap E_2 \cap \cdots \cap E_k) \qquad (4.45)$$

EXACT SYSTEM RELIABILITY

By introducing

$$W_1 = \sum_{j=1}^{k} P(E_j)$$

$$W_2 = \sum_{i<j} P(E_i \cap E_j)$$

$$\vdots$$

$$W_k = P(E_1 \cap E_2 \cap \cdots \cap E_k)$$

(4.45) may be written

$$Q_0 = W_1 - W_2 + W_3 - \cdots + (-1)^{k+1} W_k$$
$$= \sum_{j=1}^{k} (-1)^{j+1} W_j \qquad (4.46)$$

Example 4.5
The minimal cut sets of the bridge structure in Figure 4.11 are

$$K_1 = \{1,2\}, \quad K_2 = \{4,5\}, \quad K_3 = \{1,3,5\}, \quad K_4 = \{2,3,4\}$$

Let B_i denote that component i is failed, $i = 1, 2, 3, 4, 5$. According to (4.46) the unreliability Q_0 of the bridge structure is

$$Q_0 = W_1 - W_2 + W_3 - W_4$$

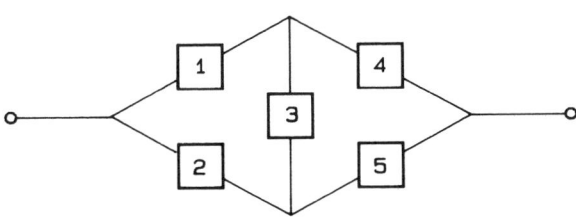

Figure 4.11 The bridge structure

where

$$W_1 = \sum_{j=1}^{4} P(E_j)$$
$$= P(B_1 \cap B_2) + P(B_4 \cap B_5) + P(B_1 \cap B_3 \cap B_5) + P(B_2 \cap B_3 \cap B_4)$$
$$= q_1 q_2 + q_4 q_5 + q_1 q_3 q_5 + q_2 q_3 q_4$$

$$W_2 = \sum_{i<j} P(E_i \cap E_j)$$
$$= P(E_1 \cap E_2) + P(E_1 \cap E_3) + P(E_1 \cap E_4) + P(E_2 \cap E_3)$$
$$+ P(E_2 \cap E_4) + P(E_3 \cap E_4)$$
$$= P(B_1 \cap B_2 \cap B_4 \cap B_5)$$
$$+ P(B_1 \cap B_2 \cap B_1 \cap B_3 \cap B_5)$$
$$+ P(B_1 \cap B_2 \cap B_2 \cap B_3 \cap B_4)$$
$$+ P(B_4 \cap B_5 \cap B_1 \cap B_3 \cap B_5)$$
$$+ P(B_4 \cap B_5 \cap B_2 \cap B_3 \cap B_4)$$
$$+ P(B_1 \cap B_3 \cap B_5 \cap B_2 \cap B_3 \cap B_4)$$
$$= q_1 q_2 q_4 q_5 + q_1 q_2 q_3 q_5 + q_1 q_2 q_3 q_4 + q_1 q_3 q_4 q_5$$
$$+ q_2 q_3 q_4 q_5 + q_1 q_2 q_3 q_4 q_5$$

Similarly

$$W_3 = 4 q_1 q_2 q_3 q_4 q_5$$

and

$$W_4 = q_1 q_2 q_3 q_4 q_5$$

Hence the system unreliability is

$$Q_0 = W_1 - W_2 + W_3 - W_4$$
$$= q_1 q_2 + q_4 q_5 + q_1 q_3 q_5 + q_2 q_3 q_4 - q_1 q_2 q_4 q_5 - q_1 q_2 q_3 q_5$$
$$- q_1 q_3 q_4 q_5 - q_2 q_3 q_4 q_5 + 2 q_1 q_2 q_3 q_4 q_5$$

ooo OOO ooo □

Example 4.5 shows that when using the general addition theorem (4.45), we have to calculate the probability of a large number of terms that later cancel each other. An alternative approach has been proposed by Satyanarayana and Prabhakar (1978). The idea behind their method is—with the help of graph theoretical arguments—to leave out the canceling terms at an early stage without having to calculate them. Their method has been computerized in a program called TRAP (topological reliability analysis program).

A number of alternatives to the inclusion-exclusion principle have been proposed. One of these alternatives is the ERAC (exact reliability/availability calculation) algorithm which was developed by Aven (1986). The ERAC algorithm has been implemented in CAFTAN, the fault tree module of the CARA program.

Calculating the exact value of a system's unreliability Q_0 by means of (4.46) may be cumbersome and time-consuming, even when the system is relatively simple. In such cases one may sometimes be content with an approximative value of the system's unreliability.

One way of determining approximate values of the system unreliability Q_0 utilizes the following result based on inclusion-exclusion:

$$Q_0 \leq W_1$$
$$W_1 - W_2 \leq Q_0$$
$$Q_0 \leq W_1 - W_2 + W_3 \tag{4.47}$$
$$\vdots$$

It can be shown (see Feller 1968, pp. 98–102) that

$$(-1)^{j-1} Q_0 \leq (-1)^{j-1} \sum_{v=1}^{j} (-1)^{v-1} W_v, \quad j = 1, 2, \ldots, k \tag{4.48}$$

Equation (4.47) may give the impression that the differences between the consequtive upper and lower bounds are monotonically decreasing. This is, however, not true in general.

In practice (4.47) is used the following way: Successively one determines upper and lower bounds for Q_0, proceeding downward in (4.47) until one obtains bounds that are sufficiently close. The first upper bound for the system unreliability at time t, $Q_0(t)$ is, according to (4.47),

$$Q_0(t) \leq W_1 = \sum_{j=1}^{k} \check{Q}_j(t) \tag{4.49}$$

According to (4.37),

$$Q_0(t) \leq 1 - \prod_{j=1}^{k}(1 - \check{Q}_j(t)) \qquad (4.50)$$

It is easy to verify that

$$1 - \prod_{j=1}^{k}(1 - \check{Q}_j(t)) \leq \sum_{j=1}^{k} \check{Q}_j(t) \qquad (4.51)$$

Hence the right-hand side of (4.50) is a more accurate approximation to the true value of $Q_0(t)$ than (4.49)

Example 4.6
Reconsider the bridge structure in Example 4.5 and assume that all the component unreliabilities q_i are equal to 0.05. By introducing these q_i's in the expression for the W_i's in Example 4.5, we obtain

$$W_1 = 5250 \cdot 10^{-6}$$
$$W_2 = 3156 \cdot 10^{-6}$$
$$W_3 = 1.25 \cdot 10^{-6}$$
$$W_4 = 0.31 \cdot 10^{-6}$$

From (4.47) we get

$$Q_0 \leq W_1 \approx 5250 \cdot 10^{-6} = 0.5250\%$$
$$Q_0 \geq W_1 - W_2 \approx 5218.4 \cdot 10^{-6} = 0.5218\%$$

Hence from the first two inequalities of (4.47) we know that

$$0.5218\% \leq Q_0 \leq 0.5250\%$$

For many applications this precision may be sufficient. If not, we proceed and calculate the next inequality:

$$Q_0 \leq W_1 - W_2 + W_3 \approx 5219.69 \cdot 10^{-6} \approx 0.5220\%$$

Now we know that Q_0 is bounded by

$$0.5218\% \leq Q_0 \leq 0.5220\%$$

The exact value is

$$Q_0 = W_1 - W_2 + W_3 - W_4 = 5219.38 \cdot 10^{-6} \approx 0.5219\%$$

By comparison, the upper bound obtained by (4.50) is equal to

$$1 - \prod_{j=1}^{k} (1 - \check{Q}_j) = 0.00524249 \approx 0.5242\% \qquad \square$$

ooo OOO ooo

A number of computer programs for reliability and fault tree analysis are based on the inclusion-exclusion principle. Among these is the fundamental KITT-code (kinetic tree theory) (see Vesely and Narum, 1970).

The inclusion-exclusion principle may also be applied to the minimal path sets $P_1, P_2, \ldots, P_p$. Let F_j denote the event that the components in the minimal path set P_j are all functioning; $j = 1, 2, \ldots, p$. In this case the system reliability $p_S = 1 - Q_0$ is

$$p_S = P\left(\bigcup_{j=1}^{p} F_j\right)$$

and

$$P(F_j) = \prod_{i \in P_j} p_i$$

where p_i is the reliability of component i for $i = 1, 2, \ldots, n$. Successive upper and lower bounds of the system reliability p_S may be derived in the same way as we dealt with (4.44).

4.6 REDUNDANCY

In some structures, single units (components, subsystems) may be of much greater importance for the system's ability to function than in others. If, for example, a single unit is operating in series with the rest of the system, failure of this single unit implies that the system fails. Two ways of ensuring higher system reliability in such situations are (1) use units with very high reliability in these critical places in the system, or (2) introduce *redundancy* in these places (i.e., introduce one or more reserve units). The type of redundancy obtained by replacing the important unit with two or more units operating in parallel is called *active redundancy*. These units then share the load right from the start until one of them fails.

The reserve units can also be kept in standby in such a way that the first of them is activated when the ordinary unit fails, the second is activated when the first reserve unit fails, and so on. If the reserve units carry no load in the waiting period before activation (and therefore cannot fail in this period), the redundancy is called *passive*. In the waiting period such a unit is said to be in *cold* standby. If the standby units carry a weak load in the waiting period (and therefore might fail in this period), the redundancy is called *partly loaded*. In the following sections we will illustrate these types of redundancy by considering some simple examples.

Passive Redundancy, Perfect Switching, No Repairs

Consider the standby system in Figure 4.12. The system functions in the following way: Unit 1 is put into operation at time $t = 0$. When it fails, unit 2 is activated. When it fails, unit 3 is activated, and so forth. The unit that is in operation is called the *active* unit, while the units standing by ready to take over are called *standby* or *passive* units. When unit n fails, the system fails.

Here we assume that the switch S functions perfectly and that units cannot fail while they are passive. Let T_i denote the time to failure of unit i, for $i = 1, 2, \ldots, n$. The lifetime T of the whole standby system is then

$$T = \sum_{i=1}^{n} T_i$$

The mean time to system failure MTTF$_S$ is obviously

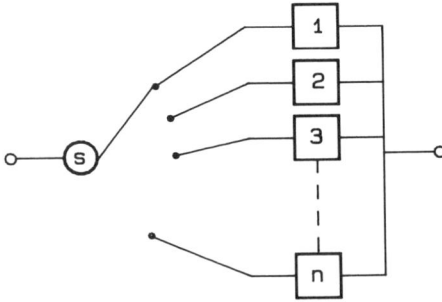

Figure 4.12 Standby system with n units

$$\text{MTTF}_S = \sum_{i=1}^{n} \text{MTTF}_i$$

where MTTF_i denotes the mean time to failure of unit i, $i = 1, 2, \ldots, n$.

The exact distribution of the lifetime T can only be determined in some very special cases. Such a special case occurs when $T_1, T_2, \ldots, T_n$ are independent and exponentially distributed with failure rate λ. According to (2.24), T is gamma distributed with parameters n and λ. The survivor function of the system is then

$$R_S(t) = \sum_{k=0}^{n-1} \frac{(\lambda t)^k}{k!} e^{-\lambda t} \tag{4.52}$$

If we have only one standby unit such that $n = 2$, the survivor function is

$$R_S(t) = e^{-\lambda t} + \frac{\lambda t}{1!} e^{-\lambda t} = (1 + \lambda t) e^{-\lambda t} \tag{4.53}$$

If we have two standby units (i.e., $n = 3$), the survivor function is

$$R_S(t) = e^{-\lambda t} + \frac{\lambda t}{1!} e^{-\lambda t} + \frac{(\lambda t)^2}{2!} e^{-\lambda t} = \left(1 + \lambda t + \frac{(\lambda t)^2}{2}\right) e^{-\lambda t} \tag{4.54}$$

If we are unable to determine the exact distribution of T, we have to be content with an approximate expression for the distribution. Assume, for example, that the lifetimes $T_1, T_2, \ldots, T_n$ are independent and identically distributed with mean time to failure μ and variance σ^2. According to Lindeberg-Levy's central limit theorem (Dudewicz and Mishra 1988, p. 316), when $n \to \infty$, T will be asymptotically normally distributed with mean $n\mu$ and variance $n\sigma^2$.

In this case the survivor function of the system may be approximated by

$$R_S(t) = P\left(\sum_{i=1}^{n} T_i > t\right) = 1 - P\left(\sum_{i=1}^{n} T_i \leq t\right)$$

$$= 1 - P\left(\frac{\sum_{i=1}^{n} T_i - n\mu}{\sigma \sqrt{n}} \leq \frac{t - n\mu}{\sigma \sqrt{n}}\right) \approx \Phi\left(\frac{n\mu - t}{\sigma \sqrt{n}}\right)$$

where $\Phi(\cdot)$ denotes the distribution function of the standard normal distribution $\mathcal{N}(0, 1)$.

Cold Standby, Imperfect Switch, No Repairs

Here we will restrict ourselves to considering the simplest case with $n = 2$ units. Figure 4.13 shows a standby system with an active unit (unit 1) and a unit in *cold* standby (unit 2). The active unit is under surveillance by a switch, which activates the standby unit when the active unit fails.

Let us furthermore assume that the active unit has constant failure rate λ_1. When the active unit fails, the switch will activate the standby unit. The probability that this switching is successful is denoted by $1 - p$. The failure rate of unit 2 in standby position is assumed to be negligible. When the standby unit is activated, its failure rate is λ_2. The three units operate independently. No repairs are carried out. In addition we assume that the only way in which the switch S can fail is by not activating the standby unit when the active unit fails. In many practical applications the switching will be performed by a human operator. The probability p of unsuccessful activation of the standby unit will often include the probability of not being able to start the standby unit.

The system is able to survive the interval $(0, t]$ in two *disjoint* ways:

1. Unit 1 does *not* fail in $(0, t]$ (i.e., $T_1 > t$).
2. Unit 1 fails in a time interval $(\tau, \tau + d\tau]$, where $0 < \tau < t$. The switch S is able to activate unit 2. Unit 2 is activated at time τ and does not fail in the time interval $(\tau, t]$.

Let T denote the time to system failure. Events 1 and 2 are clearly disjoint. Hence the survivor function of the system $R_S(t) = P(T > t)$ will be the sum of the probability of the two events.

The probability of event 1 is

$$P(T_1 > t) = e^{-\lambda_1 t}$$

Next, consider event 2: Unit 1 fails in $(\tau, \tau + d\tau]$ with probability $f_1(\tau) d\tau = \lambda_1 e^{-\lambda_1 \tau} d\tau$. The switch S is able to activate unit 2 with probability $1 - p$.

Unit 2 does *not* fail in $(\tau, t]$ with probability $e^{-\lambda_2(t-\tau)}$. Since unit 1 may fail at any point of time τ in $(0, t]$, the survivor function of the system is when $\lambda_1 \neq \lambda_2$,

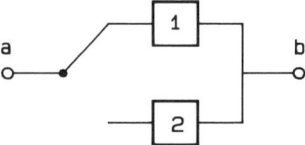

Figure 4.13 Standby system with two units

REDUNDANCY

$$R_S(t) = e^{-\lambda_1 t} + \int_0^t (1-p)e^{-\lambda_2(t-\tau)}\lambda_1 e^{-\lambda_1 \tau}\, d\tau$$

$$= e^{-\lambda_1 t} + (1-p)\lambda_1 e^{-\lambda_2 t}\int_0^t e^{-(\lambda_1-\lambda_2)\tau}\, d\tau$$

$$= e^{-\lambda_1 t} + \frac{(1-p)\lambda_1}{\lambda_1-\lambda_2}e^{-\lambda_2 t} - \frac{(1-p)\lambda_1}{\lambda_1-\lambda_2}e^{-\lambda_1 t} \qquad (4.55)$$

When $\lambda_1 = \lambda_2 = \lambda$, we get

$$R_S(t) = e^{-\lambda t} + \int_0^t (1-p)e^{-\lambda(t-\tau)}\lambda e^{-\lambda \tau}\, d\tau$$

$$= e^{-\lambda t} + (1-p)\lambda e^{-\lambda t}\int_0^t d\tau$$

$$= e^{-\lambda t} + (1-p)\lambda t e^{-\lambda t} \qquad (4.56)$$

The MTTF$_S$ for this system is

$$\text{MTTF}_S = \int_0^\infty R_S(t)\, dt = \frac{1}{\lambda_1} + \frac{(1-p)\lambda_1}{\lambda_1-\lambda_2}\left(\frac{1}{\lambda_2} - \frac{1}{\lambda_1}\right)$$

$$= \frac{1}{\lambda_1} + (1-p)\frac{1}{\lambda_2} \qquad (4.57)$$

This result applies for all values of λ_1 and λ_2.

Example 4.7
Consider the standby system in Figure 4.13, composed of two identical pumps each with constant failure rate $\lambda = 10^{-3}$ failures/hour. The probability p that the switch S will fail to activate (switch over and start) the standby pump has been estimated to 1.5%.

The survivor function of the pump system at time t is, according to (4.56),

$$R_S(t) = (1 + (1-p)\lambda t)e^{-\lambda t} \qquad (4.58)$$

Hence the probability that this system survives 1000 hours is

$$R_S(1000) = 0.7302$$

The mean time to system failure is

$$\text{MTTF}_S = \frac{1}{\lambda}(1 + (1-p)) = 1985 \text{ hours} \qquad \square$$

Partly Loaded Redundancy, Imperfect Switch, No Repairs

Consider the same standby system as the one in Figure 4.13, but change the assumptions so that unit 2 carries a certain load before it is activated. Let λ_0 denote the failure rate of unit 2 while in partly loaded standby. The system is able to survive the interval $(0, t]$ in two disjoint ways:

1. Unit 1 does *not* fail in $(0, t]$ (i.e., $T_1 > t$).

2. Unit 1 fails in a time interval $(\tau, \tau + d\tau]$, where $0 < \tau < t$. The switch S is able to activate unit 2. Unit 2 does not fail in $(0, \tau]$, is activated at time τ, and does not fail in $(\tau, t]$.

Let T denote the time to system failure. The survivor function of the system, $R_S(t) = P(T > t)$, will be the sum of the probabilities for the two events, since they are disjoint.

Consider event 2: Unit 1 fails in $(\tau, \tau + d\tau]$ with probability $f_1(\tau)d\tau = \lambda_1 e^{-\lambda_1 \tau}d\tau$. The switch S is able to activate unit 2 with probability $(1-p)$. Unit 2 does not fail in $(0, \tau]$ in partly loaded standby with probability $e^{-\lambda_0 \tau}$, and unit 2 does not fail in $(\tau, t]$ in active state with probability $e^{-\lambda_2(t-\tau)}$.

Since unit 1 may fail at any point of time τ in $(0, t]$, the survivor function of the system becomes

$$\begin{aligned} R_S(t) &= e^{-\lambda_1 t} + \int_0^t (1-p)e^{-\lambda_0 \tau}e^{-\lambda_2(t-\tau)}\lambda_1 e^{-\lambda_1 \tau}\,d\tau \\ &= e^{-\lambda_1 t} + \frac{(1-p)\lambda_1}{\lambda_0 + \lambda_1 - \lambda_2}(e^{-\lambda_2 t} - e^{-(\lambda_0+\lambda_1)t}) \end{aligned} \qquad (4.59)$$

where we have assumed that $(\lambda_1 + \lambda_0 - \lambda_2) \neq 0$.

When $(\lambda_1 + \lambda_0 - \lambda_2) = 0$, the survivor function becomes

$$R_S(t) = e^{-\lambda_1 t} + (1-p)\lambda_1 t e^{-\lambda_2 t} \qquad (4.60)$$

The mean time to system failure is

$$\text{MTTF}_S = \frac{1}{\lambda_1} + \frac{(1-p)\lambda_1}{\lambda_1 + \lambda_0 - \lambda_2}\left(\frac{1}{\lambda_2} - \frac{1}{\lambda_1 + \lambda_0}\right)$$

$$= \frac{1}{\lambda_1} + (1-p)\frac{\lambda_1}{\lambda_2(\lambda_1 + \lambda_0)} \qquad (4.61)$$

This result applies for all values of λ_0, λ_1, and λ_2. (In this section we have tacitly made certain assumptions about independence. These assumptions will not be discussed thoroughly here.)

An introduction to standby redundancy is given by Trivedi (1982), Billinton and Allan (1983), and Endrenyi (1978). A more detailed discussion is presented by Ravichandran (1990). The concept of redundancy is also discussed in Chapter 6, where we use Markov models to study repairable as well as nonrepairable standby systems.

4.7 REPAIRABLE SYSTEMS

So far we have assumed that components are discarded after the first failure. This assumption has been made for mathematical convenience and simplicity. In real life, components and systems are usually replaced or repaired after a failure. Usually one distinguishes between two types of maintenance: corrective maintenance and preventive maintenance.

Corrective maintenance is usually called *repair*; it is carried out after a component has failed. The purpose of corrective maintenance is to bring the component back to a functioning state as soon as possible. In some cases the corrective maintenance involves replacement of one or more components.

Preventive maintenance seeks to reduce the probability of failure of a component. It may involve lubrication, small adjustments, or replacing components or parts of components that are beginning to wear out. Periodic testing and maintenance based on condition monitoring are also regarded as preventive maintenance. In this section we will confine ourselves to two very simple maintenance models: replacement after failure and periodic testing/replacement.

Replacement after Failure

We will discuss the following situation: A component is put into operation and is functioning at time $t = 0$. Whenever the component fails, it is replaced by a new component of the same type or repaired to an "as-good-as-new" condition. We then get a sequence of lifetimes or *up-times* $T_1, T_2, \ldots$ for the component.

We will assume that $T_1, T_2, \ldots$ are independent and identically distributed, with distribution function $F_T(t) = P(T_i \leq t)$, $i = 1, 2, \ldots$, and mean time to failure MTTF. When a component fails, it will be "down" for a certain period, which is called the *downtime* or *repair time* of the component. We will assume that the repair times $D_1, D_2, \ldots$ are independent and identically distributed with distribution function $F_D(t) = P(D_i \leq t)$ for $i = 1, 2, \ldots$ and mean time to repair MTTR. Finally, we will assume that $T_i + D_i$ for $i = 1, 2, \ldots$ are independent. The MTTR usually comprises a number of elements such as access time, diagnosis time, replacement time, and checkout time. These elements are further discussed by Smith (1988).

As before, the state of a component is given by the state variable:

$$X(t) = \begin{cases} 1 & \text{if component is functioning at time } t \\ 0 & \text{if component is under repair at time } t \end{cases}$$

The state variable $X(t)$ is illustrated in Figure 4.14.

Definition 4.1. The availability $A(t)$ at time t of a repairable component is equal to the probability that the component is functioning at time t:

$$A(t) = P(X(t) = 1) \tag{4.62}$$

□

Note that if the component is not repaired, then $A(t) = R(t)$. $A(t)$ will generally depend on the life distribution $F_T(t)$ and the distribution of the repair time $F_D(t)$.

Definition 4.2. The unavailability $\overline{A}(t)$ at time t of a repairable component is equal to the probability that the component is in a failed state at time t:

$$\overline{A}(t) = 1 - A(t) = P(X(t) = 0) \tag{4.63}$$

□

When determining the availability $A(t)$, it is often appropriate to use renewal theory. A brief introduction to this theory is given in Chapter 7.

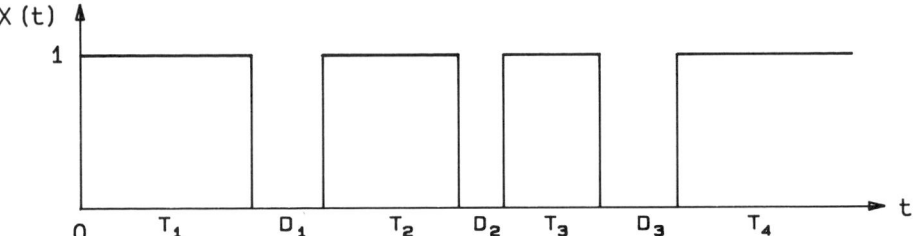

Figure 4.14 States of a repairable component

Example 4.8

Consider a repairable component where the up-times are independent and exponentially distributed with failure rate λ. The downtimes are assumed to be independent and exponentially distributed with parameter μ. The mean time to repair is thus

$$\text{MTTR} = \frac{1}{\mu}$$

The parameter μ is often called the repair rate.

In Chapter 7 it is shown in (7.96) that the availability $A(t)$ of the component is

$$A(t) = \frac{\mu}{\lambda + \mu} + \frac{\lambda}{\lambda + \mu} e^{-(\lambda+\mu)t} \qquad (4.64)$$

The availability $A(t)$ is illustrated in Figure 4.15. Note that the availability $A(t)$ will approach a constant when $t \to \infty$.

$$\lim_{t \to \infty} A(t) = \frac{\mu}{\lambda + \mu} = \frac{1/\lambda}{1/\lambda + 1/\mu} = \frac{\text{MTTF}}{\text{MTTF} + \text{MTTR}} \qquad (4.65)$$

When the component is not repaired (i.e., when $\mu = 0$), then the availability is equal to the survivor function $R(t)$:

$$A(t) = R(t) = e^{-\lambda t} \qquad \text{when } \mu = 0 \qquad \square$$

Definition 4.3. The average availability $A_{av}(0, \tau)$ in the time interval $(0, \tau]$ is defined as

$$A_{av}(0, \tau) = \frac{1}{\tau} \int_0^\tau A(t)\, dt \qquad (4.66)$$

$\square$

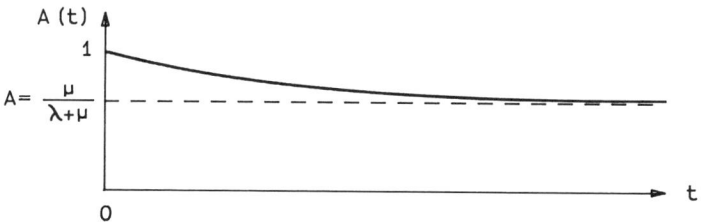

Figure 4.15 Availability $A(t)$ of a component with failure rate λ and repair rate μ

$A_{av}(0,\tau)$ may be interpreted as the mean proportion of $(0,\tau]$ where the component is able to function.

Example 4.9
Consider a time interval of 1000 hours, and let the average availability of a certain component in this interval be 0.95. We would then expect the component to be functionable 950 of those 1000 hours. Note that the availability does not tell anything about *how many* times the component will fail in this interval. □

ooo OOO ooo

If we in (4.66) let $\tau \to \infty$, we get the average availability A_{av} of a component over an infinite time interval. A_{av} may also be interpreted as the mean proportion of time, where the component is functioning.

Now suppose that we have observed a component until repair n is completed. Then we have obtained the lifetimes $T_1, T_2, \ldots, T_n$ and the repair times $D_1, D_2, \ldots, D_n$. According to the law of large numbers then, under relatively general assumptions (e.g., see Dudewicz and Mishra 1988, p. 302)

$$\frac{1}{n} \sum_{i=1}^{n} T_i \xrightarrow{P} E(T) = \text{MTTF} \quad \text{when } n \to \infty$$

$$\frac{1}{n} \sum_{i=1}^{n} D_i \xrightarrow{P} E(D) = \text{MTTR} \quad \text{when } n \to \infty \qquad (4.67)$$

The proportion of the time in which the component has been functioning is

$$\frac{\sum_{i=1}^{n} T_i}{\sum_{i=1}^{n} T_i + \sum_{i=1}^{n} D_i} = \frac{(1/n)\sum_{i=1}^{n} T_i}{(1/n)\sum_{i=1}^{n} T_i + (1/n)\sum_{i=1}^{n} D_i} \qquad (4.68)$$

Heuristically we may expect that the right-hand side of (4.68) will tend to

$$\frac{E(T)}{E(T) + E(D)} \quad \text{as } n \to \infty \qquad (4.69)$$

In the reliability literature, (4.69) is often used as definition of the *average availability* A_{av}:

$$A_{av} \stackrel{def}{=} \frac{E(T)}{E(T) + E(D)} = \frac{\text{MTTF}}{\text{MTTF} + \text{MTTR}} \qquad (4.70)$$

REPAIRABLE SYSTEMS

Definition 4.4. The limiting availability A is defined by

$$A = \lim_{t \to \infty} A(t) \tag{4.71}$$

when the limit exists. □

For most life- and repair-time distributions, $A(t)$ will converge rapidly toward A. Usually it is difficult to determine $A(t)$ exactly. In such situations it may suffice to use A as an approximation for $A(t)$.

It can be shown that if the limiting availability exists, then

$$A = A_{av} = \frac{\text{MTTF}}{\text{MTTF} + \text{MTTR}} \tag{4.72}$$

In the reliability literature A is often used as notation both for the limiting availability A and the average avaliability A_{av}.

Note

A condition for the existence of the limiting availability is that the distribution of the sum $(T_i + D_i)$ not be periodic; namely $(T_i + D_i)$ cannot be a discrete variable with values set $(c, 2c, 3c, \ldots)$ for some value of c, $i = 1, 2, \ldots$.

In preventive maintenance the components are often replaced at regular intervals so that the distribution of $(T_i + D_i)$ is periodic. In such situations formula (4.72) must be used with care. (This is discussed further in Chapter 7.)

Example 4.10

A machine with mean time to failure, MTTF = 1000 hours, and mean repair time MTTR = 5 hours has (limiting) availability

$$A = \frac{\text{MTTF}}{\text{MTTF} + \text{MTTR}} = \frac{1000}{1000 + 5} = 0.995$$

Hence this machine will on the average function 99.5% of the time. The unavailability is thus 0.5%, which corresponds to approximately 44 hours of downtime per year. □

Example 4.11

Consider a component with independent up-times with constant failure rate λ. The downtimes are independent and identically distributed with mean MTTR. Since usually MTTR $\ll$ MTTF the (limiting) unavailability of the component is approximately

$$\overline{A} = \frac{\text{MTTR}}{\text{MTTF} + \text{MTTR}} = \frac{\lambda \cdot \text{MTTR}}{1 + \lambda \cdot \text{MTTR}} \approx \lambda \cdot \text{MTTR} \qquad (4.73)$$

This approximation is often used in hand calculations (see Fussell 1975). □

ooo OOO ooo

When planning supplies of spare parts, it is of interest to know how many failures that may be expected in a given time interval. Let $W(t)$ denote the mean number of repairs carried out in the time interval $(0, t]$. Obviously $W(t)$ will then depend on the life distribution and the distribution of repair times. It is often difficult to find an exact expression for $W(t)$ (see Chapter 7). When t is relatively large, however, the following approximation may be used:

$$W(t) \approx \frac{t}{\text{MTTF} + \text{MTTR}} \qquad (4.74)$$

System Availability

Consider a system with n components and structure function $\phi(X(t))$. If the state variables $X_1(t), X_2(t), \ldots, X_n(t)$ are independent random variables, the system availability $A_S(t)$ can be determined by the procedure described in Section 4.2:

$$A_S(t) = E(\phi(X(t))) \qquad (4.75)$$

Example 4.12
The system in Figure 4.16 has structure function

$$\phi(X(t)) = X_1(t)(X_2(t) + X_3(t) - X_2(t)X_3(t)) \qquad (4.76)$$

The three components are assumed to fail and be repaired independent of each other. We want to determine the average (limiting) availability of the system.

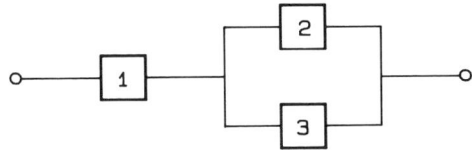

Figure 4.16 Reliability block diagram—Example 4.12

REPAIRABLE SYSTEMS

The MTTF's and MTTR's of the three components are listed below, together with the component availabilities calculated by

$$A_i = \frac{\text{MTTF}_i}{\text{MTTF}_i + \text{MTTR}_i} \quad \text{for} \quad i = 1, 2, 3$$

i	MTTF_i (hours)	MTTR_i (hours)	A_i
1	1000	10	0.990
2	500	10	0.980
3	500	10	0.980

The average availability of the system is

$$A_S = A_1(A_2 + A_3 - A_2 A_3) \approx 0.9896$$

The average system unavailability is thus $\overline{A}_S \approx 0.0104$ which corresponds to approximately 91 hours of downtime per year. □

Approximation Formulas

Consider a system with n components and k minimal cut sets $K_1, K_2, \ldots, K_k$. As described on page 109 the system may be represented by a series structure of the k minimal cut parallel structures. Furthermore the system may be represented by a reliability block diagram or a corresponding fault tree. The system unavailability $\overline{A}_S(t)$ corresponds to the TOP event probability $Q_0(t)$.

The TOP event probability (system unavailability) may be approximated either by the upper bound approximation (4.37),

$$Q_0(t) \approx 1 - \prod_{j=1}^{k} (1 - \check{Q}_j(t)) \tag{4.77}$$

or by the somewhat cruder approximation (4.39),

$$Q_0(t) \approx \sum_{j=1}^{k} \check{Q}_j(t) \tag{4.78}$$

The average (limiting) system unavailability is thus approximately

$$Q_0 \approx 1 - \prod_{j=1}^{k} (1 - \check{Q}_j) \tag{4.79}$$

or

$$Q_0 \approx \sum_{j=1}^{k} \check{Q}_j \qquad (4.80)$$

where $\check{Q}_j$ denotes the average (limiting) unavailability of the minimal cut parallel structure corresponding to the minimal cut set $K_j, j = 1, 2, \ldots, k$. For brevity, this structure will be called the *minimal cut parallel structure* K_j.

In the rest of this section we will assume that component i has constant failure rate λ_i, mean time to repair MTTR_i, and constant repair rate $\mu_i = 1/\text{MTTR}_i$ for $i = 1, 2, \ldots, n$. Furthermore we assume that $\lambda_i \ll \mu_i$ for all $i = 1, 2, \ldots, n$.

As shown in Example 4.11, the average unavailability $\overline{A}_i = q_i$ of component i may be approximated by $q_i \approx \lambda_i \cdot \text{MTTR}_i$, such that

$$\check{Q}_j \approx \prod_{i \in K_j} \lambda_i \cdot \text{MTTR}_i \qquad (4.81)$$

The TOP event probability (system unavailability) is thus approximately

$$Q_0 \approx 1 - \prod_{j=1}^{k} \left(1 - \prod_{i \in K_j} \lambda_i \cdot \text{MTTR}_i \right) \qquad (4.82)$$

or

$$Q_0 \approx \sum_{j=1}^{k} \prod_{i \in K_j} \lambda_i \cdot \text{MTTR}_i \qquad (4.83)$$

Example 4.12 (cont.)
For the structure in Example 4.12, we get

$$\lambda_1 \cdot \text{MTTR}_1 = 0.010$$
$$\lambda_2 \cdot \text{MTTR}_2 = 0.020$$
$$\lambda_3 \cdot \text{MTTR}_3 = 0.020$$

The minimal cut sets are $K_1 = \{1\}$ and $K_2 = \{2, 3\}$, and the system unavailability $\overline{A}_S = Q_0$ is, according to (4.83),

REPAIRABLE SYSTEMS

$$Q_0 \approx \lambda_1 \cdot \text{MTTR}_1 + \lambda_2 \cdot \text{MTTR}_2 \cdot \lambda_3 \cdot \text{MTTR}_3$$
$$= 0.010 + 0.020 \cdot 0.020 = 0.0104$$

which is the same as the result we obtained in Example 4.12 by exact calculation. □

ooo OOO ooo

Cut Set Information

Consider a *specific* minimal cut parallel structure K_j, that is, for a specified j for $j = 1, 2, \ldots, k$. As before we assume that the components fail and are repaired independent of each other. When all the components of the cut set K_j are in a failed state, we say that we have a *cut set failure*. Due to the assumption of independence and the "memoryless" property of the exponential distribution, the mean time to repair the minimal cut parallel structure K_j (MTTR$_j$) is the mean of the minimum of the repair times of the individual components in K_j. Hence

$$\text{MTTR}_j = \frac{1}{\sum_{i \in K_j} \mu_i} \qquad (4.84)$$

After some time the cut parallel structure will "forget" its initial state, and the availability of the structure will approach a steady state availability. This is further discussed, and verified, in Section 6.5.

The mean time between cut set failures MTBF$_j$ will thus approach a constant. The expected frequency of cut set failures ω_j may be defined as

$$\omega_j = \frac{1}{\text{MTBF}_j} \qquad (4.85)$$

The expected frequency ω_j thus denotes the mean number of cut set (K_j) failures per time unit. In Chapter 7 the expected frequency ω_j of cut set failures is called the *rate of occurrence of failures* (ROCOF) of the cut set K_j. In Section 6.5 the expected frequency, ω_j, of cut set failures is determined by Markov methods [see (6.46)]:

$$\omega_j \approx \left(\prod_{i \in K_j} \frac{\lambda_i}{\mu_i} \right) \cdot \left(\sum_{i \in K_j} \mu_i \right) \qquad (4.86)$$

Example 4.13

Consider a minimal cut parallel structure K_j of three independent components with the following failure rates and repair rates:

168 SYSTEMS OF INDEPENDENT COMPONENTS

Component i	Failure rate λ_i (hours)$^{-1}$	Repair rate μ_i (hours)$^{-1}$
1	$5 \cdot 10^{-3}$	$2 \cdot 10^{-1}$
2	$2 \cdot 10^{-4}$	10^{-1}
3	$4 \cdot 10^{-4}$	10^{-1}

The expected frequency of cut set failures is, according to (4.86),

$$\omega_j \approx 4 \cdot 10^{-10}(2 \cdot 10^2) = 8 \cdot 10^{-8}$$

The mean time between cut set failures is

$$\text{MTBF}_j = \frac{1}{\omega_j} \approx 1.250 \cdot 10^7 \text{ hours}$$

Note that MTBF$_j$ also includes the mean downtime of the cut parallel structure. The downtime is, however, usually negligible compared to the up-time. □

ooo OOO ooo

Assume that all components are functioning at time $t = 0$. It is not straightforward to determine the mean time to the first failure of the parallel structure. It may, however, be accomplished by Markov methods, as described in Chapter 6.

Since we know that all components are functioning at time $t = 0$, the mean time to the first failure is obviously greater than the steady state mean time between failures. In most cases the difference is, however, not significant. This is illustrated in the two simple examples below.

Example 4.14
Consider a minimal cut parallel structure K_j with two independent components with the same failure rate λ and the same repair rate μ. In Example 6.3 it is shown that the mean time to the *first* system failure (MTTF$_j$) of this structure is

$$\text{MTTF}_j = \frac{3}{2\lambda} + \frac{\mu}{2\lambda^2} \qquad (4.87)$$

Note that $3/2\lambda$ is the mean time to failure of the structure when no repair is carried out ($\mu = 0$).

REPAIRABLE SYSTEMS

The expected frequency of system failures is, according to (4.86),

$$\omega \approx \lambda^2 \cdot \frac{2}{\mu} = \frac{2\lambda^2}{\mu}$$

The mean time between cut set failures is thus

$$\text{MTBF}_j \approx \frac{1}{\omega} = \frac{\mu}{2\lambda^2} \qquad (4.88)$$

When $\lambda \ll \mu$, we observe that $\text{MTTF}_j \approx \text{MTBF}_j$. For example, if $\lambda = 5 \cdot 10^{-4}$ (hours)$^{-1}$ and $\mu = 10^{-1}$ (hours)$^{-1}$, then

$$\text{MTBF}_j = 2.00 \cdot 10^5 \text{ hours}$$
$$\text{MTTF}_j = 2.03 \cdot 10^5 \text{ hours} \qquad \square$$

Example 4.15
Consider a minimal cut parallel structure K_j of three independent components with the same failure rate λ and the same repair rate μ. According to Problem 4 in Section 6.8 the mean time to the *first* system failure (MTTF_j) of this structure is

$$\text{MTTF}_j = \frac{11}{6\lambda} + \frac{7\mu}{6\lambda^2} + \frac{\mu^2}{3\lambda^3} \qquad (4.89)$$

Note that $11/6\lambda$ is the mean time to failure of the structure when no repair is carried out ($\mu = 0$).

The expected frequency of system failures is, according to (4.86),

$$\omega \approx \lambda^3 \cdot \frac{3}{\mu^2} = \frac{3\lambda^3}{\mu^2}$$

The mean time between cut set failures is thus

$$\text{MTBF}_j \approx \frac{1}{\omega} = \frac{\mu^2}{3\lambda^3} \qquad (4.90)$$

When $\lambda \ll \mu$, we observe that MTTF_j is of the same order of magnitude as

MTBF$_j$. For example, if $\lambda = 5 \cdot 10^{-4}$ (hours)$^{-1}$ and $\mu = 10^{-1}$ (hours)$^{-1}$, then

$$\text{MTBF}_j = 2.67 \cdot 10^7 \text{ hours}$$
$$\text{MTTF}_j = 3.17 \cdot 10^7 \text{ hours} \qquad \square$$

The difference between MTTF$_j$ and MTBF$_j$ will obviously increase with the order of the cut set.

System Information

Consider a system of n independent components with constant failure rates λ_i and constant repair rates μ_i, such that $\lambda_i \ll \mu_i$ for $i = 1, 2, \ldots, n$. As before, we assume that the system is represented as a series structure of the k minimal cut parallel structures K_j for $j = 1, 2, \ldots, k$. The expected frequency, ω_j of failures of the minimal cut parallel structure K_j is given by (4.86).

If we assume that all the k minimal cut parallel structures are independent and that all downtimes are negligible, then the expected frequency of system failures would be

$$\omega_S = \sum_{j=1}^{k} \omega_j \qquad (4.91)$$

In general, this formula is not correct because (1) the minimal cut parallel structures are usually not independent, and (2) the downtimes of the minimal cut parallel structures are often not negligible. For a system with very high availability, (4.91) is a good approximation for the expected frequency ω_S of system failures. More accurate approximation formulas are presented, for example, by Fussell (1975) and Henley and Kumamoto (1981). They show that the expected frequency of system failures may be approximated by

$$\omega_S \approx \sum_{j=1}^{k} \frac{\partial Q_0}{\partial \check{Q}_j} \omega_j$$

The upper bound approximation for the system unavailability is

$$Q_0 \approx 1 - \prod_{j=1}^{k} (1 - \check{Q}_j)$$

such that

REPAIRABLE SYSTEMS

$$\frac{\partial Q_0}{\partial \check{Q}_j} \approx \prod_{\substack{i=1 \\ i \neq j}}^{k} (1 - \check{Q}_i) \approx 1 - \sum_{\substack{i=1 \\ 1 \neq j}}^{k} \check{Q}_i$$

The mean time between system failures MTBF_S in the steady state situation is approximately

$$\text{MTBF}_S \approx \frac{1}{\omega_S} \qquad (4.92)$$

The mean downtime MTTR_j of K_j per cut set failure is given by (4.84). The mean system downtime per system is thus approximately (see (6.55))

$$\text{MTTR}_S \approx \frac{\sum_{j=1}^{k} \omega_j \text{MTTR}_j}{\sum_{j=1}^{k} \omega_j} \qquad (4.93)$$

Periodic Testing/Replacement

A component is put into operation at time $t = 0$. The component is tested and, if necessary, repaired or replaced after regular time intervals of length τ. After a test (repair) the component is considered to be "as good as new."

As an example, let us assume that the component is a fire detector system. The fire detector system has two main failure modes: "fail to function" (FTF) and "false alarm" (FA). The failure mode FTF is normally only detected during tests. If an FTF failure occurs within a test interval and no fire breaks out in the same interval, the failure will remain undetected until the next test. The FTF failure is sometimes called a *hidden* failure.

The theory described in this section applies to components with the same properties as the fire detector system. Examples of such components include gas detectors, pressure detector systems, and safety valves. These are passive components that are installed as a protection or barrier against specified risks. The most important failure mode occurs when the component looses its ability to function as a barrier. This failure mode is called "fail to function" (FTF) and is always a hidden failure.

The state variable $X(t)$ of a component with respect to FTF failures is shown in Figure 4.17.

Let T denote the time to FTF–failure of the component, with distribution function $F_T(t)$ and probability density $f_T(t)$. Since the component is assumed to be "as good as new" after each test, all the test intervals will be "equal" from a stochastic point of view. Consider therefore only the first test interval $(0, \tau]$. Let T_1 be the part of this time interval where the component is able to function as a barrier (i.e., an FTF failure has not occurred). Let D_1 be the part of $(0, \tau]$

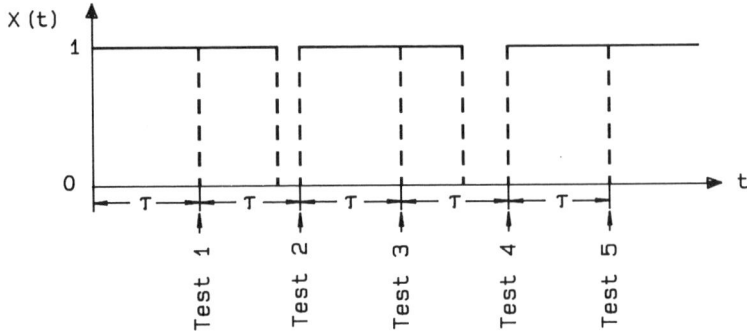

Figure 4.17 State $X(t)$ of a periodically tested component with respect to FTF failures

where the component is in a failed state (i.e., an FTF failure is present but is not detected). Hence

$$T_1 + D_1 = \tau$$

Furthermore

$$T_1 = \min\{T, \tau\}$$

The distribution function of T_1 is given by

$$F_{T_1}(t) = P(T_1 \leq t) = \begin{cases} 0 & \text{for } t < 0 \\ F_T(t) & \text{for } 0 \leq t < \tau \\ 1 & \text{for } t \geq \tau \end{cases} \quad (4.94)$$

A sketch of $F_{T_1}(t)$ is given in Figure 4.18. Note that the distribution function of T_1 is composed of a continuous part and a discrete part.

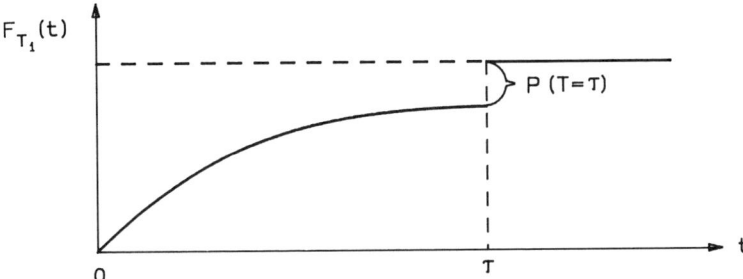

Figure 4.18 Distribution function of the up-time T_1

REPAIRABLE SYSTEMS

The survivor function of the component for $0 \leq t < \tau$ is $1 - F(t) = R_1(t)$. Since the components are assumed to be "as good as new" after each test, the test intervals $(0, \tau]$, $(\tau, 2\tau]$, ..., will be equal from a stochastic point of view. Hence the availability $A(t)$ of the component is as illustrated in Figure 4.19. Note that $A(t)$ is discontinuous for $t = n\tau$, for $n = 1, 2, \ldots$.

Now reconsider the first test interval $(0, \tau]$. The mean up-time in $(0, \tau]$ is

$$E(T_1) = \int_0^\tau t f_T(t)\, dt + \tau P(T_1 = \tau)$$
$$= \int_0^\tau t f_T(t)\, dt + \tau R(\tau) \qquad (4.95)$$

where $R(t) = P(T > t)$. $E(T_1)$ can also be determined by

$$E(T_1) = \int_0^\tau R(t)\, dt$$

Since $T_1 + D_1 = \tau$, the mean downtime in $(0, \tau]$ is given by

$$E(D_1) = \tau - E(T_1) \qquad (4.96)$$

The average availability $A_{av}(0, \tau)$ of the component in the time interval $(0, \tau]$ denotes the mean proportion of the time the component is able to function in $(0, \tau]$:

$$A_{av}(0, \tau) = \frac{E(T_1)}{\tau} = \frac{1}{\tau} \int_0^\tau R(t)\, dt \qquad (4.97)$$

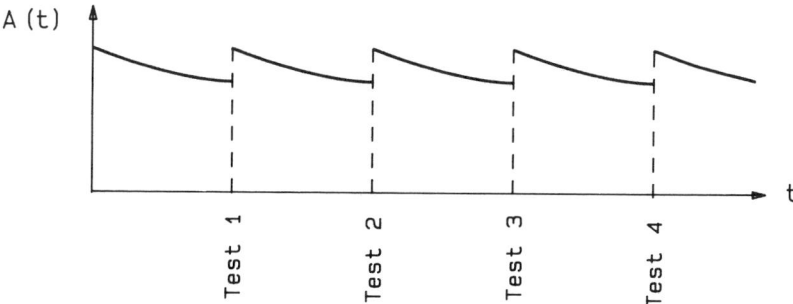

Figure 4.19 Availability $A(t)$ of a periodically tested component

Since all the test intervals have the same stochastic characteristics, the average availability in the long run A_{av} also equals $A_{av}(0,\tau)$. The average unavailability is often called the *mean fractional dead time* and is denoted by MFDT.

$$\text{MFDT} = 1 - A_{av} = \frac{1}{\tau}\left(\tau - \int_0^\tau R(t)\,dt\right) = 1 - \frac{1}{\tau}\int_0^\tau R(t)\,dt$$

$$= \frac{1}{\tau}\int_0^\tau F_T(t)\,dt \qquad (4.98)$$

Example 4.16
A fire detector, which is tested after regular intervals of length τ, has constant failure rate λ. The mean up-time in $(0, \tau]$ is then

$$E(T_1) = \int_0^\tau R(t)\,dt = \int_0^\tau e^{-\lambda t}\,dt = \frac{1}{\lambda}(1 - e^{-\lambda\tau})$$

Hence the average availability A_{av} of the fire detector is

$$A_{av} = A_{av}(0,\tau) = \frac{E(T_1)}{\tau} = \frac{1}{\lambda\tau}(1 - e^{-\lambda\tau}) \qquad (4.99)$$

The mean fractional dead time is

$$\text{MFDT} = 1 - \frac{1}{\lambda\tau}(1 - e^{-\lambda\tau}) \qquad (4.100)$$

If we replace $e^{-\lambda\tau}$ in (4.100) by its Maclaurins series, we get

$$A_{av} = \frac{1}{\lambda\tau}\left(\lambda\tau - \frac{(\lambda\tau)^2}{2} + \frac{(\lambda\tau)^3}{3!} - \frac{(\lambda\tau)^4}{4!} + \cdots\right)$$

$$= 1 - \frac{\lambda\tau}{2} + \frac{(\lambda\tau)^2}{3!} - \frac{(\lambda\tau)^3}{4!} + \cdots$$

When $\lambda\tau$ is small, then

$$A_{av} \approx 1 - \frac{\lambda\tau}{2} \qquad (4.101)$$

and

$$\text{MFDT} = 1 - A_{av} \approx \frac{\lambda \tau}{2} \qquad (4.102)$$

This approximation is often used in practical calculation.

If we assume that the failure rate is $\lambda = 5.0 \cdot 10^{-5}$ FTF failures per hour and that the test interval is $\tau = 1$ month ≈ 730 hours, the mean fractional dead time becomes

$$\text{MFDT} \approx \frac{\lambda \tau}{2} = \frac{5.0 \cdot 10^{-5} \cdot 730}{2} \approx 0.0183$$

which corresponds to approximately 160 hours per year (1 year is approximately 8760 hours). This means that we are *unprotected* on the average 160 hours per year. □

ooo OOO ooo

Assume that we have a system of n independent components with constant failure rates λ_i, $i = 1, 2, \ldots, n$. The distribution function $F_{T_i}(t)$ of the time to failure of component i is approximated by

$$F_{T_i}(t) = 1 - e^{-\lambda_i t} \approx \lambda_i t$$

Failures in a test interval are neither detected nor repaired. Hence the components may be regarded as nonrepairable within the test interval. In this case

$$q_i(t) = P(\text{component } i \text{ is in a failed state at time } t)$$
$$= F_{T_i}(t) \approx \lambda_i t$$

Let $K_1, K_2, \ldots, K_k$ be the k minimal cut sets of the system. The probability that the minimal cut parallel structure corresponding to the minimal cut set K_j is failed at time t is

$$\check{Q}_j(t) = \prod_{i \in K_j} F_i(t) \approx \prod_{i \in K_j} \lambda_i t \qquad \text{for } j = 1, 2, \ldots, k$$

The probability that the system is failed (has a hidden failure) at time t is, according to (4.49),

$$Q_0(t) = F_S(t) \approx \sum_{j=1}^{k} \check{Q}_j(t) \approx \sum_{j=1}^{k} \prod_{i \in K_j} \lambda_i t = \sum_{j=1}^{k} \left(\prod_{i \in K_j} \lambda_i \right) t^{|K_j|}$$

(4.103)

where $|K_j|$ denotes the *order* of the minimal cut set K_j, $j = 1, 2, \ldots, k$.

The mean fractional dead time MFDT of the system which is tested periodically with test interval τ is, by combining (4.98) and (4.103), approximately

$$\text{MFDT} = \frac{1}{\tau} \int_0^\tau F_s(t) \, dt \approx \sum_{j=1}^{k} \prod_{i \in K_j} \lambda_i \frac{1}{\tau} \int_0^\tau t^{|K_j|} \, dt \qquad (4.104)$$

Hence

$$\text{MFDT} \approx \sum_{j=1}^{k} \frac{1}{|K_j| + 1} \prod_{i \in K_j} \lambda_i \tau \qquad (4.105)$$

Assume now that we have a k-out-of-n system of identical and independent components with failure rate λ. A k-out-of-n system has $\binom{n}{n-k+1}$ minimal cut sets of order $(n - k + 1)$. The MFDT of the k-out-of-n system is from (4.105)

$$\text{MFDT} \approx \binom{n}{n-k+1} \frac{(\lambda \tau)^{n-k+1}}{n - k + 2} \qquad (4.106)$$

The MFDT of some simple k-out-of-n systems are listed in Table 4.4.

Now let us suppose that we test a component at time τ and find that the component is in a failed state, i.e. $X(\tau) = 0$. How large is then the expected length of time in which the component has been in a failed state in the interval $(0, \tau]$?

By using double expectation (see e.g. Dudewicz & Mishra, 1988, p. 242) we get

$$E(D_1) = E(E(D_1|X(\tau)))$$
$$= E(D_1|X(\tau) = 0) \cdot P(X(\tau) = 0) + E(D_1|X(\tau) = 1) \cdot P(X(\tau) = 1)$$

If the component is functionable at time $t = \tau$, the downtime D_1 is equal to 0. Therefore

REPAIRABLE SYSTEMS

Table 4.4 MFDT of Some k-Out-of-n Systems of Identical and Independent Components With Failure and λ and Test Interval τ

$k\backslash n$	1	2	3	4
1	$\dfrac{\lambda\tau}{2}$	$\dfrac{(\lambda\tau)^2}{3}$	$\dfrac{(\lambda\tau)^3}{4}$	$\dfrac{(\lambda\tau)^4}{5}$
2	—	$\lambda\tau$	$(\lambda\tau)^2$	$(\lambda\tau)^3$
3	—	—	$\dfrac{3\lambda\tau}{2}$	$2(\lambda\tau)^2$
4	—	—	—	$2\lambda\tau$

$$E(D_1 | X(\tau) = 1) = 0$$

Furthermore

$$P(X(\tau) = 0) = P(T \leq \tau) = F(\tau)$$

Hence

$$E(D_1) = E(D_1 | X(\tau) = 0) \cdot F(\tau)$$

By using (4.96),

$$E(D_1 | X(\tau) = 0) = \frac{E(D_1)}{F(\tau)} = \frac{1}{F(\tau)} \left(\tau - \int_0^\tau R(t)\, dt \right) \qquad (4.107)$$

Let us next determine the mean number of test intervals *until* the first failure occurs. Let C_i denote the event that the component does *not* fail in the test interval i for $i = 1, 2, \ldots$. Then

$$P(C_i) = P(T > \tau) = R(\tau)$$

Since the events $C_1, C_2, \ldots$ are independent with the same probability $p = R(\tau)$, the number of test intervals Z *until* the component fails for the first time is geometrically distributed with point probability (see Dudewicz and Mishra 1988, p. 90)

$$P(Z = z) = P(C_1 \cap C_2 \cap \cdots \cap C_z \cap C_{z+1}^c) = p^z(1 - p)$$
$$\text{for} \quad z = 0, 1, \ldots$$

The mean number of test intervals *until* the components fails is then

$$E(Z) = \sum_{z=0}^{\infty} zP(Z = z) = \frac{p}{1 - p} = \frac{R(\tau)}{F(\tau)} \qquad (4.108)$$

Let T' denote the time from the component is put into operation until its first failure. Then

$$\begin{aligned} E(T') &= \tau E(Z) + (\tau - E(D_1|X(\tau) = 0)) \\ &= \tau \frac{R(\tau)}{F(\tau)} + \tau - \frac{1}{F(\tau)} \left(\tau - \int_0^{\tau} R(t)\,dt \right) \\ &= \frac{1}{F(\tau)} \int_0^{\tau} R(t)\,dt \end{aligned} \qquad (4.109)$$

Example 4.17
If in particular the component has constant failure rate λ, then

$$E(T') = \frac{1}{F(\tau)} \int_0^{\tau} R(t)\,dt = \frac{1}{1 - e^{-\lambda\tau}} \int_0^{\tau} e^{-\lambda t}\,dt = \frac{1}{\lambda}$$

This result also follows directly from the properties of the exponential distribution. □

Now let us assume that fires occur randomly in such a way that the number of fires in a given time interval $(0, t]$ is Poisson distributed with parameter βt. β denotes the mean number of fires per time unit. We say that *a critical situation* occurs if a fire occurs while the fire detector system is in a failed state. This situation is illustrated in Figure 4.20.

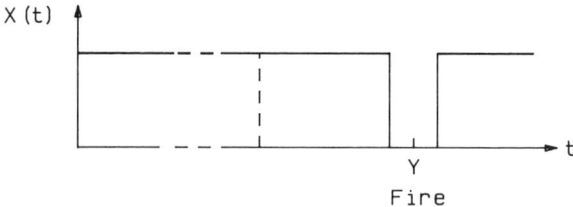

Figure 4.20 Critical situation for the fire detector system

Since all test intervals are equal from a stochastic point of view, we can confine ourselves to considering only the first test interval $(0, \tau]$. A critical situation occurs in $(0, \tau]$ if

1. The fire detector fails in an interval $(t, t + dt]$ where $t < \tau$ and simultaneously
2. At least one fire occurs in the interval $(t, \tau]$

The probability of event 1 is approximately $f(t) dt$, while the probability of event 2 is $1 - e^{-\beta(\tau-t)}$.

Thus the probability of at least one critical event $P_K(0, \tau)$ in the first test interval $(0, \tau]$ is

$$P_K(0, \tau) = \int_0^\tau (1 - e^{\beta(\tau-t)}) f(t) \, dt \qquad (4.110)$$

Usually fires are rare (i.e., β is small) so that $\beta \tau \leq 1$. In that case

$$1 - e^{-\beta(\tau-t)} \approx \beta(\tau - t)$$

Hence

$$P_K(0, \tau) \approx \int_0^\tau \beta(\tau - t) f(t) \, dt$$

By replacing $f(t)$ by $-R'(t)$ and using partial integration, we get

$$P_K(0, \tau) \approx \beta \left(\tau - \int_0^\tau R(t) \, dt \right)$$

According to (4.98),

$$\beta \cdot \tau \cdot \text{MFDT} = \beta \left(\tau - \int_0^\tau R(t) \, dt \right)$$

Hence

$$P_K(0, \tau) \approx \beta \tau \cdot \text{MFDT} \qquad (4.111)$$

Consider an interval $(0, \tau_0]$ where $\tau_0 = n\tau$ and n is a large number. Since all test intervals are equal from a stochastic point of view, the probability of experiencing at least one critical situation in $(0, \tau_0]$ is

$$P_K(0, \tau_0) = 1 - (1 - P_K(0, \tau))^n \tag{4.112}$$

Usually $P_K(0, \tau) \ll 1$. In that case

$$P_K(0, \tau_0) \approx n P_K(0, \tau) = n\beta\tau \cdot \text{MFDT}$$
$$P_K(0, \tau_0) \approx \beta\tau_0 \, \text{MFDT} \tag{4.113}$$

Let U be the number of test intervals *until* a critical situation occurs. U is then geometrically distributed

$$P(U = u) = (1 - P_K(0, \tau))^u P_K(0, \tau) \quad \text{for} \quad u = 0, 1, \ldots$$

Thus the mean number of test intervals *until* a critical situation occurs is

$$E(U) = \sum_{u=0}^{\infty} u P(U = u) = \frac{1 - P_K(0, \tau)}{P_K(0, \tau)} \tag{4.114}$$

When β is small, we may apply (4.113). Hence for small β

$$E(U) \approx \frac{1 - \beta\tau \cdot \text{MFDT}}{\beta\tau \cdot \text{MFDT}} \tag{4.115}$$

Example 4.16 (cont.)
If the fire detector system consists of only one detector with failure rate λ, from (4.102) we know that

$$\text{MFDT} \approx \frac{\lambda\tau}{2}$$

In this case the mean number of complete test intervals *until* a critical situation occurs is approximately

$$E(U) \approx \frac{1 - \beta\tau(\lambda\tau/2)}{\beta\tau(\lambda\tau/2)} = \frac{2 - \beta\lambda\tau^2}{\beta\lambda\tau^2} \tag{4.116}$$

□

Example 4.18
If the fire detector system is composed of two independent detectors in parallel, each with failure rate λ, we found in Table 4.4 that

$$\text{MFDT} \approx \tfrac{1}{3}(\lambda\tau)^2$$

In this case the mean number of test intervals *until* a critical situation occurs is approximately

$$E(U) \approx \frac{1 - \beta\tau(\lambda\tau)^2/3}{\beta\tau(\lambda\tau)^2/3} = \frac{3 - \beta\lambda^2\tau^3}{\beta\lambda^2\tau^3} \qquad (4.117)$$

□

Example 4.19
If the fire detector system is composed of three independent detectors in a 2-out-of-3 system, each with failure rate λ, we found in Table 4.4 that

$$\text{MFDT} \approx (\lambda\tau)^2$$

In that case

$$E(U) \approx \frac{1 - \beta\lambda^2\tau^3}{\beta\lambda^2\tau^3} \qquad (4.118)$$

□

Monte Carlo Simulation

The assumption of independence which is necessary for using the methods in Section 4.7 is rather restrictive. It means that if one component in a system fails and is repaired, all the other components in the system will function as normal without regard to the repair going on. For many systems this assumption will be unrealistic, and other approaches have to be used to determine the system availability.

An alternative approach based on Markov models is presented in Chapter 6. By using Markov models a wide range of dependencies can be taken into account. The Markov models are, however, rather restrictive with respect to the life- and repair-time distributions. Furthermore it is not possible to take into account any trends like seasonal effects.

A more general approach is based on *Monte Carlo (stochastic) simulation*. The simulation is carried out by generating certain random and discrete events in a computer model in order to create a realistic or "typical" lifetime scenario of the system. This approach is entirely different from the approaches discussed so far. In the Monte Carlo approach a realization of the life process is simulated on the computer and, after having observed the simulated process for some time, estimates are made of the desired measures of performance, such as average availability and the mean number of failures per time unit. Thus the simulation is treated as a series of real experiments. The *events* are simulated either with *fixed time increments* (i.e., at equidistant points of time) or with *variable time increments*. The simulation is started at a specified time t_0.

In the first case the simulator clock is incremented by a given (constant) time interval Δt. If any events have occurred in the interval $(t_0, t_0 + \Delta t)$, the new system state is recorded, and the appropriate statistics are collected. Then the time is again incremented by Δt, and so on. The variable time increment simulation is also called *next event simulation*. In this case the simulator clock is incremented by a random time interval, equal to the time interval until the next event. The two types of simulation are further discussed e.g. by Mitrani (1982). Most of the simulation programs available are based on the latter type—next event simulation.

Let T denote a random variable, not necessarily a time to failure, with distribution function $F_T(t)$, which is strictly increasing for all t, such that $F_T^{-1}(y)$ is uniquely determined for all $y \in (0, 1)$. Further let $Y = F_T(T)$. Then the distribution function $F_Y(y)$ of Y is

$$F_Y(y) = P(Y \leq y) = P(F_T(T) \leq y)$$
$$= P(T \leq F_T^{-1}(y)) = F_T(F_T^{-1}(y)) = y \quad \text{for} \quad 0 < y < 1$$

Hence $Y = F_T(T)$ has a uniform distribution over $(0, 1)$. (See also Lemma 9.1.) This implies that if a random variable Y has a uniform distribution over $(0, 1)$, then $T = F_T^{-1}(Y)$ has the distribution function $F_T(t)$.

This result can be used to generate random variables $T_1, T_2, \ldots$ with a specified distribution function $F_T(t)$ on a computer. Variables $Y_1, Y_2, \ldots$, which are uniformly distributed over $(0, 1)$, may be generated by a pseudorandom number generator. The variables $T_i = F_T^{-1}(Y_i)$ for $i = 1, 2, \ldots$ will then have distribution function $F_Y(t)$. Several alternative methods to generate random variables from specific distribution classes are discussed by Ripley (1987).

A wide range of pseudorandom number generators are available (e.g., as part of computer programs for statistical analysis). Most of these pseudorandom number generators are able to generate variables $Y_1, Y_2, \ldots$ which are approximately independent with a uniform distribution over $(0, 1)$. One of the most common types of pseudorandom generators is the *congruential generator*. This generator is defined by

$$X_i = (aX_{i-1} + c) \bmod m$$

where X_i denotes the ith pseudorandom number and a and c are integers between 0 and $m - 1$. Modulo (mod) m of $(aX_{i-1} + c)$ is the remainder when $(aX_{i-1} + c)$ is divided by m. A pseudorandom sequence $Y_1, Y_2, \ldots$ is determined from the formula above through

$$Y_i = \frac{X_i}{m} \quad \text{for} \quad i = 0, 1, 2, \ldots$$

once the *seed* X_0 is given. The congruential generators and alternative generators are discussed in detail by Ripley (1987).

If we have a repairable component, we can simulate, or imitate, a possible life history of the component by Monte Carlo simulation. First, we generate the time T_1 to the first failure according to some specified life distribution F_{T_1}. The repair or restoration time D_1 is next generated according to a specified repair time distribution F_{D_1}. The repair time distribution may, for example, depend on the season (the date) and the time of the day of the failure. The repair time may, for example, be longer for a failure that occurs during the night than for the same failure occurring during ordinary working hours. The time T_2 to the second failure is next generated according to a specified distribution F_{T_2} which may be different from F_{T_1}, and so on.

The simulation on a computer can theoretically take into account virtually all aspects and contingencies of a unit:

- Seasonal and daily variations
- Variations in loading and output
- Periodic testing and interventions into the unit
- Phased mission schemes
- Planned shutdown periods
- Interactions with other components and systems
- Dependencies between functioning times and downtimes

The life history of the component may be simulated for a specified life length (e.g., five years). The computer creates a chronological log file where all the events (failures, repairs, testing, etc.) and the time intervals between consecutive events are recorded. From this log file we are thus able to calculate the number of failures (for the various failure modes), the accumulated use of repair resources and utilities, the average availability, and so on, for the specific realization of the component's life history.

If we simulate a high number of independent life histories (with different *seeds*), we may pool the log files and obtain good estimates of the component's average availability, the average requirements for repair resources, and other parameters of interest. The standard deviation of the estimates may be assessed by standard estimation techniques. A variety of approaches to reduce the variation in the estimates are available. Variance reduction methods are discussed, for example, by Mitrani (1982) and Ripley (1987).

A similar approach may be applied to complex systems. The system model may be established in many different ways, such as

- Reliability block diagram
- Fault tree
- Flow network (e.g., see Aven 1992)
- Markov diagram (see Chapter 6)
- Petri net (e.g., see Peterson 1981)
- Hierarchical system model

As part of the system modeling we have to establish a set of decision rules for the various events and combinations of events. These rules must state which actions should be a consequence of the various events. Examples of such decision rules are

- Setting priorities between repair actions of simultaneous failures when there are limited repair resources
- Switching policies between standby units
- Deciding to replace or refurbish some additional components of the same subsystem when a component fails
- Deciding to shut down the whole subsystem after a failure of a component, until repair action of the component is completed.

To obtain estimates of satisfactory accuracy, we have to simulate a rather high number of life histories of the system. The number of replicated simulations will depend on the system's complexity and the reliabilities of the various system components. Systems with a high reliability will in general require more replications than systems with low reliability. The simulation time will be especially long when the model involves extremely rare events with extreme consequences. For complex systems we may need several thousands of replications. The simulation time will often be excessive even on a fast computer, and the log file may be very large.

A number of simulation programs have been developed for availability—and production regularity—assessment of specific systems. Three of the main simulation programs used in European oil and gas applications are listed below. All of these programs run on personal computers.

MIRIAM   A very flexible simulation program based on flow network theory (see Aven 1992). The program was designed for optimization of offshore process systems but has been applied on a wide range of systems. MIRIAM is developed by EDS Scicon Ltd. Wavendon Tower, Wavendon, Milton Keynes, Buckinghamshire MK17 8LX, England, in close cooperation with the Norwegian oil company Statoil.

PROBLEMS   **185**

MAROS   A simulation program applicable to a wide range of systems.
             MAROS (an acronym for maintainability, availability,
             reliability, and operability) produces approximately the same
             results as MIRIAM but is based on a simpler model.
             MAROS is developed by Baker Jardine and Associates
             Limited, 48a Stratford Road, London W8 6QA, England.

pcFOSP   A simulation program especially designed for subsea production
             systems for oil and gas. FOSP (an acronym for field
             operation simulation program) was initially developed by
             the oil company Shell. The program was transferred to
             SINTEF in 1990, who improved the program and made a
             PC version. pcFOSP is very user-friendly and effective; it is
             available from SINTEF Safety and Reliability, N-7034
             Trondheim, Norway.

4.8 PROBLEMS

1. Show that when the components are independent, the system reliability $p_S(t)$ may be written as in (4.5), that is, as a function of the component reliabilities $p_i(t)$, $i = 1, 2, \ldots, n$, only.

2. An old-fashioned string of Christmas tree lights has 10 bulbs connected in series. The 10 identical bulbs are assumed to have independent lifetimes with constant failure rate λ. Determine λ such that the probability that the string survives three weeks is at least 99%.

3. Consider three identical units in parallel. What is the system reliability if each unit has a reliability of 98%?

4. A system consists of five identical components connected in parallel. Determine the reliability of the components such that the system reliability is 97%.

5. A system must have a reliability of 99% How many components are required in parallel when each component has a reliability of 65%?

6. A plant has two identical and parallel process streams A and B. Each process stream has a transfer pump and a rotary filter, as shown in Figure 4.21. Both process streams have to be functioning to secure full production. (This problem is adapted from an example in Henley and Kumamoto 1981, p. 305.) To improve the system availability, it has been proposed that an extra process stream be installed, as shown in Figure 4.22.

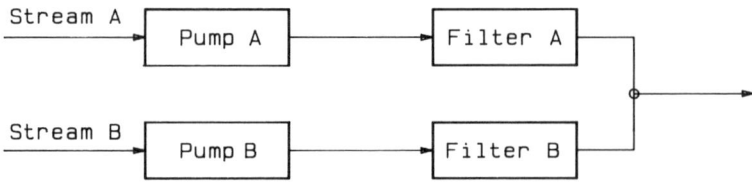

Figure 4.21 Two parallel process streams (system 1)

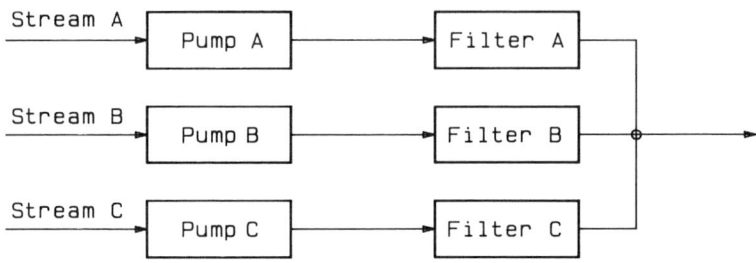

Figure 4.22 Three parallel process streams (system 2)

System 2 has full production when at least two of the three process streams are functioning (i.e., a 2-out-of-3 system). It is assumed that the pumps and the filters are functioning and repaired independent of each other. The average availability of a pump has been estimated to be 99.2% while the average availability of a filter is 96.8%.

(a) Determine the average availability with respect to full production for the two systems (Note that full production is achieved when at least two process streams are functioning.)

(b) Assume that the total cost of a pump is USD 15 per day (including installation, operation, and maintenance). The total cost of a filter is estimated to USD 60 per day. The company gets a penalty of USD 10,000 per day when the system is not able to give full production. Which of the two systems would you choose to minimize the cost?

7. Consider a series structure with two "independent" components B and C with constant failure rates $\lambda_B = 5.0 \cdot 10^{-4}$ failures per hour and $\lambda_C = 3.0 \cdot 10^{-3}$ failures per hour, respectively. The system is put into operation and is functioning at time $t = 0$. Each time the system fails, the component responsible for the failure is repaired to an "as-good-as-new" condition. When one of the components fails, the load on the other component will disappear, and this component will consequently not fail when the failed component is "down." The mean repair time of component B is $\tau_B = 5$ hours, while the mean repair time of component C is $\tau_C = 10$ hours.

(a) Show that the mean repairtime of the *system* is

$$\text{MTTR} = \frac{\lambda_B \tau_B + \lambda_C \tau_C}{\lambda_B + \lambda_C}$$

(b) Determine the average availability A of the system.

(c) Since the failure of one component prevents the other component from failing, the states of the components are not fully independent of each other. Determine the average availability A_I of the system when you assume (wrongly) that the components are fully independent, and compare with the average availability A which was found in (a). Discuss the difference between A and A_I.

8. Consider a series structure of 10 independent and identical components. Each component has a mean time to failure MTTF = 5000 hours.

   (a) Determine $\text{MTTF}_{\text{system}}$ when the components have constant failure rates.

   (b) Assume next that the life lengths of the components are Weibull distributed with shape parameter $\alpha = 2.0$. Determine $\text{MTTF}_{\text{system}}$, and compare with (a).

   (c) Assume now that you have a parallel structure of two independent and identical components with MTTF = 5000 hours. Repeat problems (a) and (b) and discuss the results.

9. Figure 4.23 illustrates a part of a smoke detection system. The system comprises two optical smoke detectors (with separate batteries) and a start relay. All components are assumed to be independent with constant failure rates:

   Smoke detector 1 and 2 $\lambda_{SD} = 2 \cdot 10^{-4}$ failures per hour
   Start relay $\lambda_{SR} = 5 \cdot 10^{-5}$ failures per hour

   The system is tested and, if necessary, repaired after time intervals of equal length $\tau = 1$ month. After each test (repair) the system is considered to

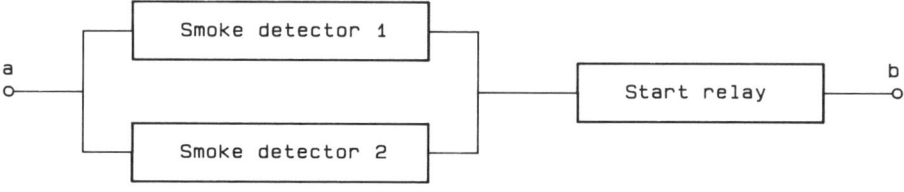

Figure 4.23 Smoke detector system (simplified)

be "as good as new." The repair time is assumed to be negligible. Fail to function (FTF) failures are only detected during tests.

(a) Determine the mean fractional deadtime (MFDT) of the system with respect to FTF failures.

(b) Determine the mean number of test intervals the system passes from $t = 0$ until the first FTF failure.

(c) Assume that you in a specific test find that the system has an FTF failure. Determine the mean time the system has been in a failed state.

(d) Assume that fires occur as a homogeneous Poisson process with intensity $\beta = 5$ fires per year. Find the probability that a fire occurs while an FTF failure of the smoke detection system is present, during a period of two years.

10. Compare the three detector systems in Examples 4.16, 4.18 and 4.19 when $\lambda = 10^{-4}$ hours^{-1}, the detectors are being tested every month, and fire occurs on the average four times per year.

11. You are planning to install a pressure sensor system on a pressure vessel. From past experience you know that the pressure sensors you are planning to use have the following constant failure rates with respect to the actual failure modes:

No signal when pressure
increases beyond pressure setting $\lambda_{FTF} = 3.10 \cdot 10^{-6}$ failures/hour
False high pressure signal $\lambda_{FA} = 3.60 \cdot 10^{-6}$ failures/hour

The pressure sensors will be connected to a programmable logic controller (PLC). The PLC transforms the incoming signals and transmits them to the emergency shutdown (ESD) system. The failure rates of the PLC are estimated to be

Does not transmit
  correct signal $\lambda_A = 0.10 \cdot 10^{-6}$ failures/hour per input
False high pressure $\lambda_B = 0.05 \cdot 10^{-6}$ failures/hour
  signal out

Four different system configurations are considered:
- One single pressure sensor (with PLC)
- Two pressure sensors in parallel
- Three pressure sensors as a 2-out-of-3 system
- Four pressure sensors as a 2-out-of-4 system

The pressure sensors and the PLC will be tested and, if necessary, repaired at the same time once a month. FTF failures will only be detected during tests. After a test (repair) all units are assumed to be "as good as new." The time required for testing and repair is assumed to be negligible.

(a) Determine the mean fractional dead time (MFDT) with respect to FTF failures for each of the four system configurations when you assume that all units are independent and that the failure rates of cables, and so on, are negligible.

(b) Determine the probability of getting at least one false alarm (FA) from each of the four system configurations during a period of one year.

(c) Which of the four system configurations would you install?

12. A process plant needs a regular supply of high-pressure steam. If the steam supply is shut down, the process must also be shut down immediately. The start-up procedures are rather time-consuming. A shutdown of the steam supply may therefore imply significant consequences. The steam producing system comprises three identical steam vessels. The three vessels are physically separated and have separate control and supply systems. The three vessels may thus be considered as independent units.

Consider first only one of the vessels. Assume that the vessel has a constant failure rate $2.02 \cdot 10^{-4}$ failures per hour when it is operated with normal capacity. Failure is in this context defined to be a spurious shutdown of the steam supply from the vessel.

(a) Find the mean time to failure (MTTF) of the vessel.

(b) Find the probability that the vessel will survive a period of four months without failure.

The failure rate of the vessel depends on the capacity the vessel is operated at. It has been estimated that

$$\lambda_p = \lambda_0 + 2p(p - 1.6)\lambda_1 \quad \text{for} \quad 0 < p \leq 1$$

where λ_p denotes the failure rate when the vessel is operated at $100p\%$ capacity, and

$$\lambda_0 = 6.5 \cdot 10^{-4} \quad \text{failures per hour}$$
$$\lambda_1 = 3.5 \cdot 10^{-4} \quad \text{failures per hour}$$

(c) Make a sketch of the failure rate λ_p as a function of p. Write especially down the failure rate when the capacity is 40%, 80%, and 100%.

Each vessel is equipped with four independent burner elements. When a burner has been started (ignited), it has a constant failure rate $\lambda_b = 5.0 \cdot 10^{-5}$ failure per hour. When a vessel is operated at 80% capacity, all the four burner elements are normally active.

(d) Consider a vessel operated at 80% capacity where all the four burner elements are active at time $t = 0$. Determine the probability that the burner system survives a period of two months without any burner element failures when no repair is carried out.

The probability that a passive burner can *not* be started has been estimated to be 3%.

(e) Determine the probability that a standby (passive) vessel may be started and survives a period of two months, when the vessel is operated at 80% capacity, and no repair is carried out.

The process needs a steam supply corresponding to 80% capacity of *one* vessel. Based on regularity arguments, two vessels are, however, normally active at 40% capacity each. The third vessel remains in *cold* standby. The changeover from passive to active operation is normally carried out without other problems than those connected to the start-up of the burner elements. When a vessel is operated at 40% capacity, only two of the four burners elements are used. Two burners are also necessary for the operation. If one of the two active vessels fails, the following procedure is used:

- The capacity of the functioning vessel is increased to 80% if possible
- The standby vessel is started, if possible, by starting two burner elements (40% capacity)

(f) Assume that one of the active vessels fails. Determine the probability that the capacity of the other active vessel can be increased from 40% to 80%, when you know that the two other burner elements in the same vessel have been passive since the last major overhaul of the vessel.

13. Figure 4.24 shows a part of a shutdown system of a process plant. There are two process sections, A and B. If a fire occurs in one of the process sections,

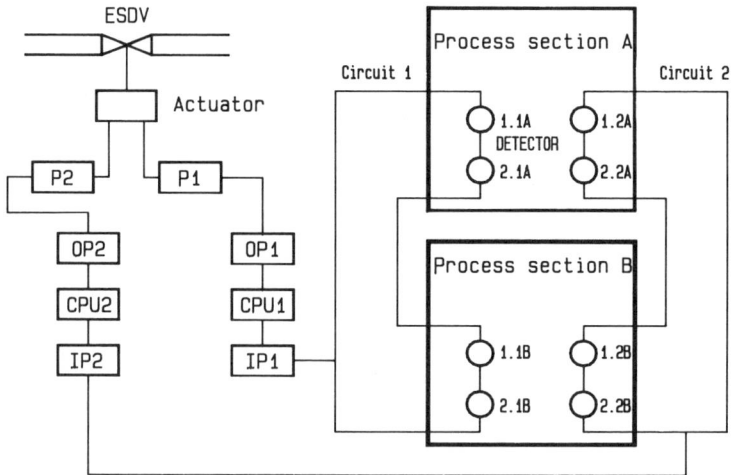

Figure 4.24 Sketch of an emergency shutdown system

the emergency shutdown (ESD) system is installed to close the emergency shutdown valve (ESDV). The ESD valve has a fail-safe hydraulic actuator. The valve is held open by hydraulic pressure. When the hydraulic pressure is bled off, the valve will close.

Each process section has two redundant detector circuits (circuit 1 and circuit 2). Each detector circuit is connected to the ESDV actuator by a pilot valve, which by signal from the detectors opens and bleeds off the hydraulic pressure in the ESDV actuator and thereby closes the ESD valve. Further each circuit comprises an input card, a central processing unit (CPU), an output card, and two fire detectors in each process section. When a fire detector is activated, the current in that circuit is broken. When the current to the input card is broken, a "message" is sent to the output card via the CPU to open the pilot valve. It is assumed that minor fires in one of the process sections cannot be detected by the fire detectors in the other process section. It is assumed that all the components are independent with constant failure rates. Each component has two different failure modes:

- Fail to function (FTF), namely no reaction when a signal is received
- False alarm

The system components, their symbols, and FTF-failure rates are listed in Table 4.5.

Table 4.5 Failure Rates for the "Fail to Function" Mode

Component	Symbol	FTF-Failure Rate λ (Failures per Hour)
ESD value	ESDV	$3.0 \cdot 10^{-6}$
Actuator	Actuator	$5.0 \cdot 10^{-6}$
Pilot value	P1, P2	$2.0 \cdot 10^{-6}$
Output card	OP1, OP2	$0.1 \cdot 10^{-6}$
Input card	IP1, IP2	$0.1 \cdot 10^{-6}$
CPU	CPU1, CPU2	$0.1 \cdot 10^{-7}$
Fire detector	1.1A, 1.2A, 2.1A, 2.2A 1.1B, 1.2B, 2.1B, 2.2B	$4.0 \cdot 10^{-6}$

(a) Construct a fault tree with respect to the TOP event: "The ESD valve does not close when a fire occurs in process section A."

Write down the extra assumptions you have to make during the fault tree construction. As seen from Table 4.5 the failure rates of the input card, the CPU, and the output card are negligible compared to the failure rates of the other components. To simplify the fault tree construction, you may therefore disregard the input/output cards and the CPU.

Show that the fault tree has the following minimal cut sets:

{Actuator}
{ESDV}
{P1, P2}
{P2, 1.1A, 2.1A}
{P1, 2.1A, 2.2A}
{1.1A, 1.2A, 2.1A, 2.2A}

All the components are tested once a month. FTF failures are normally only detected during tests. The time required for testing and, if necessary, repair is assumed to be negligible compared to the length of the testing interval. In question (b) we will assume that the testing of the various components are carried out at different and, for us, unknown times.

(b) **A.** Determine the mean fractional dead times (MFDT) for each of the relevant components.
  B. Determine the TOP event probability by the "upper bound approximation," when the basic events of the fault tree are assumed to be independent.
  C. Discuss the accuracy of the "upper bound approximation" in this case.
  D. Describe other—and more exact methods—to compute the TOP event probability. Discuss pros and cons for each of these methods.

(c) Minor fires are assumed to occur in process section A on the average two times a year, according to a homogeneous Poisson process. A *critical situation* occurs when a fire occurs at the same time as the ESD system has FTF failure (i.e., when the TOP event is present). Find the probability of at least one such a *critical situation* during a period of ten years.

(d) Next consider the subsystem comprising the two fire detectors $1.1A$ and $2.1A$. Determine the mean fractional dead time (MFDT) of this subsystem when the detectors are tested:

  (i) Once every third months at different and, for us, unknown time points
  (ii) At the same time once every third month
  (iii) By "staggered testing," where detector $1.1A$ is tested once every third month and detector $2.1A$ is also tested once every third month but always one month later than detector $1.1A$.

Which of these testing regimes would you prefer (give pros and cons). Explain why the MFDT in case i is different from the MFDT in case ii.

PROBLEMS

193

(e) Do you consider the suggested system structure to be optimal with respect to avoid "false alarm" failures? Suggest an improved structure and discuss possible positive and negative properties of this structure.

14. A downhole safety valve (DHSV) is placed in the oil/gas production tubing on offshore production platforms—approximately 50–100 meters below the sea floor. The valve is held open by hydraulic pressure through a 1/16″ hydraulic pipeline from the platform. When the hydraulic pressure is bled off, the valve will close by spring force. The valve is thus *fail-safe close*. The valve is the last barrier against blowouts in case of an emergency situation on the platform. It is very important that the valve is functioning as a safety barrier, and the valve is therefore tested at regular intervals.

There are two main types of DHSVs: wireline retrievable (WR) valves and tubing retrievable (TR) valves. WR valves are locked in a landing nipple in the tubing and may be installed and retrieved by a wireline operation from the platform. A TR valve is an integrated part of the tubing. To retrieve a TR valve, the tubing has to be pulled. Here we will consider a WR valve. When the WR valve fails, it will be retrieved by a wireline operation, and a new valve of the same type will be installed in the same nipple.

The DHSV is tested once a month. During the testing, which requires approximately 1.5 hours, the production has to be closed down. The mean time to repair a failure is estimated to be 9 hours.

The DHSV has four main failure modes:

FTC	Fail to close on command
LCP	Leakage in closed position
FTO	Fail to open on command
PC	Premature closure

The failure modes FTC and LCP are critical with respect to safety. The failure modes FTO and PC are noncritical with respect to safety but will stop the production. The three failure modes FTC, LCP, and FTO may only be detected during testing, while PC failures are detected at once since the production from the well closes down.

The following failure mode distribution has been discovered:

FTC	15%
LCP	20%
FTO	15%
PC	50%

The mean time between valve failures (with respect to all failure modes) has been estimated to 44 months.

If a critical failure is detected during a test, the well will be unsafe during approximately 1/3 of the repair time. If a noncritical failure is detected, the well will be safe during these operations.

(a) Determine the mean time between FTC failures of a valve.

(b) Determine the probability that a valve survives a test interval without any failure.

(c) Find the mean fractional dead time (MFDT) with respect to safety (the time required for testing and repair shall be taken into account). Discuss the complications encountered in this calculation due to PC failures.

(d) Find the mean proportion of time the production is shut down due to DHSV testing and failures.

(e) Assume now that an emergency situation occurs on the platform on the average once every 50 platform years—which requires that the DHSV must be closed. A *critical situation* occurs for such an emergency situation when the DHSV is not functioning as a safety barrier. Compute the mean time between this type of *critical situations*.

(f) Consider a platform with 20 production wells, with a DHSV in each well. In an emergency situation all the wells have to be closed down. With the same assumptions as above, determine the mean time between *critical situations* on the platform.

CHAPTER 5

Component Importance

From the preceding chapters it should be obvious that the reliability importance of a component in a system depends on two factors:

- The location of the component in the system
- The reliability of the component in question

Nevertheless, it would be of value to the designer, the reliability analyst, as well as the repairperson to have a quantitative measure of the reliability importance of each component. Such a measure would, for example, identify the components that, by being improved, would increase the reliability of the system the most or, by means of a list, tell the repairperson in which order to check the components that may have caused the system failure.

A number of different measures of reliability importance of a component have been defined. We will be content with looking at four of these measures. Several other measures are defined and described by Lambert (1975), Natvig (1979), and Henley and Kumamoto (1981).

Consider a system with n independent components, and let $p_i(t)$ denote the reliability of component i at time t, for $i = 1, 2, \ldots, n$. Then, according to (4.5), the system reliability at time t is a function of $p_1(t), p_2(t), \ldots, p_n(t)$ only, and we denote it by

$$h(\boldsymbol{p}(t)) = h(p_1(t), p_2(t), \ldots, p_n(t))$$

5.1 BIRNBAUM'S MEASURE

Birnbaum (1969) proposed the following measure of reliability importance:

Definition 5.1. Birnbaum's measure of importance of component i at time t is

$$I^B(i|t) = \frac{\partial h(\boldsymbol{p}(t))}{\partial p_i(t)} \quad \text{for} \quad i = 1, 2, \ldots, n \quad (5.1)$$

□

Birnbaum's measure is thus obtained by partial differentiation of the system reliability with respect to $p_i(t)$. This approach is well known from classical sensitivity analysis.

By using the fault tree notation introduced in Section 4.4,

$$q_i(t) = 1 - p_i(t) \quad \text{for} \quad i = 1, 2, \ldots, n$$
$$Q_0(t) = 1 - h(\boldsymbol{p}(t)) \quad (5.2)$$

we observe that (5.1) may be written

$$I^B(i|t) = \frac{\partial Q_0(t)}{\partial q_i(t)} \quad \text{for} \quad i = 1, 2, \ldots, n \quad (5.3)$$

If $I^B(i|t)$ is large, a small change in the reliability of component i will result in a comparatively large change in the system reliability at time t. In the following examples, without loss of generality, let $p_1(t) \geq p_2(t) \geq p_3(t)$. In these examples we will, for the sake of brevity, leave out t and write p_i instead of $p_i(t)$, $i = 1, 2, \ldots, n$.

Example 5.1
Consider a series structure, such as in Figure 5.1, with two independent components where at a given time t, the components have the following reliabilities:

$$p_1 = 0.98$$
$$p_2 = 0.96$$

The system reliability at time t is

$$h(p_1, p_2) = p_1 \cdot p_2 = 0.9408$$

Birnbaum's measure of the reliability importance for components 1 and 2, respectively, then becomes

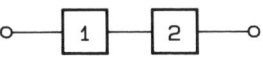

Figure 5.1 Series structure

BIRNBAUM'S MEASURE

$$I^B(1|t) = \frac{\partial h(p_1,p_2)}{\partial p_1} = p_2 = 0.96$$

$$I^B(2|t) = \frac{\partial h(p_1,p_2)}{\partial p_2} = p_1 = 0.98 \qquad \square$$

Example 5.1 illustrates the fact that in a series structure the component with the lowest reliability is the most important one according to Birnbaum's measure. This corresponds well with our intuition. A series structure can be compared with a chain. We know that a chain is never stronger than the weakest link in the chain. The weakest link is therefore the most important.

Example 5.2
Consider a parallel structure, such as in Figure 5.2, with two independent components. Let the component reliabilities at time t be as in Example 5.1. Then the system reliability at time t is

$$h(p_1,p_2) = p_1 + p_2 - p_1 \cdot p_2 = 0.9992$$

Birnbaum's measure of reliability importance for components 1 and 2, respectively, is

$$I^B(1|t) = \frac{\partial h(p_1,p_2)}{\partial p_1} = 1 - p_2 = 0.04$$

$$I^B(2|t) = \frac{\partial h(p_1,p_2)}{\partial p_2} = 1 - p_1 = 0.02 \qquad \square$$

Example 5.2 illustrates that in a parallel structure, the component with the highest reliability is the most important one according to Birnbaum's measure.

Example 5.3
Consider a 2-out-of-3 structure of three independent components, such as in Figure 5.3, with component reliabilities (at time t)

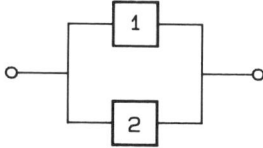

Figure 5.2 Parallel structure

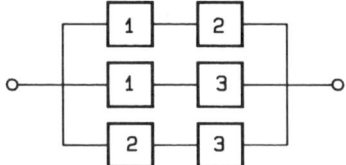

Figure 5.3 A 2-out-of-3 structure

$$p_1 = 0.98$$
$$p_2 = 0.96$$
$$p_3 = 0.94$$

Then the system reliability (at time t) is

$$h(\boldsymbol{p}) = p_1 p_2 + p_1 p_3 + p_2 p_3 - 2 p_1 p_2 p_3 = 0.9957$$

Birnbaum's measure of reliability importance of components 1, 2, and 3, respectively, becomes

$$I^B(1|t) = \frac{\partial h(\boldsymbol{p})}{\partial p_1} = p_2 + p_3 - 2 p_2 p_3 = 0.0952$$

$$I^B(2|t) = \frac{\partial h(\boldsymbol{p})}{\partial p_2} = p_1 + p_3 - 2 p_1 p_3 = 0.0776$$

$$I^B(3|t) = \frac{\partial h(\boldsymbol{p})}{\partial p_3} = p_1 + p_2 - 2 p_1 p_2 = 0.0584$$

Hence in this particular case

$$I^B(1|t) > I^B(2|t) > I^B(3|t)$$

In words, Birnbaum's measure decreases with decreasing reliability. □

ooo OOO ooo

Let us now see whether we can find some connection between the structural importance (see Section 3.5, p. 109–113) and the reliability importance of component i, $i = 1, 2, \ldots, n$, using Birnbaum's measure. By pivotal decomposition the structure function $\phi(\boldsymbol{X}(t))$ at time t can be written as (3.18),

$$\begin{aligned}\phi(\boldsymbol{X}(t)) &= X_i(t) \cdot \phi(1_i, \boldsymbol{X}(t)) + (1 - X_i(t)) \cdot \phi(0_i, \boldsymbol{X}(t)) \\ &= X_i(t) \cdot [\phi(1_i, \boldsymbol{X}(t)) - \phi(0_i, \boldsymbol{X}(t))] + \phi(0_i, \boldsymbol{X}(t))\end{aligned} \quad (5.4)$$

If we assume that the components are independent, the system reliability becomes

$$h(\boldsymbol{p}(t)) = p_i(t) \cdot E[\phi(1_i, X(t))] + (1 - p_i(t)) \cdot E[\phi(0_i, X(t))]$$

Now let $h(1_i, \boldsymbol{p}(t)) = E[\phi(1_i, X(t))]$ and $h(0_i, \boldsymbol{p}(t)) = E[\phi(0_i, X(t))]$. Hence

$$\begin{aligned} h(\boldsymbol{p}(t)) &= p_i(t) \cdot h(1_i, \boldsymbol{p}(t)) + (1 - p_i(t)) \cdot h(0_i, \boldsymbol{p}(t)) \\ &= p_i(t) \cdot [h(1_i, \boldsymbol{p}(t)) - h(0_i, \boldsymbol{p}(t))] - h(0_i, \boldsymbol{p}(t)) \end{aligned} \quad (5.5)$$

Birnbaum's measure of the reliability importance of component i at time t can thus be written

$$I^B(i|t) = \frac{\partial h(\boldsymbol{p}(t))}{\partial p_i(t)} = h(1_i, \boldsymbol{p}(t)) - h(0_i, \boldsymbol{p}(t)) \quad (5.6)$$

where $h(1_i, \boldsymbol{p}(t))$ denotes the system reliability when it is known that component i is functioning at time t, and $h(0_i, \boldsymbol{p}(t))$ denotes the system reliability when component i is in a failed state at time t. This procedure of determining Birnbaum's measure is used in some computer programs, for example, in CAFTAN, the fault tree module of the CARA (ref.) program.

Since $(\phi(1_i, X(t)) - \phi(0_i, X(t)))$ can only take on the values 0 and 1, (5.6) can be written

$$I^B(i|t) = P(\phi(1_i, X(t)) - \phi(0_i, X(t)) = 1) \quad (5.7)$$

This is to say that $I^B(i|t)$ is equal to the probability that $(1_i, X(t))$ is a *critical path vector* for component i at time t (see Definition 3.6, p. 110).

When $(1_i, X(t))$ is a critical path vector for component i, we often, for the sake of brevity, say that component i is critical for the system. The state of component i will, in other words, be decisive for whether or not the system functions at time t.

Note that the fact that component i is critical for the system tells nothing about the state of component i. The statement concerns the other components of the system only.

Consider the series structure in Figure 5.1. There component 1 is critical for the system if, and only if, component 2 is functioning. If component 2 is in a failed state, it makes no difference to the system whether or not component 1 is functioning. Similarly in the parallel structure in Figure 5.2, component 1 is critical for the system if, and only if, component 2 is failed.

An alternative definition of Birnbaum's measure is thus:

Definition 5.2. Birnbaum's measure of reliability importance of component i at time t is equal to the probability that the system is in such a state at time t that component i is critical for the system. □

Note that this definition of reliability importance also can be used when the components are *dependent*.

If the reliabilities $p_j(t) = 1/2$ for all $j \neq i$, the different realizations of the stochastic vector

$$(\cdot_i, X(t)) = (X_1(t), \ldots, X_{i-1}(t), \cdot, X_{i+1}(t), \ldots, X_n(t))$$

all have the probability

$$\frac{1}{2^{n-1}}$$

since the $X_i(t)$'s are assumed to be independent. Then

$$\begin{aligned} I^B(t) &= E(\phi(1_i, X(t)) - \phi(0_i, X(t))) \\ &= \sum_{(\cdot_i, x)} (\phi(1_i, x) - (\phi(0_i, x)) \cdot P(X(t) = x) \\ &= \frac{1}{2^{n-1}} \sum_{(\cdot_i, x)} (\phi(1_i, x) - \phi(0_i, x)) \\ &= \frac{\eta_\phi(i)}{2^{n-1}} = B_\phi(i) \end{aligned} \qquad (5.8)$$

where $\eta_\phi(i)$ is defined as in (3.16).

Thus we have shown that when all the component reliabilities $p_j(t) = \frac{1}{2}$ for $j \neq i$, then Birnbaum's measure of reliability importance of component i and his measure of structural importance for component i coincide:

$$B_\phi(i) = I^B(t)|_{p_j(t)=1/2, j \neq i} = \left.\frac{\partial h(p(t))}{\partial p_i(t)}\right|_{p_j(t)=1/2, j \neq i} \qquad (5.9)$$

Equation (5.9) is often used to calculate structural importance.

5.2 CRITICALITY IMPORTANCE

Let $C(1_i, X(t))$ denote the event that the system at time t is in a state where component i is critical. According to (5.7) the probability of this event is equal

CRITICALITY IMPORTANCE

to Birnbaum's measure of reliability importance of component i at time t:

$$P(C(1_i, X(t))) = I^B(i|t) \qquad (5.10)$$

As before, we assume that all the components in the system are independent. The event $C(1_i, X(t))$ will then, as shown in Section 5.1, be independent of the state of component i at time t.

The probability that component i is critical for the system and simultaneously is failed at time t, is then

$$P(C(1_i, X(t)) \cap (X_i(t) = 0)) = I^B(i|t) \cdot (1 - p_i(t)) \qquad (5.11)$$

Definition 5.3. The measure "criticality importance" $I^{CR}(i|t)$ is the probability that component i is critical for the system and is failed at time t, given that the system is failed at time t. □

$I^{CR}(i|t)$ is in other words the probability that component i has caused system failure, given that the system is failed at time t. When component i is repaired, the system will start functioning again.

Hence we have

$$I^{CR}(i|t) = P(C(1_i, X(t)) \cap (X_i(t) = 0 | \phi(X(t)) = 0)) \qquad (5.12)$$

Since the event $C(1_i, X(t)) \cap (X(t) = 0)$ implies that $\phi(X(t)) = 0$, we get

$$I^{CR}(i|t) = \frac{P(C(1_i, X(t)) \cap (X_i(t) = 0))}{P(\phi(X(t)) = 0)}$$

By using (5.11), we get

$$I^{CR}(i|t) = \frac{I^B(i|t) \cdot (1 - p_i(t))}{1 - h(\mathbf{p}(t))} \qquad (5.13)$$

By using the fault tree notation (5.2), we observe that $I^{CR}(i|t)$ may be written

$$I^{CR}(i|t) = \frac{I^B(i|t) \cdot q_i(t)}{Q_0(t)} \qquad (5.14)$$

Example 5.1 (cont.)
For the series structure at time t we have

$$I^{CR}(1|t) = \frac{I^B(1|t) \cdot (1-p_1)}{1-p_1 p_2} = 0.3243$$

$$I^{CR}(2|t) = \frac{I^B(2|t) \cdot (1-p_2)}{1-p_1 p_2} = 0.6622$$

such that

$$I^{CR}(1|t) < I^{CR}(2|t)$$

This agrees with the ranking we obtained by using Birnbaum's measure of importance. □

Example 5.2 (cont.)
For the parallel structure at time t, we get

$$I^{CR}(1|t) = \frac{I^B(1|t) \cdot (1-p_1)}{1-p_1-p_2+p_1 p_2} = \frac{(1-p_1)(1-p_2)}{(1-p_1)(1-p_2)} = 1.000$$

$$I^{CR}(2|t) = \frac{I^B(2|t) \cdot (1-p_2)}{1-p_1-p_2+p_1 p_2} = \frac{(1-p_1)(1-p_2)}{(1-p_1)(1-p_2)} = 1.000$$

Hence

$$I^{CR}(2|t) = I^{CR}(1|t)$$

This result seems reasonable. If a parallel structure is failed, it will start functioning again irrespective of which of the components we repair. Accordingly all the components in a parallel structure should have the same importance when we use the criticality importance measure. □

Example 5.3 (cont.)
In the 2-out-of-3 structure we get

$$I^{CR}(1|t) = \frac{I^B(1|t) \cdot (1-p_1)}{1-p_1 p_2 - p_1 p_3 - p_2 p_3 + 2 p_1 p_2 p_3} = 0.4428$$

Correspondingly

$$I^{CR}(2|t) = 0.7219$$

$$I^{CR}(3|t) = 0.8149$$

and hence

$$I^{CR}(1|t) < I^{CR}(2|t) < I^{CR}(3|t)$$

Note that this ranking does *not* correspond to the ranking we obtained by using Birnbaum's measure of importance. □

5.3 VESELY-FUSSELL'S MEASURE

W. Vesely and J. B. Fussell suggested the following measure of reliability importance of component i:

Definition 5.4. Vesely-Fussell's measure of importance $I^{VF}(i|t)$ is the probability that at least one minimal cut that contains component i is failed at time t, given that the system is failed at time t. □

We say that a minimal cut is failed when all the components in the minimal cut set are failed.

Vesely-Fussell's measure takes into account the fact that a component may contribute to system failure without being critical. The component contributes to system failure when a minimal cut, containing the component, is failed.

Consider a system with k minimal cut sets $K_1, K_2, \ldots, K_k$. According to (3.12), at time t, the system can then be represented logically by a series structure of k minimal cut parallel structures $\kappa_1(X(t)), \ldots, \kappa_k(X(t))$. The system is failed if and only if at least one of the k minimal cuts is failed. Note that the same component may be a member of *several* different minimal cut sets.

We introduce the following notation:

$D_i(t)$ At least one minimal cut set that contains component i is failed at time t
$C(t)$ The system is failed at time t
m_i The number of minimal cut sets that contain component i
$E_j^i(t)$ Minimal cut set j among those containing component i is failed at time t for $i = 1, 2, \ldots, n$ and $j = 1, 2, \ldots, m_i$

Then

$$I^{VF}(i|t) = P(D_i(t)|C(t)) = \frac{P(D_i(t) \cap C(t))}{P(C(t))}$$

but since $D_i(t)$ implies $C(t)$,

$$I^{VF}(i|t) = \frac{P(D_i(t))}{P(C(t))} \tag{5.15}$$

Since the event $D_i(t)$ occurs if at least one of the events $E_j^i(t)$ occurs, $j = 1, 2, \ldots, m_i$, $i = 1, 2, \ldots, n$:

$$D_i(t) = E_1^i(t) \cup E_2^i(t) \cup \cdots \cup E_{m_i}^i(t)$$

Since the components are assumed to be independent, then

$$P(C(t)) = P(\phi(X(t)) = 0) = 1 - h(p(t))$$

and

$$P(E_j^i(t)) = P(\kappa_j^i(X(t)) = 0) = \prod_{\ell \in K_j^i} (1 - p_\ell) \tag{5.16}$$

where K_j^i denotes the jth minimal cut set that contains component i and $\kappa_j^i(X(t))$ is the corresponding cut parallel structure.

Since the same component may be a member of several cut sets, the events $E_j^i(t)$, $j = 1, 2, \ldots, m_i$, are usually not disjoint. For the same reason the events $E_j^i(t)$, $j = 1, 2, \ldots, m_i$, will not in general be independent, even if all the components are independent.

If the components are independent or associated (see Chapter 8), the cut parallel structures $\kappa_j^i(X(t))$ will also be associated. It can then be shown that the following inequality is valid: [compare with (4.37)]:

$$P(D_i(t)) \leq 1 - \prod_{j=1}^{m_i} (1 - P(E_j^i(t))) \tag{5.17}$$

When the component reliabilities are high, (5.17)—with equality sign—will be approximately correct. Then

$$I^{VF}(i|t) \approx \frac{1 - \prod_{j=1}^{m_i}(1 - P(E_j^i(t)))}{1 - h(\boldsymbol{p}(t))} \qquad (5.18)$$

A somewhat cruder approximation is

$$I^{VF}(i|t) \approx \frac{\sum_{j=1}^{m_i} P(E_j^i(t))}{1 - h(\boldsymbol{p}(t))} \qquad (5.19)$$

By using the fault tree notation (5.2) and

$$\check{Q}_j^i(t) = P(E_j^i(t)) = \prod_{\ell \in K_j^i} q_\ell(t)$$

from formulas (5.18) and (5.19) we obtain

$$I^{VF}(i|t) \approx \frac{1 - \prod_{j=1}^{m_i}(1 - \check{Q}_j^i(t))}{Q_0(t)} \qquad (5.20)$$

and

$$I^{VF}(i|t) \approx \frac{\sum_{j=1}^{m_i} \check{Q}_j^i(t)}{Q_0(t)} \qquad (5.21)$$

For complex systems Vesely-Fussell's measure is considerably easier to calculate by hand than Birnbaum's measure and the criticality importance measure.

Example 5.1 (cont.)
In this series structure each of the two components constitutes its own cut sets, and we have

$$P(D_1(t)) = 1 - p_1 = 0.02$$

$$P(D_2(t)) = 1 - p_2 = 0.04$$

while

$$P(C(t)) = 1 - h(p_1, p_2) = 1 - p_1 p_2 = 0.0592$$

Vesely-Fussell's measure is then for components 1 and 2, respectively,

$$I^{VF}(1|t) = \frac{P(D_1(t))}{P(C(t))} = \frac{1-p_1}{1-p_1 p_2} = 0.3378$$
$$I^{VF}(2|t) = \frac{P(D_2(t))}{P(C(t))} = \frac{1-p_2}{1-p_1 p_2} = 0.6757$$

Hence

$$I^{VF}(1|t) < I^{VF}(2|t)$$

which agrees with the ranking we obtained by using the two other measures. □

Example 5.2 (cont.)
In this parallel structure the system itself constitutes a minimal cut set, that is, $D_1(t) = D_2(t) = C(t)$. Hence

$$I^{VF}(1|t) = I^{VF}(2|t) = 1$$

Hence, by use of Vesely-Fussell's measure, all the components in a parallel structure will be of equal importance. □

Example 5.3 (cont.)
The 2-out-of-3 structure represented as a series structure of the minimal cut parallel structures can be illustrated by the reliability block diagram in Figure 5.4. This structure has three minimal cut sets where each component enters in two cut sets.

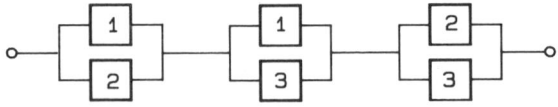

Figure 5.4 The 2-out-of-3 structure as a series structure of minimal cut parallel structures

Thus we have

$$P(D_1(t)) = P(E_1^1(t) \cup E_2^1(t))$$
$$= P(E_1^1(t)) + P(E_2^1(t)) - P(E_1^1(t) \cap E_2^1(t))$$
$$= q_1(t)q_2(t) + q_1(t)q_3(t) - q_1(t)q_2(t)q_3(t) \approx 0.0020$$
$$P(D_2(t)) = P(E_1^2(t) \cup E_2^2(t))$$
$$= P(E_1^2(t)) + P(E_2^2(t)) - P(E_1^2(t) \cap E_2^2(t))$$
$$= q_1(t)q_2(t) + q_2(t)q_3(t) - q_1(t)q_2(t)q_3(t) \approx 0.0032$$
$$P(D_3(t)) = P(E_1^3(t) \cup E_2^3(t))$$
$$= P(E_1^3(t)) + P(E_2^3(t)) - P(E_1^3(t) \cap E_2^3(t))$$
$$= q_1(t)q_3(t) + q_2(t)q_3(t) - q_1(t)q_2(t)q_3(t) \approx 0.0036$$

In Example 5.3 we found that $P(C(t)) = 1 - h(\boldsymbol{p}(t)) = 1 - 0.9957 = 0.0043$. Vesely-Fussell's measure of reliability importance of components 1, 2, and 3, respectively, become

$$I^{VF}(1|t) = \frac{P(D_1(t))}{P(C(t))} = \frac{0.0020}{0.0043} \approx 0.4651$$
$$I^{VF}(2|t) = \frac{P(D_2(t))}{P(C(t))} = \frac{0.0032}{0.0043} \approx 0.7442$$
$$I^{VF}(3|t) = \frac{P(D_3(t))}{P(C(t))} = \frac{0.0036}{0.0043} \approx 0.8372$$

We have here determined Vesely-Fussell's measure exactly. One of the approximation formulas could also be used.

As an illustration, we will calculate Vesely-Fussell's measure for components 1, 2, and 3, respectively, according to the approximation formula (5.21):

$$I^{VF}(1|t) \approx \frac{\check{Q}_1^1(t) + \check{Q}_2^1(t)}{Q_0(t)} = \frac{q_1(t)q_2(t) + q_1(t)q_3(t)}{Q_0(t)} \approx 0.4651$$
$$I^{VF}(2|t) \approx \frac{\check{Q}_1^2(t) + \check{Q}_2^2(t)}{Q_0(t)} = \frac{q_1(t)q_2(t) + q_2(t)q_3(t)}{Q_0(t)} \approx 0.7442$$
$$I^{VF}(3|t) \approx \frac{\check{Q}_1^3(t) + \check{Q}_2^3(t)}{Q_0(t)} = \frac{q_1(t)q_3(t) + q_2(t)q_3(t)}{Q_0(t)} \approx 0.8372$$

As is clear, the four first decimals are correct. The approximation seems appropriate, at least for this simple example. □

When Vesely-Fussell's measure is to be calculated by hand, the formula

(5.21) is normally used. The formula is simple to use and at the same time gives a good approximation when the component reliabilities are high.

5.4 IMPROVEMENT POTENTIAL

Consider a system with reliability $h(p(t))$ at time t. In some cases it may be of interest to know how much the system reliability increases if component i ($i = 1, 2, \ldots, n$) is replaced by a perfect component, that is, a component such that $p_i(t) = 1$. The difference between $h(1_i, p(t))$ and $h(p(t))$ is sometimes called the *improvement potential* with respect to component i and denoted by $I^{IP}(i|t)$.

Definition 5.5. The improvement potential with respect to component i at time t is

$$I^{IP}(i|t) = h(1_i, p(t)) - h(p(t)) \quad \text{for} \quad i = 1, 2, \ldots, n \quad (5.22)$$

□

From (5.5) we note that $h(p(t))$ is a linear function of $p_i(t)$ for each i, $i = 1, 2, \ldots, n$. Hence it is straightforward to verify (see Figure 5.5) that Birnbaum's measure of importance is

$$I^B(i|t) = \frac{h(1_i, p(t)) - h(p(t))}{1 - p_i(t)} \quad (5.23)$$

Then the improvement potential with respect to component i may be written

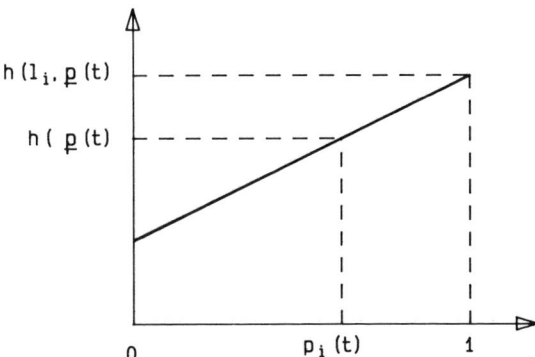

Figure 5.5 Birnbaum's measure

IMPROVEMENT POTENTIAL

$$I^{IP}(i|t) = I^{B}(i|t) \cdot (1 - p_i(t)) \quad (5.24)$$

or, by using the fault tree notation (5.2),

$$I^{IP}(i|t) = I^{B}(i|t) \cdot q_i(t) \quad (5.25)$$

The improvement potential may also be expressed by the criticality importance measure $I^{CR}(i|t)$ by using (5.13),

$$I^{IP}(i|t) = I^{CR}(i|t) \cdot (1 - h(\mathbf{p}(t))) \quad (5.26)$$

or, by using fault tree notation,

$$I^{IP}(i|t) = I^{CR}(i|t) \cdot Q_0(t) \quad (5.27)$$

Example 5.1 (cont.)
For the series system at time t we have

$$I^{IP}(1|t) = I^{B}(1|t) \cdot (1 - p_1(t)) = 0.0192$$

$$I^{IP}(2|t) = I^{B}(2|t) \cdot (1 - p_2(t)) = 0.0392$$

such that

$$I^{IP}(1|t) < I^{IP}(2|t)$$

This illustrates the fact that in a series structure the improvement potential in system reliability is highest for the component with the lowest reliability. □

Example 5.2 (cont.)
For the parallel system at time t we have

$$I^{IP}(1|t) = I^{B}(1|t) \cdot (1 - p_1(t)) = 0.0008$$

$$I^{IP}(2|t) = I^{B}(2|t) \cdot (1 - p_2(t)) = 0.0008$$

such that

$$I^{IP}(1|t) = I^{IP}(2|t)$$

This illustrates the fact that in a parallel system all components have the same improvement potential. □

Example 5.3 (cont.)
For the 2-out-of-3 system at time t we have

$$I^{IP}(1|t) = I^B(1|t) \cdot (1 - p_1(t)) = 0.0019$$
$$I^{IP}(2|t) = I^B(2|t) \cdot (1 - p_2(t)) = 0.0031$$
$$I^{IP}(3|t) = I^B(3|t) \cdot (1 - p_3(t)) = 0.0035$$

such that

$$I^{IP}(1|t) < I^{IP}(2|t) < I^{IP}(3|t)$$

Hence in a 2-out-of-3 system the improvement potential in system reliability is highest for the component with the lowest reliability. □

5.5 A BRIEF COMPARISON

As mentioned in the introduction to Chapter 5, a number of different measures of reliability importance have been suggested. We have discussed only four of these, namely Birnbaum's measure, the critical importance measure, Vesely-Fussell's measure, and the improvement potential. To illustrate these measures, we have used each of them in the three situations in Examples 5.1, 5.2, and 5.3. Let us for comparison put these results together into Table 5.1.

In Table 5.1 we have also ranked the components according to their reliability importance determined by the four measures. The examples show that the measures may lead to different rankings. This was to be expected, since the measures are defined differently. When analyzing a specific system, one must choose the measure that is of greatest interest for the situation at hand.

To identify the component that should be improved to increase system reliability, Birnbaum's measure and the improvement potential are normally the most appropriate. On the other hand, to identify the component that has the largest probability of being the cause of system failure, the critical importance measure or the Vesely-Fussell's measure is the most appropriate. These two measures usually lead to approximately the same result. They may also be used to set up a repairperson's checklist.

A number of objections have been raised to each of these four measures. Birnbaum's measure, for example, is criticized for not taking into account the costs of improving the individual components. With limited economic resources, one may in certain situations attain the "best solution" from an eco-

Table 5.1 Measures and Ranking (in Parantheses) of Reliability Importance

Structure	Component	Birnbaum	Criticality Importance	Vesely-Fussell	Improvement Potential
Series (Example 5.1)	1	0.9600 (2)	0.3242 (2)	0.3378 (2)	0.0192 (2)
	2	0.9800 (1)	0.6622 (1)	0.6757 (1)	0.0392 (1)
Parallel (Example 5.2)	1	0.0400 (1)	1.000 (1)	1.000 (1)	0.0008 (1)
	2	0.0200 (2)	1.000 (1)	1.000 (1)	0.0008 (1)
2-out-of-3 (Example 5.3)	1	0.0952 (1)	0.4428 (3)	0.4651 (3)	0.0019 (3)
	2	0.0776 (2)	0.7219 (2)	0.7442 (2)	0.0031 (2)
	3	0.0584 (3)	0.8149 (1)	0.8372 (1)	0.0035 (1)

nomic point of view by improving a component that has comparatively low Birnbaum importance. Note that all four measures must be calculated at a fixed time. At different times one may get different rankings of the reliability importance of the components.

The four measures may be used for systems that are composed of repairable as well as nonrepairable components. However, the measures are often difficult to interpret when components are repairable.

5.6 PROBLEMS

1. Prove that Definition 5.1 is identical to Definition 5.3.

2. Show that if a 2-out-of-3 structure of independent components has component reliabilities $p_1 \geq p_2 \geq p_3$, then
   (a) If $p_0 \geq 0.5$, $I^B(1|t) \geq I^B(2|t) \geq I^B(3|t)$.
   (b) If $p_1 \leq 0.5$, $I^B(1|t) \leq I^B(2|t) \leq I^B(3|t)$.

3. Find the structural importance of component 1 in the 2-out-of-3 structure in Example 5.3 by using (5.6).

4. Find the reliability importance and structural importance of component 7 in Example 4.1.

5. Find the critical importance for component 7 in Example 4.1.

6. Find the reliability importance for component 7 in Example 4.1 by using Vesely-Fussell's measure.

7. Consider the nonrepairable structure in Figure 5.6.

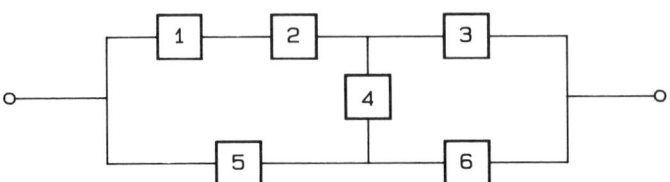

Figure 5.6 Reliability block diagram—Problem 7

(a) Determine the structure function.

(b) Assume the components to be independent. Determine the reliability importance according to Birnbaum's measure of components 2 and 4 when $p_i = 0.99$ for $i = 1, 2, \ldots, 6$.

8. Consider the nonrepairable structure in Figure 5.7

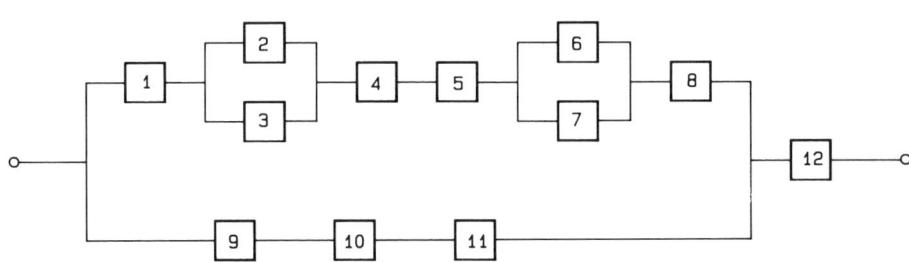

Figure 5.7 Reliability block diagram—Problem 8

(a) Show that the corresponding structure function may be written

$$\phi(X) = [X_1 \cdot (X_2 + X_3 - X_2X_3)X_4X_5(X_6 + X_7 - X_6X_7)X_8 \\ + X_9X_{10}X_{11} - X_1(X_2 + X_3 - X_2X_3)X_4X_5(X_6 \\ + X_7 - X_6X_7)X_8X_9X_{10}X_{11}]X_{12}$$

(b) Determine the system reliability when the different component reliabilities are given as:

$$p_1 = 0.970 \quad p_5 = 0.920 \quad p_9 = 0.910$$
$$p_2 = 0.960 \quad p_6 = 0.950 \quad p_{10} = 0.930$$
$$p_3 = 0.960 \quad p_7 = 0.959 \quad p_{11} = 0.940$$
$$p_4 = 0.940 \quad p_8 = 0.900 \quad p_{12} = 0.990$$

    (c) Determine the reliability importance of component 8 by using Birnbaum's measure and the criticality importance measure.

    (d) Similarly determine the reliability importance of component 11, using the same measures as in (c). Compare and comment on the results obtained.

9. Let (C, ϕ) be a coherent structure of n independent components with state variables $X_1, X_2, \ldots, X_n$. Consider the following modular decomposition of (C, ϕ):

    (i) $C = \bigcup_{j=1}^{r} A_j$ where $A_i \cap A_j = \emptyset$ for $i \neq j$

    (ii) $\phi(x) = \omega(\chi_1(x^{A_1}), \chi_2(x^{A_2}), \ldots, \chi_r(x^{A_r}))$

Assume that $k \in A_j$, and show that

- the Birnbaum measure of importance of component k is equal to the product of
- the Birnbaum measure of importance of module j relative to the system, and
- the Birnbaum measure of importance of component k relative to module j.

Is the same relation valid for the other three measures?

CHAPTER 6

Markov Models

6.1 BASIC CONCEPTS

The models in the preceding chapters are all based on the assumption that the components can be in one of two possible states: *a functioning state or a failed state*. We have also seen that the models are rather static and not well suited for analysis of repairable systems.

In this chapter we introduce the *state-space method* for system reliability evaluation. A system is described by its states and by the possible transitions between these states. The system states and the possible transitions are illustrated by a *state-space diagram*, which is also known as a *Markov diagram*. An example of such a diagram is given later in Figure 6.2. The various system states are defined by the states of the components comprising the system. By the state–space method the components are not restricted to having only two possible states. The components may have a number of different states such as functioning, derated, in standby, completely failed, and under maintenance. The various failure modes may also be defined as states. The transitions between the states are caused by various mechanisms and activities such as failures, repairs, replacements, and switching operations.

The state-space method is thus not restricted to only two possible states of the components. The method can be used to model rather complicated repair and switching strategies. Common cause failures may also be modeled by the state-space method. Common cause failures and modeling of such failures are further discussed in Chapter 8.

The number of system states, however, increases rapidly with the size and complexity of the system. The state-space method therefore is suitable only for relatively small systems.

Example 6.1
Consider the parallel structure of two components in Figure 6.1. The structure is functioning when at least one of the components is functioning. Since each of

BASIC CONCEPTS

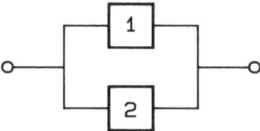

Figure 6.1 Parallel structure of two components

the components has two possible states, the parallel structure has four possible states. These states are listed in Table 6.1.

When the parallel structure is in state 0, the system is failed. In the states 1 and 2 the parallel structure is operating with only one component in function. The functioning component may in these states be exposed to a higher stress level than in state 3 where both components are sharing the load.

It is assumed that the following corrective maintenance strategy is adopted: When a component fails, a repair action is initiated to bring this component back to its initial functioning state. After the repair is completed, the component is assumed to be *as good as new*.

The possible transitions between the four system states are illustrated in the state–space diagram in Figure 6.2. In this system we disregard the possibility of common cause failures. Thus a transition between state 3 and state 0 is assumed to be impossible during a time interval of length Δt. Observe also that there are no transitions between state 2 and state 1 during the same interval. Such a transition would imply two "simultaneous" events; failure of one component and repair of the other. When drawing the diagram, Δt is assumed to be very short.

□

Table 6.1 Possible States of a System of Two Components

System State	State of Component 2	State of Component 1	Comment
3	1	1	Both components functioning
2	1	0	Component 2 functioning, component 1 in failed state
1	0	1	Component 1 functioning, component 2 in failed state
0	0	0	Both components in failed state

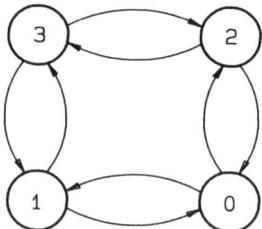

Figure 6.2 State-space diagram of the parallel structure in Example 6.1

6.2 MARKOV PROCESSES

In this chapter the state variable of the system at time t is denoted by $X(t)$. The system is assumed to have $(r + 1)$ possible states numbered $0, 1, 2, \ldots, r$. The event $\{X(t) = j\}$ means that the system at time t is in state j, $j = 0, 1, 2, \ldots, r$. The probability of this event is denoted by

$$P_j(t) = P(X(t) = j) \quad \text{for} \quad j = 0, 1, 2, \ldots, r$$

The system is assumed to start in a specified state, say, state i, at time $t = 0$. The transitions between the states may be described by a stochastic process $\{X(t); t \geq 0\}$. For a general introduction to stochastic processes, see, for example, Cox and Miller (1965), Taylor and Karlin (1984), or Ross (1983).

Many systems have transitions that can approximately be described by a stochastic process with the *Markov property*: Given that a system is in state i at time t [i.e., $X(t) = i$], the future states $X(t+v)$ do not depend on the previous states $X(u)$, $u < t$. In other words, when its present state is known, the probability of any particular future behavior of the process is not altered by additional knowledge about its past behavior.

In probabilistic terms the Markov property is defined by

$$\begin{aligned}P(X(t + v) = j | X(t) &= i; X(u) = x(u); 0 \leq u < t) \\&= P(X(t + v) = j | X(t) = i) \\&\quad \text{for all possible } x(u); 0 \leq u < t\end{aligned} \quad (6.1)$$

A stochastic process satisfying the Markov property is called a *Markov process*. The conditional probabilities

$$P(X(t + v) = j | X(t) = i) \quad \text{for} \quad i,j = 0, 1, 2, \ldots, r$$

are called the *transition probabilities* of the Markov process.

In order to obtain an easy-to-use mathematical model, we introduce another simplification: stationary transition probabilities. If the transition probability

MARKOV PROCESSES

$P(X(t + v) = j | X(t) = i)$ does not depend on the time t but only on the time interval v for the transition, the transition probabilities are said to be stationary

$$P(X(t + v) = j | X(t) = i) = P_{ij}(v) \quad \text{for} \quad t, v > 0; \quad i, j = 0, 1, \ldots, r$$

(6.2)

A Markov process with stationary (or steady state) transition probabilities is often called a process with *no memory*.

A Markov process with stationary transition probabilities is often convenient and realistic for modeling technical systems, and we will keep within the framework of such a process. In the following, when we say that the process $\{X(t); t \geq 0\}$ is a Markov process, this implies that it satisfies the Markov property and has stationary transition probabilities. The Poisson point process, see Section 2.5, is an example of a Markov process.

The transition probabilities satisfy

$$P_{ij}(t) \geq 0 \quad \text{for} \quad t > 0 \quad (6.3\text{a})$$

$$\sum_{j=0}^{r} P_{ij}(t) = 1 \quad \text{for} \quad t > 0 \quad (6.3\text{b})$$

$$P_{ij}(t + v) = \sum_{k=0}^{r} P_{ik}(t) \cdot P_{kj}(v) \quad \text{for} \quad t, v > 0 \quad (6.3\text{c})$$

Property (6.3c) is known as the Chapman-Kolmogorov equation and follows from the Markov property and the rule for total probability (e.g., see Dudewicz and Mishra 1988, p. 41).

In the same way as the failure rate was defined in Section 2.3, we can define the transition rate from state i to state j as

$$a_{ij} = \lim_{\Delta t \to 0} \frac{P(X(t + \Delta t) = j | X(t) = i)}{\Delta t}$$

assuming that this limit exists. Further, by using (6.2),

$$a_{ij} = \lim_{\Delta t \to 0} \frac{P_{ij}(\Delta t)}{\Delta t} = \dot{P}_{ij}(0) \quad (6.4)$$

where the following notation for the time derivative is introduced:

$$\dot{P}_{ij}(t) = \frac{d}{dt} P_{ij}(t)$$

Since the transition rate a_{ij} from state i to state j is constant, the time T_{ij} the system is staying in state i until transition to state j is exponentially distributed with parameter a_{ij} (see Section 2.6).

6.3 STATE EQUATIONS

Now let Δt be a positive number so small that we can disregard the possibility of more than one transition in a time interval of length Δt. From (6.3c) we have

$$P_{ij}(t + \Delta t) = \sum_{k=0}^{r} P_{ik}(t) \cdot P_{kj}(\Delta t)$$

$$= \sum_{\substack{k=0 \\ k \neq j}}^{r} P_{ik}(t) \cdot P_{kj}(\Delta t) + P_{ij}(t) \cdot P_{jj}(\Delta t) \qquad (6.5)$$

The probability $P_{jj}(\Delta t)$ is according to property (6.3b),

$$P_{jj}(\Delta t) = 1 - \sum_{\substack{k=0 \\ k \neq j}}^{r} P_{jk}(\Delta t)$$

$P_{jj}(\Delta t)$ is thus the probability that the process does *not* leave state j in a time interval of length Δt.

Hence

$$P_{ij}(t + \Delta t) = \sum_{\substack{k=0 \\ k \neq j}}^{r} P_{ik}(t) \cdot P_{kj}(\Delta t) + P_{ij}(t) \cdot \left[1 - \sum_{\substack{k=0 \\ k \neq j}}^{r} P_{jk}(\Delta t) \right]$$

and

$$P_{ij}(t + \Delta t) - P_{ij}(t) = -P_{ij}(t) \sum_{\substack{k=0 \\ k \neq j}}^{r} P_{jk}(\Delta t) + \sum_{\substack{k=0 \\ k \neq j}}^{r} P_{ik}(t) \cdot P_{kj}(\Delta t)$$

After dividing by Δt, letting $\Delta t \to 0$ and using (6.4), we get the *state equations*

STATE EQUATIONS

$$\dot{P}_{ij}(t) = -P_{ij}(t) \cdot \sum_{\substack{k=0 \\ k \neq j}}^{r} a_{jk} + \sum_{\substack{k=0 \\ k \neq j}}^{r} P_{ik}(t) \cdot a_{kj} \quad (6.6)$$

Let us now assume that $X(0) = i$, namely that the process is in state i at time $t = 0$. This can be expressed as

$$P_i(0) = P(X(0) = i) = 1$$
$$P_k(0) = P(X(0) = k) = 0 \quad \text{for} \quad k \neq i$$

From now on we will assume that the initial state is known. Then we can simplify the notation of the state equations (6.6) by omitting the index i:

$$\dot{P}_j(t) = -P_j(t) \sum_{\substack{k=0 \\ k \neq j}}^{r} a_{jk} + \sum_{\substack{k=0 \\ k \neq j}}^{r} P_k(t) \cdot a_{kj}$$
$$P_i(0) = 1, \quad P_k(0) = 0 \quad \text{for} \quad k \neq i \quad (6.7)$$

The transition rates a_{ij} may be written as a matrix A, which is called the *transition rate matrix*:

$$A = \begin{bmatrix} -a_{00} & a_{10} & a_{20} & \cdots & a_{r0} \\ a_{01} & -a_{11} & a_{21} & \cdots & a_{r1} \\ a_{02} & a_{12} & -a_{22} & \cdots & a_{r2} \\ \vdots & \vdots & \vdots & \cdots & \vdots \\ a_{0r} & a_{1r} & a_{2r} & \cdots & -a_{rr} \end{bmatrix} \quad (6.8)$$

where we have introduced the following notation for the diagonal elements:

$$a_{jj} = \sum_{\substack{k=0 \\ k \neq j}}^{r} a_{jk} \quad (6.9)$$

The state equations (6.7) may now be written in matrix form

$$\begin{bmatrix} -a_{00} & a_{10} & a_{20} & \cdots & a_{r0} \\ a_{01} & -a_{11} & a_{21} & \cdots & a_{r1} \\ a_{02} & a_{12} & -a_{22} & \cdots & a_{r2} \\ \vdots & \vdots & \vdots & \cdots & \vdots \\ a_{0r} & a_{1r} & a_{2r} & \cdots & -a_{rr} \end{bmatrix} \cdot \begin{bmatrix} P_0(t) \\ P_1(t) \\ P_2(t) \\ \vdots \\ P_r(t) \end{bmatrix} = \begin{bmatrix} \dot{P}_0(t) \\ \dot{P}_1(t) \\ \dot{P}_2(t) \\ \vdots \\ \dot{P}_r(t) \end{bmatrix}$$

which may be abbreviated to

$$A \cdot P(t) = \dot{P}(t) \qquad (6.10)$$

Observe that a_{jk}, $k = 0, 1, \ldots, j-1, j+1, \cdots, r$, are transition rates *from* state j to the other states. We will call them *departure rates* from state j. Then according to (6.9), a_{jj} is the sum of the departure rates from state j.

When the process enters state j, the system will stay in this state a time T_j, which is exponentially distributed with parameter a_{jj}. The mean staying time in state j is thus

$$E(T_j) = \frac{1}{a_{jj}} \qquad \text{for } j = 0, 1, 2, \ldots, r. \qquad (6.11)$$

From (6.8) and (6.9) we observe that the sums of the columns of the transition rate matrix A add up to zero. Consequently the matrix is singular, and the state equations (6.10) do not have a unique solution. However, we know that the system must be in one of the $(r+1)$ states:

$$\sum_{j=0}^{r} P_j(t) = 1 \qquad (6.12)$$

By combining the state equations (6.10) and (6.12) and the known initial state ($P_i(0) = 1$), we are now able to compute all the state probabilities $P_j(t)$ for $j = 0, 1, 2, \ldots, r$. (Conditions for existence and uniqueness of the solutions are discussed, for example, by Cox and Miller 1965.)

6.4 TIME-DEPENDENT SOLUTION

The state equations (6.10) are seen to be a set of linear, first-order differential equations. The easiest and most widely used method to solve such equations is by Laplace transforms. An introduction to Laplace transforms is given in Appendix D.

The Laplace transform of the state probability $P_j(t)$ is denoted by $P_j^*(s)$,

TIME-DEPENDENT SOLUTION   221

and the Laplace transform of the time derivative of $P_j(t)$ is, according to Appendix D,

$$\mathcal{L}(\dot{P}_j(t)) = sP_j^*(s) - P_j(0) \quad \text{for } j = 0, 1, 2, \ldots, r$$

The Laplace transform of the state equations (6.10) is thus

$$\begin{bmatrix} -a_{00} & a_{10} & \cdots & a_{r0} \\ a_{01} & -a_{11} & \cdots & a_{r1} \\ \vdots & \vdots & & \vdots \\ a_{0r} & a_{1r} & \cdots & -a_{rr} \end{bmatrix} \cdot \begin{bmatrix} P_0^*(s) \\ P_1^*(s) \\ \vdots \\ P_r^*(s) \end{bmatrix} = \begin{bmatrix} sP_0^*(s) \\ sP_1^*(s) \\ \vdots \\ sP_r^*(s) \end{bmatrix} - \begin{bmatrix} P_0(0) \\ P_1(0) \\ \vdots \\ P_r(0) \end{bmatrix} \quad (6.13)$$

We assume that the initial state of the system is known:

$$P_i(0) = 1, \quad P_j(0) = 0 \quad \text{when } j \neq i$$

By summing all the $(r+1)$ equations in (6.13), we obtain

$$s\sum_{i=0}^{r} P_i^*(s) = 1$$

$$\sum_{i=0}^{r} P_i^*(s) = \frac{1}{s}$$

Since $1/s$ is the Laplace transform of a function that is equal to 1 for $t \geq 0$, this is equivalent to

$$\sum_{i=0}^{r} P_i(t) = 1.$$

By introducing the Laplace transforms, we have reduced the differential equations to a set of linear equations. The Laplace transforms $P_j^*(s)$ may now be computed from (6.13). Afterward the state probabilities $P_j(t)$ may be determined from the inverse Laplace transforms.

The matrix equation (6.10) may also be solved by an alternative procedure. If $\mathbf{P}(t)$ were a one-dimensional function $P(t)$ of t and $\mathbf{A}$ a constant A, then the solution of (6.10) would be

$$P(t) = e^{At} = \sum_{k=0}^{\infty} \frac{(At)^k}{k!}$$

Hence it seems reasonable to consider

$$P(t) = \sum_{k=0}^{\infty} A^k \cdot \frac{t^k}{k!} \qquad (6.14)$$

as a possible solution for $P(t)$, where $A^0 = I$ denotes the identity matrix. It can be shown (e.g., see Cox and Miller 1965, p. 182) that (6.14) is a valid solution of (6.10) under rather mild conditions. Equation (6.14) is often a computationally convenient way of approximating $P(t)$.

Example 6.2
Consider a single component. The component can be in two states:

> State 1   The component is functioning
> State 0   The component is in a failed state

Transition from state 1 to state 0 means that the component fails, and transition from state 0 to state 1 means that the component is repaired. The transition rate a_{10} is thus the failure rate of the component, and the transition rate a_{01} is the repair rate of the component. In this example we will use the following notation

> $a_{10} = \lambda$ The failure rate of the component
> $a_{01} = \mu$ The repair rate of the component

A consequence of the memoryless property of the Markov process is that all transition times are exponentially distributed. In this case the time to repair of the component is exponentially distributed with rate μ. The mean time to repair (MTTR) is thus equal to $1/\mu$.

The state-space diagram of the single component is illustrated in Figure 6.3. The state equations for this simple system is

$$\begin{bmatrix} -\mu & \lambda \\ \mu & -\lambda \end{bmatrix} \cdot \begin{bmatrix} P_0(t) \\ P_1(t) \end{bmatrix} = \begin{bmatrix} \dot{P}_0(t) \\ \dot{P}_1(t) \end{bmatrix}$$

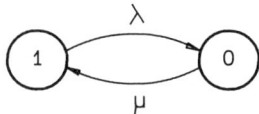

Figure 6.3 State-space diagram for a single component (function-repair cycle)

TIME-DEPENDENT SOLUTION

The component is assumed to be functioning at time $t = 0$:

$$P_1(0) = 1, \quad P_0(0) = 0$$

According to (6.13) the Laplace transform of the state equations is

$$\begin{bmatrix} -\mu & \lambda \\ \mu & -\lambda \end{bmatrix} \cdot \begin{bmatrix} P_0^*(s) \\ P_1^*(s) \end{bmatrix} = \begin{bmatrix} sP_0^*(s) \\ sP_1^*(s) - 1 \end{bmatrix}$$

Thus

$$-\mu P_0^*(s) + \lambda P_1^*(s) = sP_0^*(s) \qquad (6.15a)$$
$$\mu P_0^*(s) - \lambda P_1^*(s) = sP_1^*(s) - 1 \qquad (6.15b)$$

By adding (6.15a) and (6.15b), we get

$$sP_0^*(s) + sP_1^*(s) = 1$$

Thus

$$P_0^*(s) = \frac{1}{s} - P_1^*(s)$$

By inserting this expression for $P_0^*(s)$ into (6.15b), we obtain

$$\frac{\mu}{s} - \mu P_1^*(s) - \lambda P_1^*(s) = sP_1^*(s) - 1$$

$$P_1^*(s) = \frac{1}{\lambda + \mu + s} + \frac{\mu}{s} \cdot \frac{1}{\lambda + \mu + s}$$

To find the inverse Laplace transform, we rewrite this expression as

$$P_1^*(s) = \frac{\lambda}{\lambda + \mu} \cdot \frac{1}{\lambda + \mu + s} + \frac{\mu}{\lambda + \mu} \cdot \frac{1}{s} \qquad (6.16)$$

According to Appendix D the inverse Laplace transform of (6.16) is

$$P_1(t) = \frac{\lambda}{\lambda + \mu} e^{-(\lambda + \mu)t} + \frac{\mu}{\lambda + \mu} \qquad (6.17)$$

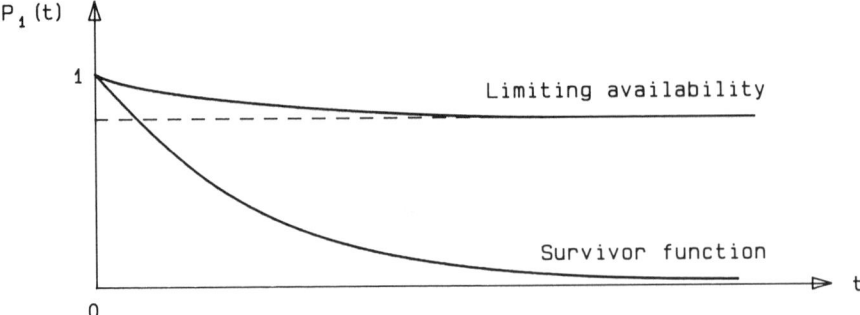

Figure 6.4 The availability and the survivor function of a single component

$P_1(t)$ denotes the probability that the component is functioning at time t, that is, the availability $A(t)$ of the component (see Example 4.8 p. 161).

Since the unavailability $P_0(t)$ of the component is equal to $1 - P_1(t)$,

$$P_0(t) = -\frac{\lambda}{\lambda + \mu} e^{-(\lambda + \mu)t} + \frac{\lambda}{\lambda + \mu} \qquad (6.18)$$

The limiting availability $A = \lim_{t \to \infty} A(t) = \lim_{t \to \infty} P_1(t)$ is, from (6.17),

$$A = P_1 = \lim_{t \to \infty} P_1(t) = \frac{\mu}{\lambda + \mu} \qquad (6.19)$$

The mean time to failure (MTTF) is equal to $1/\lambda$, and the mean time to repair (MTTR) is equal to $1/\mu$.

Thus the limiting availability may be written

$$A = P_1 = \frac{\text{MTTF}}{\text{MTTF} + \text{MTTR}} \qquad (6.20)$$

When there is no repair ($\mu = 0$), the availability is $P_1(t) = e^{-\lambda t}$, which coincides with the survivor function of the component. The availability $P_1(t)$ is illustrated in Figure 6.4. □

6.5 ASYMPTOTIC SOLUTION

In many applications only the long-run (steady state) probabilities are of interest—that is, the values of $P_j(t)$ when $t \to \infty$. In Example 6.2 the state probabilities $P_j(t)$ ($j = 0, 1$) approached a steady state P_j when $t \to \infty$. The same steady state value would have been found irrespective of whether the system started in the operating state or in the failed state.

Convergence toward steady state probabilities is assumed of the Markov processes we are studying in this chapter. A state j is said to be *reachable* from state i if for some $t > 0$ the transition rate $a_{ij} > 0$. The process is said to be *irreducible* if every state is reachable from every other state.

Theorem 6.1. For an irreducible Markov process, the limits

$$\lim_{t \to \infty} P_j(t) = P_j \quad \text{for } j = 0, 1, 2, \ldots, r$$

always exist and are independent of the initial state of the process (at time $t = 0$).

(See Cox and Miller 1965, sec. 4.4.) Hence a process that has been running for a long time has lost its dependency of its initial state $X(0)$. The process will converge to a process where the probability of being in state j is

$$P_j = P_j(\infty) = \lim_{t \to \infty} P_j(t) \quad \text{for } j = 0, 1, \ldots, r.$$

These asymptotic probabilities are often called the *steady state probabilities* for the process.

If $P_j(t)$ tends to a constant value when $t \to \infty$, then

$$\lim_{t \to \infty} \dot{P}_j(t) = 0 \quad \text{for } j = 0, 1, \ldots, r$$

The steady state probabilities $P_0, \ldots, P_r$ must therefore satisfy the matrix equation

$$\begin{bmatrix} -a_{00} & a_{10} & \cdots & a_{r0} \\ a_{01} & -a_{11} & \cdots & a_{r1} \\ \vdots & \vdots & & \vdots \\ a_{0r} & a_{1r} & \cdots & -a_{rr} \end{bmatrix} \cdot \begin{bmatrix} P_0 \\ P_1 \\ \vdots \\ P_r \end{bmatrix} = \begin{bmatrix} 0 \\ 0 \\ \vdots \\ 0 \end{bmatrix} \quad (6.21)$$

which may be abbreviated to

$$A \cdot P = 0$$

where, as before,

$$\sum_{j=0}^{r} P_j = 1$$

To calculate the steady state probabilities, $P_0, P_1, \ldots, P_r$ of such a process, we use r of the $r+1$ linear algebraic equation from the matrix equation (6.21), and in addition the fact that the sum of the state probabilities always is equal to 1. The initial state of the process has no influence on the steady state probabilities.

System Performance Characteristics

Visit Frequency

The visit frequency[1] ν_j to state j is defined as the expected number of visits to (arrivals into, or departures from) state j per unit time, computed over a long period of time. The visit frequency may be computed as follows: The (unconditional) probability of a departure from state j to state k in the time interval $(t, t + \Delta t]$ is

$$P((X(t + \Delta t) = k) \cap (X(t) = j))$$
$$= P(X(t + \Delta t) = k | X(t) = j) \cdot P(X(t) = j) = P_{jk}(\Delta t) \cdot P_j(t)$$

The frequency of departures from state j to state k is thus

$$\nu_{jk}^{dep}(t) = \lim_{\Delta t \to 0} \frac{P((X(t+\Delta t) = k) \cap (X(t) = j))}{\Delta t} = P_j(t) \cdot a_{jk}$$

In the steady state situation

$$\nu_{jk}^{dep} = P_j a_{jk}$$

and the total frequency of departures from state j in the steady state situation becomes

$$\nu_j^{dep} = P_j \sum_{\substack{k=0 \\ k \neq j}}^{r} a_{jk} = a_{jj} P_j \qquad (6.22)$$

We may also consider arrivals to state j. By an argument similar to the one

---

[1] In some textbooks (e.g. Endrenyi, 1978) the visit frequency is called the *state frequency*.

ASYMPTOTIC SOLUTION

used above, the frequency of arrivals to state j from state k in the steady state situation is

$$\nu_{kj}^{arr} = P_k a_{kj}$$

Hence the total frequency of arrivals to state j in the steady state situation is

$$\nu_j^{arr} = \sum_{\substack{k=0 \\ k \neq j}}^{r} P_k a_{kj} \qquad (6.23)$$

The matrix equation (6.21) implies that

$$a_{jj} P_j = \sum_{\substack{k=0 \\ k \neq j}}^{r} a_{kj} P_k \quad \text{for} \quad j = 0, 1, \ldots, r \qquad (6.24)$$

Thus in the steady state situation the frequency of departures from state j is equal to the frequency of arrivals (visit frequency) to state j. Then the visit frequency to state j is

$$\nu_j = \sum_{\substack{k=0 \\ k \neq j}}^{r} a_{kj} P_k = a_{jj} P_j \quad \text{for} \quad j = 0, 1, \ldots, r \qquad (6.25)$$

Mean Duration of a Visit
When the process arrives a state j, the system will stay in this state a time T_j until the process departures from that state, $j = 0, 1, \ldots, r$. The total departure rate from state j is

$$a_{jj} = \sum_{\substack{k=0 \\ k \neq j}}^{r} a_{jk} \quad \text{for} \quad j = 0, 1, \ldots, r$$

The departures are illustrated in Figure 6.5.

Since the departure rate is constant, the duration of a stay in state j is exponentially distributed with parameter a_{jj}. The mean duration of a stay in state j is thus

$$\theta_j = \frac{1}{a_{jj}} \quad \text{for} \quad j = 0, 1, \ldots, r \qquad (6.26)$$

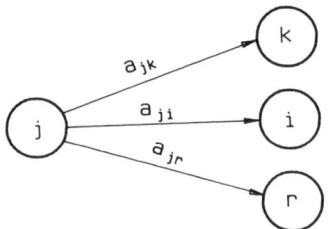

Figure 6.5 Departure rates for state j

By combining (6.25) and (6.26), we obtain

$$v_j = a_{jj} \cdot P_j = \frac{P_j}{\theta_j}$$
$$P_j = v_j \theta_j \qquad (6.27)$$

The mean proportion of time P_j the system is spending in state j is thus equal to the visit frequency to state j multiplied by the mean duration of a visit in state j for $j = 0, 1, \ldots, r$.

System Availability

Let $S = \{0, 1, \ldots, r\}$ be the set of all possible states of a system. Some of these states represent system functioning according to some specified criteria. Let B denote the subset of states in which the system is functioning, and let $F = S - B$ denote the states in which the system is failed.

The average, or long-term, *availability* of the system is the mean proportion of time when the system is functioning; that is, its state is a member of B. The average system availability A_s is thus defined as

$$A_s = \sum_{j \in B} P_j \qquad (6.28)$$

In the following we will omit the term average and call A_s the system availability.

The system unavailability $1 - A_s$ is then

$$1 - A_s = \sum_{j \in F} P_j \qquad (6.29)$$

The unavailability $1 - A_s$ of the system is the mean proportion of time when the system is in a failed state.

Frequency of System Failures

The frequency ω_F of system failures is defined as the expected number of visits to (arrivals into, or departures from) a failed state (in F) per unit time, computed over a long period of time.

Mean Duration of a System Failure

The mean duration θ_F of a system failure is defined as the mean time from the system enters into a failed state (F) until it is repaired/restored and brought back into a functioning state (B). Analogous with (6.27) it is obvious that the system unavailability $1 - A_s$ is equal to the frequency of system failures multiplied by the mean duration of a system failure. Hence

$$1 - A_s = \omega_F \cdot \theta_F \qquad (6.30)$$

Mean Time between System Failures

The mean time between system failures MTBF$_s$ is the mean time between consecutive transitions from a functioning state (B) into a failed state (F). The MTBF$_s$ may be computed from the frequency of system failures by

$$\text{MTBF}_s = \frac{1}{\omega_F} \qquad (6.31)$$

Mean Functioning Time until System Failure

The mean functioning time (up-time) until system failure $E(U)_s$ is the mean time from a transition from a failed state (F) into a functioning state (B) until the first transition back to a failed state (F). It is obvious that

$$\text{MTBF}_s = E(U)_s + \theta_F \qquad (6.32)$$

Note the difference between the mean functioning time (up-time) and the mean time to system failure MTTF$_S$. The MTTF$_S$ is discussed in Section 6.6, and it is normally calculated as the mean time until system failure when the system initially is in a *specified* functioning state.

Independent Components in Parallel

Reconsider the parallel structure of two independent components in Example 6.1. The following transition rates are assumed:

$$\begin{aligned}
a_{32} = a_{10} = \lambda_1 &\quad \text{Failure rate of component 1} \\
a_{31} = a_{20} = \lambda_2 &\quad \text{Failure rate of component 2} \\
a_{23} = a_{01} = \mu_1 &\quad \text{Repair rate of component 1} \\
a_{13} = a_{02} = \mu_2 &\quad \text{Repair rate of component 2}
\end{aligned}$$

The steady state equations (6.21) are

$$\begin{bmatrix} -(\mu_1+\mu_2) & \lambda_1 & \lambda_2 & 0 \\ \mu_1 & -(\lambda_1+\mu_2) & 0 & \lambda_2 \\ \mu_2 & 0 & -(\lambda_2+\mu_1) & \lambda_1 \\ 0 & \mu_2 & \mu_1 & -(\lambda_1+\lambda_2) \end{bmatrix} \cdot \begin{bmatrix} P_0 \\ P_1 \\ P_2 \\ P_3 \end{bmatrix} = \begin{bmatrix} 0 \\ 0 \\ 0 \\ 0 \end{bmatrix}$$

(6.33)

The steady state probabilities P_j for $j = 0, 1, 2, 3$ may be found from the equations

$$-(\mu_1 + \mu_2)P_0 + \lambda_1 P_1 + \lambda_2 P_2 = 0 \quad (6.34\text{a})$$
$$\mu_1 P_0 - (\lambda_1 + \mu_2)P_1 + \lambda_2 P_3 = 0 \quad (6.34\text{b})$$
$$\mu_2 P_0 - (\lambda_2 + \mu_1)P_2 + \lambda_1 P_3 = 0 \quad (6.34\text{c})$$
$$P_0 + P_1 + P_2 + P_3 = 1 \quad (6.34\text{d})$$

The solution is

$$\begin{aligned}
P_0 &= \frac{\lambda_1 \lambda_2}{(\lambda_1 + \mu_1)(\lambda_2 + \mu_2)} \\
P_1 &= \frac{\lambda_2 \mu_1}{(\lambda_1 + \mu_1)(\lambda_2 + \mu_2)} \\
P_2 &= \frac{\lambda_1 \mu_2}{(\lambda_1 + \mu_1)(\lambda_2 + \mu_2)} \\
P_3 &= \frac{\mu_1 \mu_2}{(\lambda_1 + \mu_1)(\lambda_2 + \mu_2)}
\end{aligned} \quad (6.35)$$

Now, for $i = 1, 2$, let:

$$q_i = \frac{\lambda_i}{\lambda_i + \mu_i} = \frac{\text{MTTR}_i}{\text{MTTF}_i + \text{MTTR}_i}$$
$$p_i = \frac{\mu_i}{\lambda_i + \mu_i} = \frac{\text{MTTF}_i}{\text{MTTF}_i + \text{MTTR}_i}$$

where $\text{MTTR}_i = 1/\mu_i$ is the mean time to repair of component i, and $\text{MTTF}_i = 1/\lambda_i$ is the mean time to failure of component i ($i = 1, 2$). q_i thus denotes the average, or limiting, unavailability of component i, while p_i denotes the average (limiting) availability of component i ($i = 1, 2$).

The steady state probabilities may thus be written as

$$P_0 = q_1 q_2$$
$$P_1 = p_1 q_2$$
$$P_2 = q_1 p_2$$
$$P_3 = p_1 p_2 \qquad (6.36)$$

In this very simple example the results in (6.36) may also be obtained by direct reasoning:

$P_0 = P$ (unit 1 is failed) $\cdot P$ (unit 2 is failed) $= q_1 q_2$
$P_1 = P$ (unit 1 is functioning) $\cdot P$ (unit 2 is failed) $= p_1 q_2$
$P_2 = P$ (unit 1 is failed) $\cdot P$ (unit 2 is functioning) $= q_1 p_2$
$P_3 = P$ (unit 1 is functioning) $\cdot P$ (unit 2 is functioning) $= p_1 p_2$

Note that the steady state probabilities can be interpreted as the mean proportion of time the system stays in the state concerned. For example,

$$P_1 = 0.005 \left(\frac{\text{year}}{\text{year}} \right) = 0.005 \cdot 8760 \left(\frac{\text{hours}}{\text{year}} \right) \approx 44 \left(\frac{\text{hours}}{\text{year}} \right)$$

In the long run the system will stay in state 1 for 44 hours per year. This does *not* mean that state 1 occurs on average once per year and lasts for 44 hours each time. The visit frequencies and the mean durations of the visits, however, can be calculated from the formulas on page 226–229.

Mean Duration of the Visits
From (6.26), we get

$$\theta_0 = \frac{1}{\mu_1 + \mu_2}$$
$$\theta_1 = \frac{1}{\lambda_1 + \mu_2}$$
$$\theta_2 = \frac{1}{\lambda_2 + \mu_1}$$
$$\theta_3 = \frac{1}{\lambda_1 + \lambda_2} \qquad (6.37)$$

Visit Frequency
From (6.27) and (6.37), we get

$$\begin{aligned} \nu_0 &= P_0(\mu_1 + \mu_2) \\ \nu_1 &= P_1(\lambda_1 + \mu_2) \\ \nu_2 &= P_2(\lambda_2 + \mu_1) \\ \nu_3 &= P_3(\lambda_1 + \lambda_2) \end{aligned} \qquad (6.38)$$

The parallel structure is functioning when at least one of its two components is functioning. When the system is in state 1, 2, or 3, the system is functioning, while state 0 corresponds to system failure.

The average system unavailability is

$$1 - A_s = P_0 = q_1 q_2 \qquad (6.39)$$

and the average system availability is

$$A_s = P_1 + P_2 + P_3 = 1 - q_1 q_2$$

The frequency of system failures ω_F is equal to the visit frequency to state 0, which is

$$\omega_F = \nu_0 = P_0(\mu_1 + \mu_2) = (1 - A_s) \cdot (\mu_1 + \mu_2) \qquad (6.40)$$

The mean duration of a system failure θ_F is in this case equal to the mean duration of a stay in state 0. Thus

$$\theta_F = \theta_0 = \frac{1}{\mu_1 + \mu_2} = \frac{1 - A_s}{\omega_F} \qquad (6.41)$$

For a parallel structure of n independent components, the above results may be generalized as follows: For system unavailability,

$$1 - A_s = \prod_{i=1}^{n} q_i = \prod_{i=1}^{n} \frac{\lambda_i}{\lambda_i + \mu_i} \qquad (6.42)$$

For frequency of system failures,

$$\omega_F = (1 - A_s) \cdot \sum_{i=1}^{n} \mu_i \qquad (6.43)$$

ASYMPTOTIC SOLUTION

For mean duration of a system failure,

$$\theta_F = \frac{1}{\sum_{i=1}^{n} \mu_i} \qquad (6.44)$$

The mean functioning time (up-time) $E(U)_P$ of the parallel structure can be determined from

$$1 - A_s = \frac{\theta_F}{\theta_F + E(U)_P}$$

Hence

$$E(U)_P = \frac{\theta_F A_s}{1 - A_s} = \frac{1 - \prod_{i=1}^{n} \lambda_i/(\lambda_i + \mu_i)}{\prod_{i=1}^{n} \lambda_i/(\lambda_i + \mu_i) \cdot \sum_{j=1}^{n} \mu_j} \qquad (6.45)$$

When the component availabilities are very high: (i.e., $\lambda_i \ll \mu_i$ for all $i = 1, 2, \ldots, n$), then

$$\frac{\lambda_i}{\lambda_i + \mu_i} = \frac{\lambda_i \text{MTTR}_i}{1 + \lambda_i \text{MTTR}_i} \approx \lambda_i \text{MTTR}_i$$

The frequency ω_F of system failures can now be approximated as

$$\omega_F = (1 - A_s) \cdot \sum_{i=1}^{n} \mu_i = \prod_{i=1}^{n} \frac{\lambda_i}{\lambda_i + \mu_i} \cdot \sum_{j=1}^{n} \mu_j$$

$$\approx \prod_{i=1}^{n} \lambda_i \text{MTTR}_i \cdot \sum_{j=1}^{n} \frac{1}{\text{MTTR}_j} \qquad (6.46)$$

For two components (6.46) reduces to

$$\omega_F \approx \lambda_1 \lambda_2 \cdot (\text{MTTR}_1 + \text{MTTR}_2) \qquad (6.47)$$

For three components (6.46) reduces to

$$\omega_F \approx \lambda_1 \lambda_2 \lambda_3 \cdot (\text{MTTR}_1 \cdot \text{MTTR}_2 + \text{MTTR}_1 \cdot \text{MTTR}_3 + \text{MTTR}_2 \cdot \text{MTTR}_3)$$

Independent Components in Series

Consider a series structure of two independent components. The states of the system and the transition rates are as defined on page 230. The state-space diagram of the series structure is shown in Figure 6.6. The corresponding steady state equations are equal to those found for the parallel structure on pages 230–231.

The average availability A_s of the structure is equal to P_3, which was found in (6.35) to be

$$A_s = P_3 = \frac{\mu_1 \mu_2}{(\lambda_1 + \mu_1)(\lambda_2 + \mu_2)} = p_1 \cdot p_2 \qquad (6.48)$$

where

$$p_i = \frac{\mu_i}{\lambda_i + \mu_i} \quad \text{for} \quad i = 1, 2$$

The frequency of system failures ω_F is the same as the frequency of visits to state 3. Thus

$$\omega_F = \nu_3 = P_3 \cdot (\lambda_1 + \lambda_2) = A_s \cdot (\lambda_1 + \lambda_2) \qquad (6.49)$$

The mean duration of a system failure θ_F is equal to

$$\theta_F = \frac{1 - A_s}{\omega_F} \qquad (6.50)$$

For a series structure of n independent components the above results can be

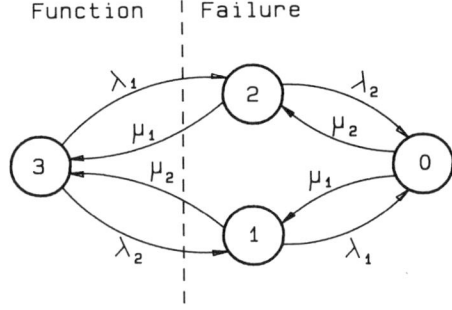

Figure 6.6 Partitioning the state-space diagram of a series structure of two independent components

generalized as follows: For system availability

$$A_s = \prod_{i=1}^{n} p_i = \prod_{i=1}^{n} \frac{\mu_i}{\lambda_i + \mu_i} \tag{6.51}$$

For frequency of system failures

$$\omega_F = A_s \sum_{i=1}^{n} \lambda_i \tag{6.52}$$

For mean duration of a system failure

$$\begin{aligned}\theta_F &= \frac{1 - A_s}{\omega_F} \\ &= \frac{1}{\sum_{i=1}^{n} \lambda_i} \frac{1 - A_s}{A_s} \\ &= \frac{1 - \prod_{i=1}^{n} \mu_i/(\lambda_i + \mu_i)}{\prod_{i=1}^{n} \mu_i/(\lambda_i + \mu_i) \cdot \sum_{j=1}^{n} \lambda_j}\end{aligned} \tag{6.53}$$

When all the component availabilities are very high such that $\lambda_i \ll \mu_i$ for all i, then $A_s \approx 1$ and the frequency of system failures is approximately

$$\omega_F \approx \sum_{i=1}^{n} \lambda_i \tag{6.54}$$

which is the same as the failure rate of nonrepairable series structure of n independent components.

The mean duration of a system failure θ_F may be approximated as

$$\begin{aligned}\theta_F &= \frac{1}{\sum_{i=1}^{n} \lambda_i} \frac{1 - A_s}{A_s} = \frac{1}{\sum_{i=1}^{n} \lambda_i} \left(\frac{1}{A_s} - 1\right) = \frac{1}{\sum_{i=1}^{n} \lambda_i} \left(\prod_{i=1}^{n} \frac{1}{p_i} - 1\right) \\ &= \frac{1}{\sum_{i=1}^{n} \lambda_i} \left(\prod_{i=1}^{n} \left(1 + \frac{\lambda_i}{\mu_i}\right) - 1\right) \approx \frac{1}{\sum_{i=1}^{n} \lambda_i} \left(1 + \sum_{i=1}^{n} \frac{\lambda_i}{\mu_i} - 1\right) \\ &= \frac{\sum_{i=1}^{n} \lambda_i/\mu_i}{\sum_{i=1}^{n} \lambda_i} = \frac{\sum_{i=1}^{n} \lambda_i \text{MTTR}_i}{\sum_{i=1}^{n} \lambda_i}\end{aligned} \tag{6.55}$$

where $\text{MTTR}_i = 1/\mu_i$ as before is the mean time to repair component i, $i = 1, 2, \ldots, n$. Equation (6.55) is a commonly used approximation for the mean duration of a failure in series structures of high reliability.

Series Structure of Components Where Failure of One Component Prevents Failure of the Other

Consider a series structure of two components. When one of the components fails, the other component is immediately taken out of operation until the failed component is repaired.[2] After a component is taken out of operation, it is no longer exposed to stress, and we can therefore assume that it will not fail. This dependence between the failures prevents a simple solution by direct reasoning, as was possible in Example 6.3. This system has three possible states, which are described in Table 6.2.

The following transition rates are assumed:

$a_{21} = \lambda_1$ Failure rate of component 1
$a_{20} = \lambda_2$ Failure rate of component 2
$a_{12} = \mu_1$ Repair rate of component 1
$a_{02} = \mu_2$ Repair rate of component 2

The state-space diagram of the series structure is illustrated in Figure 6.7.

---

[2]The same model is discussed by Barlow and Proschan (1975, pp. 194–201) in a more general context that does not assume constant failure and repair rates.

Table 6.2 Possible States of a Series Structure of Two Components Where Failure of One Component Prevents Failure of the Other

System State	State of Component 2	State of Component 1	Comment
2	1	1	Both components functioning
1	1	0	Component 2 functioning, component 1 taken out of operation
0	0	1	Component 1 functioning, component 2 taken out of operation

ASYMPTOTIC SOLUTION

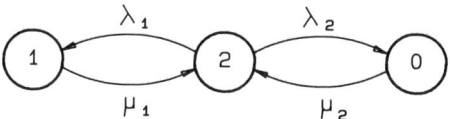

Figure 6.7 State-space diagram of a series structure of two components where failure of one component prevents failure of the other component

The steady state equations for this system are

$$\begin{bmatrix} -\mu_2 & 0 & \lambda_2 \\ 0 & -\mu_1 & \lambda_1 \\ \mu_2 & \mu_1 & -(\lambda_1+\lambda_2) \end{bmatrix} \cdot \begin{bmatrix} P_0 \\ P_1 \\ P_2 \end{bmatrix} = \begin{bmatrix} 0 \\ 0 \\ 0 \end{bmatrix} \quad (6.56)$$

The steady state probabilities may be found from the equations

$$-\mu_2 P_0 + \lambda_2 P_2 = 0 \quad (6.57a)$$
$$-\mu_1 P_1 + \lambda_1 P_2 = 0 \quad (6.57b)$$
$$P_0 + P_1 + P_2 = 1 \quad (6.57c)$$

The solution of (6.57) is

$$P_2 = \frac{\mu_1 \mu_2}{\lambda_1 \mu_2 + \lambda_2 \mu_1 + \mu_1 \mu_2} = \frac{1}{1 + (\lambda_1/\mu_1) + (\lambda_2/\mu_2)} \quad (6.58a)$$

$$P_1 = \frac{\lambda_1}{\mu_1} P_2 \quad (6.58b)$$

$$P_0 = \frac{\lambda_2}{\mu_2} P_2 \quad (6.58c)$$

Since the series structure is only functioning when both components are functioning (state 2), the average system availability is

$$A_s = P_2 = \frac{\mu_1 \mu_2}{\lambda_1 \mu_2 + \lambda_2 \mu_1 + \mu_1 \mu_2} = \frac{1}{1 + (\lambda_1/\mu_1) + (\lambda_2/\mu_2)}$$

Observe that in this case the availability of the series structure is *not* equal to the product of the component availabilities.

The mean durations of the stays in each state are

$$\theta_2 = \frac{1}{\lambda_1 + \lambda_2}$$

$$\theta_1 = \frac{1}{\mu_1}$$

$$\theta_0 = \frac{1}{\mu_2}$$

The frequency of system failures ω_F is the same as the frequency of visits to state 2

$$\omega_F = \nu_2 = P_2(\lambda_1 + \lambda_2) = A_s(\lambda_1 + \lambda_2) \qquad (6.59)$$

The mean duration of a system failure θ_F is

$$\begin{aligned}\theta_F &= \frac{1 - A_s}{\omega_F} = \frac{1}{\lambda_1 + \lambda_2} \cdot \frac{1 - A_s}{A_s} \\ &= \frac{1}{\mu_1} \cdot \frac{\lambda_1}{\lambda_1 + \lambda_2} + \frac{1}{\mu_2} \cdot \frac{\lambda_2}{\lambda_1 + \lambda_2}\end{aligned} \qquad (6.60)$$

Equation (6.60) may also be written

$$\theta_F = \text{MTTR}_1 \cdot P(\text{Component 1 fails}|\text{system failure})$$
$$+ \text{MTTR}_2 \cdot P(\text{Component 2 fails}|\text{system failure})$$

This formula is obvious since the duration of a system failure will be equal to the repair time of component 1 when component 1 fails and equal to the repair time of component 2 when component 2 fails.

The mean time between system failures MTBF_s is

$$\begin{aligned}\text{MTBF}_s = \text{MTTF}_s + \theta_F &= \frac{1}{\lambda_1 + \lambda_2} + \frac{1}{\mu_1} \frac{\lambda_1}{\lambda_1 + \lambda_2} + \frac{1}{\mu_2} \frac{\lambda_2}{\lambda_1 + \lambda_2} \\ &= \frac{1 + (\lambda_1/\mu_1) + (\lambda_2/\mu_2)}{\lambda_1 + \lambda_2}\end{aligned} \qquad (6.61)$$

The frequency of system failures may also be expressed as

$$\omega_F = \frac{1}{\text{MTBF}_s} = (\lambda_1 + \lambda_2) \cdot \frac{1}{1 + \dfrac{\lambda_1}{\mu_1} + \dfrac{\lambda_2}{\mu_2}} = A_s \cdot (\lambda_1 + \lambda_2)$$

For a series structure of n components the preceding results can be generalized as follows: For system availability

$$A_s = \frac{1}{1 + \sum_{i=1}^{n} (\lambda_i/\mu_i)} \qquad (6.62)$$

For mean time to system failure

$$\text{MTTF}_S = \frac{1}{\sum_{i=1}^{n} \lambda_i} \qquad (6.63)$$

For mean duration of a system failure

$$\theta_F = \sum_{i=1}^{n} \frac{1}{\mu_i} \cdot \frac{\lambda_i}{\sum_{j=1}^{n} \lambda_j} = \frac{1}{\sum_{j=1}^{n} \lambda_j} \cdot \sum_{i=1}^{n} \frac{\lambda_i}{\mu_i} \qquad (6.64)$$

For frequency of system failures

$$\omega_F = A_s \cdot \sum_{i=1}^{n} \lambda_i = \frac{\sum_{i=1}^{n} \lambda_i}{1 + \sum_{i=1}^{n} (\lambda_i/\mu_i)} \qquad (6.65)$$

6.6 MEAN TIME TO FAILURE

Absorbing States

All of the processes we have studied so far in this chapter have been irreducible, which means that every state is reachable from every other state. We will now introduce Markov processes with *absorbing states*. An absorbing state is a state that, once entered, cannot be left until the system starts a new mission. The popular saying is that the system is *trapped* in an absorbing state.

Example 6.3
Consider a parallel system of two independent and identical components with failure rate λ. When one of the components fails, it is repaired. The repair time is assumed to be exponentially distributed with repair rate μ. When both components have failed, the system is considered to have failed and no recovery is possible. Let the number of functioning components denote the state of the system. The state space is thus $\{0, 1, 2\}$, and state 0 is an absorbing state. The state-space diagram of the system is given in Figure 6.8.

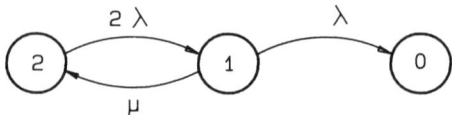

Figure 6.8 State-space diagram for a parallel system with two identical components

We assume that both components are functioning (state 2) at time $t = 0$. That is $P_2(0) = 1$. The state equations of this system are thus

$$\begin{bmatrix} 0 & \lambda & 0 \\ 0 & -(\lambda+\mu) & 2\lambda \\ 0 & \mu & -2\lambda \end{bmatrix} \cdot \begin{bmatrix} P_0(t) \\ P_1(t) \\ P_2(t) \end{bmatrix} = \begin{bmatrix} \dot{P}_0(t) \\ \dot{P}_1(t) \\ \dot{P}_2(t) \end{bmatrix} \quad (6.66)$$

Since state 0 is an absorbing state, all the transition rates from this state are equal to zero. Thus the elements of the column corresponding to the absorbing state are all equal to zero.

Since the transition rate matrix does not have full rank, we may remove one of the three equations without loosing any information about $P_0(t)$, $P_1(t)$, and $P_2(t)$. In this case we remove the first of the three equations. This is accomplished by removing the first row of the matrix. Hence we get

$$\begin{bmatrix} 0 & -(\lambda+\mu) & 2\lambda \\ 0 & \mu & -2\lambda \end{bmatrix} \cdot \begin{bmatrix} P_0(t) \\ P_1(t) \\ P_2(t) \end{bmatrix} = \begin{bmatrix} \dot{P}_1(t) \\ \dot{P}_2(t) \end{bmatrix}$$

Since all the elements of the first column of the matrix are equal to zero, $P_0(t)$ will "disappear" in the solution of the equations. We may therefore reduce the matrix equations to

$$\begin{bmatrix} -(\lambda+\mu) & 2\lambda \\ \mu & -2\lambda \end{bmatrix} \cdot \begin{bmatrix} P_1(t) \\ P_2(t) \end{bmatrix} = \begin{bmatrix} \dot{P}_1(t) \\ \dot{P}_2(t) \end{bmatrix} \quad (6.67)$$

The matrix

$$\begin{bmatrix} -(\lambda+\mu) & 2\lambda \\ \mu & -2\lambda \end{bmatrix}$$

has full rank if $\lambda > 0$. Therefore (6.67) determines $P_1(t)$ and $P_2(t)$. $P_0(t)$ may thereafter be found from $P_0(t) = 1 - P_1(t) - P_2(t)$. This solution of the reduced matrix equations (6.67) is identical to the solution of the initial matrix equations

MEAN TIME TO FAILURE   241

(6.66). The reduced matrix is seen to be obtained by deleting the row and the column corresponding to the absorbing state.

Since state 0 is absorbing and reachable from the other states, it is obvious that

$$\lim_{t \to \infty} P_0(t) = 1$$

The Laplace transforms of the reduced matrix equations are

$$\begin{bmatrix} -(\lambda + \mu) & 2\lambda \\ \mu & -2\lambda \end{bmatrix} \cdot \begin{bmatrix} P_1^*(s) \\ P_2^*(s) \end{bmatrix} = \begin{bmatrix} sP_1^*(s) \\ sP_2^*(s) - 1 \end{bmatrix}$$

since the system is assumed to be in state 2 at time $t = 0$. Thus

$$-(\lambda + \mu)P_1^*(s) + 2\lambda P_2^*(s) = sP_1^*(s)$$
$$\mu P_1^*(s) - 2\lambda P_2^*(s) = sP_2^*(s) - 1$$

Solving for $P_1^*(s)$ and $P_2^*(s)$, we get (see Appendix D)

$$P_1^*(s) = \frac{2\lambda}{s^2 + (3\lambda + \mu)s + 2\lambda^2}$$
$$P_2^*(s) = \frac{\lambda + \mu + s}{s^2 + (3\lambda + \mu)s + 2\lambda^2}$$

Let $R(t)$ denote the survivor function of the system. Since the system is functioning as long as the system is either in state 2 or in state 1, the survivor function is equal to

$$R(t) = P_1(t) + P_2(t) = 1 - P_0(t)$$

The Laplace transform of $R(t)$ is thus

$$R^*(s) = P_1^*(s) + P_2^*(s) = \frac{3\lambda + \mu + s}{s^2 + (3\lambda + \mu)s + 2\lambda^2} \qquad (6.68)$$

The survivor function $R(t)$ may now be determined by inverting the Laplace transform, or we may consider $P_0(t) = 1 - R(t)$ which denotes the distribution function of the time T_s to system failure. The Laplace transform of $P_0(t)$ is

$$P_0^*(s) = \frac{1}{s} - P_1^*(s) - P_2^*(s) = \frac{2\lambda^2}{s[s^2 + (3\lambda + \mu)s + 2\lambda^2]}$$

Let $f_s(t)$ denote the probability density function of the time T_s to system failure, that is, $f_s(t) = dP_0(t)/dt$. The Laplace transform of $f_s(t)$ is thus

$$f_s^*(s) = sP_0^*(s) - P_0(0) = \frac{2\lambda^2}{s^2 + (3\lambda + \mu)s + 2\lambda^2} \quad (6.69)$$

The denominator of (6.69) can be written

$$s^2 + (3\lambda + \mu)s + 2\lambda^2 = (s - k_1)(s - k_2)$$

where

$$k_1 = \frac{-(3\lambda + \mu) + \sqrt{\lambda^2 + 6\lambda\mu + \mu^2}}{2}$$

$$k_2 = \frac{-(3\lambda + \mu) - \sqrt{\lambda^2 + 6\lambda\mu + \mu^2}}{2}$$

The expression for $f_s^*(s)$ can be rearranged so that

$$f_s^*(s) = \frac{2\lambda^2}{k_1 - k_2} \left(\frac{1}{s + k_2} - \frac{1}{s + k_1} \right)$$

By inverting this transform, we get

$$f_s(t) = \frac{2\lambda^2}{k_1 - k_2} (e^{-k_2 t} - e^{-k_1 t})$$

The mean time to system failure MTTF$_S$ is now given by (the integration is left to the reader as an exercise)

$$\text{MTTF}_S = \int_0^\infty t f_s(t)\, dt = \frac{3}{2\lambda} + \frac{\mu}{2\lambda^2} \quad (6.70)$$

Note that the MTTF$_S$ of a two-component parallel system, without any repair (i.e., $\mu = 0$), is equal to $3/2\lambda$. The repair facility thus increases the MTTF$_S$ by $\mu/2\lambda^2$. □

ooo OOO ooo

MEAN TIME TO FAILURE 243

Survivor Function

As discussed in Section 6.5, page 228, the set of states S of a system may be grouped in a set B of functioning states and a set $F = S - B$ of failed states. In the present section we will assume that the failed states are absorbing states.

Consider a system that is in a specified functioning state at time $t = 0$. The survivor function $R(t)$ determines the probability that a system does not leave the set B of functioning states during the time interval $(0, t]$. The survivor function is thus

$$R(t) = \sum_{j \in B} P_j(t) \qquad (6.71)$$

The Laplace transform of the survivor function is

$$R^*(s) = \sum_{j \in B} P_j^*(s)$$

Mean Time to System Failure

The mean time to system failure MTTF$_S$ may, according to Section 2.4, be determined by

$$\text{MTTF}_S = \int_0^\infty R(t)\,dt \qquad (6.72)$$

The Laplace transform of $R(t)$ is given by

$$R^*(s) = \int_0^\infty R(t)e^{-st}\,dt \qquad (6.73)$$

The MTTF$_S$ of the system can be determined from (6.73) by inserting $s = 0$. Thus

$$R^*(0) = \int_0^\infty R(t)\,dt = \text{MTTF}_S \qquad (6.74)$$

Example 6.3 (Cont.)
The Laplace transform of the survivor function for the two–component parallel system was in (6.68) found to be

$$R^*(s) = \frac{3\lambda + \mu + s}{s^2 + (3\lambda + \mu)s + 2\lambda^2}$$

By introducing $s = 0$, we get

$$\text{MTTF}_S = R^*(0) = \frac{3\lambda + \mu}{2\lambda^2} = \frac{3}{2\lambda} + \frac{\mu}{2\lambda^2}$$

which is in accordance with (6.70). □

ooo OOO ooo

Assume now that we are considering a system represented by a Markov process with state space $S = \{0, 1, 2, \ldots, r\}$. Assume further that the states 0 and 2 are absorbing. Thus the state equations are given by

$$\begin{bmatrix} 0 & a_{10} & 0 & a_{30} & \cdots & a_{r0} \\ 0 & -a_{11} & 0 & a_{31} & \cdots & a_{r1} \\ 0 & a_{12} & 0 & a_{32} & \cdots & a_{r2} \\ \vdots & \vdots & \vdots & \vdots & \cdots & \vdots \\ 0 & a_{1r} & 0 & a_{3r} & \cdots & -a_{rr} \end{bmatrix} \cdot \begin{bmatrix} P_0(t) \\ P_1(t) \\ P_2(t) \\ \vdots \\ P_r(t) \end{bmatrix} = \begin{bmatrix} \dot{P}_0(t) \\ \dot{P}_1(t) \\ \dot{P}_2(t) \\ \vdots \\ \dot{P}_r(t) \end{bmatrix} \quad (6.75)$$

Let us assume that we were only interested in deriving the survivor function $R(t)$ of the system. It is indifferent for our problem whether the system is absorbed in state 0 or in state 2. These two states may therefore be pooled and considered as a single state. Since the transition rate matrix A does not have full rank, we may remove the equation (row) corresponding to this pooled state. This corresponds to deleting the first and the third row from the matrix A. Since all the elements of the first and the third column are equal to zero, $P_0(t)$ and $P_2(t)$ will disappear from the the equations:

$$\begin{bmatrix} -a_{11} & a_{31} & \cdots & a_{r1} \\ -a_{13} & -a_{33} & \cdots & a_{r3} \\ \vdots & \vdots & \cdots & \vdots \\ a_{1r} & a_{3r} & \cdots & -a_{rr} \end{bmatrix} \cdot \begin{bmatrix} P_1(t) \\ P_3(t) \\ \vdots \\ P_r(t) \end{bmatrix} = \begin{bmatrix} \dot{P}_1(t) \\ \dot{P}_3(t) \\ \vdots \\ \dot{P}_r(t) \end{bmatrix} \quad (6.76)$$

We see that the reduced matrix is obtained by deleting the rows and the columns corresponding to the absorbing states (in this example 0 and 2). This reduction of the transition rate matrix is further discussed by Billinton and Allan (1983, p. 217) who use the term *truncated* matrix instead of reduced matrix.

By introducing the Laplace transforms of the state equations (6.76) we obtain

MEAN TIME TO FAILURE   245

$$\begin{bmatrix} -a_{11} & a_{31} & \cdots & a_{r1} \\ -a_{13} & -a_{33} & \cdots & a_{r3} \\ \vdots & \vdots & \cdots & \vdots \\ a_{1r} & a_{3r} & \cdots & -a_{rr} \end{bmatrix} \cdot \begin{bmatrix} P_1^*(s) \\ P_3^*(s) \\ \vdots \\ P_r^*(s) \end{bmatrix} = \begin{bmatrix} sP_1^*(s) \\ sP_3^*(s) \\ \vdots \\ sP_r^*(s) \end{bmatrix} - \begin{bmatrix} P_1(0) \\ P_3(0) \\ \vdots \\ P_r(0) \end{bmatrix} \quad (6.77)$$

To obtain (6.77), we used the fact that (see Appendix D)

$$\mathscr{L}[\dot{P}_j(t)] = sP_j^*(s) - P_j(0) \quad \text{for } j = 1, 3, \ldots, r$$

As before, we assume that the intial state of the system at time $t = 0$ is known:

$$P_i(0) = 1$$
$$P_j(0) = 0 \quad \text{for } j \neq i$$

By inserting $s = 0$ in (6.77), we obtain

$$\begin{bmatrix} -a_{11} & a_{31} & \cdots & a_{r1} \\ -a_{13} & -a_{33} & \cdots & a_{r3} \\ \vdots & \vdots & \cdots & \vdots \\ -a_{1i} & a_{3i} & \cdots & a_{ri} \\ \vdots & \vdots & \cdots & \vdots \\ a_{1r} & a_{3r} & \cdots & -a_{rr} \end{bmatrix} \cdot \begin{bmatrix} P_1^*(s) \\ P_3^*(s) \\ \vdots \\ P_i^*(s) \\ \vdots \\ P_r^*(s) \end{bmatrix} = \begin{bmatrix} 0 \\ 0 \\ \vdots \\ -1 \\ \vdots \\ 0 \end{bmatrix} \quad (6.78)$$

Equation (6.78) can now be solved for $P_j^*(0)$ for $j = 1, 3, \ldots, r$ and the mean time to system failure MTTF$_S$ is from (6.74) determined by

$$\text{MTTF}_S = R^*(0) = \sum_{j \neq 0, 2} P_j^*(0) \quad (6.79)$$

Example 6.4

Reconsider the parallel structure of two independent components in Example 6.3, but now assume that the two components have different failure rates λ_1 and λ_2 and different repair rates μ_1 and μ_2. In this case we have to consider four different states as defined in Example 6.1 (Table 6.1). Furthermore the system is assumed to start out at time $t = 0$ in state 3 with both components functioning. The system is functioning as long as at least one of the components is functioning. The set B of functioning states is thus $\{1, 2, 3\}$. The system fails when both components are in a failed state, state 0.

Let us assume that we are primarily interested in determining the mean time to system failure MTTF$_S$. We therefore define state 0 to be an absorbing state, and set all departure rates from state 0 equal to zero.

Then the state equations become

$$\begin{bmatrix} 0 & \lambda_1 & \lambda_2 & 0 \\ 0 & -(\lambda_1+\mu_2) & 0 & \lambda_2 \\ 0 & 0 & -(\lambda_2+\mu_1) & \lambda_1 \\ 0 & \mu_2 & \mu_1 & -(\lambda_1+\lambda_2) \end{bmatrix} \cdot \begin{bmatrix} P_0(t) \\ P_1(t) \\ P_2(t) \\ P_3(t) \end{bmatrix} = \begin{bmatrix} \dot{P}_0(t) \\ \dot{P}_1(t) \\ \dot{P}_2(t) \\ \dot{P}_3(t) \end{bmatrix}$$

and the survivor function is

$$R(t) = P_1(t) + P_2(t) + P_3(t)$$

We now reduce the matrix equations by removing the row and the column corresponding to the absorbing state (state 0) and take Laplace transforms:

$$\begin{bmatrix} -(\lambda_1+\mu_2) & 0 & \lambda_2 \\ 0 & -(\lambda_2+\mu_1) & \lambda_1 \\ \mu_2 & \mu_1 & -(\lambda_1+\lambda_2) \end{bmatrix} \cdot \begin{bmatrix} P_1^*(0) \\ P_2^*(0) \\ P_3^*(0) \end{bmatrix} = \begin{bmatrix} 0 \\ 0 \\ -1 \end{bmatrix}$$

This means that

$$P_1^*(0) = \frac{\lambda_2}{\lambda_1+\mu_2} P_3^*(0) \qquad (6.80)$$

$$P_2^*(0) = \frac{\lambda_1}{\lambda_2+\mu_1} P_3^*(0) \qquad (6.81)$$

$$\left(\frac{\mu_2 \lambda_2}{\lambda_1+\mu_2} + \frac{\mu_1 \lambda_1}{\lambda_2+\mu_1} - (\lambda_1+\lambda_2) \right) P_3^*(0) = -1 \qquad (6.82)$$

The last equation leads to

$$P_3^*(0) = \frac{1}{\lambda_1 \lambda_2 [1/(\lambda_1+\mu_2) + 1/(\lambda_2+\mu_1)]} \qquad (6.83)$$

MEAN TIME TO FAILURE

Finally

$$\begin{aligned}\text{MTTF}_S &= R^*(0) = P_1^*(0) + P_2^*(0) + P_3^*(0) \\ &= \frac{\lambda_2/(\lambda_1+\mu_2) + \lambda_1/(\lambda_2+\mu_1) + 1}{\lambda_1\lambda_2[1/(\lambda_1+\mu_2) + 1/(\lambda_2+\mu_1)]}\end{aligned} \quad (6.84)$$

Then $P_1^*(0)$ and $P_2^*(0)$ are determined by inserting (6.83) in (6.80) and (6.81), respectively.

Some Special Cases

1. Non-repairable system ($\mu_1 = \mu_2 = 0$),

$$\text{MTTF}_S = \frac{(\lambda_2/\lambda_1) + (\lambda_1/\lambda_2) + 1}{\lambda_1 + \lambda_2}$$

When the two components have identical failure rates, $\lambda_1 = \lambda_2 = \lambda$, this expression is reduced to

$$\text{MTTF}_S = \frac{3}{2}\frac{1}{\lambda}$$

2. The two components have identical failure rates and identical repair rates ($\lambda_1 = \lambda_2 = \lambda$ and $\mu_1 = \mu_2 = \mu$). Then

$$\text{MTTF}_S = \frac{3}{2\lambda} + \frac{\mu}{2\lambda^2}$$

which is in accordance with Example 6.3. □

Example 6.5: *Common Cause Failures*
This example illustrates how a Markov model can be used to describe dependent failures. Dependent and common cause failures are further discussed in Chapter 8.

Consider a parallel structure of two identical components. The components may fail due to aging or other inherent defects. Such failures occur independent of each other with failure rate λ_I. The components are repaired independent of each other with repair rate μ.

An external event may occur that causes all functioning components to fail at the same time. Failures caused by the external event are called *common cause failures*. The external events occur with rate λ_C which is denoted the common cause failure rate.

The states of the system are named according to the number of components functioning. Thus the state space is $\{0, 1, 2\}$. The state space diagram of the parallel system with common cause failures is shown in Figure 6.9.

The corresponding state equations are

$$\begin{bmatrix} -2\mu & \lambda_C + \lambda_I & \lambda_F \\ 2\mu & -(\lambda_I + \lambda_C + \mu) & 2\lambda_I \\ 0 & \mu & -(2\lambda_I + \lambda_C) \end{bmatrix} \cdot \begin{bmatrix} P_0(t) \\ P_1(t) \\ P_2(t) \end{bmatrix} = \begin{bmatrix} \dot{P}_0(t) \\ \dot{P}_1(t) \\ \dot{P}_2(t) \end{bmatrix}$$

Assume that we are interested in determining the mean time to system failure MTTF$_S$. Since the system fails as soon as it enters state 0, we define state 0 as an absorbing state, and remove the row and the column from the transition rate matrix corresponding to state 0.

As before, we assume that the system is in state 2 (both components are functioning) at time $t = 0$. By introducing Laplace transforms, we get the following matrix equations:

$$\begin{bmatrix} -(\lambda_I + \lambda_C + \mu) & 2\lambda_I \\ \mu & -(2\lambda_I + \lambda_C) \end{bmatrix} \cdot \begin{bmatrix} P_1^*(0) \\ P_2^*(0) \end{bmatrix} = \begin{bmatrix} 0 \\ -1 \end{bmatrix}$$

The solutions are

$$P_1^*(0) = \frac{2\lambda_I}{(2\lambda_I + \lambda_C)(\lambda_I + \lambda_C) + \lambda_C \mu}$$

$$P_2^*(0) = \frac{\lambda_I + \lambda_C + \mu}{(2\lambda_I + \lambda_C)(\lambda_I + \lambda_C) + \lambda_C \mu}$$

and the mean time to system failure is

$$\text{MTTF}_S = P_2^*(0) + P_1^*(0) = \frac{3\lambda_I + \lambda_C + \mu}{(2\lambda_I + \lambda_C)(\lambda_I + \lambda_C) + \lambda_C \mu} \tag{6.85}$$

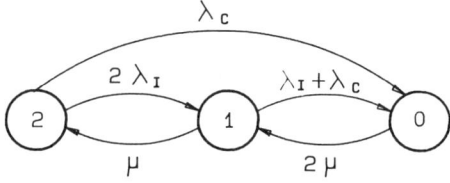

Figure 6.9 Markov diagram for a parallel system with two components exposed to common cause failures

Define a common cause factor β by

$$\beta = \frac{\lambda_C}{\lambda_C + \lambda_I} = \frac{\lambda_C}{\lambda}$$

Here $\lambda = \lambda_C + \lambda_I$ is the total failure rate of a component, and the factor β denotes the fraction of common cause failures among all failures of a component. To investigate how the common cause factor β affects the MTTF$_S$, we insert β and λ into (6.85) and get

$$\begin{aligned} \text{MTTF}_S &= \frac{3(1-\beta)\lambda + \beta\lambda + \mu}{(2(1-\beta)\lambda + \beta\lambda)\lambda + \beta\lambda\mu} \\ &= \frac{3 - 2\beta\lambda + \mu}{(2-\beta)\lambda^2 + \beta\lambda\mu} = \frac{1}{\lambda}\frac{\lambda(3-2\beta) + \mu}{(2-\beta)\lambda + \beta\mu} \end{aligned} \quad (6.86)$$

This equation may now be used to study how the MTTF$_S$ depends on the common cause factor β.

Let us consider two simple cases.

1. $\beta = 0$ (i.e., only *independent failures*, $\lambda = \lambda_I$):

$$\text{MTTF}_S = \frac{3}{2\lambda_I} + \frac{\mu}{2\lambda_I^2}$$

which is what we obtained in Example 6.3.

2. $\beta = 1$ (i.e., all failures are common cause failures, $\lambda = \lambda_C$):

$$\text{MTTF}_S = \frac{1}{\lambda_C}\frac{\lambda_C + \mu}{\lambda_C + \mu} = \frac{1}{\lambda_C}$$

The last result is evident. Only common cause failure are occurring and they affect both components simultaneously with failure rate λ_C. This β–factor model is further discussed in Chapter 8. □

Example 6.6: Sharing a Common Load

Consider a parallel system with two identical components. The components share a common load. If one component fails, the other component has to carry the whole load, and the failure rate of this component is assumed to increase immediately when the load is increased. Thus the failures of the two components are dependent. In Chapter 8 this type of dependency is referred to as *cascading failures.* The following failure rates are assumed:

λ_n = failure rate at normal load (i.e., when both components are functioning)
λ_f = failure rate at full load (i.e., when one of the components is failed)

Let μ_h denote the repair rate of a component when only one component has failed, and let μ_f denote the repair rate of a component when both components have failed. Let the number of components that are functioning denote the state of the system. The state space is thus $\{0, 1, 2\}$. The state–space diagram of the system is given in Figure 6.10.

The state equations are

$$\begin{bmatrix} -2\mu_f & \lambda_f & 0 \\ 2\mu_f & -(\mu_h + \lambda_f) & 2\lambda_n \\ 0 & \mu_h & -2\lambda_n \end{bmatrix} \cdot \begin{bmatrix} P_0(t) \\ P_1(t) \\ P_2(t) \end{bmatrix} = \begin{bmatrix} \dot{P}_0(t) \\ \dot{P}_1(t) \\ \dot{P}_2(t) \end{bmatrix}$$

The system fails when both components fail (i.e., in state 0). To determine the mean time to system failure MTTF$_S$, we define state 0 as an absorbing state and remove the row and the column corresponding to this state from the transition rate matrix. If we assume that the system starts out at time $t = 0$ with both components functioning (state 2), and take Laplace transforms with $s = 0$, we get

$$\begin{bmatrix} -(\mu_h + \lambda_f) & 2\lambda_n \\ \mu_h & -2\lambda_n \end{bmatrix} \cdot \begin{bmatrix} P_1^*(0) \\ P_2^*(0) \end{bmatrix} = \begin{bmatrix} 0 \\ -1 \end{bmatrix}$$

The solution is

$$P_1^*(0) = \frac{1}{\lambda_f}$$

$$P_2^*(0) = \frac{\lambda_f + \mu_h}{2\lambda_n \lambda_f}$$

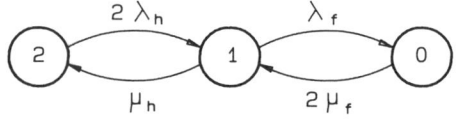

Figure 6.10 Parallel system with two components sharing a common load

The survivor function is $R(t) = P_1(t)+P_2(t)$, and the mean time to system failure is thus

$$\text{MTTF}_S = R^*(0) = P_1^*(0) + P_2^*(0) = \frac{2\lambda_n + \lambda_f + \mu_h}{2\lambda_n \lambda_f} \qquad (6.87)$$

□

6.7 STANDBY SYSTEMS

Standby systems were introduced in Section 4.6 where the survivor function $R(t)$ and the mean time to failure MTTF_S were determined for some simple nonrepairable standby systems. In the present section we will discuss some simple two-unit repairable standby systems using Markov models. The two-unit system considered is illustrated in Figure 6.11. Unit A is initially (at time $t = 0$) the operating unit and S is the sensing and changeover device.

A standby system can be operated and repaired in a number of different ways:

- The standby unit may be cold or partly loaded
- The changeover device may have several failure modes, like "fail to switch," "spurious switching," and "disconnect"
- Failure of the standby unit may be hidden (nondetectable) or detectable.

In the present section a few operation and repair modes of a standby system are illustrated. Generalizations to more complex systems and operational modes are often straightforward, at least in theory. The computations may, however, require a computer.

Parallel System with Cold Standby and Perfect Switching

Since the standby unit is passive, it is assumed not to fail in the standby state. The switching is assumed to be perfect. Then failure of the active unit is detected immediately, and the standby unit is activated with probability 1. The failure rate of unit i in operating state is denoted λ_i for $i = A, B$. When the active unit has failed, a repair action is initiated immediately. The time to

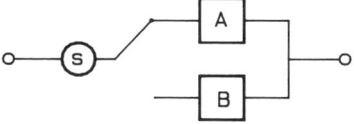

Figure 6.11 Two-unit standby system

repair is exponentially distributed with repair rate μ_i for $i = A, B$. When a repair action is completed, the unit is placed in standby state.

The possible states of the system are listed in Table 6.3 where O denotes operating state, S denotes standby state, and F denotes failed state. System failure occurs when the operating unit fails before repair of the other unit is completed. The failed state of the system is thus state 0 in Table 6.3. When both units have failed, they are repaired simultaneously, and the system is thus brought back to state 4. The repair rate in this case is denoted μ. The state-space diagram of the standby system is illustrated in Figure 6.12.

The state equations are

$$\begin{bmatrix} -\mu & \lambda_A & 0 & \lambda_B & 0 \\ 0 & -(\lambda_A + \mu_B) & \lambda_B & 0 & 0 \\ 0 & 0 & -\lambda_B & \mu_A & 0 \\ 0 & 0 & 0 & -(\lambda_B + \mu_A) & \lambda_A \\ \mu & \mu_B & 0 & 0 & -\lambda_A \end{bmatrix} \cdot \begin{bmatrix} P_0(t) \\ P_1(t) \\ P_2(t) \\ P_3(t) \\ P_4(t) \end{bmatrix} = \begin{bmatrix} \dot{P}_0(t) \\ \dot{P}_1(t) \\ \dot{P}_2(t) \\ \dot{P}_3(t) \\ \dot{P}_4(t) \end{bmatrix}$$

(6.88)

The time-dependent and steady state probabilities may be determined from (6.88) according to the procedures described in Section 6.4 and Section 6.5, respectively.

The survivor function $R(t)$ and the mean time to failure $MTTF_S$ of the system can be determined by considering the failed state of the system (state 0) to be an absorbing state. Suppose that the initial state at $t = 0$ is state 4. By deleting the row and the column of the transition rate matrix corresponding to the absorbing state 0 and taking Laplace transforms (with $s = 0$), we obtain

$$\begin{bmatrix} -(\lambda_A + \mu_B) & \lambda_B & 0 & 0 \\ 0 & -\lambda_B & \mu_A & 0 \\ 0 & 0 & -(\lambda_B + \mu_A) & \lambda_A \\ \mu_B & 0 & 0 & -\lambda_A \end{bmatrix} \cdot \begin{bmatrix} P_1^*(0) \\ P_2^*(0) \\ P_3^*(0) \\ P_4^*(0) \end{bmatrix} = \begin{bmatrix} 0 \\ 0 \\ 0 \\ -1 \end{bmatrix} \quad (6.89)$$

Table 6.3 Possible State of a Two-Unit Parallel System with Cold Standby and Perfect Switching

System State	State of Unit A	State of Unit B
4	O	S
3	F	O
2	S	O
1	O	F
0	F	F

STANDBY SYSTEMS

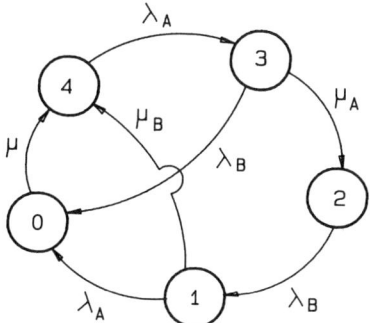

Figure 6.12 State-space diagram of a two-unit parallel system with cold standby and perfect switching

The solution is

$$P_2^*(0) = \frac{\lambda_A + \mu_B}{\lambda_B} P_1^*(0)$$

$$P_3^*(0) = \frac{\lambda_A + \mu_B}{\mu_A} P_1^*(0)$$

$$P_4^*(0) = \frac{\lambda_B + \mu_A}{\lambda_A} P_3^*(0)$$

$$= \frac{(\lambda_A + \mu_B)(\lambda_B + \mu_A)}{\lambda_A \mu_A} P_1^*(0)$$

$$= \frac{1 + \mu_B P_1^*(0)}{\lambda_A}$$

Thus

$$P_1^*(0) = \frac{\mu_A}{\lambda_A \lambda_B + \lambda_A \mu_A + \lambda_B \mu_B}$$

The mean time to failure MTTF$_S$ of the system is now, according to (6.74),

$$\text{MTTF}_S = R^*(0) = P_1^*(0) + P_2^*(0) + P_3^*(0) + P_4^*(0)$$

$$= \frac{1}{\lambda_A} + \frac{1}{\lambda_B} + \frac{\mu_A}{\lambda_B} \left(\frac{1}{\lambda_B} - \frac{1}{\lambda_B + \mu_A + \frac{\lambda_B}{\lambda_A}\mu_B} \right) \quad (6.90)$$

For a nonrepairable system, $\mu_A = \mu_B = 0$. Then

$$\text{MTTF}_S = \frac{1}{\lambda_A} + \frac{1}{\lambda_B}$$

which is an obvious result.

Parallel System with Cold Standby and Perfect Switching (Unit A is the Main Operating Unit)

Reconsider the standby system on page 251, but assume that unit A is the main operating unit. This means that unit B is only used when A is in a failed state and under repair. Unit A will thus be put into operation again as soon as the repair action is completed. System failure occurs when the operating unit B fails before repair of unit A is completed. The failed state of the system is thus state 0 in Table 6.3. When both units have failed, they are repaired simultaneously and brought back to state 4. The repair rate in this case is denoted μ. State 1 and state 2 in Table 6.3 are therefore irrelevant states for this system. The state–space diagram of this system is illustrated in Figure 6.13.

The state equations are:

$$\begin{bmatrix} -\mu & \lambda_B & 0 \\ 0 & -(\lambda_B + \mu_A) & \lambda_A \\ \mu & \mu_A & -\lambda_A \end{bmatrix} \cdot \begin{bmatrix} P_0(t) \\ P_3(t) \\ P_4(t) \end{bmatrix} = \begin{bmatrix} \dot{P}_0(t) \\ \dot{P}_3(t) \\ \dot{P}_4(t) \end{bmatrix} \qquad (6.91)$$

The steady state probabilities are determined by

$$\begin{bmatrix} -\mu & \lambda_B & 0 \\ 0 & -(\lambda_B + \mu_A) & \lambda_A \\ \mu & \mu_A & -\lambda_A \end{bmatrix} \cdot \begin{bmatrix} P_0 \\ P_3 \\ P_4 \end{bmatrix} = \begin{bmatrix} 0 \\ 0 \\ 0 \end{bmatrix}$$

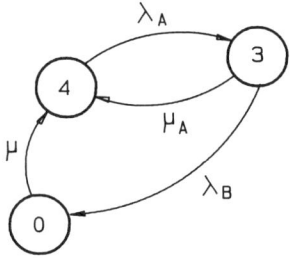

Figure 6.13 State-space diagram of a two-unit parallel system with cold standby and perfect switching (unit A is the main operating unit)

STANDBY SYSTEMS

and

$$P_0 + P_3 + P_4 = 1$$

The solution is

$$P_0 = \frac{\lambda_A \lambda_B}{\lambda_A \lambda_B + \lambda_A \mu + \lambda_B \mu + \mu \mu_A}$$

$$P_3 = \frac{\lambda_A \mu}{\lambda_A \lambda_B + \lambda_A \mu + \lambda_B \mu + \mu \mu_A}$$

$$P_4 = \frac{\lambda_B \mu + \mu \mu_A}{\lambda_A \lambda_B + \lambda_A \mu + \lambda_B \mu + \mu \mu_A}$$

where P_j is the mean proportion of time the system is spending in state j for $j = 0, 3, 4$.

The frequency of system failures ω_F is in this case equal to the visit frequency to state 0:

$$\omega_F = \nu_0 = \frac{P_0}{\mu}$$

The MTTF$_S$ of the system is determined as on page 253. By deleting the row and the column of the transition rate matrix in (6.91) and taking Laplace transforms (with $s = 0$), we obtain

$$\begin{bmatrix} -(\lambda_B + \mu_A) & \lambda_A \\ \mu_A & -\lambda_A \end{bmatrix} \cdot \begin{bmatrix} P_3^*(0) \\ P_4^*(0) \end{bmatrix} = \begin{bmatrix} 0 \\ -1 \end{bmatrix}$$

The solution is

$$P_3^*(0) = \frac{1}{\lambda_B}$$

$$P_4^*(0) = \frac{1}{\lambda_A} + \frac{\mu_A}{\lambda_A \lambda_B}$$

The mean time to failure of the system is thus

$$\text{MTTF}_S = R^*(0) = P_3^*(0) + P_4^*(0) = \frac{1}{\lambda_A} + \frac{1}{\lambda_B} + \frac{\mu_A}{\lambda_A \cdot \lambda_B} \quad (6.92)$$

The mean time to repair of the system is

$$\text{MTTR}_S = \frac{1}{\mu}$$

The average availability A of the system is

$$A = \frac{\text{MTTF}_S}{\text{MTTF}_S + \text{MTTR}_S} = \frac{1/\lambda_A + 1/\lambda_B + \mu_A/(\lambda_A\lambda_B)}{1/\lambda_A + 1/\lambda_B + \mu_A/(\lambda_A\lambda_B) + 1/\mu}$$

Parallel System with Cold Standby and Imperfect Switching (Unit A Is the Main Operating Unit)

Reconsider the standby system on page 254 but assume that the switching is no longer perfect. When the active unit A fails, the standby unit B will be activated properly with probability $(1 - p)$. The probability p may also include a "fail to start" probability of the standby unit. The state–space diagram of the system is illustrated in Figure 6.14. From state 4 the system may show a transition to state 3 with rate $(1 - p)\lambda_A$ and to state 0 with rate $p\lambda_A$

The steady state probabilities are determined by

$$\begin{bmatrix} -\mu & \lambda_B & p\lambda_A \\ 0 & -(\lambda_B + \mu_A) & (1-p)\lambda_A \\ \mu & \mu_A & -\lambda_A \end{bmatrix} \cdot \begin{bmatrix} P_0 \\ P_3 \\ P_4 \end{bmatrix} = \begin{bmatrix} 0 \\ 0 \\ 0 \end{bmatrix} \quad (6.93)$$

and

$$P_0 + P_3 + P_4 = 1$$

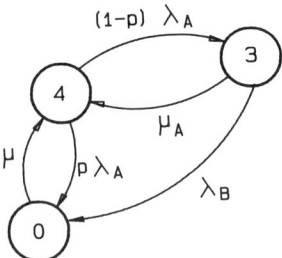

Figure 6.14 State-space diagram of a two-unit parallel system with cold standby and imperfect switching (Unit A is the main operating unit)

STANDBY SYSTEMS

The solution is

$$P_0 = \frac{\lambda_A \lambda_B + p\lambda_A \mu_A}{\lambda_A \lambda_B + p\lambda_A \mu_A + (1-p)\lambda_A \mu + \lambda_B \mu + \mu \mu_A}$$

$$P_3 = \frac{\lambda_A \mu (1-p)}{\lambda_A \lambda_B + p\lambda_A \mu_A + (1-p)\lambda_A \mu + \lambda_B \mu + \mu \mu_A}$$

$$P_4 = \frac{\lambda_B \mu + \mu \mu_A}{\lambda_A \lambda_B + p\lambda_A \mu_A + (1-p)\lambda_A \mu + \lambda_B \mu + \mu \mu_A}$$

The MTTF_S can be determined from

$$\begin{bmatrix} -(\lambda_B + \mu_A) & (1-p)\lambda_A \\ \mu_A & -\lambda_A \end{bmatrix} \cdot \begin{bmatrix} P_3^*(0) \\ P_4^*(0) \end{bmatrix} = \begin{bmatrix} 0 \\ -1 \end{bmatrix}$$

which leads to

$$P_3^*(0) = \frac{1-p}{\lambda_B + p\mu_A}$$

$$P_4^*(0) = \frac{\lambda_B + \mu_A}{\lambda_A(\lambda_B + p\mu_A)}$$

Thus

$$\mathrm{MTTF}_S = R^*(0) = P_3^*(0) + P_4^*(0) = \frac{(1-p)\lambda_A + \lambda_B + \mu_A}{\lambda_A(\lambda_B + p\mu_A)} \qquad (6.94)$$

Parallel System with Partly Loaded Standby and Perfect Switching (Unit *A* Is the Main Operating Unit)

Reconsider the standby system on page 254, but assume that the standby unit *B* may fail in standby mode and have a hidden failure when activated. The failure rate of unit *B* in standby mode is denoted λ_B^s and is normally less than the corresponding failure rate during operation. In addition to the transition in Figure 6.13, this system may also have transitions from state 4 to state 1 (in Table 6.3) and from state 1 to state 0. The state-space diagram is illustrated in Figure 6.15.

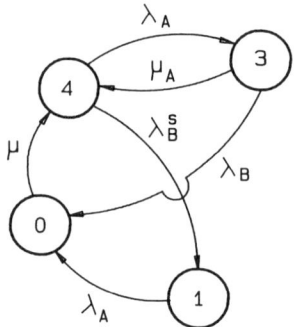

Figure 6.15 State-space diagram of a two-unit parallel system with partly loaded standby and perfect switching (unit A is the main operating unit)

The steady state probabilities are determined by

$$\begin{bmatrix} -\mu & \lambda_A & \lambda_B & 0 \\ 0 & -\lambda_A & 0 & \lambda_B^s \\ 0 & 0 & -(\lambda_B + \mu_A) & \lambda_A \\ \mu & 0 & \mu_A & -(\lambda_A + \lambda_B^s) \end{bmatrix} \cdot \begin{bmatrix} P_0(t) \\ P_1(t) \\ P_3(t) \\ P_4(t) \end{bmatrix} = \begin{bmatrix} 0 \\ 0 \\ 0 \\ 0 \end{bmatrix} \quad (6.95)$$

and

$$P_0 + P_1 + P_3 + P_4 = 1$$

The MTTF_S can be determined from

$$\begin{bmatrix} -\lambda_A & 0 & \lambda_B^s \\ 0 & -(\lambda_B + \mu_A) & \lambda_A \\ 0 & \mu_A & -(\lambda_A + \lambda_B^s) \end{bmatrix} \cdot \begin{bmatrix} P_1^*(0) \\ P_3^*(0) \\ P_4^*(0) \end{bmatrix} = \begin{bmatrix} 0 \\ 0 \\ -1 \end{bmatrix}$$

$$P_1^*(0) = \frac{\lambda_B^s(\lambda_B + \mu_A)/\lambda_A}{\lambda_A \lambda_B + \lambda_B \lambda_B^s + \lambda_B^s \mu_A}$$

$$P_3^*(0) = \frac{\lambda_A}{\lambda_A \lambda_B + \lambda_B \lambda_B^s + \lambda_B^s \mu_A}$$

$$P_1^*(0) = \frac{\lambda_B + \mu_A}{\lambda_A \lambda_B + \lambda_B \lambda_B^s + \lambda_B^s \mu_A}$$

PROBLEMS

Thus

$$\mathrm{MTTF}_S = R^*(0) = P_1^*(0) + P_3^*(0) + P_4^*(0)$$

$$= \frac{\left(\dfrac{\lambda_B^s}{\lambda_A} + 1\right)(\lambda_B + \mu_A) + \lambda_A}{\lambda_A \lambda_B + \lambda_B \lambda_B^s + \lambda_B^s \mu_A} \tag{6.96}$$

6.8 PROBLEMS

1. A fail-safe valve has two main failure modes: premature/spurious closure (PC) and fail to close (FTC), with constant failure rates

$$\lambda_{PC} = 10^{-3} \quad \text{PC failures per hour}$$
$$\lambda_{FTC} = 2 \cdot 10^{-4} \quad \text{FTC failures per hour}$$

The mean time to repair a PC failure is assumed to be 1 hour, while the mean time to repair an FTC failure is 24 hours. The repair times are assumed to be exponentially distributed.

(a) Explain why the operation of the valve may be described by a Markov process with three states. Establish the state-space diagram and the state equations for this process.

(b) Calculate the average availability of the valve, and the mean time between failures.

2. Two identical pumps are operated as a parallel system. During normal operation both pumps are functioning. When the first pump fails, the other pump has to do the whole job alone with a higher load than when both pumps are in operation. The pumps are assumed to have constant failure rates:

$\lambda_H = 1.5 \cdot 10^{-4}$ failure per hour (the failure rate when the pumps are sharing the load, i.e., "half load")

$\lambda_F = 3.5 \cdot 10^{-4}$ failure per hour (the failure rate at "full load" when one of the pumps are in a failed state)

Both pumps may fail at the same time due to some external stresses (called *common cause failure*; see Chapter 8). The failure rate with respect to common cause failures has been estimated to be $\lambda^f = 3.0 \cdot 10^{-5}$ common cause failures per hour. This type of external stresses affects the system at a rate λ^f irrespective of how many of its units that are functioning. The common cause failure rate must therefore be added to the "independent" failure rate also when only one of the pumps is functioning.

Repair is initiated as soon as one of the pumps fails. The mean downtime of a pump has been estimated to be 15 hours. When both pumps are in a failed state, the whole process system has to be shut down. In this case the system will not be put into operation again until both pumps have been repaired. The mean downtime, when both pumps have failed, has been estimated to be 25 hours.

(a) Establish a state-space (Markov) diagram for the system consisting of the two pumps.

(b) Write down the state equations for the system in matrix format.

(c) Explain what is meant by the steady state probabilities, and determine the steady state probabilities for each of the states of the pump system.

(d) Determine the percentage of time when

   (i) Both the pumps are functioning

   (ii) Only one of the pumps is functioning

   (iii) Both pumps are in a failed state.

(e) Determine the mean number of pump repairs that are necessary during a period of 5 years.

(f) How many times must we expect to have a total pump failure (i.e., both pumps in a failed state at the same time) during a period of 5 years.

3. The water chlorination system of a small town has two separate pipelines, each with a pump that supplies chlorine to the water at prescribed rates. The two pumps are denoted A and B, respectively. During normal operation both pumps are functioning and thus are sharing the load. In this case each pump is operated on approximately 60% of its capacity (cap% = 0.60). When one of the pumps fails, the corresponding pipeline is closed down, and the other pump has to supply clorine at a higher rate. In this case the single pump is operated at full capacity (cap% = 1.00). We assume that the pumps have the following constant failure rates:

$$\lambda_{cap\%} = cap\% \cdot 6.3 \quad \text{failures per year}$$

Assume that the probability of common cause failures is negligible. Repair is initiated as soon as one of the pumps fails. The mean time to repair a pump has been estimated to be eight hours, and the pump is put into operation again as soon as the repair is completed. Repairs are carried out independent of each other (i.e., maintenance crew is not a limiting factor). If both pumps are in a failed state, unchlorinated water will be supplied to the customers. Both pumps are assumed to be functioning at time $t = 0$.

(a) Define the possible system states, and establish a state space (Markov) diagram for the system.

(b) Write down the corresponding state equations on matrix format.

(c) Determine the steady state probabilities for each of the system states.

(d) Determine the mean number of pump repairs during a period of 3 years.

(e) Determine the percentage of time exactly one of the pumps is in a failed state.

(f) Determine the mean time to the first system failure (i.e., the mean time until unchlorinated water is supplied to the customers for the first time after time $t = 0$).

(g) Determine the percentage of time unchlorinated water is supplied to the customers.

4. Consider a parallel structure of three independent and identical components with failure rate λ and repair rate μ. The components are repaired independently. All the three components are assumed to be functioning at time $t = 0$.

   (a) Establish the state-space diagram and the state equations for the parallel structure.

   (b) Show that the mean time to the first system failure MTTF is given by

   $$\text{MTTF} = \frac{11}{6\lambda} + \frac{7\mu}{6\lambda^2} + \frac{\mu^2}{3\lambda^3} \qquad (6.97)$$

5. Consider a parallel structure of four independent and identical components with failure rate λ and repair rate μ. The components are repaired independently. All the three components are assumed to be functioning at time $t = 0$.

   (a) Establish the state-space diagram and the state equations for the parallel structure.

   (b) Determine the mean time to the first system failure MTTF.

   (c) Is it possible to find a general formula for a parallel structure of n components?

6. Figure 6.16 illustrates a standby system of two compressors and a switching unit. When the active compressor A fails, the standby compressor B is to be put into operation. Compressor B is assumed not to fail while in passive state (i.e., cold standby).

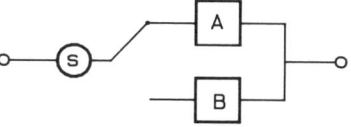

Figure 6.16 Standby system of two compressors

The probability of successful changeover to the standby compressor B is estimated to be $(1 - p)$. The probability p of unsuccessful changeover also includes the fail-to-start probability for the standby compressor. Compressor A is the main compressor. When we have a system failure (i.e., when both compressors are in a failed state), both compressors are repaired simultaneously, and the system is not started until both compressors have been repaired. The system is always started up again with compressor A as the active compressor. Common cause failures are considered to be negligible.

(a) Define the possible system states, and establish a state space (Markov) diagram.

(b) The following input data are assumed

Failure rate for compressor A: $\lambda_A = 5.0 \cdot 10^{-4}$ failures per hour
Failure rate for compressor B: $\lambda_B = 2.0 \cdot 10^{-3}$ failures per hour
Mean time to repair compressor A: $1/\mu_A = 25$ hours
Mean time to repair both the compressors: $1/\mu = 35$ hours
Probability of unsuccessful changeover: $p = 0.03$

Establish the state equations for the system and determine the steady state probabilities.

(c) Determine the average availability A_{av} for the system.

(d) How many compressor repairs may we anticipate over a period of 5 years?

(e) How many system failures may we anticipate over a period of 5 years?

(f) Determine the mean time to the first system failure, when both compressors are functioning at time $t = 0$ with compressor A as the operating compressor.

(g) Assume next that we are considering an alternative system where both compressors are operated as an ordinary parallel system (i.e., with active redundancy). All the input data are as above, except that when the two compressors share the load, their failure rates are reduced with 20%. The mean time to repair compressor B is assumed to be 20 hours. Establish the state-space (Markov) diagram for the alternative system, and compute the availability A_{av}. Discuss the result.

(h) Determine the mean time to the first system failure for the alternative system of (g).

7. Consider a system that is subject to two types of repair. Initially the system has a constant failure rate λ_1. When the system fails for the first time, a partial repair is performed to restore the system to the functioning state. This partial repair is not perfect, and the failure rate λ_2 after this partial repair is therefore larger than λ_1. After the system fails the second time, a thorough repair is performed that restores the system to an "as good as new" condition. The third repair will be a partial repair, and so on. Let μ_1 denote the constant repair rate of a partial repair, and let μ_2 be the constant repair rate of a complete repair ($\mu_1 > \mu_2$). Assume that the system is put into operation at time $t = 0$ in an "as good as new" condition.

(a) Establish the state-space diagram and the state equations for this process.

(b) Determine the steady state probabilities of the various states.

CHAPTER 7

Counting Processes

7.1 INTRODUCTION

In this chapter we consider a repairable system (subsystem or component) that is put into operation at time $t = 0$. The first failure of the system will occur at time S_1. When the system has failed, it will be replaced or restored to a functioning state. The repair time is assumed to be so short that it may be neglected. The second failure will occur at time S_2, and so on. We thus get a sequence of failure times $S_1, S_2, S_3, \ldots$. Let T_j for $j = 1, 2, \ldots$ be the time between failure $j - 1$ and failure j, where S_0 is taken to be 0. T_j will be called the *interarrival time j* for $j = 1, 2, \ldots$.

Throughout this chapter t denotes a specified point of time, irrespective whether t is *calendar time* (a realization of S_j) or *local time* (a realization of an interarrival time T_j). We hope that this convention will not confuse the reader. The sequence of interarrival times, $T_1, T_2, T_3, \ldots$ will generally not be independent and identically distributed—unless the system is replaced upon failure, or restored to an "as good as new" condition, and the environmental and operational conditions remain constant throughout the whole period.

A random variable of special interest is $N(t)$, the number of failures in the time interval $(0, t]$. The process $\{N(t), t \geq 0\}$ is called a stochastic process, or more specifically a *counting process*. A precise definition of a counting process is given below (from Ross 1983, p. 31).

Definition 7.1. A stochastic process $\{N(t), t \geq 0\}$ is said to be a counting process if $N(t)$ satisfies:

1. $N(t) \geq 0$
2. $N(t)$ is integer valued
3. If $s < t$ then $N(s) \leq N(t)$
4. For $s < t$, $[N(t) - N(s)]$ represents the number of failures that have occurred in the interval $(s, t]$ □

An excellent introduction to stochastic, and counting processes is presented by Ross (1983). A more advanced presentation is given by Cox and Isham (1980).

Example 7.1

The following failure times (calendar time in days) are presented by Ascher and Feingold (1984, p. 79). The data set is recorded from time $t = 0$ until 7 failures have been recorded during a total time of 410 (days). The data come from a single system, and the repair times are assumed to be negligible. This means that the system is assumed to be functioning again almost immediately after a failure is encountered.

Number of failures $N(t)$	Calendar time S_j	Interarrival time T_j
0	0	0
1	177	177
2	242	65
3	293	51
4	336	43
5	368	32
6	395	27
7	410	15

The data are illustrated in Figure 7.1. The interarrival times are seen to become shorter with time. The system seems to be deteriorating, and failures tend to become more frequent. A system with this property is called a *sad* system by Ascher and Feingold (1984), for obvious reasons. A system with the opposite property, where failures become less frequent with operating time, is called a *happy* system.

The number of failures $N(t)$ may also be illustrated as a function of (calendar) time t, in Figure 7.2. Note that by definition $N(t)$ is constant between failures and jumps (a height of 1 unit) at the failure times S_i for $i = 1, 2, \ldots$. It is thus sufficient to plot the jumping points $(S_i, N(S_i))$ for $i = 1, 2, \ldots$. The plot is called an $N(t)$ plot, or a Nelson-Aalen plot (see p. 317).

Note that $N(t)$ as a function of t will tend to be a convex function when the system is *sad*. In the same way $N(t)$ will tend to be a concave function of t when the system is *happy*. If $N(t)$ is (approximately) linear, the system is steady; that is, the interarrival times will have the same expected length. In Figure 7.2 $N(t)$ is clearly seen to be convex. Thus the system is *sad*. □

Example 7.2: Compressor Failure Data

Failure time data for a specific compressor at a Norwegian process plant have been collected as part of a student thesis at the Norwegian Institute of Technology. All compressor failures in the time period from 1968 until 1989 have been recorded. In this period a total of 321 failures occurred. Ninety of these

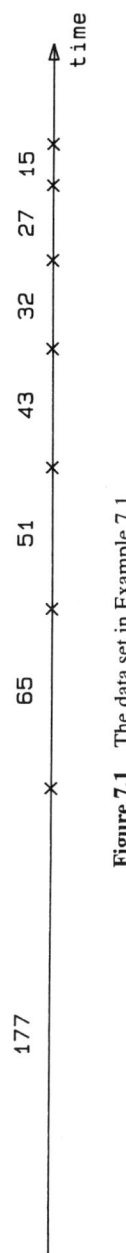

Figure 7.1 The data set in Example 7.1

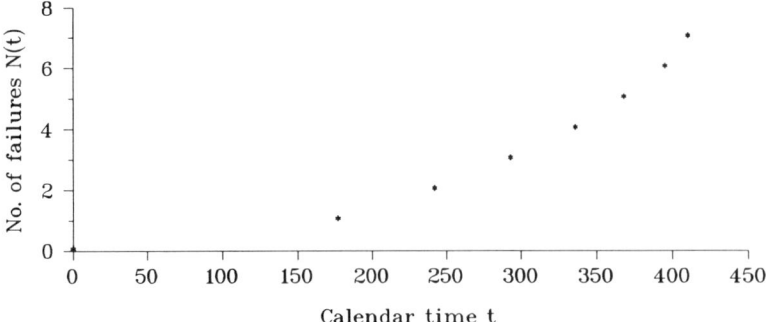

Figure 7.2 Number of failures $N(t)$ as a function of time for the data in Example 7.1

failures were critical failures, and 231 failures were noncritical. In this context a critical failure is defined to be a failure causing compressor downtime. Noncritical failures may be corrected without the compressor having to close down. The majority of the noncritical failures were instrument failures, and failures of the seal oil system and the lubrication oil system.

As above, let $N(t)$ denote the number of compressor failures in the time interval $(0, t]$. An $N(t)$ plot with respect to all the 321 failures is presented in Figure 7.3. In the figure the time t denotes the *operating* time, which means that the downtimes caused by compressor failures and process shutdowns are not included.

We observe that the $N(t)$ plot is approximately linear. The rate of compres-

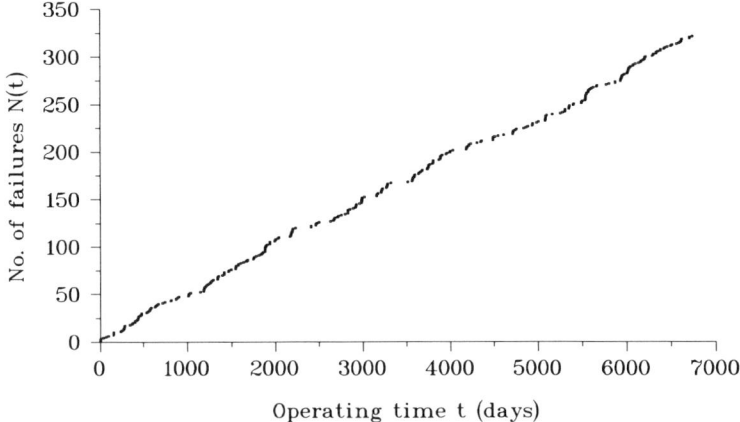

Figure 7.3 Number of compressor failures $N(t)$ as a function of time (days) with respect to both critical and noncritical failures (totally 321 failures)

INTRODUCTION

Table 7.1 Failure Times (Operating Days) in Chronological Order

1.0	4.0	4.5	92.0	252.0	277.0
277.5	284.5	374.0	440.0	444.0	475.0
536.0	568.0	744.0	884.0	904.0	1017.5
1288.0	1337.0	1338.0	1351.0	1393.0	1412.0
1413.0	1414.0	1546.0	1546.5	1575.0	1576.0
1666.0	1752.0	1884.0	1884.2	1884.4	1884.6
1884.8	1887.0	1894.0	1907.0	1939.0	1998.0
2178.0	2179.0	2188.5	2195.5	2826.0	2847.0
2914.0	3156.0	3156.5	3159.0	3211.0	3268.0
3276.0	3277.0	3321.0	3566.5	3573.0	3594.0
3640.0	3663.0	3740.0	3806.0	3806.5	3809.0
3886.0	3886.5	3892.0	3962.0	4004.0	4187.0
4191.0	4719.0	4843.0	4942.0	4946.0	5084.0
5084.5	5355.0	5503.0	5545.0	5545.2	5545.5
5671.0	5939.0	6077.0	6206.0	6206.5	6305.0

sor failures therefore seems to be constant, and the failure process is neither *sad* nor *happy* when regarding the pooled set of critical and noncritical failures.

From a production regularity point of view, the critical failures are the most important, since these failures are causing process shutdown. The operating times (in days) at which the 90 critical failures occurred are listed in Table 7.1.

An $N(t)$ plot with respect to the 90 critical failures is presented in Figure 7.4. In this case the $N(t)$ plot is slightly concave, which indicates a *happy* system. The time between critical failures hence seems to increase with the time in

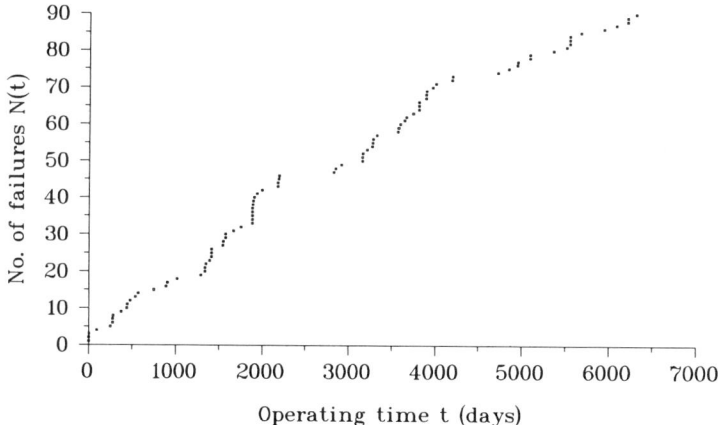

Figure 7.4 Number of critical compressor failures $N(t)$ as a function of time (days), totaling 90 failures

operation. Several failures have also occurred within short intervals. This may indicate that the failures are dependent or that the maintenance crew has not been able to correct the failures properly at the first attempt. Hence the times between failures (interarrival times) seem not to be independent and identically distributed.

Since the $N(t)$ plot of all failures is approximately linear, and the $N(t)$ plot of the critical failures is concave, indicating a decreasing rate of critical failures, the rate of noncritical failures has to be increasing. A more thorough analysis of the complete data set shows that this is correct. Both the instrumentation system and the seal oil system show an increasing rate with respect to noncritical failures. □

An analysis of life data from a repairable system should always be started by establishing an $N(t)$ plot. If $N(t)$ as a function of the time t is nonlinear, any methods based on the assumption of independent and identically distributed times between failures are obviously not appropriate. It is, however, not certain that such methods are appropriate even if the $N(t)$ plot is very close to a straight line. The interarrival times, for example, may be strongly correlated. Methods to check whether the interarrival times are correlated or not, are discussed, for example, by Ascher and Feingold (1984), and Bendell and Walls (1985). The $N(t)$ plot is further discussed in Section 7.4.

Some Basic Definitions

When considering counting processes, the following concepts are of special interest:

- *Independent increments.* A counting process $\{N(t), t \geq 0\}$ is said to have independent increments if for $0 < t_1 < \cdots < t_k, k = 2, 3, \ldots$ $[N(t_1) - N(0)]$, $[N(t_2) - N(t_1)], \ldots, [N(t_k) - N(t_{k-1})]$ are all independent random variables. In that case the number of failures in an interval is not influenced by the number of failures in any strictly earlier interval (i.e., with no overlap). This means that even if the system has experienced an unusual high number of failures in a certain time interval, these failures will not influence the distribution of future failures.
- *Stationary increments.* A counting process is said to have stationary increments if for any two disjoint time points $t > s \geq 0$ and any constant $c > 0$, the random variables $[N(t) - N(s)]$ and $[N(t+c) - N(s+c)]$ are identically distributed. This means that the distribution of the number of events in a time interval depends only on the length of the interval and not on the interval's distance from the origin.
- *Stationary process.* A counting process is said to be stationary (or homogeneous) if it has stationary increments.
- *Nonstationary increments.* A counting process is said to be nonstationary

INTRODUCTION

(or nonhomogeneous) if it is neither stationary nor eventually becomes stationary.

- *Regular process.* A counting process is said to be regular (or orderly) if

$$P(N(t + \Delta t) - N(t) \geq 2) = o(\Delta t) \qquad (7.1)$$

when Δt is small, and $o(\Delta t)$ denotes a function of Δt with the property that $\lim_{\Delta t \to 0} o(\Delta t)/\Delta t = 0$. In practice this means that the system will not experience two or more failures simultaneously.

- *ROCOF.* The rate of occurrence of failures (ROCOF) of a counting process at time t is defined as

$$w(t) = W'(t) = \frac{d}{dt} E(N(t)) \qquad (7.2)$$

where $W(t) = E(N(t))$ denotes the mean number of failures in the interval $(0, t]$. Thus

$$w(t) = W'(t) = \lim_{\Delta t \to 0} \frac{E(N(t + \Delta t) - N(t))}{\Delta t} \qquad (7.3)$$

and when Δt is small,

$$\begin{aligned} w(t) &\approx \frac{E(N(t + \Delta t) - N(t))}{\Delta t} \\ &= \frac{\text{Mean number of failures in } (t, t + \Delta t]}{\Delta t} \end{aligned}$$

Thus a natural estimator of $w(t)$ is

$$\hat{w}(t) = \frac{\text{Number of failures in } (t, t + \Delta t]}{\Delta t} \qquad (7.4)$$

for some suitable Δt. It follows that the ROCOF, $w(t)$, may be regarded as the mean number of failures per time unit at time t.

When we are dealing with a *regular* process, the probability of 2 or more failures in $(t, t + \Delta t]$ is negligible when Δt is small. Thus for small Δt we may assume that

$$N(t + \Delta t) - N(t) = 0 \text{ or } 1.$$

Thus the mean number of failures in $(t, t + \Delta t]$ is approximately equal to the probability of failure in $(t, t + \Delta t]$, and

$$w(t) \approx \frac{\text{Probability of failure in } (t, t + \Delta t]}{\Delta t} \qquad (7.5)$$

Hence $w(t)\Delta t$ can be interpreted as the probability of failure in the time interval $(t, t + \Delta t]$. Observe also that

$$E(N(t_0)) = W(t_0) = \int_0^{t_0} w(t) \, dt \qquad (7.6)$$

- *Time between failures.* We have denoted the time T_j between failure $j-1$ and failure j, for $j = 1, 2, \ldots$ the interarrival times. For a general counting process the interarrival times will neither be identically distributed nor independent. Hence the mean time between failures, $\text{MTBF}_j = E(T_j)$, will in general be a function of j and $T_1, T_2, \ldots, T_{j-1}$.
- *Forward recurrence time.* The forward recurrence time $Y(t)$ is the time to the next failure measured from an arbitrary point of time t. Thus $Y(t) = S_{N(t)+1} - t$. The forward recurrence time is also called the *remaining lifetime* or the *excess life*. The forward recurrence time is illustrated in Figure 7.5.

Three Types of Counting Processes

In this chapter three common types of counting processes are discussed.

1. Homogeneous Poisson processes (HPP)
2. Renewal processes
3. Nonhomogeneous Poisson processes (NHPP)

The homogeneous Poisson process (HPP) was introduced in Section 2.5. In the HPP model all the interarrival times are independent and exponentially distributed with the same parameter (failure rate) λ.

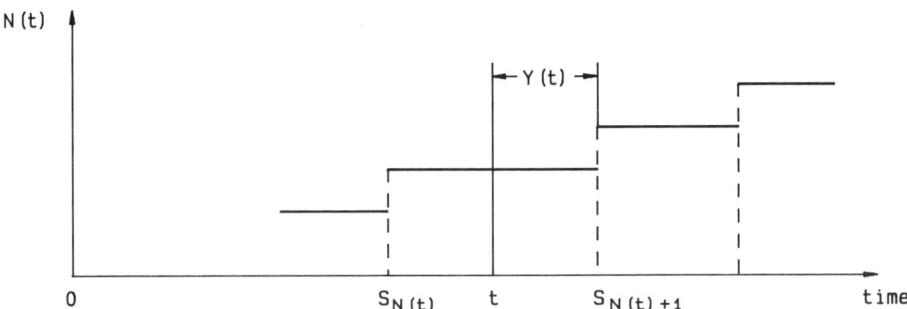

Figure 7.5 The forward recurrence time $Y(t)$

The renewal process as well as the nonhomogeneous Poisson process (NHPP) are generalizations of the HPP, both having the HPP as a special case. A renewal process is a counting process where the interarrival times are independent and identically distributed with an arbitrary distribution. Upon failure the component is thus replaced or restored to an "as good as new" condition. The analysis of observed interarrival times from a renewal process is discussed in detail in Chapter 9.

The NHPP differs from the HPP only in that the rate of occurrences of failures varies with time rather than being a constant. This implies that for an NHPP model the interarrival times are neither independent nor identically distributed. The NHPP is often used to model repairable systems that are subject to a *minimal repair* strategy, with negligible repair times. Minimal repair means that a failed system is restored just back to functioning state. After a minimal repair the system continues as if nothing had happened. The likelihood of system failure is the same immediately before and after a failure. A minimal repair thus restores the system to an "as bad as old" condition. The minimal repair strategy is discussed, for example, by Aven (1983), Ascher and Feingold (1984), and Akersten (1991) who gives a detailed list of relevant references on this subject.

The renewal process and the NHPP represent two extreme types of repair: replacement to an "as good as new" condition and replacement to "as bad as old" (minimal repair), respectively. Most repair actions are, however, somewhere between these extremes, and are often called *imperfect repair*. A number of different models have been proposed for imperfect repair. One of the best-known models is described by Brown and Proschan (1983). Further references are given by Akersten (1991). In the Brown-Proschan imperfect repair model the system is restored to an "as good as new" condition with probability p and to an "as bad as old" condition with probability $1 - p$. Each repair is thus a "weighted average" of replacement and a minimal repair. A high number of research projects are presently carried out within modeling and analysis of systems subject to imperfect repair. The topic is not further discussed in this book, and the reader is referred to the research literature, for example, the references cited in Akersten (1991).

7.2 HOMOGENEOUS POISSON PROCESSES

Definition and Basic Results

The homogeneous Poisson process (HPP) was introduced in Section 2.5. The HPP may be defined in a number of different ways. An alternative to the definition in Section 2.5 is the following definition from Ross (1983, p. 31).

Definition 7.2. The counting process $\{N(t), t \geq 0\}$ is said to be a homogeneous Poisson process, HPP, having rate λ, for $\lambda > 0$, if

1. $N(0) = 0$
2. The process has independent increments
3. The number of events in any interval of length t is Poisson distributed with mean λt. That is, for all $s, t > 0$,

$$P(N(t+s) - N(s) = n) = \frac{(\lambda t)^n}{n!} e^{-\lambda t} \quad \text{for} \quad n = 0, 1, 2, \ldots \quad (7.7)$$

□

Note that it follows from property 3 that an HPP has stationary increments and also that $E(N(t)) = \lambda t$, which explains why λ is called the rate of the process.

An alternative, and identical definition of the HPP is according to Ross (1983, p. 32):

Definition 7.3. The counting process $\{N(t), t \geq 0\}$ is said to be a homogeneous Poisson process, HPP, having rate λ, for $\lambda > 0$, if

1. $N(0) = 0$
2. The process has stationary and independent increments
3. $P(N(\Delta t) = 1) = \lambda \Delta t + o(\Delta t)$
4. $P(N(\Delta t) \geq 2) = o(\Delta t)$

□

These two alternative definitions of the HPP are presented to clarify the analogy to the definition of the nonhomogeneous Poisson process (NHPP), which is presented in Section 7.4.

The main features of the HPP were established in Section 2.5. In the present context we consider six important features:

1. The HPP is a regular (orderly) counting process with independent and stationary increments.
2. The rate of occurrence of failures (ROCOF) of the HPP is constant and independent of time,

$$w(t) = \lambda \quad \text{for all } t \geq 0 \quad (7.8)$$

3. The number of failures in the interval $(t_1, t_2]$ is Poisson distributed with mean $\lambda(t_2 - t_1)$,

$$P(N(t_2) - N(t_1) = n) = \frac{[\lambda(t_2 - t_1)]^n}{n!} e^{-\lambda(t_2 - t_1)}$$
$$\text{for all } t_2 > t_1 > 0 \quad (7.9)$$

4. The mean number of failures in the time interval $(t_1, t_2]$ is

$$W(t_2) - W(t_1) = E(N(t_2) - N(t_1)) = \lambda(t_2 - t_1) \tag{7.10}$$

5. The interarrival times $T_1, T_2, \ldots$ are independent and identically distributed exponential random variables having mean $1/\lambda$ (see Ross 1983, p. 35).

6. The arrival time of the rth failure $S_r = \sum_{j=1}^{r} T_j$ has a gamma distribution with parameters r and λ. Its probability density function is

$$f(t) = \frac{(\lambda t)^{r-1}}{(r-1)!} \lambda e^{-\lambda t} \quad \text{for} \quad t \geq 0 \tag{7.11}$$

Other features of the HPP are presented and discussed, for example, by Ross (1983), Thompson (1988), and Ascher and Feingold (1984).

Compound Poisson Processes

Consider a Poisson process $\{N(t), t \geq 0\}$ with rate λ. To failure (event) i let there be associated an independent random variable V_i for $i = 1, 2, \ldots$. The variable V_i may, for example, be the consequence (economic loss) associated to failure i. The variables $V_1, V_2, \ldots$ are furthermore assumed to be independent random variables with common distribution function

$$F_V(v) = P(V \leq v)$$

The cumulative consequence at time t is hence

$$Z(t) = \sum_{i=1}^{N(t)} V_i \quad \text{for} \quad t \geq 0 \tag{7.12}$$

The process $\{Z(t), t \geq 0\}$ is called a *compound Poisson process*. Compound Poisson processes are discussed, for example, by Ross (1983, p. 48) and Taylor and Karlin (1984, p. 200). The same model is called a *cumulative damage model* by Barlow and Proschan (1975, p. 91). To determine the mean value of $Z(t)$, we need the following important theorem:

Theorem 7.1 (Wald's Equation). Let $X_1, X_2, X_3, \ldots$ be independent and identically distributed random variables with finite mean μ. Further let N be a stochastic integer variable so that the event $(N = n)$ is independent of $X_{n+1}, X_{n+2}, \ldots$ for all $n = 1, 2, \ldots$. Then

$$E\left(\sum_{i=1}^{N} X_i\right) = E(N) \cdot \mu \qquad (7.13)$$

A proof of Wald's equation may be found, for example, in Barlow and Proschan (1975, p. 170).

Let $E(V_i) = \nu$ and $\text{var}(V_i) = \tau^2$. According to Wald's equation

$$E(Z(t)) = \nu\lambda t \quad \text{and} \quad \text{var}(Z(t)) = \lambda(\nu^2 + \tau^2)t$$

Assume now that the consequences V_i are all positive, that is, $P(V_i > 0) = 1$ for all i, and that a total system failure occurs as soon as $Z(t) > c$ for some specified critical value c. Let T_c denote the time to system failure. Note that $T_c > t$ if and only if $Z(t) \leq c$.

Let $V_0 = 0$, then

$$\begin{aligned}
P(T_c > t) &= P(Z(t) \leq c) = P\left(\sum_{i=0}^{N(t)} V_i \leq c\right) \\
&= \sum_{n=0}^{\infty} P\left(\sum_{i=0}^{n} V_i \leq c \mid N(t) = n\right) \frac{(\lambda t)^n}{n!} e^{-\lambda t} \\
&= \sum_{n=0}^{\infty} \frac{(\lambda t)^n}{n!} e^{-\lambda t} F_V^{(n)}(c) \qquad (7.14)
\end{aligned}$$

where $F_V^{(n)}(v)$ denotes the distribution function of $\sum_{i=0}^{n} V_i$, and the last equality is due to the fact that $N(t)$ is independent of $V_1, V_2, \ldots$.

The mean time to total system failure is thus

$$\begin{aligned}
E(T_c) &= \int_0^{\infty} P(T_c > t)\,dt \\
&= \sum_{n=0}^{\infty} \left(\int_0^{\infty} \frac{(\lambda t)^n}{n!} e^{-\lambda t}\,dt\right) F_V^{(n)}(c) \\
&= \frac{1}{\lambda} \sum_{n=0}^{\infty} F_V^{(n)}(c) \qquad (7.15)
\end{aligned}$$

In the special case where $V_1, V_2, \ldots$ are independent and exponentially distributed with parameter ρ, the sum $\sum_{i=1}^{n} V_i$ has a gamma distribution with parameters ρ and n [see Section 2.7 (2.25)]:

$$F_V^{(n)}(v) = 1 - \sum_{k=0}^{n-1} \frac{(\rho v)^k}{k!} e^{-\rho v}$$

$$= \sum_{k=n}^{\infty} \frac{(\rho v)^k}{k!} e^{-\rho v}$$

$$\sum_{n=0}^{\infty} F_V^{(n)}(c) = \sum_{n=0}^{\infty} \sum_{k=n}^{\infty} \frac{(\rho c)^k}{k!} e^{-\rho c}$$

$$= \sum_{k=0}^{\infty} \sum_{n=0}^{k} \frac{(\rho c)^k}{k!} e^{-\rho c}$$

$$= \sum_{k=0}^{\infty} (1 + k) \frac{(\rho c)^k}{k!} e^{-\rho c}$$

$$= 1 + \rho c$$

Hence, when the consequences $V_1, V_2, \ldots$ are exponentially distributed with parameter ρ, the mean time to total system failure is

$$E(T_c) = \frac{1 + \rho c}{\lambda} \qquad (7.16)$$

The distribution of the time T_c to total system failure is by Barlow and Proschan (1975, p. 94) shown to be an IFRA distribution for *any* distribution $F_V(v)$. (IFRA distributions are discussed in Section 2.16.)

7.3 RENEWAL PROCESSES

Introduction and Definitions

Renewal theory had its origin in the study of strategies for replacement of technical components, but later it was developed as a general theory within stochastic processes. This chapter gives a summary of some main aspects of renewal theory which are of particular interest in reliability analysis. This includes formulas for calculation of exact availability and mean number of failures within a given time interval. The latter can, for example, be used to determine optimal allocation of spare parts.

The following example will be used to illustrate the situations to be considered: A component is put into operation and is functioning at time $t = 0$. When the component fails, it is replaced by a new component of the same type, or restored to an "as good as new" condition. When this component fails, it is again replaced, and so on. The replacement time is considered to be negligible. We then have a sequence of lifetimes $T_1, T_2, T_3, \ldots$, which are assumed to be independent and identically distributed random variables with distribution function

$$F_T(t) = P(T_i \leq t) \quad \text{for} \quad t > 0, \quad i = 1, 2, \ldots$$

A process of this type is called an *ordinary renewal process*. The events observed (e.g., component failures) are called *renewals*, while the time intervals between consecutive events are called *renewal periods* or *interarrival times*. $F_T(t)$ is called the underlying distribution of the renewal process.

In some situations the lifetime T_1 of the first component has a distribution function $F_{T_1}(t)$, which is different from the life distribution function $F_T(t)$ of the subsequent components. Such a process is called a *modified renewal process*. This may, for example, be the case when the component at time $t = 0$ is not new.

The following variables associated with the renewal process are of particular interest:

1. The time until the rth renewal S_r,

$$S_r = T_1 + T_2 + \cdots + T_r = \sum_{i=1}^{r} T_i \tag{7.17}$$

2. The number of renewals in the time interval $(0, t]$,

$$N(t) = \max \{r; S_r \leq t\} \tag{7.18}$$

$N(t)$ is for example important when the number of spare parts is to be decided. The number of renewals in an arbitrary time interval $(t_1, t_2]$ is equal to $N(t_2) - N(t_1)$.

3. The renewal function

$$W(t) = E(N(t)) \tag{7.19}$$

Thus $W(t)$ is the mean number of renewals in the time interval $(0, t]$.

4. The renewal density

$$w(t) = \frac{d}{dt} W(t) \tag{7.20}$$

Note that the renewal density coincides with the ROCOF defined in Sec-

RENEWAL PROCESSES

tion 7.1 (see Ascher and Feingold 1984 for a thorough discussion of these concepts). The mean number of renewals in the time interval $(t_1, t_2]$ is thus

$$W(t_2) - W(t_1) = \int_{t_1}^{t_2} w(t)\,dt \qquad (7.21)$$

The relation between the renewal periods T_i and the number of renewals $N(t)$, the renewal process, is illustrated in Figure 7.6. The properties of renewal processes are discussed in detail by Cox (1962).

Distribution of S_r

It is often very difficult to determine the exact distribution of $S_r = \sum_{i=1}^{r} T_i$. Let

$$\begin{aligned} F^{(r)}(t) &= P(S_r \le t) = P(T_1 + T_2 + \cdots + T_r \le t) \\ &= P(S_{r-1} + T_r \le t) \end{aligned}$$

To determine the distribution $F^{(r)}(t)$, we will make use of the following definition:

Definition 7.4. Let T_1 and T_2 be independent life lengths with (for the sake of simplicity) continuous distribution functions F_1 and F_2 and corresponding densities f_1 and f_2. Introduce $T = T_1 + T_2$. Then the distribution function of $T = T_1 + T_2$ is called the *convolution* of F_1 and F_2,

$$F_T(t) = \int_0^t F_1(t - x)\,dF_2(x) = \int_0^t F_2(t - x)\,dF_1(x) \qquad (7.22)$$

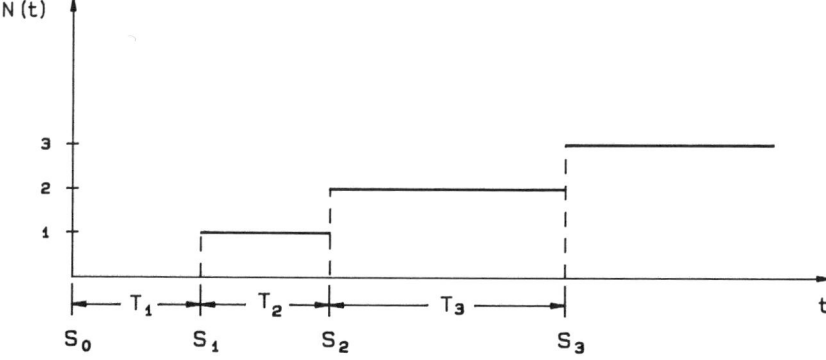

Figure 7.6 Relation between the renewal periods T_i and the number of renewals $N(t)$

If F_1 and F_2 are such that $dF_T(t)/dt$ can be obtained through differentiation under the integral sign, then the density function of $T = T_1 + T_2$ is called the *convolution* of f_1 and f_2,

$$f_T(t) = \int_0^t f_1(t-x) f_2(x)\, dx = \int_0^t f_2(t-x) f_1(x)\, dx \qquad (7.23)$$

The convolution is often denoted by use of an $*$: $F = F_1 * F_2$ and $f = f_1 * f_2$. □

Since S_{r-1} and T_r are independent, the distribution function $F^{(r)}(t)$ is the convolution of the distributions of S_{r-1} and T_r which, according to Definition 7.4, is

$$F^{(r)}(t) = \int_0^t F^{(r-1)}(t-x)\, dF_T(x) \qquad (7.24)$$

When the renewal periods are continuously distributed with probability density function $f_T(t)$, then the probability density function of S_r is

$$f^{(r)}(t) = \int_0^t f^{(r-1)}(t-x) f_T(x)\, dx \qquad (7.25)$$

By successive integration of (7.25) for $r = 2, 3, 4, \ldots$, the probability density function of S_r for a specified value of r can, in principle, be found:

$$f^{(2)}(t) = \int_0^t f^{(1)}(t-x) f_T(x)\, dx \quad \text{where} \quad f^{(1)}(t) = f_{T_1}(t)$$
$$\vdots \qquad (7.26)$$
$$f^{(r)}(t) = \int_0^t f^{(r-1)}(t-x) f_T(x)\, dx$$

Alternatively the distribution of S_r can be found by using Laplace transforms. The Laplace transform (see Appendix D) of the function $f(x)$ for $x > 0$ is

$$\mathscr{L}(f(x)) = f^*(s) = \int_0^\infty e^{-sx} f(x)\, dx \qquad (7.27)$$

When $f(x)$ is the probability density function of a nonnegative random variable

RENEWAL PROCESSES 279

X, then

$$\mathscr{L}(f(x)) = f^*(s) = E(e^{-sX}) \tag{7.28}$$

Since the renewal periods $T_1, T_2, \ldots$ are independent random variables, then

$$\mathscr{L}(f^{(r)}(t)) = f^{*(r)}(s) = E(e^{-s\sum_{i=1}^{r} T_i}) = E(e^{-sT_1}) \cdot \prod_{i=2}^{r} E(e^{-sT_i})$$

such that

$$f^{*(r)}(s) = f^*_{T_1}(s) \cdot (f^*_T(s))^{r-1} \tag{7.29}$$

The probability density function of S_r can now, at least in principle, be determined from the inverse Laplace transform of (7.29):

$$f^{(r)}(t) = \mathscr{L}^{-1}(f^{*(r)}(s)) \tag{7.30}$$

Extensive tables of inverse Laplace transforms are available; for example, see the Bateman manuscript project (1954).

Example 7.3: Homogeneous Poisson Process
Consider an ordinary renewal process where the renewal periods $T_1, T_2, \ldots$ are independent and exponentially distributed with the probability density function

$$f_T(t) = \begin{cases} \lambda e^{-\lambda t} & \text{for } t > 0, \lambda > 0 \\ 0 & \text{otherwise} \end{cases} \tag{7.31}$$

This renewal process is thus an HPP. Based on the results from Section 7.2, we obtain the necessary renewal facts:

1. The time until the rth renewal S_r has a gamma distribution with probability density function

$$f^{(r)}(t) = \frac{\lambda^r}{(r-1)!} t^{r-1} e^{-\lambda t} \quad \text{for } t \geq 0 \tag{7.32}$$

2. The number of renewals in $(0, t]$ is Poisson distributed with probability function

$$P(N(t) = r) = \frac{(\lambda t)^r}{r!} e^{-\lambda t} \quad \text{for} \quad r = 0, 1, 2, \ldots$$

3. The renewal function $W(t)$ is

$$W(t) = E(N(t)) = \lambda t$$

4. The renewal density is

$$w(t) = \frac{d}{dt} W(t) = \lambda$$

When the renewal periods represent the life lengths of components, the renewal density is identical to the failure rate λ.

The distribution of S_r may also be found by using Laplace transforms. In this case the Laplace transform is (see Appendix D)

$$f_T^*(s) = \int_0^\infty e^{-st} f_T(t)\, dt = \frac{\lambda}{\lambda + s} \tag{7.33}$$

The Laplace transform of the density $f^{(r)}(t)$ of S_r is therefore

$$f^{*(r)}(s) = (f_T^*(s))^r = \left(\frac{\lambda}{\lambda + s}\right)^r \tag{7.34}$$

This can easily be inverted (see Appendix D), and we get

$$f^{(r)}(t) = \frac{\lambda^r}{(r-1)!} t^{r-1} e^{-\lambda t} \quad \text{for} \quad t \geq 0 \tag{7.35}$$

$\square$

Asymptotic Distribution of S_r

Let $T_1, T_2, T_3, \ldots$ be independent and identically distributed, and assume that $E(T) = \mu$ and that $\text{var}(T) = \sigma^2$ exists. Then according to the central limit theorem (see Dudewicz and Mishra 1988, p. 315) $S_r = \sum_{i=1}^r T_i$ is asymptotically normally distributed $(r\mu, r\sigma^2)$. Hence

$$S_r = \sum_{i=1}^r T_i \quad \text{is approximately} \quad \mathcal{N}(r\mu, r\sigma^2) \quad \text{when } r \text{ is large}$$

RENEWAL PROCESSES

and

$$\frac{S_r - r\mu}{\sigma\sqrt{r}} \text{ is approximately } \mathcal{N}(0, 1) \text{ when } r \text{ is large.} \quad (7.36)$$

Thus, when r is large, the following approximation may be used:

$$F^{(r)}(t) = P(S_r \leq t) \approx \Phi\left(\frac{t - r\mu}{\sigma\sqrt{r}}\right) \quad (7.37)$$

where $\Phi(\cdot)$ denotes the distribution function of the standard normal distribution $\mathcal{N}(0, 1)$.

The Distribution of $N(t)$

$N(t)$ is equal to the number of renewals in the time interval $(0, t]$. From the definition of $N(t)$ and S_r, it follows that

$$N(t) \geq r \Leftrightarrow S_r \leq t \quad (7.38)$$

In words, the number of renewals in $(0, t]$ is greater than or equal to r if and only if renewal r occurs no later than time t. This implies that

$$P(N(t) \geq r) = P(S_r \leq t) = F^{(r)}(t) \quad (7.39)$$

and

$$\begin{aligned} P(N(t) = r) &= P(N(t) \geq r) - P(N(t) \geq r + 1) \\ &= F^{(r)}(t) - F^{(r+1)}(t) \end{aligned} \quad (7.40)$$

For large values of r we can apply (7.37) and obtain

$$P(N(t) = r) \approx \Phi\left(\frac{t - r\mu}{\sigma\sqrt{r}}\right) - \Phi\left(\frac{t - (r+1)\mu}{\sigma\sqrt{r+1}}\right) \quad (7.41)$$

Takács (1956) derived the following alternative approximation formula which is valid when t is large:

$$P(N(t) \leq r) \approx \Phi\left(\frac{r - (t/\mu)}{\sqrt{t\sigma^2/\mu^3}}\right) \qquad (7.42)$$

The Renewal Function

The renewal function, $W(t)$ can be expressed by

$$W(t) = \sum_{r=1}^{\infty} F^{(r)}(t) \qquad (7.43)$$

This can be shown in the following way:

$$W(t) = E(N(t)) = \sum_{r=0}^{\infty} r \cdot P(N(t) = r) = \sum_{r=0}^{\infty} r \cdot (F^{(r)}(t) - F^{(r+1)}(t))$$

$$= \sum_{r=1}^{\infty} rF^{(r)}(t) - \sum_{r=0}^{\infty} rF^{(r+1)}(t) = \sum_{r=1}^{\infty} rF^{(r)}(t) - \sum_{r=1}^{\infty} (r-1)F^{(r)}(t)$$

$$= \sum_{r=1}^{\infty} rF^{(r)}(t) - \sum_{r=1}^{\infty} rF^{(r)}(t) + \sum_{r=1}^{\infty} F^{(r)}(t) = \sum_{r=1}^{\infty} F^{(r)}(t)$$

An integral equation for $W(t)$ may be obtained by combining (7.43) and (7.24):

$$W(t) = F_{T_1}(t) + \sum_{r=2}^{\infty} F^{(r)}(t) = F_{T_1}(t) + \sum_{r=1}^{\infty} F^{(r+1)}(t)$$

$$= F_{T_1}(t) + \sum_{r=1}^{\infty} \int_0^t F^{(r)}(t - x) \, dF_T(x)$$

$$= F_{T_1}(t) + \int_0^t \sum_{r=1}^{\infty} F^{(r)}(t - x) \, dF_T(x)$$

$$= F_{T_1}(t) + \int_0^t W(t - x) \, dF_T(x) \qquad (7.44)$$

This equation is known as the *fundamental renewal equation* and can sometimes be solved for $W(t)$.

RENEWAL PROCESSES   283

For an *ordinary* renewal process (7.44) can also be derived by a more direct argument. By conditioning on the time T_1 of the first renewal, we obtain

$$W(t) = E(N(t)) = E(E(N(t)|T_1))$$
$$= \int_0^\infty E(N(t)|T_1 = x) \, dF_{T_1}(x) \qquad (7.45)$$

where

$$E(N(t)|T_1 = x) = \begin{cases} 0 & \text{for } t < x \\ 1 + W(t-x) & \text{for } t \geq x \end{cases} \qquad (7.46)$$

If the first renewal occurs at time x for $x \leq t$, the process starts over again from this point of time. The mean number of renewals in $(0, t]$ is thus 1 plus the mean number of renewals in $(x, t]$, which is $W(t-x)$.

Combining the two equations (7.45) and (7.46) yields for the ordinary renewal process

$$W(t) = \int_0^t (1 + W(t-x)) \, dF_T(x)$$
$$= F_T(t) + \int_0^t W(t-x) \, dF_T(x)$$

The Renewal Density

When $F_T(t)$ has the density $f_T(t)$ and $F_{T_1}(t)$ has the density $f_{T_1}(t)$, then

$$w(t) = \frac{d}{dt} W(t) = \frac{d}{dt} \sum_{r=1}^\infty F^{(r)}(t) = \sum_{r=1}^\infty f^{(r)}(t) \qquad (7.47)$$

By differentiating (7.44) w.r.t. t, we get

$$w(t) = f_{T_1}(t) + \int_0^t w(t-x) f_T(x) \, dx \qquad (7.48)$$

According to Appendix D the Laplace transform of (7.48) is

$$w^*(s) = f_{T_1}^*(s) + w^*(s) \cdot f_T^*(s)$$

Hence

$$w^*(s) = \frac{f_{T_1}^*(s)}{1 - f_T^*(s)} \tag{7.49}$$

and

$$W^*(s) = \frac{w^*(s)}{s} = \frac{f_{T_1}^*(s)}{s(1 - f_T^*(s))} \tag{7.50}$$

If the densities $f_{T_1}(t)$ and $f_T(t)$ are known, one can, at least in principle, also find the renewal function, that is, the mean number of renewals in $(0, t]$ from

$$W(t) = W(N(t)) = \mathscr{L}^{-1}\left(\frac{f_{T_1}^*(s)}{s(1 - f_T^*(s))}\right) \tag{7.51}$$

In the case of an *ordinary* renewal process

$$f_T^*(s) = \frac{w^*(s)}{1 + w^*(s)} \tag{7.52}$$

If the renewal density $w(t)$ of an ordinary renewal process is known, it is sometimes possible to determine the probability density function $f_T(t)$ uniquely from (7.52).

Example 7.3 (cont.)
Reconsider the homogeneous Poisson process in Example 7.3 where the renewal periods $T_1, T_2, \ldots$ are independent and exponentially distributed with parameter λ. Then according to (7.33),

$$f_T^*(s) = \frac{\lambda}{\lambda + s}$$

Hence the Laplace transform of the renewal density $w(t)$ is

$$w^*(s) = \frac{f_T^*(s)}{1 - f_T^*(s)} = \frac{\lambda}{s}$$

This implies that the renewal density is

RENEWAL PROCESSES

$$w(t) = \mathscr{L}^{-1}(w^*(s)) = \mathscr{L}^{-1}\left(\frac{\lambda}{s}\right) = \lambda \quad \text{for} \quad t > 0$$

Thus the renewal function is

$$W(t) = \int_0^t w(u)\,du = \lambda t \qquad \square$$

Example 7.4
Consider an ordinary renewal process where the renewal periods $T_1, T_2, T_3, \ldots$ are independent and Weibull distributed with shape parameter α and scale parameter λ. In this case the renewal function $W(t)$ cannot be deduced directly from (7.51). Smith and Leadbetter (1963) have, however, shown that $W(t)$ can be expressed as an infinite, absolutely convergent series where the terms can be found by a simple recursive procedure. They show that $W(t)$ can be written

$$W(t) = \sum_{k=1}^{\infty} \frac{(-1)^{k-1} \cdot A_k \cdot (\lambda t)^{k\alpha}}{\Gamma(k\alpha + 1)} \tag{7.53}$$

By introducing this expression for $W(t)$ in the fundamental renewal equation, the constants A_k; $k = 1, 2, 3, \ldots$ can be determined. The calculation, which is quite comprehensive, leads to the following *recursion* formula:

$$\begin{aligned} A_1 &= \gamma_1 \\ A_2 &= \gamma_2 - \gamma_1 A_1 \\ A_3 &= \gamma_3 - \gamma_1 A_2 - \gamma_2 A_1 \\ &\vdots \\ A_n &= \gamma_n - \sum_{j=1}^{n-1} \gamma_j A_{n-j} \\ &\vdots \end{aligned} \tag{7.54}$$

where

$$\gamma_n = \frac{\Gamma(n\alpha + 1)}{n!} \quad \text{for} \quad n = 1, 2, \ldots \tag{7.55}$$

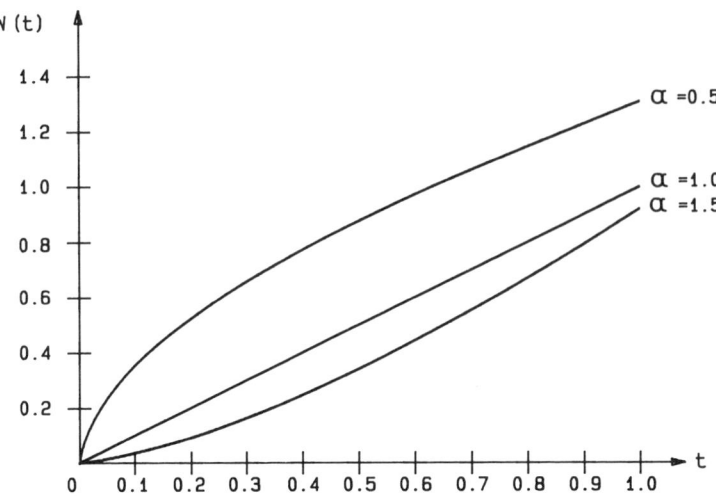

Figure 7.7 The renewal function for Weibull distributed renewal periods with $\lambda = 1$ and $\alpha = 0.5$, $\alpha = 1$, and $\alpha = 1.5$

For $\alpha = 1$ the Weibull distribution (see Figure 7.7) is an exponential distribution with parameter λ. In this case

$$\gamma_n = \frac{\Gamma(n+1)}{n!} = 1 \quad \text{for} \quad n = 1, 2, \ldots$$

This leads to

$$A_1 = 1$$
$$A_n = 0 \quad \text{for} \quad n \geq 2$$

The renewal function is thus according to (7.53),

$$W(t) = \frac{(-1)^0 A_1 \lambda t}{\Gamma(2)} = \lambda t$$

which agrees with the result in Example 7.3. □

Bounds for the Renewal Function

An exact expression for the renewal function $W(t)$ is often difficult to determine. Bounds and approximation formulas may therefore be useful. We will here content ourselves to considering an *ordinary* renewal process.

Obviously

$$\max T_i \leq \sum_{j=1}^{r} T_j = S_r$$

Therefore

$$F^{(r)}(t) = P(S_r \leq t) \leq P(\max T_i \leq t)$$

that is,

$$F^{(r)}(t) \leq (P(T_i \leq t))^r = (F_T(t))^r \qquad (7.56)$$

Since the renewal process is assumed to be ordinary, then

$$F_T^{(1)}(t) = F_T(t)$$

Hence

$$F_T(t) = F_T^{(1)}(t) \leq \sum_{r=1}^{\infty} F^{(r)}(t) \leq \sum_{r=1}^{\infty} (F_T(t))^r = \frac{F_T(t)}{1 - F_T(t)}$$

According to (7.43),

$$W(t) = \sum_{r=1}^{\infty} F^{(r)}(t)$$

Hence

$$F_T(t) \leq W(t) \leq \frac{F_T(t)}{1 - F_T(t)} \qquad (7.57)$$

Note that these bounds for $W(t)$ are only valid for *ordinary* renewal processes.

For an ordinary renewal process, $T_1, T_2, \ldots$ are independent and identically distributed with finite mean μ. Since the number of renewals up to time t is $N(t)$, then $t \leq S_{N(t)+1}$. Consequently $t \leq E(S_{N(t)+1})$.

According to Wald's equation (7.13) then

$$t \leq E(S_{N(t)+1}) = E\left(\sum_{i=1}^{N(t)+1} T_i\right)$$
$$= \mu \cdot (E(N(t)) + 1) = \mu \cdot (W(t) + 1)$$

Hence

$$W(t) \geq \frac{t}{\mu} - 1 \qquad (7.58)$$

A number of sharper upper and lower bounds of the renewal function $W(t)$ can be constructed. These bounds often depend on assumptions about the underlying distribution $F_T(t)$. A brief survey of some upper and lower bounds of $W(t)$ is given in Table 7.2, see also Siegel and Wünsche (1979). In Table 7.2 the conditions for the limits to be valid are stated together with a reference to where the limits are derived.

Table 7.2 Upper and Lower Bounds of the Renewal Function

Lower Bound	Condition	Reference	Upper Bound	Condition	Reference
$F_T(t)$	—	(7.57)	$\dfrac{F_T(t)}{1 - F_T(t)}$	$F_T(t) < 1$	(7.57)
$\dfrac{t}{\mu} - 1$	$\mu < \infty$	(7.58)	$\dfrac{t}{\mu} + \dfrac{\mu_2}{2\mu^2} - 1$	$\mu_2 < 1$	Lorden (1970)
$\dfrac{t}{\mu} - F_e(t)$	$\mu < \infty$	Butterworth and Marshall (1964)	$\dfrac{2t}{m_0} - 1$	$m_0 > 1$	Berowkow (1972)
$\dfrac{t}{\mu}$	NWUE[a]	Barlow and Proschan (1975)	$\dfrac{t}{\mu}$	NBUE[b]	Barlow and Proschan (1975)
$\dfrac{t}{\mu} - F_e(t) + F_T(t)$	NWUE	Siegel and Wünsche (1979)	$\dfrac{t}{\mu} - F_e(t) + F_T(t)$	NBUE	Siegel and Wünsche (1979)

[a]New worse than used in expectation.
[b]New better than used in expectation.

RENEWAL PROCESSES

The following notation is used in Table 7.2:

$$\mu = E(T_i)$$
$$\mu_2 = E(T_i^2)$$
$$m_0 = F_T^{-1}\left(\frac{1}{2}\right) \quad \text{[the median of } F_T(t)\text{]}$$
$$F_e(t) = \frac{1}{\mu}\int_0^t (1 - F_T(u))\,du$$

The concepts NBUE (new better than used in expectation) and NWUE (new worse than used in expectation) are defined in Section 2.16.

Asymptotic Properties of the Renewal Process

Here we will give a brief survey of some asymptotic properties of a renewal process. A more comprehensive survey can be found in Smith (1958), Cox (1962), Prabhu (1965), or Ross (1983).

For an ordinary renewal process where the underlying distribution is exponential with parameter λ (i.e., a homogeneous Poisson process), $W(t) = \lambda t$ (see Example 7.3). Hence

$$\frac{W(t)}{t} = \frac{1}{\mu}$$

where $\mu = 1/\lambda = E(T_i)$ is the mean lifetime. According to the elementary renewal theorem stated below, this relation holds asymptotically when $t \to \infty$ even in the general case.

Theorem 7.2 (The Elementary Renewal Theorem)

$$\lim_{t \to \infty} \frac{W(t)}{t} = \frac{1}{\mu} \tag{7.59}$$

where $\mu = E(T_i) < \infty$. The boundary is interpreted as 0 when $\mu = \infty$.

A proof of the elementary renewal theorem may, for example, be found in Ross (1983, p. 61).

When the renewals correspond to component failures, the mean number of failures in $(0, t]$ is thus approximately

$$E(N(t)) \approx \frac{t}{\mu} = \frac{t}{\text{MTTF}} \quad \text{when } t \text{ is large}$$

Definition 7.5. A nonnegative random variable is said to have a *lattice* (or periodic) distribution if there exists a number $d \geq 0$ such that

$$\sum_{n=0}^{\infty} P(X = nd) = 1$$

In words, X has a lattice distribution if X can only take on values that are integral multiples of some nonnegative number d. □

The following important result was shown by Blackwell (1948):

Theorem 7.3 (Blackwell's Theorem). For a renewal process with underlying distribution $F_T(t)$ which is a nonlattice distribution,

$$\lim_{t \to \infty} (W(t + \alpha) - W(t)) = \frac{\alpha}{\mu} \quad \text{for} \quad \alpha > 0 \quad (7.60)$$

The limit is equal to 0 when $\mu = \infty$.

A proof of this fundamental theorem may be found in Feller (1966, Chapter XI). Blackwell's theorem states that if $F_T(t)$ is nonlattice, then the mean number of renewals in an interval of length α, far from the origin, is approximately α/μ.

Blackwell's theorem has been generalized by Smith (1958):

Theorem 7.4 (Key Renewal Theorem). For a renewal process with underlying distribution $F_T(t)$ which is a nonlattice distribution, then

$$\lim_{t \to \infty} \int_0^t Q(t - x) \, dW(x) = \frac{1}{\mu} \int_0^\infty Q(u) \, du \quad (7.61)$$

where $Q(t)$ is a nonnegative, nonincreasing function that is Riemann integrable over $(0, \infty)$.

By introducing $Q(t) = \alpha^{-1}$ for $0 < t \leq \alpha$ and $Q(t) = 0$ otherwise, in Theorem 7.4, we get Blackwell's theorem.

Let

$$F_e(t) = \frac{1}{\mu} \int_0^t (1 - F_T(u)) \, du$$

By using $Q(t) = 1 - F_e(t)$ in the key renewal theorem, we get

RENEWAL PROCESSES

$$\lim_{t \to \infty} \left(W(t) - \frac{t}{\mu} \right) = \frac{\mu_2}{2\mu^2} - 1 \qquad (7.62)$$

if $E(T_i^2) = \mu_2 < \infty$.

For the renewal density $w(t)$ the following asymptotic properties are valid (from Smith 1958):

Theorem 7.5. For a renewal process where there exists a $p > 1$ such that $|f_T(t)|^p$ is Riemann integrable,

$$\lim_{t \to \infty} w(t) = \frac{1}{\mu} \qquad (7.63)$$

The limit is interpreted as 0 when $\mu = \infty$.

Example 7.5: Age Replacement

Under an *age replacement* policy a component is replaced upon failure or at age t_0, whichever comes first. Consider an ordinary renewal process where the component is subject to age replacement at age t_0 which is nonrandom. Let T denote the potential time to failure of the component. T is assumed to be continuous with distribution function $F_T(t)$ and density $f_T(t)$. The distribution function $F_R(t)$ of the time T_R between replacements (renewals) is

$$F_R(t) = \begin{cases} F_T(t) & \text{for } t < t_0 \\ 1 & \text{for } t \geq t_0 \end{cases}$$

Then the mean time between replacements/renewals (MTBR) is

$$\text{MTBR} = \int_0^{t_0} t f_T(t) \, dt + t_0 \cdot P(T \geq t_0) = \int_0^{t_0} (1 - F_T(t)) \, dt \qquad (7.64)$$

which is less than the mean time to failure, MTTF, of the component.

Now let $Y_1, Y_2, \ldots$ denote the times between the actual consecutive failures. This may be represented as a renewal process where the renewals are the actual failures. The renewal periods Y_i are composed of a random number N_i of time periods of length t_0 (corresponding to replacements without failure) plus a last-time period in which the component fails at an age Z_i, which is less than t_0.

Thus

$$Y_i = N_i \cdot t_0 + Z_i \qquad \text{for } i = 1, 2, \ldots$$

The distribution of N_i is given by

$$P(N_i \geq k) = (1 - F_T(t_0))^k \quad \text{for} \quad k = 0, 1, 2, \ldots$$

Hence

$$P(N_i = k) = P(N_i \geq k) - P(N_i \geq k+1) = F_T(t_0)(1 - F_T(t_0))^k$$

The mean number of replacements without failure is thus

$$E(N_i) = \sum_{k=0}^{\infty} k \cdot P(N_i = k) = F_T(t_0) \sum_{k=0}^{\infty} k \cdot (1 - F_T(t_0))^k$$

$$= \frac{1 - F_T(t_0)}{F_T(t_0)}$$

The distribution of Z_i is

$$P(Z_i \leq t) = P(T \leq t | T \leq t_0) = \frac{F_T(t)}{F_T(t_0)} \quad \text{for} \quad 0 \leq t \leq t_0$$

Hence

$$E(Z_i) = \int_0^{t_0} \left(1 - \frac{F_T(t)}{F_T(t_0)}\right) dt = \frac{1}{F_T(t_0)} \int_0^{t_0} (F_T(t_0) - F_T(t)) dt$$

The mean time between actual failures is thus

$$E(Y_i) = t_0 \cdot E(N_i) + E(Z_i)$$

$$= \frac{1}{F_T(t_0)} \left[t_0(1 - F_T(t_0)) + \int_0^{t_0} (F_T(t_0) - F_T(t)) dt \right]$$

$$= \frac{1}{F_T(t_0)} \int_0^{t_0} (1 - F_T(t)) dt \quad (7.65)$$

Let us suppose that each replacement, whether planned or not, costs c and that each failure incurs an additional penalty of K. The long-run mean cost per unit time as a function of the replacement age t_0 is thus

$$C(t_0) = \frac{c}{\text{MTBR}} + \frac{K}{E(Y_i)} = \frac{c + K \cdot F_T(t_0)}{\int_0^{t_0} (1 - F_T(t)) dt}$$

Age and Remaining Lifetime

The *age* $Z(t)$ of a unit which is operating at time t, is defined as

$$Z(t) = \begin{cases} t & \text{for } N(t) = 0 \\ t - S_{N(t)} & \text{for } N(t) > 0 \end{cases} \qquad (7.66)$$

The *remaining lifetime* $Y(t)$ of a unit which is in operation at time t is given as

$$Y(t) = S_{N(t)+1} - t \qquad (7.67)$$

The age $Z(t)$ and the remaining lifetime $Y(t)$ are illustrated in Figure 7.8. The remaining lifetime is also called the *residual life*, the *excess life*, or the *forward recurrence time* (e.g., see Ross 1983; Ascher and Feingold 1984). Note that $Y(t) > x$ is equivalent to no renewal in the time interval $(t, t+x]$.

When the distributions of $Z(t)$ and $Y(t)$ are to be determined, the following lemma is useful:

Lemma 7.1. If

$$g(t) = h(t) + \int_0^t g(t-x)\,dF(x) \qquad (7.68)$$

where the functions h and F are known while g is unknown, then

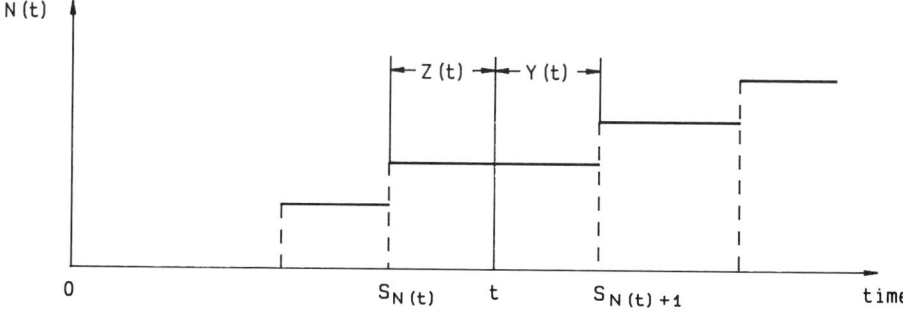

Figure 7.8 Age $Z(t)$ and the remaining lifetime $Y(t)$

$$g(t) = h(t) + \int_0^t h(t-x)\, dW_F(x) \qquad (7.69)$$

where

$$W_F(x) = \sum_{r=1}^{\infty} F^{(r)}(x)$$

Note that (7.69) is a generalization of the fundamental renewal equation (7.44).

Proof
We will prove Lemma 7.1 when $F(x)$ is differentiable. (The proof is similar when $F(x)$ is not differentiable.) By introducing

$$f(x) = \frac{d}{dx} F(x)$$

we get

$$\int_0^t g(t-x)\, dF(x) = \int_0^t g(t-x) f(x)\, dx$$

which expresses the convolution of $g(x)$ and $f(x)$ (see Definition 7.4). Equation (7.68) may thus be written

$$g = h + g * f$$

By taking Laplace transforms of (7.68) we get

$$g^*(s) = h^*(s) + g^*(s) \cdot f^*(s)$$

Hence

$$g^*(s) = \frac{h^*(s)}{1 - f^*(s)} = h^*(s)\left(1 + \frac{f^*(s)}{1 - f^*(s)}\right)$$
$$= h^*(s) + h^*(s) \cdot \frac{f^*(s)}{1 - f(s)}$$

But, according to (7.49),

RENEWAL PROCESSES

$$\frac{f^*(s)}{1-f^*(s)} = w_F^*(s)$$

Therefore

$$g^*(s) = h^*(s) + h^*(s) \cdot w_F^*(s)$$

By taking the inverse Laplace transform we get

$$g(t) = h(t) + \int_0^t h(t-x)\,dW_F(x) \qquad \square$$

Theorem 7.6. Let $Y(t)$ denote the remaining lifetime of a unit which is in operation at time t. Then

$$P(Y(t) \le y) = F_T(t+y) - \int_0^t (1 - F_T(t+y-x))\,dW_F(x) \qquad (7.70)$$

If $F_T(t)$ is a nonlattice distribution, then

$$\lim_{t \to \infty} P(Y(t) \le y) = \frac{1}{\mu} \int_0^y (1 - F_T(x))\,dx \qquad (7.71)$$

Assume that a renewal process has been "running" for a very long time and that we start observing the process at time $t = 0$. Let T_1 denote the time to the first renewal after time $t = 0$. T_1 is thus the remaining lifetime of the unit which is in operation at time $t = 0$. According to (7.71) the distribution of T_1 is

$$F_{T_1}(t) = \frac{1}{\mu} \int_0^t (1 - F_T(x))\,dx$$

Note that this distribution is equal to $F_e(t)$ in Table 7.2.

Proof (See Ross 1970, p. 44)
Let

$$P(t) = P(Y(t) > y)$$

By conditioning with respect to the first renewal period T_1,

$$P(t) = \int_0^\infty P(Y(t) > y | T_1 = x) \, dF_T(x)$$

Since the process starts again at time T_1,

$$P(Y(t) > y | T_1 = x) = \begin{cases} P(t-x) & \text{if } x \le t \\ 0 & \text{if } t < x \le t+y \\ 1 & \text{if } x > t+y \end{cases}$$

Thus

$$P(t) = \int_0^t P(t-x) \, dF(x) + \int_{t+y}^\infty dF_T(x)$$
$$= 1 - F_T(t+y) + \int_0^t P(t-x) \, dF_T(x)$$

Here we can make use of Lemma 7.1 with

$$h(t) = 1 - F_T(t+y) \quad \text{and} \quad g(t) = P(t)$$

We then get

$$P(t) = 1 - F_T(t+y) + \int_0^t (1 - F_T(t+y-x)) \, dW_{F_T}(x)$$

To show the last part of the theorem, we can use Theorem 7.4, the key renewal theorem, with $Q(t) = 1 - F_T(t+y)$.

$$\lim_{t \to \infty} P(t) = 1 - \lim_{t \to \infty} F_T(t+y) + \lim_{t \to \infty} \int_0^t Q(t-x) \, dW_{F_T}(x)$$
$$= \frac{1}{\mu} \int_0^\infty (1 - F_T(t+y)) \, dt = \frac{1}{\mu} \int_y^\infty (1 - F_T(t)) \, dt$$

or

RENEWAL PROCESSES

$$\lim_{t \to \infty} P(Y(t) \le y) = 1 - \frac{1}{\mu} \int_y^\infty (1 - F_T(t)) \, dt$$

$$= \frac{1}{\mu} \int_0^y (1 - F_T(t)) \, dt$$

since

$$\mu = \int_0^\infty (1 - F_T(t)) \, dt \qquad \square$$

When the age $Z(0)$ of the component that is in operation at time $t = 0$, is greater than 0, we have a modified renewal process. The distribution of the remaining lifetime $Y(t)$ in this case is

$$P(Y(t) \le y) = F_{T_1}(t + y) - \int_0^t (1 - F_T(t + y - x)) \, dW(x) \qquad (7.72)$$

where $W(x)$ is the renewal function of the modified renewal process (e.g., see Ross 1970, p. 48, or Prabhu 1965, p. 173).

The distribution of the age $Z(t)$ can be derived by starting with

$$Z(t) > z \Leftrightarrow \text{no renewals in } (t - z, t)$$
$$\Leftrightarrow Y(t - z) > z$$

Therefore

$$P(Z(t) > z) = P(Y(t - z) > z)$$

From Theorem 7.6 we get

Theorem 7.7. The distribution of the age $Z(t)$ is given by

$$P(Z(t) \le z) = \begin{cases} F_T(t) - \int_0^{t-z} (1 - F_T(t - x)) \, dW_{F_T}(x) & \text{for } z < t \\ 1 & \text{for } z \ge t \end{cases}$$

$$(7.73)$$

When $F_T(t)$ is a nonlattice distribution, then

$$\lim_{t \to \infty} P(Z(t) \leq z) = \frac{1}{\mu} \int_0^z (1 - F_T(x))\,dx$$

Definition 7.6. A stationary renewal process is a modified renewal process where the first renewal period has distribution function

$$F_1(t) = \frac{1}{\mu} \int_0^t (1 - F_T(x))\,dx$$

while the underlying distribution of the other renewal periods is $F_T(t)$. □

As pointed out in Cox (1962, p. 28) the stationary renewal process has a simple physical interpretation. Suppose that a renewal process is started at time $t = -\infty$ but that the process is not observed before time $t = 0$. Then the first renewal period observed T_1 is the remaining lifetime of the component in operation at time $t = 0$. According to Theorem 7.6 (7.70) the distribution function of T_1 is $F_1(t)$. A stationary renewal process is called an *equilibrium renewal process* by Cox (1962). In Ascher and Feingold (1984) the stationary renewal process is called a renewal process with *asynchronous sampling*, while an ordinary renewal process is called a renewal process with *synchronous sampling*.

Example 7.6
Consider an ordinary renewal process where the renewal periods $T_1, T_2, \ldots$ have a gamma distribution with probability density function

$$f_T(t) = \lambda^2 t e^{-\lambda t} \quad \text{for } t > 0, \lambda > 0$$

The time until the rth renewal S_r is now gamma distributed (see Section 2.7) with probability density function

$$f^{(r)}(t) = \frac{\lambda^{2r}}{(2r-1)!} t^{2r-1} e^{-\lambda t} \quad \text{for } t > 0$$

The renewal density is according to (7.47),

RENEWAL PROCESSES

$$w(t) = \sum_{r=1}^{\infty} f^{(r)}(t) = \lambda e^{-\lambda t} \sum_{r=1}^{\infty} \frac{(\lambda t)^{2r-1}}{(2r-1)!}$$

$$= \lambda e^{-\lambda t} \cdot \frac{e^{\lambda t} - e^{-\lambda t}}{2} = \frac{\lambda}{2}(1 - e^{-2\lambda t})$$

The renewal function is

$$W(t) = \int_0^t w(x)\,dx = \frac{\lambda}{2}\int_0^t (1 - e^{-2\lambda x})\,dx = \frac{\lambda t}{2} - \frac{1}{4}(1 - e^{-2\lambda t})$$

The renewal density $w(t)$ and the renewal function $W(t)$ are illustrated in Figure 7.9 for $\lambda = 1$.

The mean renewal period is

$$\mu = E(T_i) = \frac{2}{\lambda}$$

and the variance is

$$\sigma^2 = \text{var}(T_i) = \frac{2}{\lambda^2}$$

Note that when $t \to \infty$, then

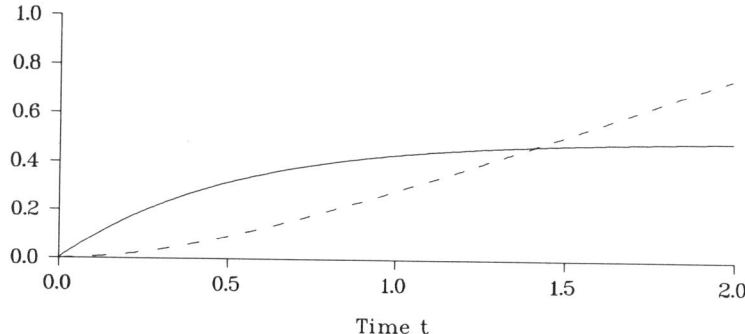

Figure 7.9 Renewal density $w(t)$ (fully drawn line) and renewal function $W(t)$ (dotted line) for Example 7.6, with $\lambda = 1$

$$w(t) \to \frac{\lambda}{2} = \frac{1}{\mu}$$

$$W(t) - \frac{t}{\mu} = W(t) - \frac{\lambda t}{2} \to -\frac{1}{4} = \frac{2/\lambda^2}{2(2/\lambda)^2} - 1 = \frac{\mu_2}{2\mu^2} - 1$$

in agreement with (7.62) and (7.63), respectively.

The distribution of the remaining lifetime $Y(t)$ in a *stationary* renewal process is from (7.71),

$$P(Y(t) \leq y) = \frac{1}{\mu} \int_0^y (1 - F_T(t))\, dt$$

In this case $F_T(t)$ is

$$F_T(t) = 1 - e^{-\lambda t} - \lambda t e^{-\lambda t}$$

Thus

$$P(Y(t) \leq y) = \frac{\lambda}{2} \int_0^y (e^{-\lambda t} + \lambda t e^{-\lambda t})\, dt = 1 - \left(1 + \frac{\lambda y}{2}\right) e^{-\lambda y}$$

The mean remaining lifetime in a stationary renewal process is

$$E(Y(t)) = \int_0^\infty (1 - P(Y(t) \leq y))\, dy$$
$$= \int_0^\infty \left(1 + \frac{\lambda y}{2}\right) e^{-\lambda y}\, dy = \frac{3}{2} \cdot \frac{1}{\lambda} \qquad \square$$

The following theorem is often useful (for a proof, see Ross 1983, p. 71):

Theorem 7.8. When the renewal periods (interarrival times) $T_1, T_2, \ldots$, have a nonlattice distribution such that $E(T_i^2) < \infty$, then

$$\lim_{t \to \infty} E(Y(t)) = \frac{E(T_i^2)}{2\mu} = \frac{\sigma^2 + \mu^2}{2\mu} \qquad (7.74)$$

If the underlying distribution function $F_T(t)$ is NBU or NWU (see Section 2.16) in an ordinary renewal process, bounds may be derived for the distribution of the remaining lifetime $Y(t)$ of the component which is in operation at time t.

Theorem 7.9. If $F_T(t)$ is NBU, then

$$P(Y(t) > y) \leq 1 - F_T(y) \qquad (7.75)$$

If $F_T(t)$ is NWU, then

$$P(Y(t) > y) \geq 1 - F_T(y) \qquad (7.76)$$

Proof of (7.75) (See Barlow and Proschan 1975, p. 169)
Let $R_T(t) = 1 - F_T(t)$. From (7.70) we get

$$P(Y(t) > y) = R_T(t + y) + \int_0^t R_T(t + y - x) \, dW_{F_T}(x)$$

If F_T is NBU, then

$$P(Y(t) > y) \leq R_T(t) \cdot R_T(y) + R_T(y) \int_0^t R_T(t - x) \, dW_{F_T}(x)$$

$$= R_T(y)(R_T(t + 0) + \int_0^t R_T(t + 0 - x) \, dW_{F_T}(x))$$

$$= R_T(y) \cdot P(Y(t) \geq 0) = R_T(y)$$

Equation (7.76) is proved by an analogous argument. □

Superimposed Renewal Processes

Consider a series structure of n independent components that is put into operation at time $t = 0$. All the n components are assumed to be new at time $t = 0$. When a component fails, it is replaced with a new component of the same type, or restored to an "as good as new" condition. Each component will thus produce an ordinary renewal process. The n components will generally be different, and the renewal processes will therefore have different underlying distributions.

The process formed by the union of all the failures is called a *superimposed renewal process* (SRP). The n individual renewal processes and the superimposed renewal process are illustrated in Figure 7.10.

In general, the SRP will *not* be a renewal process. However, it has been shown, for example, by Drenick (1960), that superposition of an infinite number of independent *equilibrium* renewal processes is a homogeneous Poisson process (HPP). Many systems are composed of a large number of components in series. Drenick's result is often used as a justification for assuming the time between system failures to be exponentially distributed.

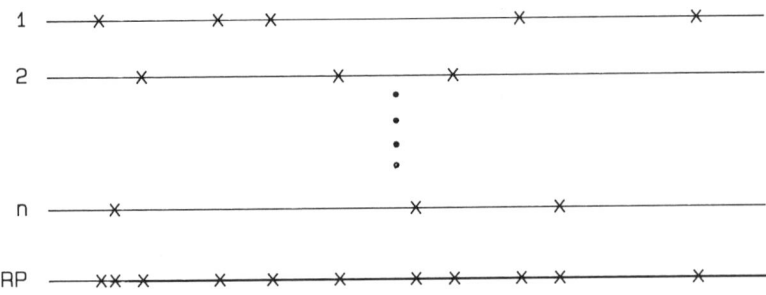

Figure 7.10 Superimposed renewal process

The superimposed renewal process is further discussed, for example, by Cox and Isham (1980), Ascher and Feingold (1984), and Thompson (1988).

Renewal Reward Processes

Consider an ordinary renewal process $\{N(t), t \geq 0\}$. The interarrival times $T_1, T_2, \ldots$ are independent and identically distributed with distribution function $F_T(t)$. To renewal i there is associated a reward V_i for $i = 1, 2, \ldots$. The rewards $V_1, V_2, \ldots$ are assumed to be independent random variables with the common distribution function $F_V(v)$. This model is comparable with the compound Poisson process described on pages 273–275. The accumulated reward in the time interval $(0, t]$ is

$$V(t) = \sum_{i=1}^{N(t)} V_i \qquad (7.77)$$

Let $E(T_i) = \mu_T$ and $E(V_i) = \mu_V$. According to Wald's equation (7.13) the mean accumulated reward is

$$E(V(t)) = \mu_V \cdot E(N(t)) \qquad (7.78)$$

According to the elementary renewal theorem (7.59), when $t \to \infty$,

$$\frac{W(t)}{t} = \frac{E(N(t))}{t} \to \frac{1}{\mu_T}$$

Hence

$$\frac{E(V(t))}{t} = \frac{\mu_V \cdot E(N(t))}{t} \to \frac{\mu_V}{\mu_T} \qquad (7.79)$$

RENEWAL PROCESSES

As shown in the following theorem, the same result is true even if the reward V_i is allowed to depend on the associated interarrival time T_i for $i = 1, 2, \ldots$. The pairs (T_i, V_i) for $i = 1, 2, \ldots$ are, however, assumed to be independent and identically distributed.

Theorem 7.10. When $\mu_V < \infty$ and $\mu_T < \infty$, then, when $t \to \infty$,

$$\frac{V(t)}{t} \to \frac{\mu_V}{\mu_T} \quad \text{with probability 1} \tag{7.80}$$

$$\frac{E(V(t))}{t} \to \frac{\mu_V}{\mu_T} \tag{7.81}$$

A proof of Theorem 7.10 is given by Ross (1983, p. 78).

When t is very large,

$$V(t) \approx \mu_V \cdot \frac{t}{\mu_T}$$

which is an obvious result.

Example 7.7

Consider a component that is activated and functioning at time $t = 0$. Whenever the component fails, it is repaired. Let $T_1, T_2, \ldots$ denote the successive lifetimes of the component. Assume that the lifetimes are independent and identically distributed with distribution function $F_T(t) = P(T_i \leq t)$ and mean $E(T) = \text{MTTF}$. Likewise we assume the corresponding repair times $D_1, D_2, \ldots$ to be independent and identically distributed with distribution function $F_D(d) = P(D_i \leq d)$ and mean $E(D) = \text{MTTR}$ (mean time to repair).

If we define the completed repairs to be the renewals, we obtain an ordinary renewal process with renewal periods (interarrival times) $T_i + D_i$ for $i = 1, 2, \ldots$. The mean time between renewals is $\mu_T = \text{MTTF} + \text{MTTR}$. This process is called an *alternating renewal process* and is further described later in this section.

Let the reward V_i associated to the ith interarrival time be defined such that we earn one unit per unit of time the component is functioning in the preceding time interval. When the reward is measured in time units, $E(V_i) = \mu_V = \text{MTTF}$. The average availability $A_{av}(0, t)$ of the component in the time interval $(0, t)$ has been defined as the mean functioning time in the interval $(0, t)$. From Theorem 7.10 we get

$$A_{av}(0,t) \to \frac{\mu_V}{\mu_T} = \frac{\text{MTTF}}{\text{MTTF} + \text{MTTR}} \qquad (7.82)$$

when $t \to \infty$ which is the same result we obtained in Section 4.7 based on heuristic arguments. □

Alternating Renewal Processes

Let us reconsider the situation we discussed in Example 7.7 and also in Section 4.7. Let the state of the component be given by the binary variable

$$X(t) = \begin{cases} 1 & \text{if the component is functioning at time } t \\ 0 & \text{otherwise} \end{cases}$$

The situation is illustrated in Figure 7.11. A process of this type is called an *alternating* renewal process.

If we let the renewals be the events when a repair is completed, the renewal periods will be $Y_i = T_i + D_i$ for $i = 1, 2, \ldots$, and we have an *ordinary* renewal process, with an underlying distribution function $H(y)$ being the convolution $F_T(t) * F_D(d)$:

$$H(y) = P(Y_i \le y) = P(T_i + D_i \le y) = \int_0^y F_T(y - t) \, dF_D(t) \qquad (7.83)$$

If instead we let the renewals be the events when a failure occurs, we get a *modified* renewal process where the first renewal period Y_1 is equal to T_1 while $Y_i = D_{i-1} + T_i$ for $i = 2, 3, \ldots$. In this case the distribution function $H_1(y)$ of the first renewal period is given by

$$H_1(y) = P(Y_1 \le y) = P(T_1 \le y) = F_T(y) \qquad (7.84)$$

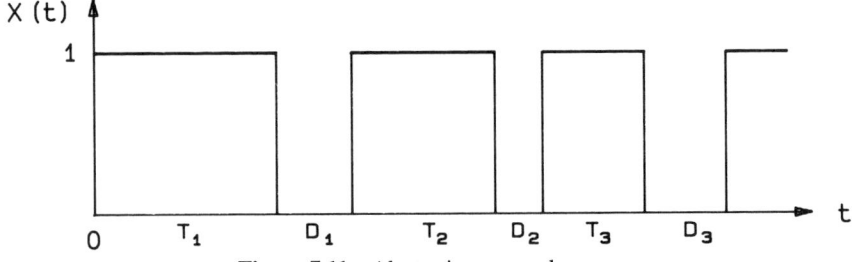

Figure 7.11 Alternating renewal process

RENEWAL PROCESSES   305

while the distribution function $H(y)$ of the other renewal periods are given by (7.83).

The *availability* at time t $A(t)$ of a component was in Section 4.7 defined as the probability that the component is functioning at time t:

$$A(t) = P(X(t) = 1) = E(X(t)) \tag{7.85}$$

The *limiting availability* A of the component is, when the limit exists, defined by

$$A = \lim_{t \to \infty} A(t) \tag{7.86}$$

The *average availability* of the component in the time interval (t_1, t_2) is defined by

$$A_{av}(t_1, t_2) = \frac{1}{t_2 - t_1} \int_{t_1}^{t_2} A(t) \, dt \tag{7.87}$$

Let $U(t_1, t_2)$ denote the fraction of the interval (t_1, t_2) where the component is functioning:

$$U(t_1, t_2) = \int_{t_1}^{t_2} X(t) \, dt$$

Then under assumptions of general nature

$$\frac{E(U(t_1, t_2))}{t_2 - t_1} = \frac{1}{t_2 - t_1} E\left(\int_{t_1}^{t_2} X(t) \, dt \right)$$
$$= \frac{1}{t_2 - t_1} \int_{t_1}^{t_2} E(X(t)) \, dt = \frac{1}{t_2 - t_1} \int_{t_1}^{t_2} A(t) \, dt \tag{7.88}$$

Hence the average availability of the component in the time interval $(t_1, t_2]$ is equal to the mean fraction of the time interval where the component is functioning.

ooo OOO ooo

Theorem 7.11. When $E(T_i + D_i) < \infty$ and the distribution $H(y)$ is a non-lattice distribution,

$$A = \lim_{t \to \infty} A(t) = \frac{\text{MTTF}}{\text{MTTF} + \text{MTTR}} \qquad (7.89)$$

Proof (See Ross 1970)
Let $Y = T_1 + D_1$ and $H(y) = P(Y \le y)$. The availability is then

$$\begin{aligned} A(t) &= P(X(t) = 1) = E(X(t)) = E(E(X(t)|Y)) \\ &= \int_0^\infty E(X(t)|Y = y)\,dH(y) \\ &= \int_0^\infty P(X(t) = 1|Y = y)\,dH(y) \end{aligned}$$

Since the component is assumed to be "as good as new" at time $Y = T_1 + D_1$, the process repeats itself from this point of time and

$$P(X(t) = 1|Y = y) = \begin{cases} A(t-y) & \text{for } t > y \\ P(T_1 > t|Y = y) & \text{for } t \le y \end{cases}$$

Therefore

$$A(t) = \int_0^t A(t-y)\,dH(y) + \int_t^\infty P(T_1 > t|Y = y)\,dH(y)$$

But, since $D_1 > 0$, then

$$\int_t^\infty P(T_1 > t|T_1 + D_1 = y)\,dH(y) = \int_0^\infty P(T_1 > t|Y = y)\,dH(y)$$
$$= P(T_1 > t) = 1 - F_{T_1}(t)$$

Hence

$$A(t) = 1 - F_{T_1}(t) + \int_0^t A(t-y)\,dH(y)$$

We may now apply Lemma 7.1, and we get

$$A(t) = 1 - F_T(t) + \int_0^t (1 - F_T(t-x))\,dW_H(x) \qquad (7.90)$$

where

$$W_H(t) = \sum_{n=1}^{\infty} H^{(n)}(t)$$

is the renewal function for a renewal process with underlying distribution $H(t)$.

Since $H(t)$ is a nonlattice distribution, Theorem 7.4, the key renewal theorem, can be used with $Q(t) = 1 - F_T(t)$, and we get

$$\int_0^t (1 - F_T(t-x))\, dW_H(x) \underset{t\to\infty}{\to} \frac{1}{E(Y)} \int_0^\infty (1 - F_T(t))\, dt = \frac{E(T)}{E(T) + E(D)}$$

Since $F_T(t) \to 1$ when $t \to \infty$, we have thus shown that

$$A = \lim_{t \to \infty} A(t) = \frac{E(T)}{E(T) + E(D)} = \frac{\text{MTTF}}{\text{MTTF} + \text{MTTR}} \qquad \square$$

Mean Number of Failures/Repairs

First, let the renewals be the events where a repair is completed. Then we have an ordinary renewal process with renewal periods $Y_1, Y_2, \ldots$ which are independent and identically distributed with distribution function (7.83).

Assume that T_i and D_i both are continuously distributed with densities $f_T(t)$ and $f_D(t)$, respectively. The probability density function of the Y_i's is then

$$h(y) = \int_0^y f_T(y-x) f_D(x)\, dx \qquad (7.91)$$

According to Appendix D the Laplace transform of (7.91) is

$$h^*(s) = f_T^*(s) \cdot f_D^*(s)$$

Let $W_1(t)$ denote the renewal function, that is, the mean number of completed repairs in the time interval $(0, t]$. According to (7.50),

$$W_1^*(s) = \frac{f_T^*(s) \cdot f_D^*(s)}{s(1 - f_T^*(s) \cdot f_D^*(s))} \qquad (7.92)$$

In this case both the T_i's and the D_i's were assumed to be continuously distributed. This, however, turns out *not* to be essential. Equation (7.92) is also valid for discrete distributions or for a mixture of discrete and continuous distributions. Thus we may use (7.28) to get

$$f_T^*(s) = E(e^{-sT_i})$$
$$f_D^*(s) = E(e^{-sD_i})$$

The number of completed repairs in $(0, t]$ can now, at least in principle, be determined for any choice of life- and repair time distributions.

ooo OOO ooo

Next let the renewals be the events where a failure occurs. In this case we get a modified renewal process. The renewal periods $Y_1, Y_2, \ldots$ are independent and

$$H_1(y) = P(Y_1 \leq y) = P(T_1 \leq y) = F_T(y)$$

while

$$H(y) = P(Y_i \leq y) = \int_0^y F_T(y - x) \, dF_D(x) \quad \text{for } i = 2, 3, \ldots$$

Let $W_2(t)$ denote the renewal function, that is, the mean number of failures in $(0, t]$ under these conditions. According to (7.50),

$$W_2^*(s) = \frac{f_T^*(s)}{s(1 - f_T^*(s) \cdot f_D^*(s))} \tag{7.93}$$

which, at least in principle, can be inverted to obtain $W_2(t)$.

Availability at a Given Point of Time

By taking Laplace transforms of (7.90), we get

$$A^*(s) = \frac{1}{s} - F_T^*(s) + \left(\frac{1}{s} - F_T^*(s)\right) \cdot w_H^*(s)$$

RENEWAL PROCESSES

Since

$$F^*(s) = \frac{1}{s} f^*(s)$$

then

$$A^*(s) = \frac{1}{s}(1 - f_T^*(s)) \cdot (1 + w_H^*(s))$$

If we have an ordinary renewal process (i.e., the renewals are the events where a repair is completed), then

$$w_H^*(s) = s W_1^*(s)$$

Hence

$$A^*(s) = \frac{1}{s}(1 - f_T^*(s)) \cdot \left(1 + \frac{f_T^*(s) \cdot f_D^*(s)}{1 - f_T^*(s) \cdot f_D^*(s)}\right)$$

that is,

$$A^*(s) = \frac{1 - f_T^*(s)}{s(1 - f_T^*(s) \cdot f_D^*(s))} \qquad (7.94)$$

The availability $A(t)$ can in principle be determined from (7.94) for any choice of life and repairtime distributions.

Example 7.8: Exponential Lifetime–Exponential Repair Time

Consider an alternating renewal process where the component lifetimes $T_1, T_2, \ldots$ are independent and exponentially distributed with failure rate λ. The corresponding downtimes are also assumed to be independent and exponentially distributed with the repair rate $\mu = 1/\text{MTTR}$. See Figure 7.12.

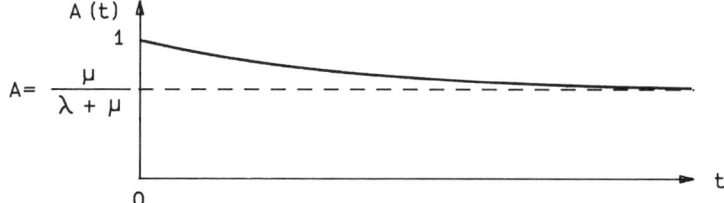

Figure 7.12 Availability of a component with exponential life- and repair time

Then

$$f_T(t) = \lambda e^{-\lambda t} \quad \text{for} \quad t > 0$$

$$f_T^*(s) = \frac{\lambda}{\lambda + s}$$

and

$$f_D(t) = \mu e^{-\mu t} \quad \text{for} \quad t > 0$$

$$f_D^*(s) = \frac{\mu}{\mu + s}$$

The availability $A(t)$ is obtained from (7.94):

$$A^*(s) = \frac{1 - \lambda/(\lambda + s)}{s[1 - (\lambda/(\lambda + s)) \cdot (\mu/(\mu + s))]}$$

$$= \frac{\mu}{\lambda + \mu} \cdot \frac{1}{s} + \frac{\lambda}{\lambda + \mu} \cdot \frac{1}{s + (\lambda + \mu)} \quad (7.95)$$

Equation (7.95) can be inverted (see Appendix D), and we get

$$A(t) = \frac{\mu}{\lambda + \mu} + \frac{\lambda}{\lambda + \mu} e^{-(\lambda + \mu)t} \quad (7.96)$$

which is the same result as we got in Section 6.4 by Markov methods.
The limiting availability is

$$A = \lim_{t \to \infty} A(t) = \frac{\mu}{\lambda + \mu} = \frac{1/\lambda}{1/\lambda + 1/\mu} = \frac{\text{MTTF}}{\text{MTTF} + \text{MTTR}}$$

By inserting $f_T^*(s)$ and $f_D^*(s)$ into (7.92), we get

$$W_1^*(s) = \frac{(\lambda/(\lambda + s)) \cdot (\mu/(\mu + s))}{s[1 - (\lambda/(\lambda + s)) \cdot (\mu/(\mu + s))]}$$

$$= \frac{\lambda \mu}{\lambda + \mu} \cdot \frac{1}{s^2} - \frac{\lambda \mu}{(\lambda + \mu)^2} \cdot \frac{1}{s} + \frac{\lambda \mu}{(\lambda + \mu)^2} \cdot \frac{1}{s + (\lambda + \mu)}$$

By inverting this expression, we get the mean number of completed repairs in the time interval $(0, t]$:

RENEWAL PROCESSES 311

$$W_1(t) = \frac{\lambda\mu}{\lambda+\mu} t - \frac{\lambda\mu}{(\lambda+\mu)^2} + \frac{\lambda\mu}{(\lambda+\mu)^2} e^{-(\lambda+\mu)t} \quad (7.97)$$

Next let us define the renewals as the events where the component fails. By substituting for f_T^* and f_D^* in (7.93), we get

$$W_2^*(s) = \frac{\lambda/(\lambda+s)}{s[1 - (\lambda/(\lambda+s)) \cdot (\mu/(\mu+s))]}$$

$$= \frac{\lambda\mu}{\lambda+\mu} \cdot \frac{1}{s^2} + \frac{\lambda^2}{(\lambda+\mu)^2} \cdot \frac{1}{s} - \frac{\lambda^2}{(\lambda+\mu)^2} \cdot \frac{1}{s+(\lambda+\mu)} \quad (7.98)$$

By inverting the right-hand side of (7.98), we get the mean number of failures in the time interval $(0, t]$ to be

$$W_2(t) = \frac{\lambda\mu}{\lambda+\mu} t + \frac{\lambda^2}{(\lambda+\mu)^2} - \frac{\lambda^2}{(\lambda+\mu)^2} e^{-(\lambda+\mu)t} \quad (7.99)$$

The corresponding renewal density is

$$w_2(t) = \frac{\lambda\mu}{\lambda+\mu} - \frac{\lambda^2}{\lambda+\mu} e^{-(\lambda+\mu)t} \quad (7.100)$$

By comparing with (7.96), we observe that

$$w_2(t) = \lambda \cdot A(t) \qquad \square$$

Example 7.9: Exponential Lifetime–Constant Repair Time
Consider an alternating renewal process where the system lifetimes $T_1, T_2, \ldots$ are independent and exponentially distributed with failure rate λ. The downtimes are assumed to be constant and equal to τ with probability 1: $P(D_i = \tau) = 1$ for $i = 1, 2, \ldots$. See Figure 7.13.

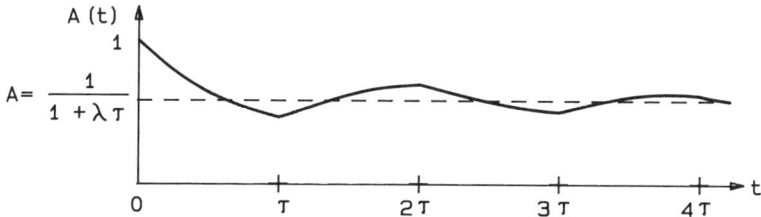

Figure 7.13 The availability of a component with exponential lifetime and constant repair time τ

The corresponding Laplace transforms are

$$f_T^*(s) = \frac{\lambda}{\lambda + s}$$
$$f_D^*(s) = E(e^{-sD}) = e^{-s\tau} \cdot P(D = \tau) = e^{-s\tau}$$

Hence the Laplace transform of the availability (7.94) becomes

$$A^*(s) = \frac{1 - \lambda/(\lambda + s)}{s[1 - (\lambda/(\lambda + s))e^{-s\tau}]} = \frac{1}{s + \lambda - \lambda e^{-s\tau}}$$
$$= \frac{1}{\lambda + s} \cdot \left[\frac{1}{1 - (\lambda/(\lambda + s))e^{-s\tau}}\right] = \frac{1}{\lambda + s} \sum_{\nu=0}^{\infty} \left(\frac{\lambda}{\lambda + s}\right)^{\nu} e^{-s\nu\tau}$$
$$= \frac{1}{\lambda} \sum_{\nu=0}^{\infty} \left(\frac{\lambda}{\lambda + s}\right)^{\nu+1} e^{-s\nu\tau} \qquad (7.101)$$

The availability then becomes

$$A(t) = \mathscr{L}^{-1}(A(s)) = \sum_{\nu=0}^{\infty} \frac{1}{\lambda} \mathscr{L}^{-1}\left[\left(\frac{\lambda}{\lambda + s}\right)^{\nu+1} e^{-s\nu\tau}\right]$$

According to Appendix D,

$$\mathscr{L}^{-1}\left[\left(\frac{\lambda}{\lambda + s}\right)^{\nu+1}\right] = \frac{\lambda^{\nu+1}}{\nu!} t^{\nu} e^{-\lambda t} = f(t)$$

$$\mathscr{L}^{-1}(e^{-s\nu\tau}) = \delta(t - \nu\tau)$$

where $\delta(t)$ denotes the Dirac delta function. Thus

$$\mathscr{L}^{-1}\left[\left(\frac{\lambda}{\lambda + s}\right)^{\nu+1} \cdot e^{-s\nu\tau}\right] = \mathscr{L}^{-1}\left[\left(\frac{\lambda}{\lambda + s}\right)^{\nu+1}\right] * \mathscr{L}^{-1}(e^{-s\nu\tau})$$
$$= \int_0^{\infty} \delta(t - \nu\tau - x) f(x)\, dx$$
$$= f(t - \nu\tau) \cdot u(t - \nu\tau)$$

where

$$u(t - \nu\tau) = \begin{cases} 1 & \text{if } t \geq \nu\tau \\ 0 & \text{if } t < \nu\tau \end{cases}$$

Hence the availability is

$$A(t) = \sum_{\nu=0}^{\infty} \frac{\lambda^\nu}{\nu!} (t - \nu\tau)^\nu e^{-\lambda(t-\nu\tau)} u(t - \nu\tau) \quad (7.102)$$

The limiting availability is then, according to Theorem 7.9,

$$A = \lim_{t \to \infty} A(t) = \frac{\text{MTTF}}{\text{MTTF} + \text{MTTR}} = \frac{1/\lambda}{(1/\lambda) + \tau} = \frac{1}{1 + \lambda\tau} \quad (7.103)$$

The Laplace transform for the renewal density is

$$w^*(s) = \frac{f_T^*(s) \cdot f_D^*(s)}{1 - f_T^*(s) \cdot f_D^*(s)} = \frac{\lambda e^{-s\tau}/(\lambda + s)}{1 - \lambda e^{-s\tau}/(\lambda + s)}$$
$$= \frac{1}{\lambda + s - \lambda e^{-s\tau}} \lambda e^{-s\tau} = \lambda \cdot A^*(s) e^{-s\tau}$$

where $A^*(s)$ is given by (7.101).
Then the renewal density becomes

$$w(t) = \lambda \mathscr{L}^{-1}(A^*(s) \cdot e^{-s\tau}) = \lambda \int_0^\infty \delta(t - \tau - x) A(x)\, dx$$

that is,

$$w(t) = \begin{cases} \lambda \cdot A(t - \tau) & \text{if } t \geq \tau \\ 0 & \text{if } t < \tau \end{cases} \quad (7.104)$$

Hence the mean number of completed repairs in the time interval $(0, t]$ for $t > \tau$ is

$$W(t) = \int_0^t w(u)\, du = \lambda \int_\tau^t A(u - \tau)\, du = \lambda \int_0^{t-\tau} A(u)\, du \quad (7.105)$$

7.4 NONHOMOGENEOUS POISSON PROCESSES

In this section the homogeneous Poisson process (HPP) is generalized by allowing the rate (intensity) of the process to be a function of time.

Introduction and Definitions

Definition 7.7. A counting process $\{N(t), t \geq 0\}$ is a nonstationary (or nonhomogeneous) Poisson process (NHPP) with intensity function $w(t)$ for $t \geq 0$, if

1. $N(0) = 0$
2. $\{N(t), t \geq 0\}$ has independent increments
3. $P(N(t + \Delta t) - N(t) \geq 2) = o(\Delta t)$, that is, the system will not experience more than one failure at the same time
4. $P(N(t + \Delta t) - N(t) = 1) = w(t)\Delta t + o(\Delta t)$ □

The basic "parameter" of the NHPP is the ROCOF function $w(t)$. This function is also called the *intensity* or the *peril rate* of the NHPP. The *cumulative intensity* of the process is

$$W(t) = \int_0^t w(u)\,du \qquad (7.106)$$

This definition of course covers the situation in which the intensity is a function of some observed explanatory variable that is a function of the time t.

It is important to note that the NHPP model does not require stationary increments. This means that failures may be more likely to occur at certain times than others, and hence the interarrival times are generally neither independent nor identically distributed. Consequently statistical techniques based on the assumption that the data are independent and identically distributed can not be applied to an NHPP.

The NHPP is often used to model trends in the interarrival times, that is, improving (*happy*) or deteriorating (*sad*) systems. It seems intuitive that a happy system will have a decreasing ROCOF function, while a sad system will have an increasing ROCOF function.

From experience one finds that many systems approximately satisfy the properties of the NHPP listed in Definition 7.7. We know from the assumption of independent increments that the number of failures in a specified interval $(t_1, t_2]$ will be independent of the failures and interarrival times prior to time t_1. A practical implication of this assumption is that the reliability is exactly the same just before a failure and immediately after the corresponding repair. This condition has been termed *minimal repair* (see Ascher and Feingold 1984, p. 51). An

NHPP clearly is not a realistic model when the failed parts to be replaced have been in operation for a long time. For the NHPP to be realistic, the parts put into service should be identical to the old ones and hence should be aged outside the system under identical conditions for the same period of time.

Now consider a system consisting of a large number of components. Suppose that a critical component fails and causes a system failure and that this component is immediately replaced by a component of the same type, thus causing a negligible system downtime. Since only a small fraction of the system is replaced, it seems natural to assume that the systems's reliability after the repair essentially is the same as immediately before the failure. In other words, the assumption of *minimal repair* is a realistic approximation. When an NHPP is used to model a repairable system, the system is treated as a *black box* in that no concern is made about how the system "looks inside."

A car is a typical example of a repairable system. Usually the operating time of a car is expressed in terms of the mileage indicated on the speedometer. Repair actions will usually not imply any extra mileage. The repair "time" is thus negligible. Many repairs are accomplished by adjustments or replacement of single components. The minimal repair assumption is therefore often applicable, and the NHPP may be accepted as a realistic model, at least as a first-order approximation.

Some Results

From the definition of the NHPP it is straightforward to verify (e.g., see Ross 1983, p. 46) that the number of failures in an interval $(t_1, t_2]$ is Poisson distributed:

$$P(N(t_2) - N(t_1) = n) = \frac{[W(t_2) - W(t_1)]^n}{n!} e^{-[W(t_2) - W(t_1)]}$$
$$\text{for} \quad n = 0, 1, 2, \ldots \quad (7.107)$$

where $W(t)$ denotes the cumulative intensity of the process (7.106). The mean number of failures in the interval $(t_1, t_2]$ is thus

$$E(N(t_2) - N(t_1)) = W(t_2) - W(t_1) = \int_{t_1}^{t_2} w(t) \, dt \quad (7.108)$$

Time to First Failure

Let T_1 denote the time from $t = 0$ until the first failure. The survivor function of T_1 is

$$R_1(t) = P(T_1 > t) = P(N(t) = 0) = e^{-W(t)} = e^{-\int_0^t w(t) \, dt} \quad (7.109)$$

Hence the failure rate (force of mortality) $z_{T_1}(t)$ of the first interarrival time T_1 is equal to the ROCOF $w(t)$ of the process. Note, however, the different meaning of the two expressions. $z_{T_1}(t)\Delta t$ approximates the probability that the *first* failure occurs in $(t, t+\Delta t]$, while $w(t)\Delta t$ approximates the probability that a failure, not necessarily the first, occurs in $(t, t + \Delta t]$.

A consequence of (7.109) is that the distribution of the first interarrival time, that is, the time from $t = 0$ until the system's first failure, will determine the ROCOF of the entire process. Thompson (1981) claims that this is a nonintuitive fact that is casting doubt on the NHPP as a realistic model for repairable systems. Use of an NHPP model implies that if we are able to estimate the failure rate (force of mortality) of the time to the *first* failure, such as for a specific type of automobiles, we at the same time have an estimate of the ROCOF of the entire life of the automobile.

Time Between Failures

Assume that the process is observed at time t_0. By using (7.107), we can express the distribution of the time Y_{t_0} until the next failure as

$$P(Y_{t_0} > t) = P(N(t_0 + t) - N(t_0) = 0) = e^{-(W(t_0+t)-W(t_0))}$$
$$= e^{-\int_{t_0}^{t_0+t} w(u)\,du} = e^{-\int_0^t w(t_0+u)\,du} \qquad (7.110)$$

Note that this result is independent of whether t_0 denotes a failure time or an arbitrary point in time.

Let t_0 denote the time at failure $k - 1$. In this case Y_{t_0} denotes the time between failure $k - 1$ and failure k (i.e., the kth interarrival time T_k). The failure rate (force of mortality) of the kth interarrival time T_k—that is, the time between failure $k - 1$ (at time t_0) and failure k—is from (7.110),

$$z_{t_0}(t) = w(t_0 + t) \qquad \text{for } t \geq 0 \qquad (7.111)$$

The mean time between failure $k - 1$ (at time t_0) and failure k is

$$E(T_k) = E(Y_{t_0}) = \int_0^\infty P(Y_{t_0} > t)\,dt$$
$$= \int_0^\infty e^{-\int_0^t w(t_0+u)\,du}\,dt \qquad (7.112)$$

Relation to the Homogeneous Poisson Process

Let $N(t)$ denote a nonhomogeneous Poisson process (NHPP) with intensity $w(t) > 0$ such that the inverse $W^{-1}(t)$ of the cumulative intensity $W(t)$ exists. Consider the process $N^*(t)$ defined by

NONHOMOGENEOUS POISSON PROCESSES

$$N^*(t) = N(W^{-1}(t)) \quad \text{for } t \geq 0$$

It is straightforward to verify (see Problem 6 in Section 7.5) that $N^*(t)$ is a homogeneous Poisson process (HPP) with rate $\lambda = 1$. An NHPP may thus be transformed into an HPP by a deterministic change of the time scale.

Let S_k denote the time until failure k for $k = 0, 1, 2, \ldots$, where $S_0 = 0$. The distribution of S_k is given by

$$P(S_k > t) = P(N(t) \leq k - 1) = P(N^*(W(t)) \leq k - 1)$$
$$= \sum_{\nu=0}^{k-1} \frac{W(t)^\nu}{\nu!} e^{-W(t)} \quad (7.113)$$

When $W(t)$ is small, this probability may be determined from standard tables of the Poisson distribution. When $W(t)$ is large, the probability may be determined by a normal approximation:

$$P(S_k > t) = P(N^*(W(t)) \leq k - 1)$$
$$= P\left(\frac{N^*(W(t)) - W(t)}{\sqrt{W(t)}} \leq \frac{k - 1 - W(t)}{\sqrt{W(t)}}\right)$$
$$\approx \Phi\left(\frac{k - 1 - W(t)}{\sqrt{W(t)}}\right) \quad (7.114)$$

The Nelson-Aalen Estimator

Assume that m different and independent NHPPs with a common ROCOF $w(t)$ have been observed. The ith process is observed in the time interval $(a_i, b_i]$, and n_i failures have occurred by time t_{ij} where $j = 1, 2, \ldots, n_i$ and $i = 1, 2, \ldots, m$. A nonparametric estimator of the cumulative intensity function $W(t) = \int_0^t w(u)\,du$ for the NHPPs in the time interval considered is given by

$$\widehat{W}(t) = \sum_{t_{ij} \leq t} \frac{1}{Y(t_{ij})} \quad (7.115)$$

where $Y(t_{ij})$ denotes the number of processes which are operating immediately before time t_{ij}. Note that $\widehat{W}(t) = 0$ for $t < \min_{ij} t_{ij}$. Note also that when there is only one sample ($m = 1$), then the Nelson–Aalen estimator coincides with $N(t)$, which is plotted as in Figure 7.2. A simple example of the Nelson-Aalen estimator for two simultaneous processes is illustrated in Figure 7.14.

The estimator (7.115) was introduced by Aalen (1975, 1978) for counting processes in general, and it generalizes the Nelson (1969) estimator, which is

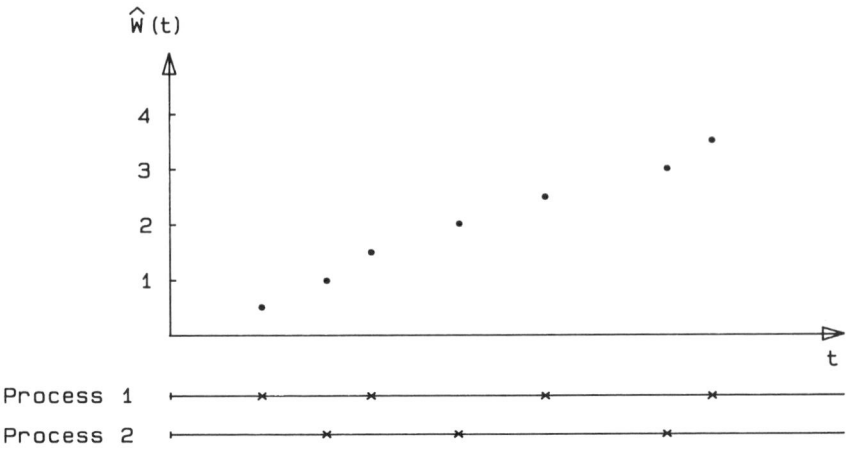

Figure 7.14 The Nelson–Aalen estimator for two simultaneous processes

further discussed in Chapter 9. It may be shown (see the discussion in Andersen and Borgan 1985) that $\widehat{W}(t)$ is an approximately unbiased estimator of $W(t)$ and that the variance can be estimated (almost unbiasedly) by:

$$\text{var}(\widehat{W}(t)) \approx \hat{\sigma}^2(t) = \sum_{t_{ij} \leq t} \frac{1}{Y(t_{ij})^2} \qquad (7.116)$$

$\widehat{W}(t)$ may further be shown (see Andersen and Borgan 1985) to be asymptotically normally distributed with mean $W(t)$ and a variance estimated by $\hat{\sigma}^2(t)$. Hence an approximate $100(1 - \alpha)\%$ pointwise confidence interval for $W(t)$, is given by

$$\widehat{W}(t) - u_{\alpha/2}\hat{\sigma}(t) \leq W(t) \leq \widehat{W}(t) + u_{\alpha/2}\hat{\sigma}(t)$$

where u_α denotes the upper $100\alpha\%$ percentile of the standard normal distribution $\mathcal{N}(0, 1)$. The Nelson–Aalen estimator is further discussed in Samset (1988).

Parametric Models

Several parametric models have been established to describe the ROCOF of an NHPP. We will discuss three of these models:

1. The power law model
2. The linear model
3. The log-linear model

All the three models may be written in the common form (see Atwood 1992)

$$w(t) = \lambda_0 g(t; \vartheta) \qquad (7.117)$$

where λ_0 is a common multiplier and $g(t; \vartheta)$ determines the shape of the ROCOF $w(t)$. The three models may be parameterized in various ways. In this section we will use the parameterization of Crowder et al. (1991), although the parameterization of Atwood (1992) may be more logical.

The Power Law Model
In the power law model the ROCOF of the NHPP is defined as

$$w(t) = \lambda \beta t^{\beta-1} \qquad \text{for} \quad \lambda > 0, \beta > 0, t \geq 0 \qquad (7.118)$$

This NHPP is sometimes referred to as a *Weibull process*, since the ROCOF has the same functional form as the failure rate (force of mortality) of the Weibull distribution. The first arrival time T_1 of this process is Weibull distributed with shape parameter β and scale parameter λ. However, according to Ascher and Feingold (1984), one should avoid the name Weibull process in this situation, since it gives the wrong impression that the Weibull distribution can be used to model trend in interarrival times of a repairable system. Hence such notation may lead to confusion.

A repairable system modeled by the Power law model is seen to be improving if $0 < \beta < 1$ and deteriorating if $\beta > 1$. If $\beta = 1$, the model reduces to an HPP. The case $\beta = 2$ is seen to give a linearly increasing ROCOF.

The power law model was first proposed by Crow (1974) based on ideas of Duane (1964). A goodness-of-fit test for the Power law model based on TTT plots (see Chapter 9) is proposed and discussed by Klefsjö and Kumar (1992).

The Linear Model
In the linear model the ROCOF of the NHPP is defined by

$$w(t) = \lambda(1 + \alpha t) \qquad \text{for} \quad \lambda > 0, t \geq 0 \qquad (7.119)$$

The linear model has been discussed by Vesely (1991) and Atwood (1992). A repairable system modeled by the linear model is deteriorating if $\alpha > 0$ and improving when $\alpha < 0$. When $\alpha < 0$, then $w(t)$ will sooner or later become less than zero. The model should only be used in time intervals where $w(t) > 0$.

The Log-Linear Model
In the log-linear model or *Cox-Lewis* model the ROCOF of the NHPP is defined by

$$w(t) = e^{\alpha+\beta t} \quad \text{for} \quad -\infty < \alpha, \beta < \infty, t \geq 0 \qquad (7.120)$$

A repairable system modeled by the log-linear model is improving if $\beta < 0$ and deteriorating if $\beta > 0$. When $\beta = 0$, the log-linear model reduces to an HPP.

The log-linear model was proposed by Cox and Lewis (1966) who used the model to investigate trends in the interarrival times between failures in air-conditioning equipment in aircrafts. The first arrival time T_1 has failure rate (force of mortality) $z(t) = e^{\alpha+\beta t}$ and hence has a truncated Gumbel distribution of the smallest extreme (i.e., a Gompertz distribution; see Section 2.14).

Parameter Estimation

The parameters of the three models described above may be estimated by standard maximum likelihood techniques. Estimators are presented, for example, by Crowder et al. (1991) and Atwood (1992).

Statistical Tests of Trend

The simple graph in Figure 7.2 clearly indicates an increasing rate of failures, that is, a deteriorating or *sad* system. The next step in an analysis of the data may be to perform a *statistical test* to find out whether the observed trend is *statistically significant* or just accidental. A number of tests have been developed for this purpose, that is, for testing the null hypothesis

H_0: "No trend" (or more precisely that the interarrival times are independent and identically exponentially distributed)

against the alternative hypothesis

H_1: "Monotonic trend" (i.e., the system is either *sad* or *happy*)

Among these are two tests that we will discuss:

1. The Laplace test
2. The Military Handbook test

These two tests are presented in detail by Ascher and Feingold (1984) and Crowder et al. (1991). It can be shown that the Laplace test is optimal when the true failure mechanism is that of a log-linear NHPP model (see Cox and Lewis 1966), while the Military Handbook test is optimal when the true failure mechanism is that of a power law NHPP model (see Bain, Engelhardt, and Wright 1985).

The Laplace Test

The test statistic for the case where the system is observed until n failures have occurred is

$$U = \frac{[\sum_{j=1}^{n-1} S_j - (S_n/2)]/(n-1)}{S_n/\sqrt{12(n-1)}} \qquad (7.121)$$

where $S_1, S_2, \ldots$ denote the failure times. For the case where the system is observed until time t_0, the test statistic is

$$U = \frac{[\sum_{j=1}^{n} S_j - (t_0/2)]/n}{t_0/\sqrt{12n}} \qquad (7.122)$$

In both cases the test statistic U is approximately standard normally $\mathcal{N}(0, 1)$ distributed when the null hypothesis H_0 is true. The value of U is seen to indicate the direction of the trend, with $U > 0$ for a *happy* system and $U < 0$ for a *sad* system. Optimal properties of the Laplace test have, for example, been investigated by Gaudoin (1992).

Military Handbook Test

The test statistic of the so-called Military Handbook test (see MIL-HDBK-189, 1981) for the case where the system is observed until n failures have occurred is

$$Z = 2 \sum_{i=1}^{n-1} \ln \frac{S_n}{S_i} \qquad (7.123)$$

For the case where the system is observed until time t_0, the test statistic is

$$Z = 2 \sum_{i=1}^{n} \ln \frac{t_0}{S_i} \qquad (7.124)$$

The asymptotic distribution of Z is in the two cases a χ^2 distribution with $2(n-1)$ and $2n$ degrees of freedom, respectively.

The hypothesis of no trend (H_0) is rejected for *small* or *large* values of Z. Low values of Z correspond to deteriorating systems, while large values of Z correspond to improving systems.

7.5 PROBLEMS

1. Consider a homogeneous Poisson process (HPP) $\{N(t), t \geq 0\}$, and let $t, s \geq 0$. Determine

$$E(N(t) \cdot N(t + s))$$

2. Consider a homogeneous Poisson process (HPP) $\{N(t), t \geq 0\}$ with rate $\lambda > 0$. Verify that

$$P(N(t) = k | N(s) = n) = \binom{n}{k}\left(\frac{t}{s}\right)^k \left(1 - \frac{t}{s}\right)^{n-k}$$

for $0 < t < s, 0 \leq k \leq n$

3. Let T_1 denote the time to the first occurrence of a homogeneous Poisson process (HPP) $\{N(t), t \geq 0\}$ with rate λ. Show that

$$P(T_1 \leq s | N(t) = 1) = \frac{s}{t} \quad \text{for } s \leq t$$

4. Let $S_1, S_2, \ldots$ be the occurrence times of a homogeneous Poisson process (HPP) $\{N(t), t \geq 0\}$ with rate λ. Assume that $N(t) = n$. Show that the random variables $S_1, S_2, \ldots, S_n$ have the joint probability density function

$$f_{S_1,\ldots,S_n | N(t)=n}(s_1, \ldots, s_n) = \frac{n!}{t^n} \quad \text{for } 0 < s_1 < \cdots < s_n \leq t$$

5. Consider a renewal process $\{N(t), t \geq 0\}$. Is it true that
   (a) $N(t) < r$ if and only if $S_r > t$?
   (b) $N(t) \leq r$ if and only if $S_r \geq t$?
   (c) $N(t) > r$ if and only if $S_r < t$?

6. Consider a nonhomogeneous Poisson process (NHPP) $\{N(t), t \geq 0\}$ where the intensity $w(t) > 0$ for all $t \geq 0$. Let

$$N^*(t) = N(W^{-1}(t))$$

Show that $\{N^*(t); t \geq 0\}$ is a homogeneous Poisson process (HPP) with rate $\lambda = 1$.

7. Consider a nonhomogeneous Poisson process (NHPP) with intensity

PROBLEMS

$$w(t) = \lambda \cdot \frac{t+1}{t} \quad \text{for } t \geq 0$$

(a) Make a sketch of $w(t)$ as a function of t.
(b) Make a sketch of the cumulative ROCOF, $W(t)$, as a function of t.

8. Consider a nonhomogeneous Poisson process (NHPP) $\{N(t), t \geq 0\}$ with intensity:

$$w(t) = \begin{cases} 6 - 2t & \text{for } 0 \leq t \leq 2 \\ 2 & \text{for } 2 < t \leq 20 \\ -18 + t & \text{for } t > 20 \end{cases}$$

(a) Make a sketch of $w(t)$ as a function of t.
(b) Make a sketch of the corresponding cumulative ROCOF, $W(t)$, as a function of t.
(c) Estimate the number of failures/events in the interval $(0, 12)$.

9. In Section 7.3 it is claimed that the superposition of independent renewal processes is generally *not* a renewal process. Explain why the superposition of independent homogeneous Poisson processes (HPP) is a renewal process. What is the renewal density of this superimposed process?

10. Atwood (1992) uses the following parameterization for the power law model, the linear model, and the log-linear model:

$w(t) = \lambda_0 (t/t_0)^\beta$ (power law model)
$w(t) = \lambda_0 [1 + \beta(t - t_0)]$ (linear model)
$w(t) = \lambda_0 e^{\beta(t - t_0)}$ (log-linear model)

(a) Discuss the meaning of t_0.
(b) Show that Atwood's parameterization is compatible with the parameterization used in Section 7.4.
(c) Show that $w(t) = \lambda_0$ when $t = t_0$ for all the three models.
(d) Show that $w(t)$ is increasing if $\beta > 0$, is constant if $\beta = 0$, and decreasing if $\beta < 0$, for all the three models.

11. Use the MIL-HDBK test described in Section 7.4 to check if the "increasing trend" of the data in Example 7.1 is significant (at a 5% level).

12. Table 7.3 shows the intervals in days between successive failures of a piece of software developed as part of a large data system. The first interval is

9, the second is 12, and so on. The data are from Jelinski and Moranda (1972).

Table 7.3 Intervals in Days Between Successive Failures of a Piece of Software

9	12	11	4	7	2	5	8	5	7
1	6	1	9	4	1	3	3	6	1
11	33	7	91	2	1	87	47	12	9
135	258	16	35						

(a) Establish the Nelson-Aalen plot ($N(t)$ plot) of the data set. Is the ROCOF increasing or decreasing?

(b) Assume that the ROCOF follows a log-linear model, and use maximum likelihood techniques to estimate the parameters of this model (e.g., formulas given by Crowder et al. 1991, sec. 8.7).

(c) Draw the estimated cumulative ROCOF in the same diagram as the Nelson-Aalen plot. Is the fit acceptable?

(d) Use the Laplace test to test whether the ROCOF is decreasing or not (use a 5% level of significance).

CHAPTER 8

Dependent Failures

8.1 INTRODUCTION

In Chapter 4 we studied situations where the n components of a system fail independently of each other. This is modeled by assuming that the state variables of the n components $X_1(t), X_2(t), \ldots, X_n(t)$ are independent random variables. This assumption considerably simplifies the modeling as well as the statistical analysis.

However, when the components of a system fail, they do not necessarily have to fail independently of each other. We may distinguish between two main types of dependence: *positive* and *negative*. If a failure of one component leads to an increased tendency for another component to fail, the dependence is said to be *positive*. If, on the other hand, the failure of one component leads to a reduced tendency for another component to fail, the dependence is called *negative*.

In practical applications, positive dependence is usually the most relevant type of dependence. Negative dependence may, however, also occur.

Dependent failures may be classified in three main groups:

1. *Common cause failures.* Common cause failures are multiple failures that are a direct result of a common or shared root cause. The root cause may be extreme environmental conditions (fire, flood, earthquake, lightning strike, etc.), failure of a piece of hardware external to the system, or a human error. The root cause is not a failure of another component in the system.

   Operation and maintenance errors are often reported to be the root cause for common cause failures. Examples include carelessness, improper adjustments and miscalibrations, human errors, and erraneous or nonadequate procedures. The number of components that fail due to the common or root cause is called the *multiplicity* of the failures.

2. *Cascading failures.* Cascading failures are multiple failures initiated by the failure of one component in the system that results in a chain reaction or "domino effect." When several components share a common load,

failure of one component may lead to increased load on the remaining components and consequently to an increased likelihood of failure.

Components may also influence each other through the internal environment. Malfunction of a component may, for example, lead to a more hostile working environment for the other components through increased pressure, temperature, humidity, and so on. Cascading failures are sometimes called *propagating* failures.

3. *Negative dependencies.* Negative dependency failures are single failures that reduce the likelihood of failures of other components. If for example an electrical fuse fails open such that the "downstream" circuit is disconnected, the load on the electrical devices in this circuit is removed and their likelihood of failure is reduced.

When a system is "down" for repair of a specific component, the load on other components is often removed and their likelihood of failure is consequently reduced during the system downtime.

Example 8.1

One of the best known accidents resulting from a common cause failure is the fire at the Browns Ferry nuclear power plant near Decatur, Alabama on March 22, 1975 (Rippon 1975). The fire started when two of the operators used a candle to check for air leaks between the cable room and one of the reactor buildings, which was kept at a negative air pressure. The candle's flame was drawn out along the conduit, and the urethane seal used where the cables penetrate the wall caught fire. The fire continued until the insulation of about 2000 cables was damaged. Among these were all the cables to the automatic emergency shutdown (ESD) systems and also the cables to all the "manually" operated valves, apart from four relief valves. With these four valves it was possible to close down the reactor so that nuclear meltdown was avoided. This accident resulted in new instructions that require cables to the different emergency shutdown systems to be put in separate conduits and prohibit the use of combustible filling (e.g., urethane foam).

ooo OOO ooo   □

A taxonomy and classification of the root causes for common cause failures is presented in Table 8.1 which is adapted from Edwards and Watson (1979). This taxonomy may help the analyst systematically to concentrate attention on the possible root causes and to judge them one at a time. Common cause failures may greatly reduce the reliability of a system, especially of systems with a high degree of redundancy. A significant research activity has therefore been devoted to this problem. A wide range of technical reports and papers describing various aspects of dependent failures have been written during the last 10 to 15 years. A special 1991 issue on "Dependent Failure Analysis" of the journal *Reliability Engineering and System Safety* (vol. 34, no. 3), gives a state of the art survey and lists a large number of references. Other valuable references include

Table 8.1 Classification of Reasons for Common Cause Failures

Engineering				Operations			
Design		Construction		Procedural		Environmental	
Functional Deficiencies	Realization Faults	Manufacture	Installation and Commissioning	Maintenance and Test	Operation	Normal Extremes	Energetic Events
Hazard undetectable	Channel dependency	Inadequate quality control	Inadequate quality control	Imperfect repair	Operator errors	Temperature Pressure Humidity	Fire Flood Weather Earthquake
Inadequate instrumentation	Common operation and protection components	Inadequate standards	Inadequate standards	Imperfect testing	Inadequate procedures	Vibration Acceleration Stress	Explosion Missiles Electrical power
Inadequate control	Operational deficiencies	Inadequate inspection	Inadequate inspection	Imperfect calibration	Inadequate supervision	Corrosion Contamination Interference	Radiation Chemical sources
	Inadequate components	Inadequate testing	Inadequate testing and commissioning	Imperfect procedures	Communication error	Radiation Static charge	
	Design errors Design limitations			Inadequate supervision			

Source: Adapted from Edwards and Watson (1979). Reproduced with the permission of AEA Technology.

Edwards and Watson (1979), Fleming et al. (1986), Mosleh et al. (1988, 1989), and Bodsberg and Hokstad (1988), who discuss the impact of dependent failures on high-reliability process shutdown systems.

8.2 HOW TO OBTAIN RELIABLE SYSTEMS

As for any reliability analysis, failure mode and effect analysis (FMEA) also provides the basic framework for the identification and investigation of dependent failures. In IEEE Std. 352 two extensions of the conventional FMEA are described in order to encompass potential interdependencies between the system components:

1. A common cause failure analysis to identify root causes and mechanisms of failures of components normally considered to be redundant
2. A cascading failure analysis to identify failures that can lead to "chain-type" events affecting several different areas or systems in a plant

Detailed procedures for the analyses are given in IEEE Std. 352.

The most important defense against accidental component failures is the use of redundancy (see Section 4.6). The fire at Browns Ferry shows, however, that redundancy itself is not enough. Even if there had been a higher number of redundant shutdown systems, all these would have been made inactive by a single error—the flame from a candle. Common cause failures can be prevented by a variety of defenses. A number of procedures and checklists have been developed to assist the engineer in creating a design that is robust against common cause failures; see, for example, Smith et al. (1988), Edwards and Watson (1979), and Bourne et al. (1981).

Some general defensive tactics to avoid common cause failures are presented by Parry (1991):[1]

- *Barriers.* Any physical impediment that tends to confine and/or restrict a potentially damaging condition.
- *Personnel training.* A program to ensure that the operators and maintainers are familiar with procedures and are able to follow them correctly during all conditions of operation.
- *Quality control.* A program to ensure that the product is in conformance with the documented design and that its operation and maintenance are according to approved procedures, standards, and regulatory requirements.
- *Redundancy.* Additional, identical, redundant components added to a sys-

---

[1] Reprinted from Reliability Engineering & System Safety, Volume 34, G. W. Parry, "Common Cause Failure Analysis: A Critique and Some Suggestions." pp. 320. Copyright 1991, with kind permission from Elsevier Science Ltd, The Boulevard, Langford Lane, Kidlington OX5 1GB, UK.

tem solely for the purpose of increasing the likelihood that a sufficient number of components will survive exposure to a given cause of failure and are available to perform a given function. This is a tactic to improve system availability, but, by definition, common cause failures decrease the positive impact of this particular tactic. Nevertheless, increased redundancy will generally still have value.

- *Preventive maintenance.* A program of applicable and effective preventive maintenance tasks designed to prevent premature failure or degradation of components.
- *Monitoring, surveillance testing, and inspection.* Monitoring via alarms, frequent tests, and/or inspections so that unannounced failures from any detectable causes are not allowed to accumulate. This includes special tests performed on redundant components in response to observed failures.
- *Procedures review.* A review of operational, maintenance, and calibration/test procedures to eliminate incorrect or inappropriate actions resulting in component or system unavailability.
- *Diversity.* The mixture of interchangeable components made by different manufacturers (equipment diversity) or the introduction of a totally redundant system with an entirely different principle of operation (functional diversity) for the express purpose of reducing the likelihood of a total loss of function that might occur because all like components are vulnerable to the same cause(s) of failure. Another form of applying this concept is diversity in staff, whereby different teams to maintain and test redundant trains. This is a tactic that specifically addresses common cause failures.

An analysis of common cause failures constitutes an essential part of reliability studies of complex high-reliability systems. A systematic approach is needed. A general approach for treating common cause failures in safety and reliability studies is presented by Mosleh et al. (1988, 1989). A more brief presentation is given by Mosleh (1991).

The first step in an analysis of common cause failures is to identify so-called common cause candidates, that is, components for which the independence assumption is suspected to be incorrect. Several checklists have been developed to aid the analyst in this task.

A systematic engineering study has to be performed on how to mitigate against each dependency. The impact of potential dependencies on the reliability of the system has to be estimated. Data on the occurrence of common cause failures must be collected, analyzed statistically, and saved for later engineering use, in particular, for the following situations:

1. Future designs of the system
2. Assessment of system reliability
3. Working out rules for system operators

8.3 MODELING OF DEPENDENT FAILURES

In situations where we have good reasons to believe that calculations based on the assumption of independent failures are adequate, the models and methods described in Chapter 4 can be used. Cascading failures and negative dependencies may, however, often be taken into account in these analyses because they arise from functional or physical interrelationships between components and subsystems that may explicitly be modeled in a fault tree/reliability block diagram or in a Markov model. Nevertheless, in a number of situations an analysis based on the assumption of independence will lead to unrealistic results and thus be of limited value for practical purposes.

At the moment large efforts are being made to develop suitable models that take into account different types of dependence. In the following sections we will give examples of some such models. In these models there are sometimes built-in approximations, but the results obtained in this way are often far more realistic than the ones obtained by assuming independence.

When analyzing reliability, one usually deals with technical systems. In some cases it may be possible to express typical features of the system mathematically and derive a model from this (mechanistic model). In other situations, when sufficient data (information) is at hand, an *empirical* model based on a sample of the data may be adequate for the analysis. In both cases the analyst must confront the possible models with data to see if they reflect the specific design features of the system. The analyst also has to be aware that the validity of a model may be limited.

8.4 SPECIAL MODELS

In this section we will discuss four models for dependent failures:

1. The "square-root" method
2. The β-factor model
3. The generalized β-factor model
4. The binomial failure rate (BFR) model

The "Square-Root" Method

In the reactor safety study WASH-1400 a simple bounding technique was used to estimate the effect of common cause failures on a system. We will illustrate this technique by a simple example.

Consider a parallel structure such as that in Figure 8.1 with two components 1 and 2, both of which may fail as a result of a common cause. Let A_i represent the situation where component i is down at time t, ($i = 1, 2$). The probability that the parallel structure is down at time t is then $P(A_1 \cap A_2)$.

SPECIAL MODELS

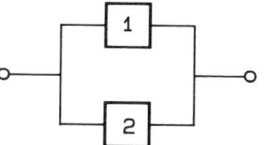

Figure 8.1 Parallel system

Since $(A_1 \cap A_2) \subseteq A_1$,

$$P(A_1 \cap A_2) \leq P(A_1)$$

Similarly

$$P(A_1 \cap A_2) \leq P(A_2)$$

Therefore

$$P(A_1 \cap A_2) \leq \min\{P(A_1), P(A_2)\} \qquad (8.1)$$

If the events A_1 and A_2 are independent, by definition

$$P(A_1 \cap A_2) = P(A_1) \cdot P(A_2)$$

If the events are positively dependent,

$$P(A_1|A_2) \geq P(A_1)$$

Hence

$$P(A_1 \cap A_2) = P(A_1|A_2) \cdot P(A_2) \geq P(A_1) \cdot P(A_2) \qquad (8.2)$$

Combining (8.1) and (8.2), we get the following result when A_1 and A_2 are positively dependent:

$$P(A_1) \cdot P(A_2) \leq P(A_1 \cap A_2) \leq \min\{P(A_1), P(A_2)\} \qquad (8.3)$$

Equation (8.3) can be written

$$P_L \leq P(A_1 \cap A_2) \leq P_U$$

where

$$P_L = P(A_1) \cdot P(A_2)$$
$$P_U = \min \{P(A_1), P(A_2)\} \qquad (8.4)$$

In WASH-1400, $P(A_1 \cap A_2)$ is calculated as the geometric mean of the lower bound P_L and the upper bound P_U:

$$P_M(A_1 \cap A_2) = \sqrt{P(A_1) \cdot P(A_2) \cdot \min(P(A_1) \cdot P(A_2))} \qquad (8.5)$$

There is, however, no proper theoretical foundation for the choice of geometric averaging of the two limits. The result further depends heavily on this somewhat arbitrary averaging.

In this section the square-root method has been given a very simple formulation to clarify the main principle of the method. A more general formulation is presented by Edwards and Watson (1979), and Harris (1986). A weakness of the square-root method is that it does not take into account the various degrees of coupling between the components. Attempts have therefore been made to develop generalizations of this method. See Harris (1986) for a thorough discussion.

Example 8.2

Consider a parallel structure of n components with a common unavailability q at a specified time t. Let A_i denote the situation that component i is down at time t. Thus we have $P(A_i) = q$ for $i = 1, \ldots, n$.

If the n components are all independent, the unavailability Q_0 of the parallel system becomes $Q_0 = q^n$. If the components are positively dependent, we can apply the square-root method with lower bound

$$P_L = \prod_{i=1}^{n} P(A_i) = q^n$$

and upper bound

$$P_U = \min \{P(A_1), \ldots, P(A_n)\} = q$$

The geometric mean is thus

$$Q_0^M = \sqrt{P_L \cdot P_U} = q^{(n+1)/2}$$

The effect of the dependency modeled by the square-root method is illustrated in Table 8.2 for some values of n and $q = 0.01$. □

Table 8.2 The Unavailability of a Parallel Structure of n Identical Components with Unavailability $q = 0.01$ Modeled by the Square-Root Method

n	Independent Components $Q_0 = q^n$	Square-root Method $Q_0^M = q^{(n+1)/2}$
1	10^{-2}	10^{-2}
2	10^{-4}	10^{-3}
3	10^{-6}	10^{-4}
4	10^{-8}	10^{-5}
5	10^{-10}	10^{-6}

The β-Factor Model

The β-factor model was introduced by Fleming (1974); it is today the most commonly used model for common cause failures. Let a system be composed of n identical components, each with constant failure rate λ. The situation is further assumed to be such that the failure of a component may be due to one of two possible causes:

1. Circumstances that concern only the component (independent of the condition of the remaining components)
2. Occurrence of an external event (independent of the system) whereby *all* the components fail at the same time

Let λ_I denote the failure rate due to failure cause of type 1, and let λ_C denote the failure rate due to failure cause of type 2. Assuming independence of the two failure causes, the total failure rate λ of the component can be written as the sum of the two failure rates

$$\lambda = \lambda_I + \lambda_C \tag{8.6}$$

Now introduce β as the "common cause factor":

$$\beta = \frac{\lambda_C}{\lambda_I + \lambda_C} = \frac{\lambda_C}{\lambda} \tag{8.7}$$

Then

$$\lambda_C = \beta\lambda \tag{8.8}$$

and

$$\lambda_I = (1 - \beta)\lambda \tag{8.9}$$

The β-factor thus denotes the relative proportion of common cause failures among all the failures of a component.

Example 8.3
Consider a parallel structure of n identical and independent components with failure rate λ. The external event that causes failure in each and every component of the system can be represented by a "hypothetical" component (C) which is in series with the rest of the system. This is illustrated in Figure 8.2.

If we assume the system to be nonrepairable and let $R_I(t)$ denote the survivor function of the identical components while $R_C(t)$ denotes the survivor function of the "hypothetical" component C, then

$$R_C(t) = e^{-\beta\lambda t} \tag{8.10}$$

The survivor function of the system can be written

$$\begin{aligned} R(t) &= [1 - (1 - R_I(t))^n] \cdot R_C(t) \\ &= [1 - (1 - e^{-(1-\beta)\lambda t})^n] \cdot e^{-\beta\lambda t} \end{aligned} \tag{8.11}$$

Figure 8.3 shows the survivor function $R(t)$ of a parallel system of $n = 4$ components for selected values of β. Note that when the common cause factor β is increased, the reliability of the system declines. □

Example 8.4
Consider the three simple systems:

1. A single component
2. A parallel structure of two identical components
3. A 2-out-of-3 system with identical components

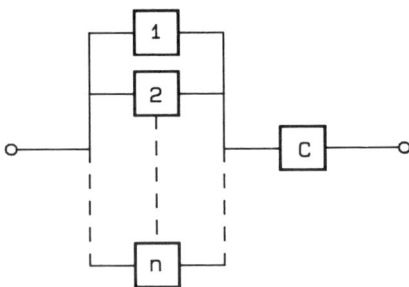

Figure 8.2 Parallel structure with common cause "component" C

SPECIAL MODELS

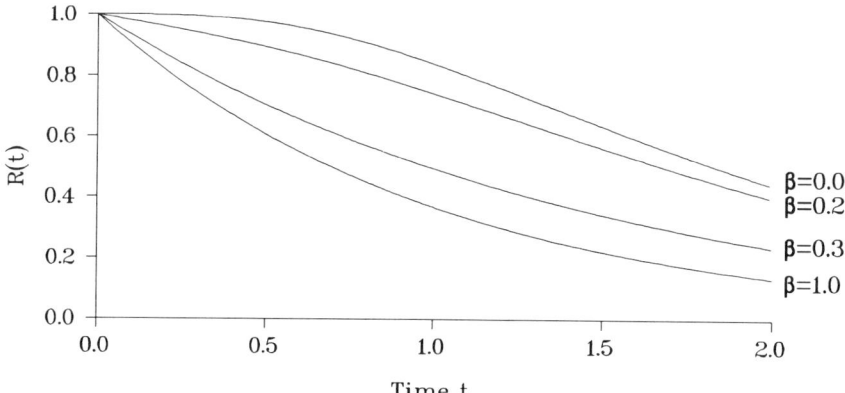

Figure 8.3 Survivor function of a parallel system of four components for some selected values of the common cause factor β, with component constant failure rate $\lambda = 1$

All the components have the same constant failure rate λ. The systems are exposed to common cause failures which may be modeled by a β-factor model.

Since common cause failures are not relevant for a single component, the survivor function and the MTTF are

$$R_{(a)}(t) = e^{-\lambda t} \quad \text{and} \quad \text{MTTF}_{(a)} = \frac{1}{\lambda}$$

The survivor function of the parallel system is, according to (8.11),

$$\begin{aligned} R_{(b)}(t) &= (2e^{-(1-\beta)\lambda t} - e^{-2(1-\beta)\lambda t}) \cdot e^{-\beta \lambda t} \\ &= 2e^{-\lambda t} - e^{-(2-\beta)\lambda t} \end{aligned}$$

Its mean time to failure is hence

$$\text{MTTF}_{(b)} = \frac{2}{\lambda} - \frac{1}{(2-\beta)\lambda}$$

The survivor function of the 2-out-of-3 system is

$$\begin{aligned} R_{(c)}(t) &= (3e^{-2(1-\beta)\lambda t} - 2e^{-3(1-\beta)\lambda t}) \cdot e^{-\beta \lambda t} \\ &= 3e^{-(2-\beta)\lambda t} - 2e^{-(3-2\beta)\lambda t} \end{aligned}$$

Its mean time to failure is hence

$$\text{MTTF}_{(c)} = \frac{3}{(2-\beta)\lambda} - \frac{2}{(3-2\beta)\lambda}$$

The MTTFs of the three simple systems are illustrated in Figure 8.4 for $\lambda = 1$. It is obvious that all three systems have the same MTTF when $\beta = 1$, namely total dependence. □

ooo OOO ooo

The β-factor model is very simple, and it is easy to understand the practical interpretation of the factor β. β is related to the degree of protection against common cause failures. Procedures for assessing the value of β have been developed for example by Humphreys (1987). He is basing the assessment of β on an evaluation of the following (weighted) factors:

Factor		Weight
Design	Separation	8
	Similarity	6
	Complexity	6
	Analysis	6
Operation	Procedures	10
	Training	5
Environment	Control	6
	Tests	4

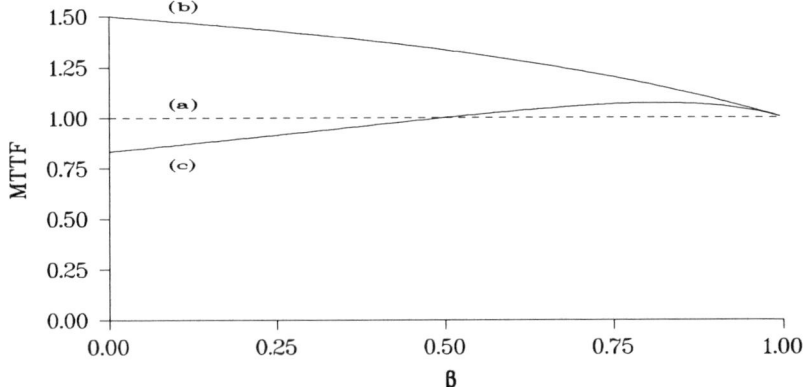

Figure 8.4 MTTF as a function of β for (*a*) a single component, (*b*) a parallel system of two identical components, and (*c*) a 2-out-of-3 system of identical components. All components have constant failure rate $\lambda = 1$

SPECIAL MODELS 337

When using the β-factor model, we assume that all system failures are either single, independent failures or common cause failures where *all* the components of the system fail simultaneously. A serious limitation of the β-factor model is that it does not allow that only a certain fraction of the components fails. The model seems quite adequate for parallel systems with two components but may not be adequate for more complex systems.

To use the β-factor model, we need an estimate of the failure rate λ, or λ_I, and an estimate of β. As mentioned above, β may be estimated based on a procedure similar to the one suggested by Humphreys (1987) or based on sound engineering judgment. Failure rates may be found in a variety of data sources (see Chapter 12). Some of the data sources present the total failure rate λ, while other sources present the independent failure rate λ_I. It is often difficult to decide whether a specific data source presents λ or λ_I. To decide, we have to study carefully how the data have been collected, from which sources, and so on. The data in MIL-HDBK 217 are, for example, mainly compiled from laboratory testing of single components. Hence only independent failure rates λ_I's are presented. The data in OREDA (1992) are, on the other hand, field data collected from maintenance files. The maintenance records do not normally distinguish between independent failures and common cause failures. Hence OREDA (1992) presents the total failure rates.

Generalized β-Factor Model

Let us consider the same situation as described above, with the change that we allow the n independent components to have different failure rates $\lambda_{I_1}, \lambda_{I_2}, \ldots, \lambda_{I_n}$. Then the total failure rate λ_i of component i can be written

$$\lambda_i = \lambda_{I_i} + \lambda_C \quad \text{for} \quad i = 1, 2, \ldots, n \tag{8.12}$$

Let us introduce

$$\lambda_{Iav} = \frac{1}{n} \sum_{i=1}^{n} \lambda_{I_i}$$

as the "average independent failure rate."

Now define the common cause part β of failures

$$\beta = \frac{\lambda_C}{\lambda_C + \lambda_{Iav}} = \frac{\lambda_C}{\lambda_C + (\sum_{i=1}^{n} \lambda_{I_i})/n} \tag{8.13}$$

Note that if $\lambda_{I_i} = \lambda_I$ for all i (identical components), (8.13) coincides with (8.7).

The survivor function of component i due to independent failures is, in this case,

$$R_{I_i}(t) = e^{-\lambda_{I_i} t} \quad \text{for} \quad i = 1, 2, \ldots, n$$

The survivor function of the corresponding common cause "component" is

$$R_C(t) = e^{-\lambda_C t}$$

For the parallel structure in Figure 8.2, we get

$$R(t) = \left(1 - \prod_{i=1}^{n} (1 - e^{-\lambda_{I_i} t})\right) e^{-\lambda_C t} \tag{8.14}$$

Note that (8.14) is identical with (8.11) when $\lambda_{I_i} = \lambda_I$ for all i.

Binomial Failure Rate Model and Its Extensions

The binomial failure rate (BFR) model was introduced by Vesely (1977) and is a simple special case of Marshall-Olkins multivariate exponential model (see Marshall and Olkin 1967). The situation under study is the following: A system is composed of n identical components. Each component can fail at random times, independently of each other, and they are all supposed to have the same failure rate λ_I. We will denote such failures as "individual failures."

Furthermore a common cause shock can hit the system with occurrence rate ν. Whenever a shock occurs, each of the n individual components is assumed to fail with probability p, independent of the states of the other components. The number Z of individual components failing as a consequence of the shock, is thus binomially distributed (n, p). The probability that the multiplicity Z of failures is equal to z is accordingly

$$P(Z = z) = \binom{n}{z} p^z (1 - p)^{n-z} \quad \text{for} \quad z = 0, 1, 2, \ldots$$

The mean number of components that fail in one shock is $E(Z) = np$.
Two conditions are furthermore assumed:

1. Shocks and individual failures occur independently of each other
2. All failures are immediately discovered and repaired, with the repair time being negligible

As a consequence the time between individual failures, in the absence of shocks, will be exponentially distributed with failure rate λ_I, and the time between

SPECIAL MODELS

shocks will be exponentially distributed with failure rate ν. The number of individual failures in any time period of length t_0 will be Poisson distributed ($\lambda_I t_0$). Similarly the number of shocks in any time period of length t_0 will be Poisson distributed (νt_0).

The component failure rate caused by shocks thus equals $p \cdot \nu$, and the total failure rate of one component equals

$$\lambda = \lambda_I + p \cdot \nu$$

By using this model, we have to estimate the independent failure rate λ_i and the two parameters ν and p. The parameter ν relates to the degree of "stress" on the system, while p is a function of the built in component protection against external shocks. Note that the binomial failure rate (BFR) model is identical to the β-factor model when the system has only two components.

This is Vesely's original BFR model. The statistical analysis of such models is discussed by Atwood (1986). Several aspects of the situation must be clarified to make the analysis possible. It may, for example, happen that ν cannot be estimated in a direct way from failure data because shocks may occur unnoticed when no component fails.

Several extensions of Vesely's BFR model have been studied. It may, for example, happen that p varies from shock to shock. One way of modeling this is to assume p to be beta distributed; that is, it has the probability density

$$f(p) = \frac{\Gamma(r+s)}{\Gamma(r)\Gamma(s)} p^{r-1}(1-p)^{s-1} \quad \text{for} \quad r > 0, s > 0.$$

Such a Bayesian approach is discussed, for example, by Hokstad (1988). He introduces a re-parameterization of the beta distribution and is able to interpret the new parameters in the context of how well the components and the system are protected against the shocks.

The assumption that the components will fail independently of each other, given that a shock has occurred, represents a rather serious limitation, and this assumption is often not satisfied in practice. The problem can, to some extent, be remedied by defining one fraction of the shocks as being "lethal" shocks, namely shocks that automatically causes all the components to fail ($p = 1$). If all the shocks are "lethal," one is back to the β-factor model. Observe that this case ($p = 1$) corresponds to the situation that there is no built-in protection against these shocks.

Situations where individual failures may occur together with nonlethal as well as lethal shocks are often realistic. Such models are, however, rather complicated, even if the nonlethal and the lethal shocks occur independently of each other. Further extensions are presented, for example, by Apostolakis and Moieni (1987).

8.5 ASSOCIATED VARIABLES

Definition

As we pointed out in Section 8.3, reliability analysis based on the assumption that state variables are independent when in fact they are dependent, may lead to results of very limited practical value. In Section 8.4 we gave several examples on how to model dependency in special situations. Now we are going to study dependence in a more general setup where we will confine ourselves to *positive* dependence.

Esary, Proschan, and Walkup (1967) define a broad class of state variables, which they call *associated*. This class includes independent variables as well as variables with a certain type of positive dependence appropriate for use in reliability theory. They note that this class cannot be *too broad* to be of any use in reliability analysis. Let us follow their argument.

The covariance of two random variables S and T, $\text{cov}(S, T)$, is commonly used to indicate how the values of S relates to the values of T. With this measure S and T are said to be positively dependent when

$$\text{cov}(S, T) = E(S \cdot T) - E(S)E(T) \geq 0 \qquad (8.15)$$

However, the class of state variables $X_1, X_2, \ldots, X_n$ with the only restriction that the X's pairwise have nonnegative covariance turns out to be too broad to be of any use in reliability analyses.

A somewhat stronger restriction would be to ask that

$$\text{cov}(f(S), g(T)) \geq 0 \qquad (8.16)$$

for all pairs of nonnegative functions f and g, nondecreasing in S and T, respectively. An even stronger requirement would be to ask that

$$\text{cov}(f(S, T), g(S, T)) \geq 0 \qquad (8.17)$$

for all pairs of functions f and g which are nondecreasing in each argument.

Equation (8.17) has a natural multivariate generalization which Esary, Proschan, and Walkup (1967) find restrictive enough to serve as a useful definition of the wanted class of state variables.

Definition 8.1. The variables $T_1, T_2, \ldots, T_n$ are associated if

$$\text{cov}(f(T_1, T_2, \ldots, T_n), g(T_1, T_2, \ldots, T_n)) \geq 0 \qquad (8.18)$$

for all functions f and g that are nondecreasing in each argument.[2] □

---

[2] $E[f(T_1, T_2, \ldots, T_n)]$, $E[g(T_1, T_2, \ldots, T_n)]$, and $E[f(T_1, T_2, \ldots, T_n) g(T_1, T_2, \ldots, T_n)]$ are all assumed to exist.

ASSOCIATED VARIABLES 341

It turns out that one, without loss of generality, can restrict the test functions f and g to be binary (Theorem 8.5). Hence we may replace Definition 8.1 by the following definition, which is easier to apply when developing the theory:

Definition 8.2. The random variables $T_1, T_2, \ldots, T_n$ (not necessarily binary) are said to be associated if, for each pair of binary functions, $\gamma(T_1, T_2, \ldots, T_n)$ and $\delta(T_1, T_2, \ldots, T_n)$, which are both nondecreasing in each argument,

$$\text{cov}(\gamma(T_1, T_2, \ldots, T_n) \delta(T_1, T_2, \ldots, T_n)) \geq 0 \qquad (8.19)$$

□

In Theorem 8.6 we show that the class of associated random variables includes the independent variables.

Some Basic Theorems

In the following pages by an associated set we will mean a set of associated variables. We now claim that the following five theorems are valid for associated variables:

Theorem 8.1. Any subset of an associated set is itself associated.

Theorem 8.2. A single random variable is always associated.

Theorem 8.3. Nondecreasing functions of associated variables are associated.

Theorem 8.4. Let $\{X_1, X_2, \ldots, X_m\}$ and $\{Y_1, Y_2, \ldots, Y_n\}$ be associated sets and $(X_1, X_2, \ldots, X_m)$ and $(Y_1, Y_2, \ldots, Y_n)$ independent random vectors. Then the set $\{X_1, X_2, \ldots, X_m, Y_1, Y_2, \ldots, Y_n\}$ is associated.

Theorem 8.5. Let $T_1, \ldots, T_n$ be associated and $f(T), g(T)$ be two functions of $T_1, \ldots, T_n$, both nondecreasing in each argument. Assume that

$$E(f(T)) < \infty, \quad E(g(T)) < \infty, \quad E(f(T) \cdot g(T)) < \infty$$

Then $\text{cov}(f(T), g(T)) \geq 0$.

Proof of Theorem 8.1

Let $\{T_1, T_2, \ldots, T_n\}$ be associated, and let $\{T_{\nu_1}, T_{\nu_2}, \ldots, T_{\nu_m}\} \in \{T_1, T_2, \ldots, T_n\}$. Furthermore let $\gamma_1(T_{\nu_1}, T_{\nu_2}, \ldots, T_{\nu_m})$ and $\delta_1(T_{\nu_1}, T_{\nu_2}, \ldots, T_{\nu_m})$ be any two binary functions of $(T_{\nu_1}, T_{\nu_2}, \ldots, T_{\nu_m})$, both nondecreasing in each argument. Since $\{T_1, T_2, \ldots, T_n\}$ is associated, then (8.19) holds for any pair

$\gamma(T_1, T_2, \ldots, T_n)$ and $\delta(T_1, T_2, \ldots, T_n)$ of binary functions, each nondecreasing in each argument. By choosing $\gamma(T_1, T_2, \ldots, T_n) = \gamma_1(T_{\nu_1}, T_{\nu_2}, \ldots, T_{\nu_m})$ and $\delta(T_1, T_2, \ldots, T_n) = \delta_1(T_{\nu_1}, T_{\nu_2}, \ldots, T_{\nu_m})$, we get cov $(\gamma_1(T_{\nu_1}, T_{\nu_2}, \ldots, T_{\nu_m}), \delta(T_{\nu_1}, T_{\nu_2}, \ldots, T_{\nu_m})) \geq 0$. □

Proof of Theorem 8.2
Let T be a random variable and $\gamma(T)$ and $\delta(T)$ be any two nondecreasing binary functions of T. Then

$$\text{cov}(\gamma(T), \delta(T)) = E(\gamma(T) \cdot \delta(T)) - E(\gamma(T)) \cdot E(\delta(T)) \qquad (8.20)$$

Since $\gamma(T)$ and $\delta(T)$ are both binary and nondecreasing, we must have either

$$\gamma(T) \geq \delta(T) \quad \text{for all } T \qquad (8.21a)$$

or

$$\delta(T) \geq \gamma(T) \quad \text{for all } T \qquad (8.21b)$$

In (8.21a)

$$\gamma(T) \cdot \delta(T) \geq \delta(T)^2 = \delta(T) \quad \text{for all } T \qquad (8.22)$$

Hence

$$E(\gamma(T) \cdot \delta(T)) \geq E(\delta(T))$$

By introducing (8.22) into (8.20), we get

$$\text{cov}(\gamma(T), \delta(T)) \geq E(\delta(T))(1 - E(\gamma(T)))$$

By definition

$$E(\delta(T)) \geq 0$$
$$0 \leq E(\gamma(T)) \leq 1 \qquad \square$$

The proof of Theorem 8.3 is left to the reader as Problem 8. For proofs of Theorem 8.4 and Theorem 8.5, see Barlow and Proschan (1975, pp. 30–31).

Theorem 8.6. If $X_1, X_2, \ldots, X_n$ are independent variables, then they are also associated.

Theorem 8.7. If $P(X_1 = X_2 = \cdots = X_n) = 1$, then $X_1, X_2, \ldots, X_n$ are associated.

Theorem 8.8. If $X_1, X_2, \ldots, X_n$ are associated and binary variables, then $(1 - X_1), (1 - X_2), \ldots, (1 - X_n)$ are also associated and binary.

The proofs of Theorem 8.6, 8.7, and 8.8 are left to the reader as Problems 9, 10, and 11 in Section 8.6.

Theorem 8.9. If the binary state variables $X_1, X_2, \ldots, X_n$ of a coherent structure ϕ are associated, then

1. The minimal path series structures $\rho_j(X_1, X_2, \ldots, X_n)$ for $j = 1, 2, \ldots, p$ are also binary and associated
2. The minimal cut parallel structures $\kappa_j(X_1, \ldots, X_n)$ for $j = 1, \ldots, k$ are binary and associated

Proof

1. Denote the minimal path sets by $P_1, P_2, \ldots, P_p$. Then

$$\rho_j(X) = \prod_{i \in P_j} X_i \quad \text{for } j = 1, 2, \ldots, p$$

Since $\rho_1(X), \rho_2(X), \ldots, \rho_p(X)$ are all binary, nondecreasing functions of the associated variables $X_1, X_2, \ldots, X_n$, they are associated according to Theorem 8.3.

2. The proof is left to the reader as Problem 12 in Section 8.6. □

System Reliability, Upper and Lower Bounds

We now raise the question: What happens if we, when determining the system reliability of a series structure and a parallel structure of associated variables, respectively, proceed as if the components were independent? Theorem 8.10 gives the answer. In common words it states:

1. For a series structure we are conservative and underestimate the system reliability
2. For a parallel structure we are overoptimistic and overestimate the reliability

Theorem 8.10. Let $X_1, X_2, \ldots, X_n$ be binary and associated. Then

$$P\left(\prod_{j=1}^{n} X_j = 1\right) \geq \prod_{j=1}^{n} P(X_j = 1) \tag{8.23}$$

that is,

$$E\left(\prod_{j=1}^{n} X_j\right) \geq \prod_{j=1}^{n} E(X_j) \tag{8.24}$$

$$P\left(\coprod_{j=1}^{n} X_j = 1\right) \leq \coprod_{j=1}^{n} P(X_j = 1) \tag{8.25}$$

Proof
In (8.23), X_1 and $\prod_{j=2}^{n} X_j$ are both binary and nondecreasing functions of the associated variables $X_1, X_2, \ldots, X_n$. Hence they are associated, according to Theorem 8.3, which means that

$$\mathrm{cov}\left(X_1, \prod_{j=2}^{n} X_j\right) \geq 0$$

that is,

$$E\left(\prod_{j=1}^{n} X_j\right) \geq E(X_1) \cdot E\left(\prod_{j=2}^{n} X_j\right)$$

In the same way we show, for example, that

$$E\left(\prod_{j=2}^{n} X_j\right) \geq E(X_2) \cdot E\left(\prod_{j=3}^{n} X_j\right)$$

Hence

$$E\left(\prod_{j=1}^{n} X_j\right) \geq \prod_{j=1}^{n} E(X_j)$$

The following statements are equivalent to (8.25):

$$P\left(\left(1 - \prod_{j=1}^{n}(1 - X_j)\right) = 1\right) \leq 1 - \prod_{j=1}^{n}(1 - P(X_j = 1))$$

$$P\left(\prod_{j=1}^{n}(1 - X_j) = 0\right) \leq 1 - \prod_{j=1}^{n} P(X_j = 0)$$

$$\prod_{j=1}^{n} P(X_j = 0) \leq 1 - P\left(\prod_{j=1}^{n}(1 - X_j) = 0\right)$$

$$\prod_{j=1}^{n} P((1 - X_j) = 1) \leq P\left(\prod_{j=1}^{n}(1 - X_j) = 1\right) \quad (8.26)$$

We now prove (8.25) by proving (8.26), which is easy.

Introduce $Y_j = 1 - X_j$ for $j = 1, 2, \ldots, n$. Then according to Theorem 8.8, $Y_1, Y_2, \ldots, Y_n$ are binary and associated. By using (8.23),

$$P\left(\prod_{j=1}^{n} Y_j = 1\right) \geq \prod_{j=1}^{n} P(Y_j = 1)$$

which is equal to (8.26). □

For a system that is neither a series nor a parallel system, it is impossible to predict whether or not we over- or underestimate the reliability by proceeding as if the state variables are independent when all we know is that they are associated. In such cases, however, rough upper and lower bounds for the system reliability can be determined, using the following theorem:

Theorem 8.11. Let ϕ be a coherent structure of n components, with associated state variables $X_1, X_2, \ldots, X_n$, and denote

$$P(X_j = 1) = p_j \quad \text{for} \quad j = 1, 2, \ldots, n.$$

Then

$$\prod_{j=1}^{n} p_j \leq P(\phi(X) = 1) \leq \coprod_{j=1}^{n} p_j \quad (8.27)$$

Proof

Since ϕ is coherent, then according to Theorem 3.2

$$\prod_{j=1}^{n} X_j \leq \phi(X) \leq \coprod_{j=1}^{n} X_j$$

implying that

$$E\left(\prod_{j=1}^{n} X_j\right) \leq E(\phi(X)) \leq E\left(\coprod_{j=1}^{n} X_j\right) \tag{8.28}$$

Since $\prod_{j=1}^{n} X_j$, $\coprod_{j=1}^{n} X_j$ and $\phi(X)$ are all binary

$$P\left(\prod_{j=1}^{n} X_j = 1\right) \leq P(\phi(X) = 1) \leq P\left(\coprod_{j=1}^{n} X_j = 1\right) \tag{8.29}$$

but, according to Theorem 8.10, (8.23) and (8.25),

$$P\left(\prod_{j=1}^{n} X_j = 1\right) \geq \prod_{j=1}^{n} P(X_j = 1) = \prod_{j=1}^{n} p_j \tag{8.30}$$

$$P\left(\prod_{j=1}^{n} X_j = 1\right) \leq \coprod_{j=1}^{n} P(X_j = 1) = \coprod_{j=1}^{n} p_j \tag{8.31}$$

and Theorem 8.11 follows from (8.29), (8.30), and (8.31). □

In common language Theorem 8.11 states that the reliability of a coherent structure of associated state variables is larger than or equal to the reliability of an imaginary series system of independent components, and smaller than or equal to an imaginary parallel system of independent components. These bounds obtained by Theorem 8.11 are, however, normally rather wide and therefore not very useful for practical purposes.

In order to find more narrow bounds for system reliability, we will look for bounds based on the minimal path series structures $\rho_1(X), \ldots, \rho_S(X)$ and the minimal cut parallel structures $\kappa_1(X), \kappa_2(X), \ldots, \kappa_k(X)$. The state variables $X_1, X_2, \ldots, X_n$ are assumed to be binary and associated as above.

The next theorem states that a lower bound may be found from a series structure of modules where the modules are the minimal cut parallel structures of the system. The upper bound is found from a parallel structure of modules where the modules are the minimal path series structures of the system.

Theorem 8.12. Let ϕ be a coherent structure of associated state variables. Let the minimal path series structures be $\rho_i(X)$ for $i = 1, 2, \ldots, p$, and the minimal cut parallel structures be $\kappa_j(X)$ for $j = 1, 2, \ldots, k$. Then

$$\prod_{j=1}^{k} P(\kappa_j(X) = 1) \leq P(\phi(X) = 1) \leq \coprod_{j=1}^{p} P(\rho_i(X) = 1) \qquad (8.32)$$

Proof
According to (3.9) and (3.12)

$$\phi(X) = \coprod_{j=1}^{p} \rho_j(X) = \prod_{j=1}^{k} \kappa_j(X)$$

Hence

$$P\left(\prod_{j=1}^{k} \kappa_j(X) = 1\right) = P(\phi(X) = 1) = P\left(\coprod_{j=1}^{p} \rho_j(X) = 1\right) \qquad (8.33)$$

According to Theorem 8.9, $\rho_1(X), \rho_2(X), \ldots, \rho_p(X)$ and $\kappa_1(X), \kappa_2(X), \ldots, \kappa_k(X)$ are binary and associated. Hence Theorem 8.10, (8.23) and (8.25) are applicable:

$$P\left(\coprod_{j=1}^{p} \rho_j(X) = 1\right) \leq \coprod_{j=1}^{p} P(\rho_j(X) = 1) \qquad (8.34)$$

$$P\left(\prod_{j=1}^{k} \kappa_j(X) = 1\right) \geq \prod_{j=1}^{k} P(\kappa_j(X) = 1) \qquad (8.35)$$

Theorem 8.12 follows from (8.33), (8.34), and (8.35). □

Corollary 8.1. Let ϕ be a coherent structure of independent state variables. Then

$$\prod_{j=1}^{k} \coprod_{i \in K_j} p_i \leq P(\phi(X) = 1) \leq \coprod_{j=1}^{p} \prod_{i \in P_j} p_i \qquad (8.36)$$

Proof
In the case of independent state variables

$$P(\rho_j(X) = 1) = \prod_{i \in P_j} p_i$$

$$P(\kappa_j(X) = 1) = \coprod_{i \in K_j} p_i \qquad (8.37)$$

The corollary follows directly from Theorem 8.12 by introducing (8.37). □

Like the bounds obtained from Theorem 8.11, the bounds obtained from Theorem 8.12 are usually too wide to be of any practical use. According to the following theorem the minimal path series structure with maximum reliability constitutes a lower bound, and the minimal cut parallel structure with minimum reliability constitutes an upper bound for the reliability of a coherent structure with associated state variables.

Theorem 8.13. Let ϕ be a coherent structure of state variables $X_1, X_2, \ldots, X_n$. Denote the minimal path sets by $P_1, P_2, \ldots, P_p$ and the minimal cut sets by $K_1, K_2, \ldots, K_k$. Then it is generally true that

$$\max_{1 \leq j \leq p} P(\min_{i \in P_j} X_i = 1) \leq P(\phi(X) = 1) \leq \min_{1 \leq j \leq k} P(\max_{i \in K_j} X_i = 1) \qquad (8.38)$$

If additionally it is assumed that $X_1, X_2, \ldots, X_n$ are associated, then

$$\max_{1 \leq j \leq p} \prod_{i \in P_j} p_i \leq P(\phi(X) = 1) \leq \min_{1 \leq j \leq k} \coprod_{i \in K_j} p_i \qquad (8.39)$$

and these bounds are at least as narrow as the bounds obtained from Theorem 8.11.

Proof
For (8.38),

$$\phi(X) = \coprod_{j=1}^{p} \prod_{i \in P_j} X_i = \prod_{j=1}^{k} \coprod_{i \in K_j} X_i$$

Since $X_1, X_2, \ldots, X_n$ are binary,

$$\prod_{i \in P_j} X_i = \min_{i \in P_j} X_i \quad \text{and} \quad \coprod_{i \in K_j} X_i = \max_{i \in K_j} X_i$$

are both binary. Hence

$$\max_{1 \leq r \leq p} \min_{i \in P_r} X_i = \phi(X) = \min_{1 \leq s \leq k} \max_{i \in K_s} X_i$$

Therefore

$$\min_{i \in P_r} X_i \leq \phi(X) \leq \max_{i \in K_s} X_i \quad \text{for } r = 1, 2, \ldots, p, \ s = 1, 2, \ldots, k$$

implying that

$$P(\min_{i \in P_r} X_i = 1) \leq P(\phi(X) = 1) \leq P(\max_{i \in K_s} X_i = 1) \quad \text{for any pair } (r, s)$$

Now (8.38) follows if we maximize over r and minimize over s. Then (8.39) follows from (8.38) if we apply Theorems 8.10, (8.23) and (8.25). Since $\prod_{i=1}^{n} p_i \leq \prod_{i \in P_j} p_i$ for all $j = 1, 2, \ldots, p$, then

$$\prod_{i=1}^{n} p_i \leq \min_{j} \prod_{i \in P_j} p_i \leq \max_{j} \prod_{i \in P_j} p_i \tag{8.40}$$

Similarly, since $\coprod_{i=1}^{n} p_i \geq \coprod_{i \in K_j} p_i$ for all $j = 1, 2, \ldots, k$, then

$$\coprod_{i=1}^{n} p_i \geq \max_{j} \coprod_{i \in K_j} p_i \geq \min_{j} \coprod_{i \in K_j} p_i \tag{8.41}$$

$\square$

Example 8.5

Reconsider the gas detector system in Example 4.1. Let the component reliabilities at time t_0 be

$$\begin{aligned} p_1 &= p_2 = p_3 = 0.93 \\ p_4 &= 0.98 \\ p_5 &= p_6 = 0.97 \\ p_7 &= p_8 = 0.90 \end{aligned} \qquad (8.42)$$

If the components are independent, we find the system reliability at time t_0 by inserting (8.42) in (4.12) and thus get

$$p_S \approx 0.90$$

Suppose that the state variables of the system are associated but not necessarily independent. Let us first determine the bounds we get for the system reliability from Theorem 8.11 (8.27).

The lower bound becomes

$$\prod_{i=1}^{8} p_i = (0.93)^3 \cdot 0.98 \cdot (0.97)^2 \cdot (0.90)^2 \approx 0.60$$

The upper bound becomes

$$\coprod_{i=1}^{8} p_i = 1 - \prod_{i=1}^{8}(1 - p_i) = 1 - (0.07)^3 \cdot (0.02 \cdot (0.03)^2 \cdot (0.10)^2 \approx 1.00$$

Hence

$$0.60 \leq p_S \leq 1.00$$

Let us next calculate the bounds for the system reliability from Theorem 8.13 (8.39). The minimal path and cut sets were found in Example 4.1.

ASSOCIATED VARIABLES 351

Minimal path sets	Minimal cut sets
$S_1 = \{1,2,4,5,6,7\}$	$K_1 = \{1,2\}$
$S_2 = \{1,2,4,5,6,8\}$	$K_2 = \{1,3\}$
$S_3 = \{1,3,4,5,6,7\}$	$K_3 = \{2,3\}$
$S_4 = \{1,3,4,5,6,8\}$	$K_4 = \{4\}$
$S_5 = \{2,3,4,5,6,7\}$	$K_5 = \{5\}$
$S_6 = \{2,3,4,5,6,8\}$	$K_6 = \{6\}$
	$K_7 = \{7,8\}$

Using the data given by (8.44), we observe that all paths S_1 to S_6 have the same reliability

$$\prod_{i \in P_j} p_i = (0.93)^2 0.98 (0.97)^2 \cdot 0.90 \approx 0.718 \quad \text{for all } j$$

Hence the lower bound is

$$\max_{1 \leq j \leq 6} \prod_{i \in P_j} p_i \approx 0.72$$

To find the upper bound, we must calculate $\coprod_{i \in K_j} p_i$ for K_1 to K_7:

Cut 1 $\quad \coprod_{i \in K_1} p_i = p_1 \amalg p_2 = 1 - (0.08)^2 \approx 0.995$

Cut 2 $\quad \coprod_{i \in K_2} p_i = 1 - (0.07)^2 \approx 0.995$

Cut 3 $\quad \coprod_{i \in K_3} p_i = 1 - (0.07)^2 \approx 0.995$

Cut 4 $\quad \coprod_{i \in K_4} p_i = p_4 \approx 0.98$

Cut 5 $\quad \coprod_{i \in K_5} p_i = p_5 \approx 0.97$

Cut 6 $\quad \coprod_{i \in K_6} p_i = p_6 \approx 0.97$

Cut 7 $\quad \coprod_{i \in K_7} p_i = 1 - (0.10)^2 \approx 0.99$

Hence the upper bound becomes

$$\min_{1 \leq j \leq 7} \coprod_{i \in K_j} p_i \approx 0.97$$

Hence application of Theorem 8.11 leads to the following bounds for p_S:

$$0.72 \leq p_S \leq 0.97 \qquad \square$$

8.6 PROBLEMS

1. Consider a parallel structure of two identical components. Let q denote the unavailability of each component.
   (a) Make a sketch of the system unavailability as a function of q when the two components are independent.
   (b) Determine the system unavailability by the square-root method, and make a sketch of this unavailability as a function of q in the same coordinate system as in (a).
   (c) Determine the difference between the system unavailability when the components are independent, and the unavailability determined by the square-root method when $q = 0.15$.

2. Consider a 2-out-of-3 system of identical components. Let q denote the unavailability of each component. Edwards and Watson (1979) states that the square-root method gives the system unavailability $\sqrt{3 \cdot q^3}$. Discuss this statement. Is it correct? See, for example, Harris (1986) for a thorough discussion.

3. Consider a 2-out-of-3 system of identical components with constant failure rate λ. The system is exposed to common cause failures which may be modeled by a β-factor model. In Figure 8.4 it is shown that the MTTF of the system has a minimum for $\beta = 0$. Determine the value of β for which MTTF attain its maximum. Explain why MTTF as a function of β has this particular shape.

4. Consider a parallel structure of n identical components with constant failure rates λ. The system is put into operation at time $t = 0$. The system is tested and, if necessary, repaired after regular time intervals of length τ. After a test (repair) the system is considered to be "as good as new." This model is further discussed in Section 4.7. The system is exposed to common cause failures, which may be modeled by a β-factor model. Let MFDT_n denote the mean fractional dead-time of a parallel structure of order n.

(a) Determine MFDT_n as a function of λ, τ, and β.

(b) Let $\lambda = 5 \cdot 10^{-5}$ failures per hour and $\tau = 3$ months, and make a sketch of MFDT_n as a function of β for $n = 2$ and $n = 3$.

(c) With the same data as in question (b), determine the difference between MFDT_2 and MFDT_3 when $\beta = 0$, and $\beta = 0.20$, respectively.

5. Reconsider the bridge structure in Example 3.5. Assume that all the five components are identical and have constant failure rate λ. The system is exposed to common cause failures which may be modeled by a β-factor model. Determine the MTTF of the bridge structure as a function of β, and make a sketch of MTTF as a function of β when $\lambda = 5 \cdot 10^{-4}$ failures per hour and no repair is carried out.

6. Consider a 2-out-of-3 structure of identical components. The system is exposed to common cause failures, which may be modeled by a binomial failure rate (BFR) model. The "individual" failure rate of the components is $\lambda_I = 5 \cdot 10^{-5}$ failures per hour. Nonlethal shocks occur with frequency $\nu = 10^{-5}$ nonlethal shocks per hour. When a nonlethal shock occurs, the components may fail independently with probability $p = 0.20$. Lethal shocks occur with frequency $\omega = 10^{-7}$ lethal shocks per hour. When a lethal shock occurs, all the three components will fail simultaneously. The lethal and the nonlethal shocks are assumed to be independent.

(a) Determine the mean time between system failures MTBF_I caused by "individual" component failures, assuming that the system is only repaired when a system failure occurs. In such a case the system is repaired to an "as good as new" condition.

(b) Determine the mean time between system failures MTBF_{NL}, assuming that the only cause of system failures is the nonlethal shocks.

(c) Determine the mean time between system failures MTBF_L, assuming that the only cause of system failures is the lethal shocks.

(d) Try to find the total mean time between system failures. Discuss the problems you encounter during this assessment.

7. Explain why (8.16) is more restrictive than (8.15), and why (8.17) is more restrictive than (8.16).

8. Prove Theorem 8.3. Hint: Let $f_j(T_1, T_2, \ldots, T_n)$ for $j = 1, 2, \ldots, n$ denote the nondecreasing functions. Introduce the variables

$$S_j = f_j(T_1, T_2, \ldots, T_n) \quad \text{for } j = 1, 2, \ldots, n$$

Theorem 8.3 will be proved if one can show that

$$\text{cov}(\gamma(S_1, S_2, \ldots, S_n), \delta(S_1, S_2, \ldots, S_n)) \geq 0$$

for any pair (γ, δ) of binary functions, nondecreasing in each argument.

9. Prove Theorem 8.6. Hint: Use Theorem 8.2 and Theorem 8.4.

10. Prove Theorem 8.7

11. Prove Theorem 8.8. Hint: Use $\text{cov}(\gamma(1 - X), \delta(1 - X)) = \text{cov}(1 - \gamma(1 - X), 1 - \delta(1 - X))$.

12. Prove Property 2 of Theorem 8.9.

13. Let $T_1, \ldots, T_n$ be associated (not necessarily binary) variables. Show that in this case

$$P(T_1 > t_1, \ldots, T_n > t_n) \geq \prod_{i=1}^{n} P(T_i > t_i) \tag{8.43}$$

and that

$$P(T_1 \leq t_1, \cdots, T_n \leq t_n) \geq \prod_{i=1}^{n} P(T_i \leq t_i) \tag{8.44}$$

Hint: Use indicator variables

$$X_{T_i}(t_i) = \begin{cases} 1 & \text{if } T_i > t_i \\ 0 & \text{otherwise} \end{cases}$$

14. Let $T_1, \ldots, T_n$ be associated (not necessarily binary) variables and show that

$$P(\min_{1 \leq i \leq n} T_i > t) \geq \prod_{i=1}^{n} P(T_i > t) \tag{8.45}$$

and that

$$P(\max_{1 \leq i \leq n} T_i > t) \leq \prod_{i=1}^{n} P(T_i > t) \tag{8.46}$$

Hint: Use the result from Problem 13.

CHAPTER 9

Life Data Analysis

9.1 INTRODUCTION

To obtain information about a particular life distribution $F(t)$ for a component/system, it is often necessary to carry out a so-called life test where n identical, numbered units of the component/system are activated in order to record their lifetimes. If the test is allowed to run until all the n components have failed and the lifetimes are recorded, the data set thus obtained is said to be *complete*.

Often we have to be satisfied with incomplete data sets. This may be because it is impractical or too expensive to wait until all the components have failed, or because individual components are "lost" for one reason or another, or because in recording lifetime we must make do with stating relatively large time intervals to which the lifetimes belong. In such situations the data set is said to be *censored*. The examples show that such censoring can be planned but also that circumstances may arise that are beyond control.

On the following pages we will first look at analysis of complete data sets and in particular at the so-called TTT plot, which can be used to decide whether the actual, unknown life distribution $F(t)$ has an increasing failure rate (IFR), a decreasing failure rate (DFR), or neither of these. Thereafter we will discuss analysis of exponentially distributed data, consider the most usual types of censoring, and see how the analysis is influenced by such censoring.

In Section 9.3 we will discuss Kaplan-Meier's estimator of the survivor function $R(t)$ and also a graphical method—hazard plotting—that can be used to decide whether or not a set of life data originates from a specified life distribution. Both these methods can be used for complete and censored data sets. We will denote the lifetime for component j by T_j for $j = 1, 2, \ldots, n$. The corresponding order statistic is denoted $(T_{(1)}, T_{(2)}, \ldots, T_{(n)})$.

Throughout this chapter we will assume that the lifetimes of the n components are stochastically independent and identically distributed with a continuous distribution function $F(t)$. Note that the lifetimes in the context of censored data set should be interpreted as the potential lifetimes of the censored components.

When considering data from repairable units, we must first verify that we have a renewal process where the lifetimes (interarrival times) are independent and identically distributed. A number of graphical methods and formal tests have been developed to check this assertion. It is always wise to start an analysis of data from repairable units by drawing a Nelson-Aalen plot as described in Figure 7.2. If the Nelson-Aalen plot is nonlinear, the methods described in this chapter should *not* be used. See, for example, Ascher and Feingold (1984) and Akersten (1991) for a more detailed discussion.

In practice, the assumption of identically distributed lifetimes corresponds to the assumption that the components are nominally identical, that is, of the same type and exposed to approximately the same environmental and operational stresses. The assumption of independence means that the components are not affected by the operation or failure of any other component in the study. The censoring mechanism must satisfy the requirement of *independent censoring*. Briefly this means that censorings occur independent of any information gained from previously failed components in the same study.

Explanatory variables (covariates, concomitant variables) are not considered in this chapter. Analysis of life data with explanatory variables is briefly discussed in Chapter 10. This topic is thoroughly discussed, for example, by Kalbfleisch and Prentice (1980), Lawless (1982), and Cox and Oakes (1984).

9.2 COMPLETE DATA SETS

Total Time on Test Plot

Assume that we activate n identical, numbered components to observe their lifetimes and obtain information about the underlying life distribution $F(t)$ which is more or less unknown. For the sake of simplicity, we will assume that we are dealing with a continuous life distribution and that $F(t)$ is strictly increasing for $F^{-1}(0) = 0 < t < F^{-1}(1)$. Further it is assumed that the distribution has finite mean μ.

Definition 9.1. The total time on test at time t, $\mathscr{T}(t)$ is defined as

$$\mathscr{T}(t) = \sum_{j=1}^{i} T_{(j)} + (n - i)t \qquad (9.1)$$

where i is such that

$$T_{(i)} \leq t < T_{(i+1)} \quad \text{for} \quad i = 0, 1, \ldots, n$$

and $T_{(0)}$ is defined to be equal to 0 and $T_{(n+1)} = +\infty$. □

COMPLETE DATA SETS

The total time on test $\mathcal{T}(t)$ denotes the total observed lifetime of the n components at time t. We assume that all the n components are put into operation at time $t = 0$ and that the observation is terminated at time t. In the time interval $(0, t]$, a number i of the components have failed. The total functioning time of these i components is $\sum_{j=0}^{i} T_j$. The remaining $n - i$ components survive the time interval $(0, t]$. The total functioning time of these $n - i$ components at time t is thus $(n - i)t$.

The total time on test at the ith failure for $i = 1, 2, \ldots, n$ is

$$\mathcal{T}(T_{(i)}) = \sum_{j=1}^{i} T_{(j)} + (n - i)T_{(i)} \qquad \text{for } i = 1, 2, \ldots, n \qquad (9.2)$$

In particular,

$$\mathcal{T}(T_{(n)}) = \sum_{j=1}^{n} T_{(j)} = \sum_{j=1}^{n} T_j$$

The total time on test at the ith failure $\mathcal{T}(T_{(i)})$ may be scaled by dividing by $\mathcal{T}(T_{(n)})$. The scaled total time on test at time t is defined as $\mathcal{T}(t)/\mathcal{T}(T_{(n)})$.

If we plot the points

$$\left(\frac{i}{n}, \frac{\mathcal{T}(T_{(i)})}{\mathcal{T}(T_{(n)})} \right) \qquad \text{for } i = 1, 2, \ldots, n$$

we obtain the so-called TTT plot of the data set.

Example 9.1
Suppose that we have activated 10 identical components and observed their lifetimes (in hours):

$$\begin{array}{ccccc} 6.3 & 11.0 & 21.5 & 48.4 & 90.1 \\ 120.2 & 163.0 & 182.5 & 198.0 & 219.0 \end{array}$$

Let us construct the TTT plot for this data set.

First we calculate the quantities we are going to need and put them in a table as done in Table 9.1. The TTT plot for this (complete) data set is shown in Figure 9.1. □

Table 9.1 TTT Estimates for the Data in Example 9.1

i	$T_{(i)}$	$\sum_{j=1}^{i} T_{(j)}$	$\sum_{j=1}^{i} T_{(j)} + (n-i)T_{(i)} = \mathscr{T}(T_{(i)})$			$\dfrac{i}{n}$	$\dfrac{\mathscr{T}(T_{(i)})}{\mathscr{T}(T_{(n)})}$
1	6.3	6.3	6.3 + 9·6.3	=	63.0	0.1	0.06
2	11.0	17.3	17.3 + 8·11.0	=	105.3	0.2	0.10
3	21.5	38.8	38.8 + 7·21.5	=	189.3	0.3	0.18
4	48.4	87.2	87.2 + 6·48.4	=	377.6	0.4	0.36
5	90.1	177.3	177.3 + 5·90.1	=	627.8	0.5	0.59
6	120.2	297.5	297.5 + 4·120.2	=	778.3	0.6	0.73
7	163.0	460.5	460.5 + 3·163.0	=	949.5	0.7	0.90
8	182.5	643.0	643.0 + 2·182.5	=	1008.0	0.8	0.95
9	198.0	841.0	841.0 + 1·198.0	=	1039.0	0.9	0.98
10	219.0	1060.0	1060.0 + 0	=	1060.0	1.0	1.00

In the introduction to this chapter we mentioned that the TTT plot can be used to decide whether the underlying life distribution $F(t)$ is IFR or DFR. To be able to do this, we need the following theorem which is stated here without proof (see Barlow and Campo 1975).

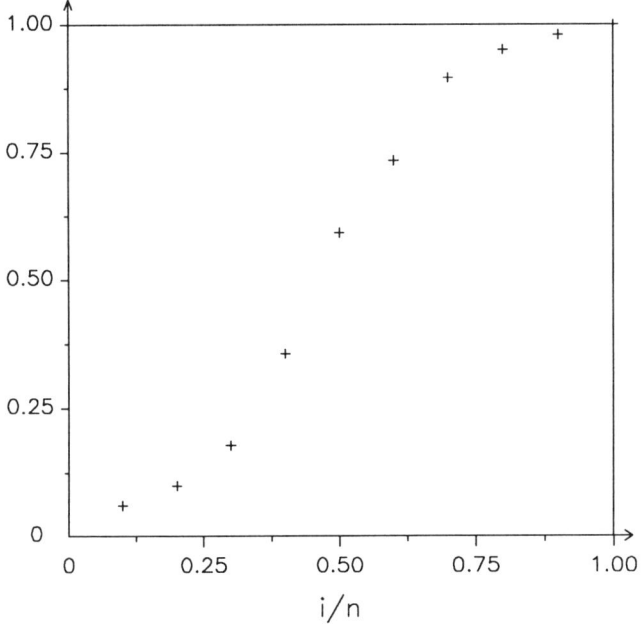

Figure 9.1 TTT plot of the data in Example 9.1

Theorem 9.1. Let $U_1, U_2, \ldots, U_{n-1}$ be independent random variables with a uniform distribution over $(0, 1]$. If the underlying life distribution is exponential, the random variables

$$\frac{\mathcal{T}(T_{(1)})}{\mathcal{T}(T_{(n)})}, \frac{\mathcal{T}(T_{(2)})}{\mathcal{T}(T_{(n)})}, \ldots, \frac{\mathcal{T}(T_{(n-1)})}{\mathcal{T}(T_{(n)})}$$

have the same joint distribution as the $(n-1)$ ordered variables $U_{(1)}, U_{(2)}, \ldots, U_{(n-1)}$.

From this theorem follows

Corollary 9.1. If the underlying life distribution is exponential, then

1. $\text{var}(\mathcal{T}(T_i)/\mathcal{T}(T_n))$ is finite
2. $E(\mathcal{T}(T_i)/\mathcal{T}(T_n)) = i/n$, for $i = 1, 2, \ldots, n$

If the underlying life distribution is exponential, it is expected for large n that

$$\frac{\mathcal{T}(T_{(i)})}{\mathcal{T}(T_{(n)})} \approx \frac{i}{n} \quad \text{for} \quad i = 1, 2, \ldots, (n-1)$$

As this is not the case for the TTT plot in Figure 9.1, we can conclude that the underlying life distribution for the data in Example 9.1 is probably not exponential.

To decide from a TTT plot whether or not the corresponding life distribution is IFR or DFR, we need a little more theory. We will be content with a heuristic argument.[1]

Let $F_n(t)$ denote the empirical distribution function of the data, defined as

$$F_n(t) = \begin{cases} 0 & \text{for } t < T_{(1)} \\ \dfrac{i}{n} & \text{for } T_{(i)} \leq t < T_{(i+1)}, \quad i = 1, 2, \ldots, (n-1) \\ 1 & \text{for } T_{(n)} \leq t \end{cases} \qquad (9.3)$$

We now claim that

$$\mathcal{T}(T_{(i)}) = n \int_0^{T_{(i)}} (1 - F_n(u)) \, du \qquad (9.4)$$

---

[1] A more rigorous treatment is found, for example, in Barlow and Campo (1975).

This assertion can be proved in the following way (remember that per definition $T_{(0)} = 0$):

$$n \int_0^{T_{(i)}} (1 - F_n(u)) \, du$$

$$= n \left[\sum_{j=1}^{i} \int_{T_{(j-1)}}^{T_{(j)}} \left(1 - \frac{j-1}{n} \right) du \right]$$

$$= \sum_{j=1}^{i} (n - j + 1)(T_{(j)} - T_{(j-1)})$$

$$= n T_{(1)} + (n - 1)(T_{(2)} - T_{(1)}) + \cdots + (n - i + 1)(T_{(i)} - T_{(i-1)})$$

$$= \sum_{j=1}^{i} T_{(j)} + (n - i) T_{(i)} = \mathcal{T}(T_{(i)})$$

We now come to the heuristic part of the argument.

First, let n equal $2m + 1$, where m is an integer. Then $T_{(m+1)}$ is the median of the data set. When $m \to \infty$, what happens to the integral

$$\int_0^{T_{(m+1)}} (1 - F_n(u)) \, du$$

When $m \to \infty$, we can expect that

$$F_n(u) \to F(u)$$

and that

$$T_{(m+1)} \to \{\text{median of } F\} = F^{-1} \left(\tfrac{1}{2} \right)$$

and therefore that

$$\frac{1}{n} \mathcal{T}(T_{(m+1)}) \to \int_0^{F^{-1}(1/2)} (1 - F(u)) \, du \qquad (9.5)$$

Next, let n equal $4m + 3$. In this case $T_{(2m+2)}$ is the median of the data, and $T_{(m+1)}$ and $T_{(3m+3)}$ are the lower and upper quartiles, respectively.

When $m \to \infty$, by the same argument as above, we can expect the following:

COMPLETE DATA SETS

$$\frac{1}{n} \mathcal{T}(T_{(m+1)}) \to \int_0^{F^{-1}(1/4)} (1 - F(u))\, du \qquad (9.6)$$

$$\frac{1}{n} \mathcal{T}(T_{(2m+2)}) \to \int_0^{F^{-1}(1/2)} (1 - F(u))\, du$$

$$\frac{1}{n} \mathcal{T}(T_{(3m+3)}) \to \int_0^{F^{-1}(3/4)} (1 - F(u))\, du$$

In addition, according to (2.11),

$$E(T) = \theta = \int_0^\infty (1 - F(u))\, du = \int_0^{F^{-1}(1)} (1 - F(u))\, du \qquad (9.7)$$

When $n \to \infty$, we can expect

$$\frac{1}{n} \sum_{i=1}^n T_i = \frac{1}{n} \mathcal{T}(T_{(n)}) \to \int_0^{F^{-1}(1)} (1 - F(u))\, du \qquad (9.8)$$

The integrals that we obtain as limits by this approach are of interest, and we will look at them more closely. They are all of the type

$$\int_0^{F^{-1}(v)} (1 - F(u))\, du \qquad \text{for } 0 \le v \le 1$$

We call this integral, $\int_0^{F^{-1}(v)} (1 - F(u))\, du$, the *total time on test (TTT) transform* of the distribution F and denote it by $H_F^{-1}(v)$.

Definition 9.2. The total time on test (TTT) transform of the distribution F is

$$H_F^{-1}(v) = \int_0^{F^{-1}(v)} (1 - F(u))\, du \qquad \text{for } 0 \le v \le 1 \qquad (9.9)$$

□

The TTT transform of the distribution F is illustrated in Figure 9.2.

It can be shown under assumptions of general nature that there is a one-to-one correspondence between a distribution $F(t)$ and its TTT transform $H_F^{-1}(v)$ (see Barlow and Campo 1975).

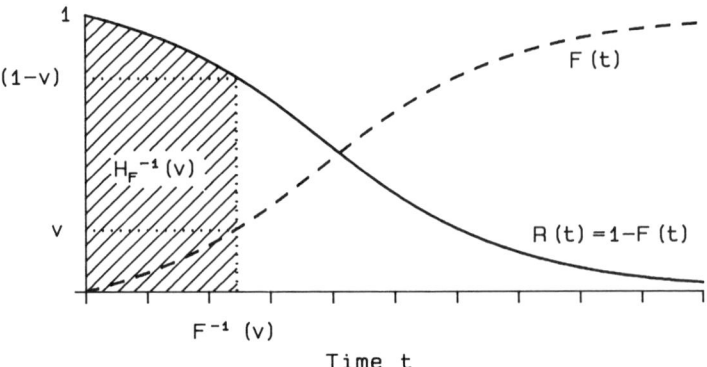

Figure 9.2 The TTT transform of the distribution F

We see from (9.9) and (9.7) that

$$H_F^{-1}(1) = \int_0^{F^{-1}(1)} (1 - F(u))\, du = \theta$$

The *scaled TTT transform* of $F(t)$ is defined as

$$\varphi_F(v) = \frac{H_F^{-1}(v)}{H_F^{-1}(1)} = \frac{1}{\theta} H_F^{-1}(v) \qquad \text{for } 0 \le v \le 1 \qquad (9.10)$$

Example 9.2. The distribution function of the exponential distribution is

$$F(t) = 1 - e^{-\lambda t} \qquad \text{for } t \ge 0,\, \lambda > 0$$

and hence

$$F^{-1}(v) = -\frac{1}{\lambda} \ln(1 - v) \qquad \text{for } 0 \le v \le 1$$

Thus the TTT transform of the exponential distribution is

COMPLETE DATA SETS

$$H_F^{-1}(v) = \int_0^{-\ln(1-v)/\lambda} e^{-\lambda u}\, du$$

$$= -\frac{1}{\lambda} e^{-\lambda u}\Big|_0^{-\ln(1-v)/\lambda}$$

$$= \frac{1}{\lambda} - \frac{1}{\lambda} e^{\lambda \ln(1-v)/\lambda}$$

$$= \frac{1}{\lambda} - \frac{1}{\lambda}(1-v) = \frac{v}{\lambda} \quad \text{for} \quad 0 \le v \le 1$$

Further

$$H_F^{-1}(1) = \frac{1}{\lambda}$$

The corresponding scaled TTT transform is therefore

$$\frac{v/\lambda}{1/\lambda} = v \quad \text{for} \quad 0 \le v \le 1 \tag{9.11}$$

The scaled TTT transform of the exponential distribution is thus a straight line from $(0,0)$ to $(1,1)$, as illustrated in Figure 9.3. □

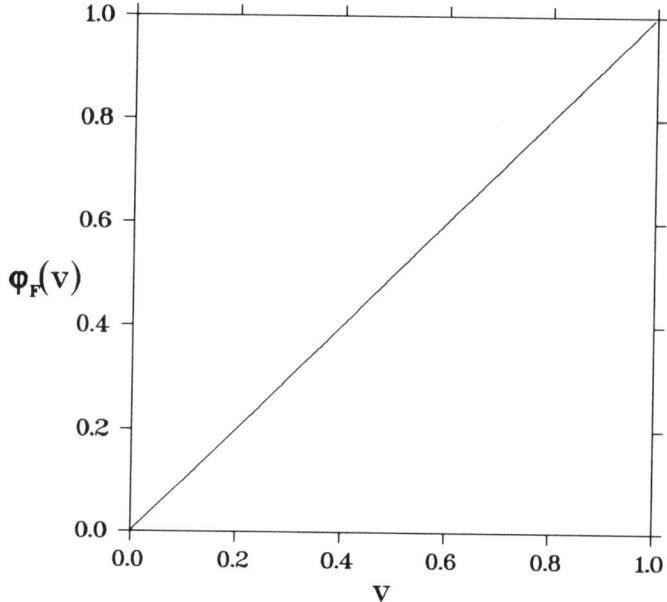

Figure 9.3 Scaled TTT transform of the exponential distribution

Example 9.3
It is usually not straightforward to determine the TTT transform of a life distribution. We will illustrate this by trying to determine the TTT transform of the Weibull distribution

$$F(t) = 1 - e^{-(\lambda t)^\alpha} \quad \text{for } t \geq 0, \lambda > 0, \alpha > 0$$

The inverse function of F is

$$F^{-1}(v) = \frac{1}{\lambda}(-\ln(1-v))^{1/\alpha} \quad \text{for } 0 \leq v \leq 1$$

According to (9.9) the TTT transform of the Weibull distribution is

$$H_F^{-1}(v) = \int_0^{F^{-1}(v)} (1 - F(u)) \, du = \int_0^{(-\ln(1-v))^{1/\alpha}/\lambda} e^{-(\lambda u)^\alpha} \, du$$

By substituting $x = (\lambda u)^\alpha$, we obtain

$$H_F^{-1}(v) = \frac{1}{\alpha\lambda} \int_0^{-\ln(1-v)} x^{1/\alpha+1} e^{-x} \, dx \quad (9.12)$$

which shows that the TTT transform of the Weibull distribution may be expressed by the incomplete gamma function. However, several approximation formulas are available.

The mean time to failure is obtained by inserting $v = 1$ in $H_F^{-1}(v)$ [see (9.7)]:

$$H_F^{-1}(1) = \frac{1}{\alpha\lambda} \int_0^\infty x^{1/\alpha+1} e^{-x} \, dx = \frac{1}{\alpha\lambda} \Gamma\left(\frac{1}{\alpha}\right) = \frac{1}{\lambda} \Gamma\left(\frac{1}{\alpha} + 1\right)$$

which coincides with the result we obtained in (2.37). Note that the scaled TTT transform of the Weibull distribution depends only on the shape parameter α and is independent of the scale parameter λ. □

Computer programs able to compute the TTT transforms of the most common life distributions and to print scaled TTT transforms are available. One of these programs is SAREPTA. Scaled TTT transforms of the Weibull distribution for some selected values of the shape parameter α are illustrated in Figure 9.4.

COMPLETE DATA SETS

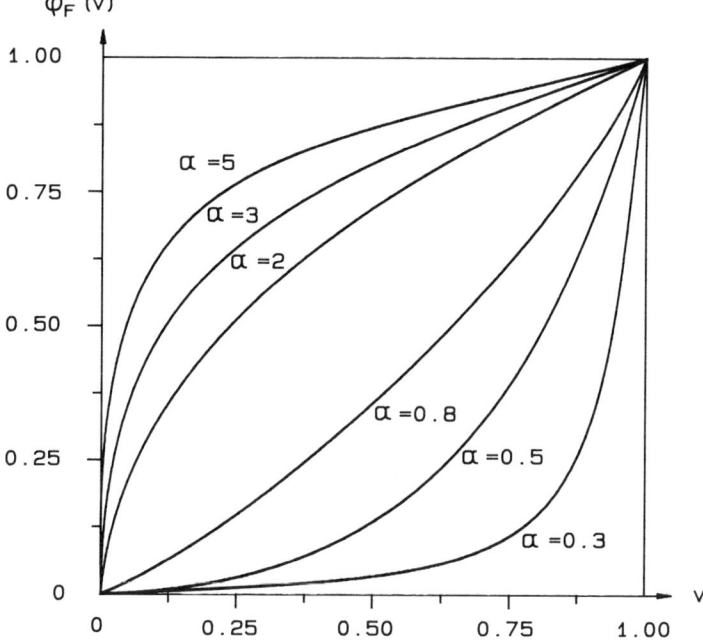

Figure 9.4 Scaled TTT transforms of the Weibull distribution for some selected values of α

We will now prove the following theorem:

Theorem 9.2. If $F(t)$ is a continuous life distribution that is strictly increasing for $F^{-1}(0) = 0 < t < F^{-1}(1)$, then

$$\frac{d}{dv} H_F^{-1}(v)\big|_{v=F(t)} = \frac{1}{z(t)} \qquad (9.13)$$

where $z(t)$ is the failure rate of the distribution $F(t)$.

Proof
Since

$$\frac{d}{dv} H_F^{-1}(v) = \frac{d}{dv} \int_0^{F^{-1}(v)} (1 - F(u))\, du$$

$$= [1 - F(F^{-1}(v))] \frac{d}{dv} F^{-1}(v) = (1 - v) \frac{1}{f(F^{-1}(v))}$$

then

$$\frac{d}{dv} H_F^{-1}(v)|_{v=F(t)} = (1 - F(t)) \cdot \frac{1}{f(t)} = \frac{1}{z(t)} \qquad \square$$

From Theorem 9.2 we can now prove

Theorem 9.3. If $F(t)$ is a continuous life distribution, strictly increasing for $F^{-1}(0) = 0 < t < F^{-1}(1)$, then

1. $F \sim \text{IFR} \Leftrightarrow H_F^{-1}(v)$ concave, $0 \leq v \leq 1$
2. $F \sim \text{DFR} \Leftrightarrow H_F^{-1}(v)$ convex, $0 \leq v \leq 1$

The arguments used to prove properties 1 and 2 are completely analogous. We therefore prove only property 1.

Proof

$$F \sim \text{IFR} \Leftrightarrow z(t) \quad \text{is nondecreasing in } t$$
$$\Leftrightarrow \frac{1}{z(t)} \quad \text{is nonincreasing in } t$$
$$\Leftrightarrow \frac{d}{dv} H_F^{-1}(v)|_{v=F(t)} \quad \text{is nonincreasing in } t$$
$$\Leftrightarrow \frac{d}{dv} H_F^{-1}(v) \quad \text{is nonincreasing in } v,$$
$$\text{since } F(t) \text{ is strictly increasing in } t$$
$$\Leftrightarrow H_F^{-1}(v) \quad \text{is concave, } 0 \leq v \leq 1 \qquad \square$$

If we are going to estimate the scaled TTT transform of $F(t)$ for different v values on the basis of the observed lifetimes, it is natural to use the estimator

$$\frac{\int_0^{F_n^{-1}(v)} (1 - F_n(u)) \, du}{\int_0^{F_n^{-1}(1)} (1 - F_n(u)) \, du} \quad \text{for } v = \frac{i}{n}, i = 1, 2, \ldots, n \qquad (9.14)$$

Introducing the notation

$$H_n^{-1}(v) = \int_0^{F_n^{-1}(v)} (1 - F_n(u)) \, du \quad \text{for } v = \frac{i}{n}, i = 1, 2, \ldots, n \qquad (9.15)$$

this estimator can be written

COMPLETE DATA SETS

$$\frac{H_n^{-1}(v)}{H_n^{-1}(1)} \quad \text{for} \quad v = \frac{i}{n}, i = 1, 2, \ldots, n \tag{9.16}$$

By comparing (9.16) with (9.10), it seems natural to call $H_n^{-1}(v)/H_n^{-1}(1)$ the *empirical*, scaled TTT transform of the distribution F.

The following theorem is useful when we wish to exploit the TTT plot to provide information about the life distribution $F(t)$:

Theorem 9.4. If $F(t)$ is a continuous life distribution function strictly increasing for $F^{-1}(0) = 0 < t < F^{-1}(1)$, then

$$\frac{H_n^{-1}(\frac{i}{n})}{H_n^{-1}(1)} = \frac{\mathcal{T}(T_{(i)})}{\mathcal{T}(T_{(n)})} \quad \text{for} \quad i = 1, 2, \ldots, n \tag{9.17}$$

where $\mathcal{T}(T_{(i)})$, as before, denotes the total time on test at time $T_{(i)}$.

Proof
According to (9.15) and (9.3), for $i = 1, 2, \ldots, n$,

$$H_n^{-1}\left(\frac{i}{n}\right) = \int_0^{F_n^{-1}(i/n)} (1 - F_n(u)) \, du$$

$$= \int_0^{T_{(i)}} (1 - F_n(u)) \, du = \frac{1}{n} \mathcal{T}(T_{(i)}) \tag{9.18}$$

while

$$H_n^{-1}(1) = \int_0^{F_n^{-1}(1)} (1 - F_n(u)) \, du$$

$$= \int_0^{\infty} (1 - F_n(u)) \, du = \frac{1}{n} \mathcal{T}(T_{(n)}) = \frac{1}{n} \sum_{i=1}^{n} T_i \tag{9.19}$$

By introducing these results in (9.16), we get (9.17). □

Hence the scaled total time on test at time $T_{(i)}$ seems to be a natural estimator of the scaled TTT transform of $F(t)$ for $v = i/n$ $i = 1, 2, \ldots, n$. One way of obtaining an estimate for the scaled TTT transform of $F(v)$ for $(i-1)/n < v < i/n$ $i = 1, 2, \ldots, n$, is by applying linear interpolation between the estimate for $v = (i-1)/n$ and $v = i/n$. On the following pages we will use this procedure.

Now suppose that we have carried out a life test as described in the introduction to this chapter. We determine $\mathcal{T}(T_{(i)})/\mathcal{T}(T_{(n)})$ for $i = 1, 2, \ldots, n$ as we did in Example 9.1, plot the points $[i/n, \mathcal{T}(T_{(i)})/\mathcal{T}(T_{(n)})]$, and join pairs of neighboring points with straight lines. The curve obtained is an estimate for $H_F^{-1}(v)/H_F^{-1}(1) = H_F^{-1}(v)/\theta$, $0 \le v \le 1$. We may now assess the shape of the curve [the estimate for $H_F^{-1}(v)$] in the light of Theorem 9.3, and in this way obtain information about the underlying distribution $F(t)$.

A plot like the one shown in Figure 9.5(a) indicates that $H_F^{-1}(v)$ is concave. The plot therefore indicates that the corresponding life distribution $F(t)$ is IFR. By the same argument the plot in Figure 9.5(b) indicates that $H_F^{-1}(v)$ is convex, so the corresponding life distribution $F(t)$ is DFR. Similarly the plot in Figure 9.5 (c) indicates that $H_F^{-1}(v)$ "is first convex" and "thereafter concave." In other words, the failure rate of the corresponding lifetime distribution has a bathtub shape. The TTT plot obtained in Example 9.1 therefore indicates that these data originate from a life distribution with bathtub-shaped failure rate.

Example 9.4
The following data from Lieblein and Zelen (1956) are the numbers of millions of revolutions to failure for each of 23 ball bearings. The original data have been put in numerical order for convenience.

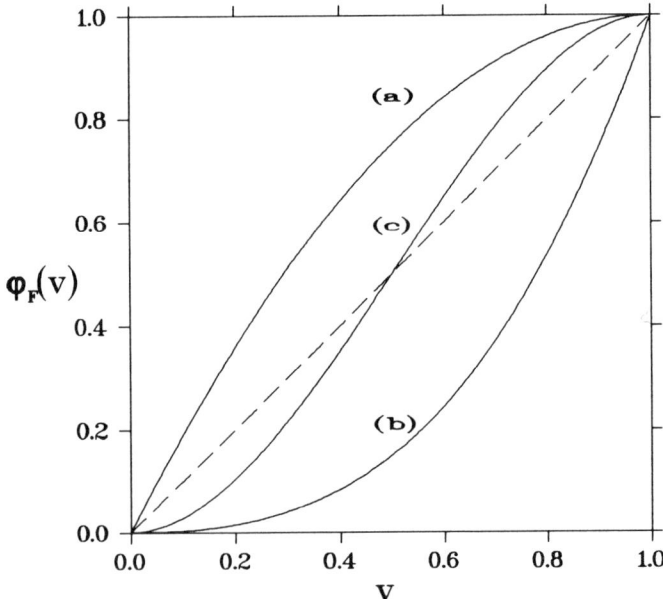

Figure 9.5 TTT plots indicating (a) increasing failure rate (IFR), (b) decreasing failure rate (DFR), and (c) Bathtub-shaped failure rate

COMPLETE DATA SETS

17.88	28.92	33.00	41.52	42.12	45.60	48.40
51.84	51.96	54.12	55.56	67.80	68.64	68.64
68.88	84.12	93.12	98.64	105.12	105.84	127.92
128.04	173.40					

The TTT plot of the ball bearing-data is presented in Figure 9.6. The TTT plot indicates an increasing failure rate. We may try to fit a Weibull distribution to the data. The Weibull parameters α and λ are estimated by the SAREPTA program to be $\hat{\alpha} = 2.10$ and $\hat{\lambda} = 1.22 \cdot 10^{-2}$. The TTT transform of the Weibull distribution with these parameters has been determined by SAREPTA and plotted as an overlay curve to the TTT plot in Figure 9.6. □

Age Replacements

A well-known application of the TTT transform and the TTT plot is the ordinary age replacement problem. Here a unit is replaced at a cost $c+k$ at *failure* or at a cost c at a *planned replacement* when the unit has reached a certain age t_0. The cost c covers the hardware and worker hour costs, while k is the extra cost incurred by the unplanned replacement, such as production loss. The problem

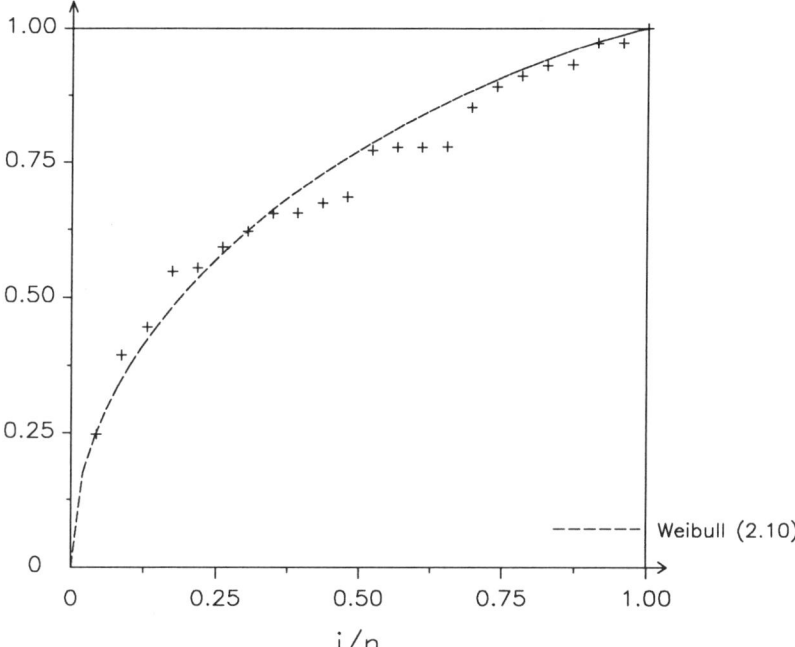

Figure 9.6 TTT plot of the ball-bearing data in Example 9.4 together with an overlay curve of the TTT transform of the Weibull distribution with shape parameter $\alpha = 2.10$ (printout from SAREPTA)

is to determine the optimal replacement age t_0 to minimize the long-term cost per unit time.

Let T denote the potential time to failure of the unit. T is assumed to have a continuous distribution function $F(t)$ with density $f(t) = F'(t)$. The mean time between replacements MTBR of the unit when the replacement age is t_0 is thus

$$\text{MTBR} = \int_0^{t_0} t f(t) \, dt + t_0 \cdot P(T > t_0) = \int_0^{t_0} (1 - F(t)) \, dt \qquad (9.20)$$

The last equality in (9.20) follows by using partial integration:

$$\int_0^{t_0} (1 - F(t)) \, dt = \int_0^{t_0} dt - \int_0^{t_0} F(t) \, dt = t_0 - \int_0^{t_0} F(t) \cdot 1 \, dt$$

$$= t_0 - \left[[F(t) \cdot t]_0^{t_0} - \int_0^{t_0} t f(t) \, dt \right]$$

$$= t_0 [1 - F(t_0)] + \int_0^{t_0} t f(t) \, dt$$

$$= \int_0^{t_0} t f(t) \, dt + t_0 \cdot P(T > t_0)$$

The cost per replacement period is equal to the replacement cost c plus the extra cost k whenever a failure occurs. The *mean* total cost per replacement period is thus

$$c + k \cdot P(\text{``failure''}) = c + k \cdot P(T < t_0) = c + k \cdot F(t_0)$$

The total cost per unit time $C(t_0)$ with replacement age t_0 is determined by

$$C(t_0) \cdot \text{MTBR} = c + k \cdot F(t_0)$$

Hence

$$C(t_0) = \frac{c + k \cdot F(t_0)}{\int_0^{t_0} (1 - F(t)) \, dt} \qquad (9.21)$$

The objective is now to determine the value of t_0 that minimizes $C(t_0)$. If the distribution function $F(t)$ and all its parameters are known, it is a straightforward task to determine the optimal value of t_0. One way to solve this problem is to apply the TTT transform. Recall that the TTT transform of $F(t)$ was defined as

COMPLETE DATA SETS

$$H_F^{-1}(v) = \int_0^{F^{-1}(v)} (1 - F(t))\, dt$$

By introducing $v = F(t_0)$, we obtain

$$H_F^{-1}(F(t_0)) = \int_0^{F^{-1}(F(t_0))} (1 - F(t))\, dt = \int_0^{t_0} (1 - F(t))\, dt$$

Formula (9.21) may now be rewritten as

$$C(t_0) = \frac{c + k \cdot F(t_0)}{H_F^{-1}(F(t_0))} = \frac{1}{H_F^{-1}(1)} \frac{c + k \cdot F(t_0)}{\varphi_F(F(t_0))}$$

where $H_F^{-1}(1)$ is the MTTF of the unit, and

$$\varphi_F(v) = \frac{H_F^{-1}(v)}{H_F^{-1}(1)}$$

is the scaled TTT transform of the distribution function $F(t)$.

The optimal value of t_0 may be determined by first finding the value $v_0 = F(t_0)$, which minimizes

$$C_1(v_0) = \frac{c + k \cdot v_0}{\varphi_F(v_0)}$$

and thereafter determining t_0 such that $v_0 = F(t_0)$. The minimizing value of v_0 may be found by setting the derivative of $C_1(v_0)$ with respect to v_0 equal to zero and solving the equation for v_0:

$$\frac{d}{v_0} C_1(v_0) = \frac{\varphi_F(v_0) \cdot k - \varphi_F'(v_0)(c + k \cdot v_0)}{\varphi_F(v_0)^2} = 0$$

This implies that

$$\varphi_F'(v_0) = \frac{\varphi_F(v_0)}{c/k + v_0} \tag{9.22}$$

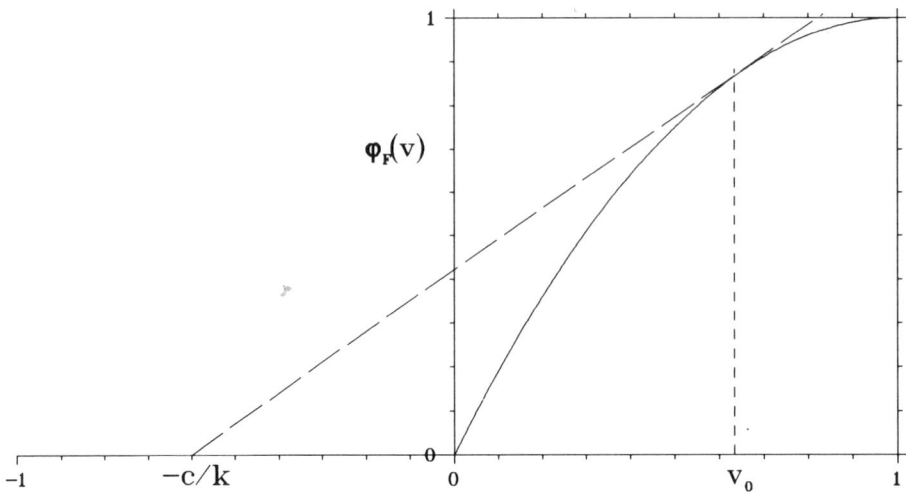

Figure 9.7 Determination of the optimal replacement age from the scaled TTT transform

The optimal value of v_0, and hence t_0, may now be determined by the following simple graphical method:

1. Draw the scaled TTT transform in a 1×1-coordinate system
2. Identify the point $(-c/k, 0)$ on the abcissa axis
3. Draw a tangent from $(-c/k, 0)$ to the TTT transform

The optimal value of v_0 can be read as the abcissa of the point where the tangent touches the TTT transform. If $v_0 = 1$, then $t_0 = \infty$, and no preventive replacements should be performed. The procedure is illustrated in Figure 9.7.

When a set of times to failure of the actual type of unit has been recorded, we may use this data set to obtain the empirical, scaled TTT transform of the underlying distribution function $F(t)$, and to draw a TTT plot. The optimal replacement age t_0 may now be determined by the same procedure as described above. This is illustrated in Figure 9.8. The procedure is further discussed, for example, by Bergman and Klefsjö (1982, 1984).

Exponentially Distributed Lifetimes

We shall now show how life data can be analyzed when the life distribution $F(t)$ is assumed to be exponential with unknown failure rate λ.

Let us first assume that we have observed a population of identical components a total time in service τ. Under the assumptions made in Section 2.5, failures will occur according to a Poisson process and the number of failures X observed during this period will have a Poisson distribution with parameter

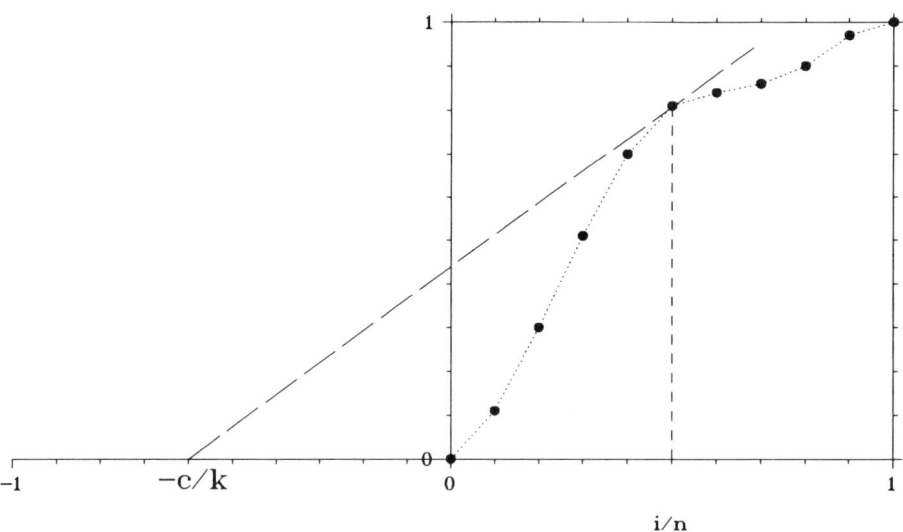

Figure 9.8 Determination of the optimal replacement age from a TTT plot

$\lambda\tau$. An unbiased estimator of the failure rate λ is thus

$$\tilde{\lambda} = \frac{X}{\tau} \qquad (9.23)$$

and a $1 - \epsilon$ confidence interval for λ is given by (e.g., see Sverdrup 1967, p. 316; Cox and Oakes 1984, p. 41.):

$$\left(\frac{1}{\tau} z_{\epsilon/2,2X}, \frac{1}{\tau} z_{1-\epsilon/2,2(X+1)} \right) \qquad (9.24)$$

where $z_{\epsilon,\nu}$ denotes the upper $100\epsilon\%$ percentile of the chi-square (χ^2) distribution with ν degrees of freedom.

In many situations it is of interest to give an upper $1 - \epsilon$ confidence limit for λ. Such a limit is obtained through the one-sided confidence interval given by

$$\left(0, \frac{1}{\tau} z_{1-\epsilon,2(X+1)} \right) \qquad (9.25)$$

Note that this interval is applicable even if no failures ($X = 0$) are observed during the total time in service τ.

When $\lambda\tau$ is large (e.g., $\lambda\tau > 15$), the Poisson distribution may be approximated by the normal distribution $\mathcal{N}(\lambda\tau, \lambda\tau)$. Then

$$\frac{X - \lambda\tau}{\sqrt{\lambda\tau}} \sim \mathcal{N}(0, 1)$$

Hence if $u_{\epsilon/2}$ denotes the $\epsilon/2$ quantile of the standard normal distribution

$$P\left(-u_{\epsilon/2} \leq \frac{X - \lambda\tau}{\sqrt{\lambda\tau}} \leq u_{\epsilon/2}\right) \approx 1 - \epsilon$$

That is,

$$P\left(\left(\frac{X - \lambda\tau}{\sqrt{\lambda\tau}}\right)^2 \leq u_{\epsilon/2}^2\right) \approx 1 - \epsilon$$

But

$$\left(\frac{X - \lambda\tau}{\sqrt{\lambda\tau}}\right)^2 \leq u_{\epsilon/2}^2 \Leftrightarrow (\lambda\tau)^2 - \lambda\tau(2X + u_{\epsilon/2}^2) + X^2 \leq 0$$

which implies that

$$P(\lambda_1(X) \leq \lambda \leq \lambda_2(X)) \approx 1 - \epsilon$$

where $\lambda_1(X)$ and $\lambda_2(X)$, respectively, denote the lower and the upper roots of the equation

$$\lambda^2 - \lambda(2X + u_{\epsilon/2}^2)/\tau + X^2/\tau^2 = 0$$

That is,

$$\lambda_1(X) = \frac{1}{\tau}\left(X + \frac{1}{2}u_{\epsilon/2}^2 - u_{\epsilon/2}\sqrt{X + \frac{1}{4}u_{\epsilon/2}^2}\right)$$

$$\lambda_2(X) = \frac{1}{\tau}\left(X + \frac{1}{2}u_{\epsilon/2}^2 + u_{\epsilon/2}\sqrt{X + \frac{1}{4}u_{\epsilon/2}^2}\right)$$

ooo OOO ooo

COMPLETE DATA SETS   375

Next, let us assume that we have recorded the actual lifetimes of n components. We denote the lifetime of component j by $T_j, j = 1, 2, \ldots, n$, and assume that $T_1, T_2, \ldots, T_n$ are independent and identically exponentially distributed with unknown failure rate λ. The likelihood function is then

$$\ell(\lambda; t_1, \ldots t_n) = \lambda^n e^{-\lambda \sum_{j=1}^{n} t_j} \qquad \lambda > 0, t_j > 0, j = 1, 2, \ldots, n \qquad (9.26)$$

The maximum likelihood estimator of λ is

$$\lambda^* = \frac{n}{\sum_{j=1}^{n} T_{(j)}} = \frac{n}{\mathscr{T}(T_{(n)})} \qquad (9.27)$$

where $\mathscr{T}(T_{(n)})$, as before, denotes the total time on test at the last failure. A brief introduction to maximum likelihood (ML) estimation is given in Appendix C.

Let us study the properties of this estimator and first find out whether it is unbiased or not. Since T_j is exponentially distributed with parameter λ, $2\lambda T_j$ will be χ^2 distributed with two degrees of freedom for $j = 1, 2, \ldots, n$ (Dudewicz and Mishra 1988, p. 276). Since the T_j's are independent, then $2\lambda \sum_{j=1}^{n} T_j$ will be χ^2 distributed with $2n$ degrees of freedom. Therefore

$$\lambda^* = \frac{n}{\sum_{j=1}^{n} T_{(j)}} = \frac{2n\lambda}{2\lambda \sum_{j=1}^{n} T_{(j)}}$$

will have the same distribution as $2n\lambda/Z$, where Z is χ^2 distributed with $2n$ degrees of freedom. Accordingly

$$E(\lambda^*) = 2n\lambda E\left(\frac{1}{Z}\right)$$

But

$$E\left(\frac{1}{Z}\right) = \int_0^\infty \frac{1}{z} \cdot \frac{1}{2^n} \frac{1}{\Gamma(n)} \cdot z^{n-1} e^{-z/2} \, dz$$
$$= \frac{1}{2(n-1)} \int_0^\infty \frac{1}{2^{n-1}\Gamma(n-1)} z^{n-2} e^{-z/2} \, dz$$
$$= \frac{1}{2(n-1)}$$

Therefore

$$E(\lambda^*) = 2n\lambda \cdot \frac{1}{2(n-1)} = \frac{n}{n-1} \cdot \lambda$$

λ^* is accordingly not unbiased. The estimator $\hat{\lambda}$ given by

$$\hat{\lambda} = \frac{n-1}{n} \cdot \lambda^* = \frac{n-1}{\sum_{j=1}^{n} T_j} = \frac{n-1}{\mathcal{T}(T_{(n)})} \tag{9.28}$$

is easily seen to be unbiased.
Let us determine $\text{var}(\hat{\lambda})$.

$$\text{var}(\hat{\lambda}) = \left(\frac{n-1}{n}\right)^2 \cdot \text{var}(\lambda^*) = 4(n-1)^2 \lambda^2 \, \text{var}\left(\frac{1}{Z}\right)$$

where Z has the same meaning as above. Now

$$\text{var}\left(\frac{1}{Z}\right) = E\left(\frac{1}{Z^2}\right) - \left[E\left(\frac{1}{Z}\right)\right]^2$$

and

$$E\left(\frac{1}{Z^2}\right) = \int_0^\infty \frac{1}{z^2} \frac{1}{2^n} \frac{1}{\Gamma(n)} z^{n-1} e^{-z/2} \, dz = \frac{1}{4(n-1)(n-2)} \tag{9.29}$$

Hence

$$\text{var}(\hat{\lambda}) = 4(n-1)^2 \lambda^2 \cdot \left(\frac{1}{4(n-1)(n-2)} - \frac{1}{4(n-1)^2}\right)$$
$$= (n-1)\lambda^2 \cdot \left(\frac{1}{n-2} - \frac{1}{n-1}\right) = \frac{\lambda^2}{n-2} \tag{9.30}$$

The estimator

$$\hat{\lambda} = \frac{n-1}{\mathcal{T}(T_{(n)})} \tag{9.31}$$

is therefore unbiased and has variance

$$\text{var}(\hat{\lambda}) = \frac{\lambda^2}{n-2}$$

ooo OOO ooo

To establish a $1-\epsilon$ confidence interval for λ, we use the fact that $2\lambda \sum_{j=1}^{n} T_j$ is χ^2-distributed with $2n$ degrees of freedom. Hence

$$P\left(z_{1-\epsilon/2,2n} \leq 2\lambda \sum_{j=1}^{n} T_j \leq z_{\epsilon/2,2n}\right) = 1 - \epsilon$$

and

$$P\left(\frac{z_{1-\epsilon/2,2n}}{2\sum_{j=1}^{n} T_j} \leq \lambda \leq \frac{z_{\epsilon/2,2n}}{2\sum_{j=1}^{n} T_j}\right) = 1 - \epsilon$$

Thus a $1-\epsilon$ confidence interval for λ is

$$\left(\frac{z_{1-\epsilon/2,2n}}{2\sum_{j=1}^{n} T_j}, \frac{z_{\epsilon/2,2n}}{2\sum_{j=1}^{n} T_j}\right) \tag{9.32}$$

ooo OOO ooo

To find out whether or not the observed lifetimes $T_1, T_2, \ldots, T_n$ support the assertion that $\lambda < \lambda_0$, we need a test for

$$H_0 : \lambda \geq \lambda_0 \quad \text{against} \quad H_1 : \lambda < \lambda_0$$

As a first step let us derive a test for

$$H_0' : \lambda = \lambda_0 \quad \text{against} \quad H_1 : \lambda < \lambda_0$$

Then it seems reasonable to reject H_0' when $\hat{\lambda} \leq k$, where k is determined to give the test the significance level ϵ:

$$P(\hat{\lambda} \leq k | H_0') \leq \epsilon$$

Now

$$\hat{\lambda} \leq k \Leftrightarrow \frac{1}{\hat{\lambda}} \geq \frac{1}{k}$$

$$\Leftrightarrow \frac{1}{(n-1)} \sum_{j=1}^{n} T_j \geq \frac{1}{k}$$

$$\Leftrightarrow 2\lambda_0 \sum_{j=1}^{n} T_j \geq \frac{2(n-1)}{k} \cdot \lambda_0$$

By introducing $2\lambda_0(n-1)/k = c$, the test can be written as

$$\text{Reject } H'_0 \quad \text{when} \quad 2\lambda_0 \sum_{j=1}^{n} T_j \geq c$$

Under H'_0, $2\lambda_0 \sum_{j=1}^{n} T_j$ is χ^2 distributed with $2n$ degrees of freedom, and accordingly c is chosen equal to $z_{\epsilon,2n}$.

Intuitively, the same test can be used when H_0 is to be tested against H_1. The test then has the power function

$$P\left(2\lambda_0 \sum_{j=1}^{n} T_j \geq z_{\epsilon,2n} | \lambda \right) = P\left(2\lambda \sum_{j=1}^{n} T_j \geq \frac{\lambda}{\lambda_0} z_{\epsilon,2n}\right)$$

$$= 1 - \Gamma_{2n}\left(z_{\epsilon,2n} \cdot \frac{\lambda}{\lambda_0}\right) \quad (9.33)$$

Here $\Gamma_{2n}(z)$ denotes the distribution function of the χ^2 distribution with $2n$ degrees of freedom and is nondecreasing in z. Accordingly $1 - \Gamma_{2n}(z_{\epsilon,2n} \cdot \lambda/\lambda_0)$ is nonincreasing in λ, and

$$P\left(2\lambda_0 \sum_{j=1}^{n} T_j \geq z_{\epsilon,2n} | \lambda\right) \leq \epsilon \quad \text{for} \quad \lambda \geq \lambda_0$$

with the equality sign valid for $\lambda = \lambda_0$.

For testing

$$H_0: \lambda \geq \lambda_0 \quad \text{against} \quad H_1: \lambda < \lambda_0$$

COMPLETE DATA SETS

we therefore use the test criterion:

$$\text{Reject } H_0 \text{ when } 2\lambda_0 \sum_{j=1}^{n} T_j \geq z_{\epsilon,2n} \qquad (9.34)$$

The power function of the test is given by (9.33), and the significance level is ϵ.

Inverse Gaussian Distributed Lifetimes

We will restrain ourselves to considering a one–sample situation where μ and λ are both positive but unknown.

ML Estimators for μ and λ

Theorem 9.5. Let $T_1, \ldots, T_k$ be independent, and let T_j for $j = 1, 2, \ldots, k$, be $IG(\mu, n_j\lambda)$ where $n_1, n_2, \ldots, n_k$ are known, positive integers, $\sum_{j=1}^{k} n_j = N$, and $\mu > 0$, $\lambda > 0$, but otherwise unknown. Then the likelihood function becomes

$$\ell(\mu, \lambda; t_1, \ldots, t_k) = \prod_{j=1}^{k} \left(\sqrt{\frac{n_j}{2\pi t_j^3}} \cdot \lambda^{1/2} \cdot \exp\left(-\frac{\lambda n_j (t_j - \mu)^2}{2\mu^2 t_j} \right) \right) \qquad (9.35)$$

and the maximum likelihood estimators of μ and λ, μ^* and λ^* are given by

$$\mu^* = \frac{1}{N} \sum_{j=1}^{k} n_j T_j, \qquad (9.36)$$

and

$$\frac{1}{\lambda^*} = \frac{1}{k} \sum_{j=1}^{k} \frac{n_j (T_j - \mu^*)^2}{\mu^{*2} T_j} = \frac{1}{k} \sum_{j=1}^{k} n_j \left(\frac{1}{T_j} - \frac{1}{\mu^*} \right). \qquad (9.37)$$

Furthermore

$$E(\mu^*) = \mu, \quad \text{var}(\mu^*) = \frac{\mu^3}{\lambda N}. \qquad (9.38)$$

$$E\left(\frac{1}{\lambda^*}\right) = \frac{k-1}{k} \cdot \frac{1}{\lambda}, \quad \text{var}\left(\frac{1}{\lambda^*}\right) = \frac{2(k-1)}{k} \cdot \frac{1}{\lambda^2} \qquad (9.39)$$

Proof. See Problem 9.4.

ooo OOO ooo □

Corollary 9.2. Let $T_1, \ldots, T_k$ be i.i.d. $IG(\mu, \lambda)$, where $\mu > 0$ and $\lambda > 0$ but otherwise unknown. Then the MLE for μ and λ, μ^* and λ^* are given by

$$\mu^* = \frac{1}{k} \sum_{j=1}^{k} T_j = \overline{T}, \qquad \frac{1}{\lambda^*} = \frac{1}{k} \sum_{j=1}^{k} \left(\frac{1}{T_j} - \frac{1}{\overline{T}} \right) \qquad (9.40)$$

The corollary follows directly from Theorem 9.4, by choosing $n_1 = n_2 = \cdots = n_k = 1$. Schrödinger proposed the estimators in (9.40) as early as 1915 and denoted them *die Wahrscheinlichste* (Schrödinger 1915).

In the following we restrain ourselves to the case where $T_1, \ldots, T_k$ are i.i.d. $IG(\mu, \lambda)$, where μ and λ are both positive but otherwise unknown. By utilizing the well-known result that the MLE of a function $g(\mu, \lambda)$ is $g(\mu^*, \lambda^*)$, where μ^* and λ^* are the MLEs of μ and λ, respectively, one gets the MLE for var(T):

$$\widehat{\text{var}(T)} = \frac{\mu^{*3}}{\lambda^*} \qquad (9.41)$$

Similarly the MLE for the survivor function $R(t; \mu, \lambda)$ is

$$R(t; \widehat{\mu^*, \lambda^*}) = \Phi\left(-\frac{\sqrt{\lambda^*}}{\mu^*} \sqrt{t} - \sqrt{\lambda^*} \cdot \frac{1}{\sqrt{t}} \right)$$
$$- e^{2(\lambda^*/\mu^*)} \Phi\left(-\frac{\sqrt{\lambda^*}}{\mu^*} \sqrt{t} - \sqrt{\lambda^*} \cdot \frac{1}{\sqrt{t}} \right) \qquad (9.42)$$

and the MLE of the failure rate $z(t; \mu, \lambda)$ is

$$z(t; \widehat{\mu^*, \lambda^*})$$
$$= \frac{\sqrt{\lambda^*/(2\pi t^3)} \exp[-\lambda^*(t-\mu^*)^2/(2\mu^{*2}t)]}{\Phi(-\sqrt{\lambda^*t}/\mu^* + \sqrt{\lambda^*}/\sqrt{t}) + \exp[2\lambda^*/\mu^*] \cdot \Phi(-\sqrt{\lambda^*t}/\mu^* - \sqrt{\lambda^*}/\sqrt{t})}$$

$$(9.43)$$

Exponentiality, Completeness, and Sufficiency

Let $T_1, \ldots, T_n$ be i.i.d. $IG(\mu, \lambda)$, where μ and λ are positive but unknown. Then the joint density of $T_1, \ldots, T_n$ can be written

$$f_{T_1,\ldots,T_n}(t_1,\ldots,t_n;\mu,\lambda)$$

$$= \left(\frac{\lambda}{2\pi}\right)^{n/2} \cdot \prod_{j=1}^{n} t_j^{-3/2} \cdot \exp\left(-\frac{\lambda}{2\mu^2} \sum_{j=1}^{n} \frac{(t_j-\mu)^2}{t_j}\right)$$

$$= \exp\left(\frac{n}{2}\ln\frac{\lambda}{2\pi} - \sum_{j=1}^{n}\ln t_j^{3/2} - \frac{\lambda}{2\mu^2}\sum_{j=1}^{n} t_j + \frac{\lambda n}{\mu} - \frac{\lambda}{2}\sum_{j=1}^{n}\frac{1}{t_j}\right)$$

$$= \exp\left(\sum_{i=1}^{2} c_i(\mu,\lambda)\cdot Y_i(t) + d(\mu,\lambda) + s(t)I_{A(T)}\right)$$

where

$$c_1(\mu,\lambda) = -\frac{1}{2\mu^2}, \quad c_2(\mu,\lambda) = -\frac{\lambda}{2}, \quad d(\mu,\lambda) = \frac{n}{2}\ln\frac{\lambda}{2\pi} + \frac{\lambda n}{\mu}$$

$$S(t) = -\sum \ln t_j^{3/2}, \quad A(t) = \{t; t_j > 0, j=1,\ldots,n\}$$

$$Y_1(t) = \sum_j t_j, \quad Y_2(t) = \sum_j \frac{1}{t_j}$$

$I_{A(t)}$ is the indicator for $A(t) \subset R^n$. Hence the $IG(\mu,\lambda)$ family constitutes a two-parameter exponential family (e.g., Bickel and Doksum 1977, p. 72), and $(\sum_{j=1}^{n} T_j, \sum_{j=1}^{n} 1/T_j)$ is a natural sufficient statistic for this family.

Since the range of $c(\mu,\lambda) = (c_1(\mu,\lambda), c_2(\mu,\lambda))$ contains an open rectangle, the statistic $(\sum_{j=1}^{n} T_j, \sum_{j=1}^{n} 1/T_i)$ is complete as well as sufficient (e.g., Bickel and Doksum 1977, p. 123.)

Theorem 9.6. Let $T_1,\ldots,T_k$ be i.i.d. $IG(\mu,\lambda)$, $\mu > 0$, $\lambda > 0$, but otherwise unknown. Then the estimators

$$\mu^* = \overline{T} \quad \text{and} \quad \frac{1}{\hat{\lambda}} = \frac{1}{(k-1)}\sum_{j=1}^{k}\left(\frac{1}{T_j} - \frac{1}{\overline{T}}\right) \tag{9.44}$$

are UMVU (uniform minimum variance unbiased) for μ and $1/\lambda$ respectively.

Proof
From Theorem 9.4 follows that μ^* and $1/\hat{\lambda}$ are unbiased esimators of μ and $1/\lambda$, respectively. Furthermore they are functions of the observations through the complete sufficient statistic $(\sum_{j=1}^{k} T_j, \sum_{j=1}^{k} 1/T_j)$. Hence they are UMVU for μ and $1/\lambda$. (e.g., Bickel and Doksum 1977, p. 122). □

Chhikara and Folks (1974) have derived UMVU estimators of $F_T(t_0; \mu, \lambda)$ in the situation where

- μ is known and λ is unknown
- μ is unknown and λ is known
- μ and λ are both unknown

These estimators can be used directly to find UMVU estimators of the survivor function $R(t; \mu, \lambda) = 1 - F_T(t; \mu, \lambda)$.

Example 9.5
The following data consisting of the times to failure, given in 1000 hours, of a new class H insulation on trial in a motorette test at 260°C are taken from Nelson (1971):

260°C: 0.600, 0.744, 0.744, 0.744, 0.912, 1.128, 1.320, 1.464, 1.608, 1.896

In his analysis Wayne Nelson assumed the lifetimes to be independent and *lognormally* distributed. For the purpose of illustration, we will instead assume them to be i.i.d. $IG(\mu, \lambda)$, where μ and λ are both positive but otherwise unknown.

We obtain the following MLE of μ and λ:

$$\mu^* = 1.116$$

$$\frac{1}{\lambda^*} = \frac{1}{10} \sum_{j=1}^{10} \left(\frac{1}{t_j} - \frac{1}{\bar{t}} \right) = 0.13113$$

$$\lambda^* = 7.62602$$

Furthermore the MLE of var(T) is

$$\widehat{\text{var}(T)} = \frac{\mu^{*3}}{\lambda^*} = 0.18226$$

As a check, let us also estimate var(T) by the common estimator

$$S_{k-1}^2 = \frac{1}{k-1} \sum_{i=1}^{k} (T_i - \bar{T})^2$$

This leads to the estimate 0.19293 of var(T), which is in good correspondence with the MLE of var(T) above. Furthermore $\lim_{t\to\infty} z(t; 260°C)$ may be estimated by $\lambda^*/2\mu^{*2} = 3.06$. In Wayne Nelson's model, this limit is *zero*.

One may now ask the question: Which one of these two limiting values appears to be the most realistic one in this situation? The choice between the lognormal and the inverse Gaussian model should then be made accordingly.

□

Analysis of Binomial Data

A large variety of situations such as games, quality control, and sex distribution may be realistically modeled by the binomial model. In all such cases the situation may be characterized the following way:

1. n independent trials are carried out.
2. Each trial results in exactly one of the same two mutually exclusive outcomes, usually denoted as success (S) and failure (F).
3. The probability p of outcome S is constant from trial to trial.

If we define a random variable X to be the number of successes (S) in the n trials, X has a binomial distribution

$$P(X = x) = \binom{n}{x} p^x (1-p)^{n-x} \quad \text{for} \quad x = 1, 2, \ldots, n \qquad (9.45)$$

The mean and variance of X are

$$E(X) = np$$
$$\text{var}(X) = np(1-p)$$

A natural estimator of p is

$$\hat{p} = \frac{X}{n}$$

A $1 - \epsilon$ confidence interval for p is given by

$$\left(\frac{x}{x + (n-x+1)f_{1-\epsilon/2, 2(n-x+1), 2x}}, \frac{(x+1)f_{1-\epsilon/2, 2(x+1), 2(n-x)}}{n - x + (x+1)f_{1-\epsilon/2, 2(x+1), 2(n-x)}} \right) \qquad (9.46)$$

where $f_{\epsilon, \nu_1, \nu_2}$ denotes the $100\epsilon\%$ percentile of the F-distribution with ν_1 and ν_2 degrees of freedom (see Sverdrup 1967, p. 289; OREDA 1992).

When np and $n(1-p)$ are both large, then

$$\frac{X-np}{\sqrt{np(1-p)}} = \frac{n(\hat{p}-p)}{\sqrt{np(1-p)}} \approx \mathcal{N}(0,1)$$

This approximation is usually good when np and $n(1-p)$ are both greater than 5. It implies that

$$P\left(-u_{\epsilon/2} \leq \frac{n(\hat{p}-p)}{\sqrt{np(1-p)}} \leq u_{\epsilon/2}\right) \approx 1-\epsilon$$

which is equivalent to

$$n(\hat{p}-p)^2 \leq u_{\epsilon/2}^2 \cdot p(1-p)$$

and

$$(n+u_{\epsilon/2}^2)p^2 - (2n\hat{p}+u_{\epsilon/2}^2)p + n\hat{p}^2 \leq 0$$

Since $n+u_{\epsilon/2}^2 > 0$, the left-hand side of the inequality above is negative when $p_1 < p < p_2$, where p_1 and p_2 are the roots of

$$(n+u_{\epsilon/2}^2)p^2 - (2n\hat{p}+u_{\epsilon/2}^2)p + n\hat{p}^2 = 0$$

These roots are easily determined to be

$$\frac{1}{2a}(-b \pm \sqrt{b^2-4ac})$$

where $a = n+u_{\epsilon/2}^2$, $b = 2n\hat{p}+u_{\epsilon/2}^2$, and $c = n\hat{p}^2$.

Hence, if p_1 and p_2 are the smaller and the larger of these two roots, respectively, then

$$p_1 < p < p_2$$

constitutes an approximate $1-\epsilon$ confidence interval for p.

9.3 CENSORED DATA SETS

We will first define some of the most usual types of censoring. As an illustrative example we will see how the analysis is affected by censoring when we are dealing with exponentially distributed data.

Four Normally Occurring Forms of Censoring

Type I Censoring
Sometimes, for economical or other reasons, a life test has to be terminated at a specified time t_0. All units are activated at time $t = 0$ and followed until failure or until time t_0 when the experiment is terminated. This is often the case in medical research. After the experiment only the lifetimes of those units that have failed before t_0 will be known exactly.

This type of censoring is called *censoring of type I*, and the information in the data set obtained then consists of $s (\leq n)$ observed ordered lifetimes:

$$T_{(1)} \leq T_{(2)} \leq \cdots \leq T_{(s)}$$

In addition one knows that $n - s$ components have survived the time t_0, and this information should also be utilized.

Since the number (S) of components that fail before time t_0 obviously is stochastic, there is a chance that none or relatively few of the components will fail before t_0. This may be a weakness of the design.

Type II Censoring
If one wants to ensure that the resulting data set contains a fixed number r observed lifetimes and furthermore wants to terminate the test as fast as possible, the design must allow for the test to terminate at the rth failure, $0 < r < n$. As before, all the units are activated at $t = 0$. The information obtained through the experiment then consists of the data set

$$T_{(1)} \leq T_{(2)} \leq \cdots \leq T_{(r)}$$

in addition to the fact that $n - r$ components have survived the time $T_{(r)}$.

In this case the number (r) of recorded failures is nonstochastic. The price for obtaining this is that the time $T_{(r)}$ to complete the experiment is stochastic. A weakness of this design is therefore that one cannot know beforehand how long the experiment will last.

Type III Censoring
Type III censoring is a combination of the first two types. The test terminates at the time that occurs first, t_0 or the rth failure (t_0 and r must both be fixed beforehand).

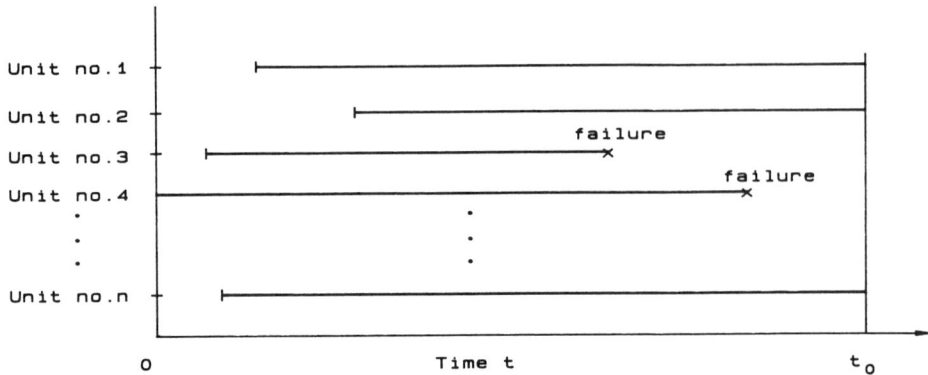

Figure 9.9 Censored data with staggered entry

Type IV Censoring

In this case n numbered identical units are activated at different given point(s) in time. If the time for censoring of unit i (S_i, $i = 1, 2, \ldots, n$) is stochastic, the censoring is said to be of type IV. This model can also be used in the following situation:

Example 9.6

Let us assume that there are compelling reasons why a test must be terminated by time t_0. Further, the activating times for the individual units are stochastic, for example, as they are in a medical experiment where patients may enter the study more or less randomly; see Figure 9.9. If we in this situation set the activating time points for the individual units to $t = 0$, then the censoring time points may be regarded as stochastic; see Figure 9.10. □

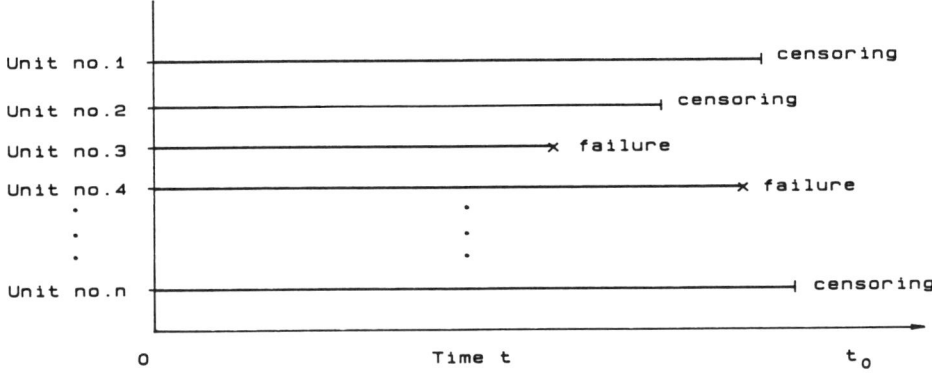

Figure 9.10 Censored data with staggered entry shifted toward $t = 0$

Example 9.7

Suppose that we are studying a series system composed of two different components A and B. If we are primarily concerned with studying to what degree failure in component A leads to system failure, we can interpret system failure caused by component B, as censoring. □

ooo OOO ooo

We will return to type IV censoring in connection with the Kaplan-Meier estimator (see p. 394).

Analysis of Type-II Censored Exponentially Distributed Data

As mentioned earlier in this section, the information in the data set consists of the recorded lifetimes

$$T_{(1)} \leq T_{(2)} \leq \cdots \leq T_{(r)}$$

in addition to the fact that $n - r$ units survived $T_{(r)}$.

According to Theorem B.3 (see Appendix B) the joint probability density for $T_{(1)} \leq T_{(2)} \leq \cdots \leq T_{(r)}$ is given by

$$f_{T_{(1)},\ldots,T_{(r)}}(t_1,\ldots,t_r) = \frac{n!}{(n-r)!} \lambda^r e^{-\lambda \sum_{j=1}^{r} t_j} \cdot e^{-\lambda t_r(n-r)}$$
$$\text{for } t_1 < t_2 < \cdots < t_r \qquad (9.47)$$

The corresponding likelihood function is therefore

$$\ell(\lambda; t_1,\ldots,t_r) = \frac{n!}{(n-r)!} \lambda^r e^{-\lambda[\sum_{j=1}^{r} t_j + (n-r)t_r]}$$
$$= \frac{n!}{(n-r)!} \lambda^r e^{-\lambda \mathcal{T}(t_r)} \qquad \text{for } 0 < t_1 < \cdots < t_r$$

and we find the maximum likelihood estimator λ_{II}^* of λ in the usual way:

$$\lambda_{II}^* = \frac{r}{\mathcal{T}(T_{(r)})} \qquad (9.48)$$

Which properties does this estimator have? Let us first see what we can say about the probability distribution of λ_{II}^*.

If D_j denotes the time interval from the $(j-1)$th to the jth failure, then

$$T_1 = D_1$$
$$T_2 = D_1 + D_2$$
$$\vdots \quad \vdots \quad \vdots$$
$$T_r = D_1 + D_2 + \cdots + D_r$$

and

$$\sum_{j=1}^{n} T_{(j)} = rD_1 + (r-1)D_2 + \cdots + D_r$$

Furthermore

$$(n-r)T_{(r)} = (n-r)(D_1 + D_2 + \cdots + D_r)$$

Therefore the total time on test at time $T_{(r)}$ is

$$\mathcal{T}(T_{(r)}) = nD_1 + (n-1)D_2 + \cdots + (n-(r-1))D_r$$
$$= \sum_{j=1}^{r}(n-(j-1))D_j$$

Introducing

$$D_j^* = (n-(j-1))D_j \quad \text{for} \quad j = 1, 2, \ldots, r$$

we know from Theorem B.4 in Appendix B that $2\lambda D_1^*, 2\lambda D_2^*, \ldots, 2\lambda D_r^*$ are independent and χ^2 distributed, each with two degrees of freedom. Hence

$$2\lambda \mathcal{T}(T_{(r)}) \quad \text{is } \chi^2 \text{distributed with } 2r \text{ degrees of freedom} \tag{9.49}$$

We can utilize this to find $E(\lambda_{II}^*)$:

$$E(\lambda_{II}^*) = E\left(\frac{r}{\mathcal{T}(T_{(r)})}\right) = 2\lambda r \cdot E\left(\frac{1}{2\lambda \mathcal{T}(T_{(r)})}\right) = 2\lambda r \cdot E\left(\frac{1}{Z}\right)$$

where Z is χ^2 distributed with $2r$ degrees of freedom. This implies that (see p. 375)

$$E\left(\frac{1}{Z}\right) = \frac{1}{2(r-1)}$$

Hence

$$E(\lambda_{II}^*) = 2\lambda r \cdot \frac{1}{2r-1} = \lambda \cdot \frac{r}{(r-1)}$$

λ_{II}^* is accordingly not unbiased. However,

$$\hat{\lambda}_{II} = \frac{(r-1)}{\mathcal{T}(T_{(r)})} \qquad (9.50)$$

is easily seen to be unbiased. As on page 376, we find that

$$\text{var}(\hat{\lambda}_{II}) = \frac{\lambda^2}{(r-2)}$$

Confidence intervals, as well as tests for standard hypotheses about λ, may now be derived from the fact that $2\lambda \cdot \mathcal{T}(T_{(r)})$ is χ^2-distributed with $2r$ degrees of freedom. The procedure is the same as the one used on page 377.

Analysis of Type-I Censored Exponentially Distributed Data

The fact that the number S of components failing before time t_0 is stochastic makes this situation more difficult to deal with from a probabilistic point of view. We will therefore confine ourselves to suggesting an intuitive estimator of λ.

First notice that the estimators for λ, derived in the case of complete data sets and of type-II censored data, both could be written as a fraction with numerator equal to "number of recorded failures -1" and denominator equal to "total time on test at the termination of the test." It seems intuitively reasonable to use the same fraction when we have type-I censoring.

In this case the number of failures is S, while the total time on test is

$$\mathcal{T}(t_0) = \sum_{j=1}^{S} T_{(j)} + (n-S)t_0$$

Hence

$$\hat{\lambda}_I = \frac{S-1}{\mathcal{T}(t_0)}$$

seems to be a reasonable estimator of λ. It can be shown that this estimator is biased for small samples. However, asymptotically it will have the same properties as $\hat{\lambda}_{II}$ has (see Mann, Schafer, and Singpurwalla 1974, p. 173).

Barlow-Proschan's Test

The TTT plot presented in Section 9.2 provides a visual method for checking whether the data indicate a constant, an increasing, a decreasing, or, for example, a bathtub-shaped failure rate. In some situations it may also be of interest to establish a formal statistical test to decide whether or not the data have a constant failure rate. Barlow and Proschan (1969) proposed a test based on the test statistic W. W is so designed that it has a tendency to become large (small) when the underlying distribution has an increasing (decreasing) failure rate. For a *complete* data set $(T_1, T_2, \ldots, T_n)$ the Barlow-Proschan statistic W simply is the sum of the scaled total time on test values at the failure times:

$$W = U_1 + U_2 + \cdots + U_n \tag{9.51}$$

where

$$U_i = \frac{\mathcal{T}(T_{(i)})}{\mathcal{T}(T_{(n)})} \quad \text{for} \quad i = 1, 2, \ldots, n \tag{9.52}$$

denote the scaled total time on test at the ith failure.

For a *censored* data set (of type IV) the test statistic W is modified as follows: Let $T_{(1)}, T_{(2)}, \ldots, T_{(k)}$ denote the ordered *failure* times in the sample of n functioning times. Note that "withdrawals" may occur between $T_{(i)}$ and $T_{(i+1)}$ and that k in general is a random variable. Let further S_i for $i = 1, 2, \ldots, k$ be the total time on test between the $(i-1)$th and the ith failure, that is, between time $X_{(i-1)}$ and time $X_{(i)}$. S_i is thus a sum with one term for each unit that was functioning in the relevant time interval, the contribution from each unit being its functioning time in that interval.

Barlow-Proschan's test statistic W for censored data is defined as

$$W = \frac{\sum_{i=1}^{k-1}(k-i)S_i}{\sum_{i=1}^{k}S_i} \tag{9.53}$$

It is straightforward to verify that (9.53) coincides with (9.51) when the data set is complete (in which case $k = n$). When the failure rate is constant, it is shown (Barlow and Proschan 1969) that W may be written

$$W = U_1 + U_2 + \cdots + U_{k-1}$$

where U_i ($i = 1, 2, \ldots, k - 1$) are independent uniform random variables on $[0, 1]$. Hence, when the failure rate is constant (λ_0), then W and $(k - 1 - W)$ have the same distribution, and

$$E_{\lambda_0}(W) = \frac{k - 1}{2} \quad \text{and} \quad \text{var}_{\lambda_0}(W) = \frac{k - 1}{12}$$

When the failure rate is constant (λ_0), and k is large, we may use the normal approximation

$$W \approx \mathcal{N}\left(\frac{k - 1}{2}, \frac{k - 1}{12}\right)$$

that is,

$$\frac{W - (k - 1)/2}{\sqrt{(k - 1)/12}} \approx \mathcal{N}(0, 1)$$

Example 9.8
Let us illustrate the use of Barlow–Proschan's test statistic W for the data set (measured in 10^4 hours):

$$\begin{array}{cccccc} 0.35 & 0.50* & 0.75* & 1.00 & 1.30 & 1.80 \\ 3.00* & 3.15* & 4.85* & 5.50 & 5.50* & 6.25* \end{array}$$

Censored times are starred (*). From these data we compute

$$\begin{aligned} S_1 &= 12 \cdot 0.35 = 4.20 \\ S_2 &= (0.50 - 0.35) + (0.75 - 0.35) + 9(1.00 - 0.35) = 6.40 \\ S_3 &= 8(1.30 - 1.00) = 2.40 \\ S_4 &= 7(1.80 - 1.30) = 3.50 \\ S_5 &= (3.00 - 1.80) + (3.15 - 1.80) + (4.85 - 1.80) \\ &\quad + 3(5.50 - 1.80) = 16.70 \end{aligned}$$

Thus

$$W = \frac{4S_1 + 3S_2 + 2S_3 + S_4}{S_1 + S_2 + S_3 + S_4 + S_5} = \frac{44.3}{33.2} = 1.33 \qquad \square$$

As an illustration, let us use Barlow–Proschan's test to test the null hypothesis (H_0): "The failure rate is constant" against the alternative hypothesis (H_1): "The failure rate is increasing on the average (IFRA)."

Barlow-Proschan's test criterion is

Reject H_0 when $W \geq w_\alpha$, where the critical value w_α is determined so as to get significance level α.

The critical values w_α are given in Table 9.2 for selected values of α and number of failures k. The table is adapted from Barlow and Proschan (1969). When $k \geq 13$, normal approximation may be used. Then

$$w_\alpha = u_\alpha \sqrt{\frac{(k-1)}{12}} + \frac{(k-1)}{2}$$

where u_α denotes the upper $100\alpha\%$ percentile of the standard normal distribution (e.g., $u_{0.05} = 1.645$).

Table 9.2 Critical Values for Barlow and Proschan's Test Statistic W

			α		
$k - 1$	0.100	0.050	0.025	0.010	0.005
2	1.553	1.684	1.776	1.859	1.900
3	2.157	2.331	2.469	2.609	2.689
4	2.753	2.953	3.120	3.300	3.411
5	3.339	3.565	3.754	3.963	4.097
6	3.917	4.166	4.376	4.610	4.762
7	4.489	4.759	4.988	5.244	5.413
8	5.056	5.346	5.592	5.869	6.053
9	5.619	5.927	6.189	6.487	6.683
10	6.178	6.504	6.781	7.097	7.307
11	6.735	7.077	7.369	7.702	7.924
12	7.289	7.647	7.953	8.302	8.535

Source: Adapted from Barlow and Proschan (1969)[1]
[1]Reproduced with the permission of the Institute of Mathematical Statistics.

Example 9.8 (Cont.)
Let us return to Example 9.8 and use Barlow and Proschan's test to check whether the failure rate is constant or increasing, with a significance level of $\alpha = 0.10$. In this case $(k - 1) = 4$, and the critical value is from Table 9.2 equal to $w_{0.10} = 2.753$. The test statistic W was computed to be $W = 1.33$, which is less than $w_{0.10} = 2.753$. There is thus *no reason to reject the null hypothesis* of constant failure rate and accept the alternative hypothesis of increasing failure rate average (IFRA). □

We may also use the test statistic W to test the null hypothesis (H_0): "The failure rate is constant" against the alternative hypothesis (H_1): "The failure rate is on the average decreasing (DFRA)." The test criterion of Barlow-Proschan's test now becomes

Reject H_0 when $(k - 1 - W) > w_\alpha$, where the critical value w_α is determined so as to give significance level α.

For the data in Example 9.8, we find that $k - 1 - W = 5 - 1 - 1.33 = 2.67$ and $w_{0.10} = 3.339$. Hence these data give *no reason to reject the null hypothesis* of constant failure rate and accept the alternative hypothesis of decreasing failure rate average (DFRA).

Barlow-Proschan's test is part of the SAREPTA program. A test of H_0: "The failure rate is constant" against Weibull alternatives is discussed by Cox and Oakes (1984, p. 43).

Nonexponentially Distributed Lifetimes

Parameter estimation in connection with nonexponential distributions is not covered in this book, except for the inverse Gaussian distribution for complete data sets. Valuable references include Mann, Schafer and Singpurwalla (1974), Lawless (1982), Cox and Oakes (1984), and Crowder et al. (1991).

Several personal computer programs for estimation of parameters in nonexponential distributions are available. Many of these programs are general purpose statistical programs that are also able to handle Kaplan-Meier and Hazard plotting, discussed later in this chapter. A survey of statistical programs for PC (DOS, Windows, and OS/2), Macintosh, and UNIX platforms is published regularly by the journal IEEE Spectrum. Advertisements and reviews of new or upgraded programs may be found, for example, in the journal *AMSTAT News* (published by the American Statistical Association). PC programs that can estimate parameters of nonexponential distributions include the following:

BMDP   A comprehensive statistical software package, including ANOVA, regression, multivariate statistics, time series, survival analysis, and more. BMDP runs on PCs, VAX, and a wide range of

	UNIX platforms. For further information contact: BMDP Statistical Software, Inc. 1440 Sepulveda Boulevard, Los Angeles, CA 90025.
SAS	A comprehensive software package with a lot of different modules. SAS runs on PCs under Windows and OS/2, VAX/VMS, and a wide range of UNIX platforms. For further information contact: SAS Institute Inc. SAS Campus Drive, Cary, NC 27513.
S-PLUS	An interactive computing environment with graphical data analysis system and object-oriented language used for exploratory data analysis, graphics, statistics, and mathematical computing. S-PLUS runs on PCs (Windows) and a wide range of UNIX platforms. For further information contact: Statistical Sciences (StatSci), 1700 Westlake Avenue, N. Suite 500, Seattle, WA 98109.
SPSS	A widely used program, especially within the social sciences. SPSS runs on PCs under Windows and on several other platforms. SPSS has a wide range of add-on modules including the QI Analyst for quality control. For further information contact: SPSS Inc. 444 N. Michigan Avenue, Chicago, IL 60611.
STATA	A versatile software program suitable for a wide range of applications. STATA also includes a module for analysis of survival data. STATA runs on PCs under DOS and OS/2, Macintosh, and a wide range of UNIX platforms. For further information contact: Computing Resource Center, 1640 Fifth Street, Santa Monica, CA 90401.
SYSTAT	A comprehensive statistical software package, ranging from basic to complex analyses. SYSTAT has a complete set of graph types, and a wide range of add-on packages for special applications, including a module for analysis of survival data called "Survival." SYSTAT runs on PCs under DOS and Windows, and Macintosh. For further information contact: SYSTAT, Inc. 1800 Sherman Avenue, Evanston, IL 60201.

Kaplan-Meier Estimator of the Survivor Function

Let $F(t)$ denote the life distribution for a certain type of units. We know the distribution to be continuous, but make no further assumptions about $F(t)$ (i.e., a nonparametric model). We want to estimate the survivor function

$$P(T > t) = R(t) = 1 - F(t) \qquad (9.54)$$

on the basis of n observed lifetimes of such units.

Let t_j denote the observed lifetime of unit $j, j = 1, 2, \ldots, n$. Then the *empirical distribution function* is defined as

CENSORED DATA SETS

$$F_n(t) = \frac{\text{(Number of lifetimes } \leq t)}{n} \tag{9.55}$$

and the corresponding empirical survivor function is

$$\hat{R}(t) = 1 - F_n(t) = \frac{\text{(Number of lifetimes } > t)}{n} \tag{9.56}$$

$\hat{R}(t)$ may obviously be used as an estimator of the survivor function $R(t)$. If all observations are distinct, $\hat{R}(t)$ is a step function that decrease by $1/n$ just before each observed failure time. A simple adjustment accommodates any ties present in the data. Note that $\hat{R}(t)$, like $F_n(t)$, is continuous from the right.

Example 9.9
Suppose that a test has been carried out as described above, $n = 16$, and the observed lifetimes are (given in months):

31.7	39.2	57.5	65.0	65.8	70.0	75.0	75.2
87.7	88.3	94.2	101.7	105.8	109.2	110.0	130.0

(The data are adapted from an example in Nelson 1972.) The empirical distribution function $F_n(t)$ is illustrated in Figure 9.11. The empirical survivor function $\hat{R}(t) = 1 - F_n(t)$ is illustrated in Figure 9.12. □

Next let us show how to estimate $R(t)$ from an incomplete data set with censoring of type IV (see p. 386). n numbered units are activated at time $t = 0$, and the censoring time for unit i, S_i is stochastic. Associated with unit

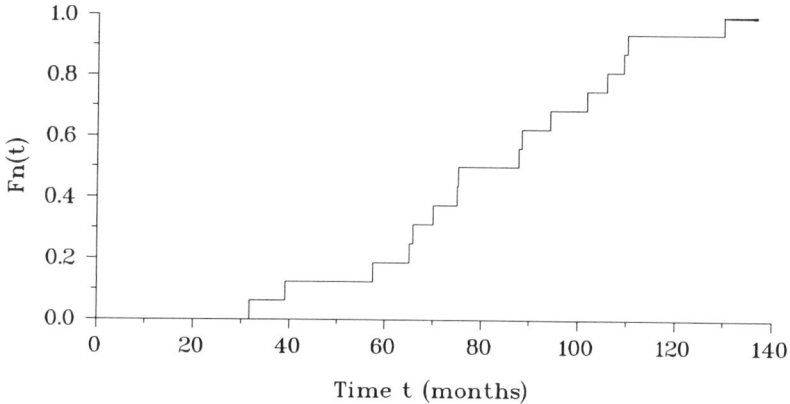

Figure 9.11 Empirical distribution function $F_n(t)$

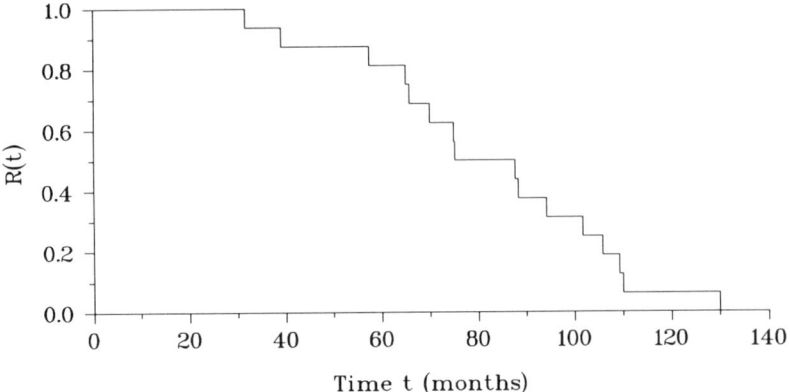

Figure 9.12 Empirical survivor function $\hat{R}(t) = 1 - F_n(t)$

i for $i = 1, 2, \ldots, n$ are two nonnegative random variables, namely the lifetime T_i, which would be observed if unit i were not exposed to censoring, called the *potential lifetime*, and the time S_i, when the unit is possibly censored. We will assume that the vectors (T_i, S_i) for $i = 1, 2, \ldots, n$, are i.i.d. with a continuous distribution. Further we assume that T_i and S_i for $i = 1, 2, \ldots, n$ are independent with continuous marginal distributions.

In this situation it is only possible to record the smaller of T_i and S_i for component i for $i = 1, \ldots, n$, though at the same time we know whether we are observing a failure or a censoring. Let us introduce

$$Y_i = \min(T_i, S_i)$$

and the indicators

$$\delta_i = \begin{cases} 1 & \text{if } T_i \leq S_i \\ 0 & \text{if } T_i > S_i \end{cases} \quad i = 1, \ldots, n$$

After the life test is terminated, we are left with the data set

$$(Y_1, \delta_1), (Y_2, \delta_2), \ldots, (Y_n, \delta_n)$$

Kaplan and Meier (1958) suggested the following estimation procedure: Fix $t > 0$. Let $t_{(1)} < t_{(2)} < \cdots < t_{(n)}$ denote the recorded functioning times, either until failure or to censoring, ordered according to size. Let J_t denote the set of all indices j where $t_{(j)} \leq t$ and $t_{(j)}$ represents a failure time. Let n_j denote the number of units, functioning and in observation immediately before time $t_{(j)}$,

CENSORED DATA SETS 397

$j = 1, 2, \ldots, n$. Then the Kaplan–Meier estimator of $R(t)$ is defined as

$$\hat{R}(t) = \prod_{j \in J_t} \frac{n_j - 1}{n_j} \qquad (9.57)$$

Before we give the motivation for this estimator, let us illustrate its use by an example.

Example 9.10
We change the situation given in Example 9.9 so that only the recorded lifetimes, which are not starred (*), in Table 9.3 represent the times to failure. In Table 9.3 we have calculated $\hat{R}(t)$ from (9.57). In Table 9.4 the Kaplan-Meier estimate is determined as a function of time. In the time interval (0, 31.7) until the first failure it is reasonable to set $\hat{R}(t) = 1$. The estimate is displayed graphically by a Kaplan-Meier plot in Figure 9.13. □

Table 9.3 Computation of the Kaplan-Meier Estimate

Rank j	Inverse Rank $n - j + 1$	Ordered Failure and Censoring Times $t_{(j)}$	$\hat{p}_j$	$\hat{R}(t_{(j)})$
0	—	—	1	1.000
1	16	31.7	$\frac{15}{16}$	0.938
2	15	39.2	$\frac{14}{15}$	0.875
3	14	57.2	$\frac{13}{14}$	0.813
4	13	65.0*	1	0.813
5	12	65.8	$\frac{11}{12}$	0.745
6	11	70.0	$\frac{10}{11}$	0.677
7	10	75.0*	1	0.677
8	9	75.2*	1	0.677
9	8	87.5*	1	0.677
10	7	88.3*	1	0.677
11	6	94.2*	1	0.677
12	5	101.7*	1	0.677
13	4	105.8	$\frac{3}{4}$	0.508
14	3	109.2*	1	0.508
15	2	110.0	$\frac{1}{2}$	0.254
16	1	130.0*	1	0.254

Note: Censoring times are starred(*).

LIFE DATA ANALYSIS

Table 9.4 Kaplan-Meier Estimate as a Function of Time

t	$\hat{R}(t)$
$0 \leq t < 31.7$	$= 1.000$
$31.7 \leq t < 39.2$	$\frac{15}{16} = 0.938$
$39.2 \leq t < 57.5$	$\frac{15}{16} \cdot \frac{14}{15} = 0.875$
$57.5 \leq t < 65.8$	$\frac{15}{16} \cdot \frac{14}{15} \cdot \frac{13}{14} = 0.813$
$65.8 \leq t < 70.0$	$\frac{15}{16} \cdot \frac{14}{15} \cdot \frac{13}{14} \cdot \frac{11}{12} = 0.745$
$70.0 \leq t < 105.8$	$\frac{15}{16} \cdot \frac{14}{15} \cdot \frac{13}{14} \cdot \frac{11}{12} \cdot \frac{10}{11} = 0.677$
$105.8 \leq t < 110.0$	$\frac{15}{16} \cdot \frac{14}{15} \cdot \frac{13}{14} \cdot \frac{11}{12} \cdot \frac{10}{11} \cdot \frac{3}{4} = 0.508$
$110.0 \leq t$	$\frac{15}{16} \cdot \frac{14}{15} \cdot \frac{13}{14} \cdot \frac{11}{12} \cdot \frac{10}{11} \cdot \frac{3}{4} \cdot \frac{1}{2} = 0.254$

We see from (9.57) and also from Figure 9.13 that $\hat{R}(t)$ is a step function, continuous from the right, that equals 1 at t = 0. $\hat{R}(t)$ drops by a factor of $(n_j - 1)/n_j$ at each failure time $t_{(j)}$. The estimator $\hat{R}(t)$ does not change at the censoring times. The effect of the censoring is, however, influencing the values of n_j and hence the size of the steps in $\hat{R}(t)$.

A slightly problematic point is that $\hat{R}(t)$ never reduces to zero when the last time $t_{(n)}$ recorded is a censored time. For this reason $\hat{R}(t)$ is usually taken to be undefined for $t > t_{(n)}$. This problem is further discussed by Kalbfleisch and Prentice (1980, p. 12).

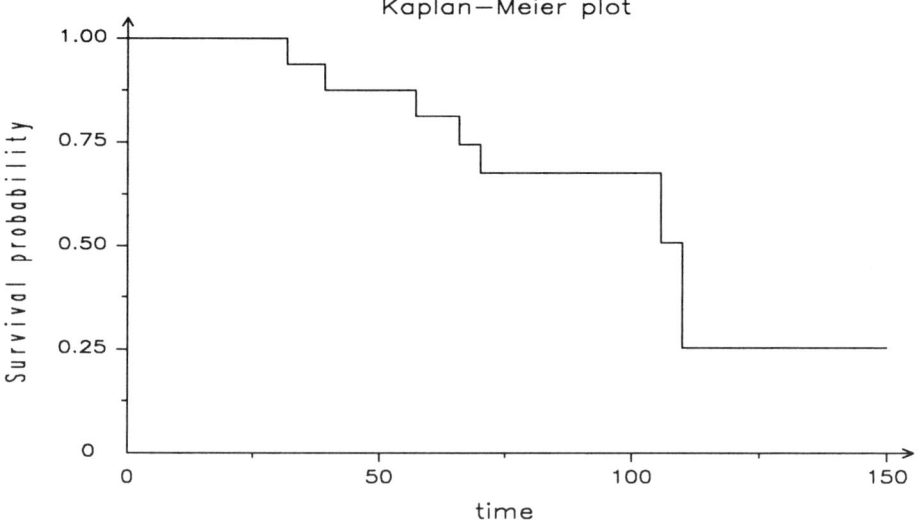

Figure 9.13 Kaplan-Meier plot for the data in Example 9.10

Justification for the Kaplan-Meier Estimator

The motivation for the Kaplan-Meier estimator is as follows[2]: Let the time period $[0, \infty)$ be divided into small time intervals $(u_j, u_{j+1}]$ for $j = 0, 1, \ldots$, where $u_0 = 0$ and the intervals are so short that, based on the continuity assumptions about T and S, we can disregard the following possibilities:

1. Two or more units fail in the *same* interval
2. One unit fails and another is censored in the *same* interval

Now let $t \in (u_m, u_{m+1}]$. Then obviously

$$R(t) = P(T > t)$$
$$= P(T > u_0) \cdot P(T > u_1 | T > u_0) \cdots P(T > t | T > u_m) \quad (9.58)$$

Since $F(t)$ is assumed to represent a continuous life distribution for all $t \geq 0$, then $P(T > u_0) = P(T > 0) = 1$.

Hence

$$R(t) = P(T > u_1 | T > u_0) \cdot P(T > u_2 | T > u_1) \cdots P(T > t | T > u_m)$$
$$= \prod_{j=0}^{m} p_j \quad (9.59)$$

where

$$p_j = P(T > u_{j+1} | T > u_j) \quad \text{for} \quad j = 0, 1, 2, \ldots, (m-1)$$
$$p_m = P(T > t | T > u_m)$$

Kaplan-Meier's idea is to estimate each single factor on the right-hand side of (9.59) and thereafter use the product of these estimators as an estimator of $R(t)$. What will now be a reasonable estimator of $p_j = P(T > u_{j+1} | T > u_j)$?

1. If neither failure nor censoring occurs in $(u_j, u_{j+1}]$, then the same number of units will be active at the start and at the end of this interval. In this case it seems reasonable to use the estimator

$$\hat{p}_j = P(T > u_{j+1} | T > u_j) = 1$$

2. Next suppose that censoring occurs in $(u_j, u_{j+1}]$. Then due to the assump-

---

[2] The following derivation is slightly different from the derivation presented by Kaplan and Meier (1958). See also Kalbfleish and Prentice (1980), Lawless (1982), and Cox and Oakes (1984) for further alternatives.

tion about short intervals, we may ignore the possibility that failure occurs in the same interval. Accordingly in this case we have recorded no failures in the interval, and it seems reasonable to use the same estimator as in condition 1.

3. But suppose that failures occur in $(u_j, u_{j+1}]$. Due to the assumption about short intervals, we may ignore the possibility of more than one failure occurring in this interval. Let n_j denote the number of units at risk (i.e., which are functioning and in observation) at the beginning of the interval $(u_j, u_{j+1}]$. The number of units at risk at the end of the same interval is then $n_j - 1$. Since $n_j - 1$ units out of n_j survive the interval $(u_j, u_{j+1}]$, a natural estimator of $p_j = P(T > u_{j+1} | T > u_j)$ is

$$\hat{p}_j = \frac{n_j - 1}{n_j} \tag{9.60}$$

A reasonable estimator of $P(T > t | T > u_m)$ is found in the same way. Thus the only intervals where the estimator $\hat{p}_j$ is different from 1 are the intervals where a failure occurs. By increasing the number of intervals such that the length of each interval, except the last, approaches zero, we see that the estimator $\hat{p}_j$ is different from 1 only "at" the failure times.

As stated above, we may partly disregard the intervals where no failures occur. To simplify the notation, we therefore redefine the n_j's:

n_j = number of units at risk (functioning and in observation) immediately before time $t_{(j)}$ for $j = 1, 2, \ldots, n$

The probabilities p_j may now be estimated for infinitesimal intervals around the $t_{(j)}$'s by

$$\hat{p}_j = \begin{cases} 1 & \text{if a censoring occurred at } t_{(j)} \\ \frac{n_j - 1}{n_j} & \text{if a failure occurred at } t_{(j)} \quad j = 1, 2, \ldots, n \end{cases} \tag{9.61}$$

and $\hat{p}_0 = 1$.

The Kaplan-Meier estimator of the survivor function $R(t)$ is then given by [see (9.59)]

$$\hat{R}(t) = \prod_{j=0}^{n} \hat{p}_j \tag{9.62}$$

Let J_t denote the set of all integers j such that $t_{(j)}$ is a failure time and $t_{(j)} \leq t$.

The Kaplan-Meier estimator[3] in (9.62) may then be written

$$\hat{R}(t) = \prod_{j \in J_t} \frac{n_j - 1}{n_j} \qquad (9.63)$$

Let $t_{(1)} \leq t_{(2)} \leq \cdots \leq t_{(n)}$ denote the recorded functioning times, until either failure or censoring, arranged by size. If two or more of these coincide, they are arranged in random order. If a failure time and a censoring time are recorded as equal, the convention is often adopted (see Cox and Oakes 1984, p. 49) that censoring times are considered to be infinitesimally larger than the failure times. This makes sense since a unit that is censored at time t almost certainly survives past t.

Some Properties of the Kaplan-Meier Estimator

A thorough discussion of the properties of the Kaplan-Meier estimator $\hat{R}(t)$ may be found in Kalbfleisch and Prentice (1980), Lawless (1982), and Cox and Oakes (1984). We will here only summarize a few properties without proofs:

1. The Kaplan-Meier estimator $\hat{R}(t)$ can be derived as a nonparametric maximum likelihood estimator (MLE). This derivation was originally given by Kaplan and Meier (1958).
2. The Kaplan-Meier estimator may be slightly modified to data sets with *ties*. If we assume that d_i units fail at time t_i for $i = 1, 2, \ldots, n$, the Kaplan-Meier estimator becomes (e.g., see Kalbfleish and Prentice 1980, p. 12):

$$\hat{R}(t) = \prod_{j \in J_t} \frac{n_j - d_j}{n_j} \qquad (9.64)$$

3. $\hat{R}(t)$ is a consistent estimator of R(t) under quite general conditions with estimated asymptotic variance (Kalbfleisch and Prentice 1980, p. 14):

$$\widehat{\text{var}}(\hat{R}(t)) = (\hat{R}(t))^2 \sum_{j \in J_t} \frac{d_j}{n_j(n_j - d_j)} \qquad (9.65)$$

Expression (9.65) is known as *Greenwood's formula*.

4. The Kaplan-Meier estimator has an asymptotic normal distribution, since it is a maximum likelihood estimator. Hence confidence limits for $R(t)$ can be determined using normal approximation. For details see Cox and Oakes (1984, pp. 51–52).

---

[3] The estimator is also called the *product limit* (PL) estimator.

Nelson's Estimator of the Cumulative Failure Rate

Let $F(t)$ denote the life distribution for a certain type of unit, and assume that the distribution is continuous with probability density $f(t)$, where $f(t) > 0$ for $t > 0$. No further assumptions are made about the distribution $F(t)$ (nonparametric model).

The failure rate is defined as

$$z(t) = \frac{f(t)}{1 - F(t)} = -\frac{d}{dt} \ln R(t) \qquad (9.66)$$

and the cumulative failure rate as

$$Z(t) = \int_0^t z(u)\, du = -\ln R(t) = -\ln[1 - F(t)] \qquad (9.67)$$

Hence

$$R(t) = e^{-Z(t)} \qquad (9.68)$$

Let us first discuss the situation where we have an incomplete data set with stochastic censoring (type IV).

A natural estimator of the cumulative failure rate $Z(t)$ is then deducted from the Kaplan-Meier estimator $\hat{R}(t)$ as

$$\hat{Z}(t) = -\ln \hat{R}(t) = -\ln \prod_{j=0}^{n} \hat{p}_j \qquad (9.69)$$

where $\hat{p}_j$ for $j = 1, 2, \ldots, n$ is defined in (9.61). If we, as before, let J_t be the set of integers such that $t_{(j)}$ is a failure time and $t_{(j)} \le t$, then (9.69) may be written

$$\hat{Z}(t) = -\ln \prod_{j=0}^{n} \hat{p}_j = -\ln \prod_{j \in J_t} \left(\frac{n_j - 1}{n_j}\right) = -\sum_{j \in J_t} \ln\left(1 - \frac{1}{n_j}\right)$$

$$= \sum_{j \in J_t} \left(\frac{1}{n_j} + \frac{1}{2n_j^2} + \cdots\right) \qquad (9.70)$$

ooo OOO ooo

An alternative estimator of $Z(t)$ is the so-called Nelson estimator, which we can

CENSORED DATA SETS

derive as follows: As before, we activate n components in order to record the corresponding lifetimes $T_1, \ldots, T_n$ and thereafter to study the characteristics of the life distribution. For the time being, let us disregard the possibility of censoring.

It turns out to be difficult to estimate the failure rate $z(t)$ without first specifying the distribution more closely. However, it is considerably easier to find a suitable estimator of the cumulative failure rate $Z(t)$. For this we need some lemmas.

Lemma 9.1. Let T be continuously distributed with strictly increasing distribution function $F(t)$. Then

1. $U = F(T)$ has a uniform distribution over the interval $(0, 1)$
2. $Z(T) = -\ln(1 - F(T))$ is exponentially distributed with parameter 1

Proof

1. Let $u \in (0, 1]$. Then

$$P(U \le u) = P(F(T) \le u) = P(T \le F^{-1}(u)) = F(F^{-1}(u)) = u$$

2. Let $z > 0$. Then

$$\begin{aligned} P(Z \le z) &= P(-\ln(1 - F(T)) \le z) \\ &= P(F(T) \le 1 - e^{-z}) = P(U \le 1 - e^{-z}) = 1 - e^{-z} \end{aligned} \quad \square$$

By using Lemma 9.1, we easily obtain

Corollary 9.3. Let $T_1, \ldots, T_n$ be independent and identically continuously distributed with strictly increasing distribution function $F(t)$. Denote the corresponding order statistic

$$T_{(1)} < T_{(2)} < \cdots < T_{(n)}$$

Let

$$Z(T_{(j)}) = -\ln[1 - F(T_{(j)})] \quad \text{for } j = 1, 2, \ldots, n \quad (9.71)$$

and replace $Z(T_{(j)})$ by $Z_{(j)}$. Then

$$Z_{(1)} < Z_{(2)} < \cdots < Z_{(n)}$$

can be interpreted as the order statistic of n independent, identical, exponentially distributed variables with parameter 1.

Lemma 9.2. Let the assumptions be the same as in Corollary 9.3, and let $Z_{(j)}$ be defined as in (9.71). Then

$$E(Z_{(j)}) = E(Z(T_{(j)})) = \frac{1}{n} + \frac{1}{n+1} + \cdots + \frac{1}{n-j+1} \qquad (9.72)$$

The proof of Lemma 9.2 may be found for example in Barlow and Proschan (1975, p. 60). From (9.72) Nelson (1969) proposed to estimate the cumulative failure rate up to time t by

$$\hat{Z}(t) = \begin{cases} 0 & \text{for } t < T_{(1)} \\ \sum_{j=1}^{r} \frac{1}{n-j+1} & \text{for } T_r \leq t < T_{(r+1)} \end{cases} \qquad (9.73)$$

where $r = 1, 2, \ldots, n-1$.

Having estimated the cumulative failure rate by $\hat{Z}(t)$, it seems natural to estimate the survivor function by

$$R^*(t) = e^{-\hat{Z}(t)} \qquad (9.74)$$

Before giving a justification for the estimator $\hat{Z}(t)$, we will illustrate its use by an example.

Example 9.11: Complete Data Set

Reconsider the situation in Example 9.9, calculate $\hat{Z}(t)$ by (9.73) for $t = t_{(j)}$ for $j = 1, \ldots, n$, and afterward the Nelson estimate $R^*(t)$ for the survivor function based on the same t values. The result is shown in Table 9.5. On the right-hand side of the table the corresponding Kaplan-Meier estimate is shown. □

ooo OOO ooo

Censored Data Sets

Assume now that we have a data set which is subject to censoring of type IV. In this situation Nelson proposed the following procedure for estimating the cumulative failure rate and the survivor function: As before, let

$$t_{(1)} \leq t_{(2)} \leq \cdots \leq t_{(n)}$$

CENSORED DATA SETS

Table 9.5 Nelson Estimator for a Complete Data Set Compared with the Kaplan-Meier Estimator

Rank j	Lifetime $t_{(j)}$	Inverse of Number at Risk $(n - j + 1)^{-1}$	Nelson Estimate $\hat{Z}(t_{(j)})$	Survivor Function Estimate	
				Nelson $R^*(t_{(j)})$	Kaplan-Meier $\hat{R}(t_{(j)})$
1	31.7	$\frac{1}{16}$	0.0625	0.939	0.938
2	39.2	$\frac{1}{15}$	0.1292	0.879	0.875
3	57.5	$\frac{1}{14}$	0.2006	0.818	0.813
4	65.0	$\frac{1}{13}$	0.2775	0.758	0.750
5	65.8	$\frac{1}{12}$	0.3608	0.697	0.688
6	70.0	$\frac{1}{11}$	0.4517	0.637	0.625
7	75.0	$\frac{1}{10}$	0.5517	0.576	0.563
8	75.2	$\frac{1}{9}$	0.6628	0.515	0.500
9	87.5	$\frac{1}{8}$	0.7878	0.455	0.448
10	88.3	$\frac{1}{7}$	0.9307	0.394	0.375
11	94.2	$\frac{1}{6}$	1.0974	0.334	0.313
12	101.7	$\frac{1}{5}$	1.2974	0.273	0.250
13	105.8	$\frac{1}{4}$	1.5474	0.213	0.188
14	109.2	$\frac{1}{3}$	1.9807	0.138	0.125
15	110.0	$\frac{1}{2}$	2.4807	0.084	0.063
16	130.0	$\frac{1}{1}$	3.4807	0.031	0.000

denote the recorded lifetimes until either failure or censoring, and they are ordered according to size.[4] Let the index ν run through the integers j, where $t_{(j)}, j = 1, 2, \ldots$, denotes the times *to failure such that* $t_{(j)} < t$.

Nelson's estimator of the cumulative failure rate is then

$$\hat{Z}(t) = \sum_{\nu} \frac{1}{n - \nu + 1} = \sum_{j \in J_t} \frac{1}{n_j} \qquad (9.75)$$

Note that the Nelson estimator $\hat{Z}(t)$ is a first-order approximation to the estimator (9.70) derived from the Kaplan-Meier estimator. Nelson's estimator of the survivor function at time t is

---

[4]If two or more of the observations coincide, they are arranged in random order.

$$R^*(t) = e^{-\hat{Z}(t)} \tag{9.76}$$

Before we give the justification for these estimators, we will use them in an example.

Example 9.12
Again consider the situation in Example 9.10, calculate $\hat{Z}(t)$ from (9.75) for the failure times $t_{(1)}$, $t_{(2)}$, $t_{(3)}$, $t_{(5)}$, $t_{(6)}$, $t_{(13)}$, and $t_{(15)}$ (hence ν runs only through the values 1, 2, 3, 5, 6, 13, and 15). Then $R^*(t)$ is determined from (9.76). The results are shown in Table 9.6. On the right-hand side of Table 9.6, the corresponding Kaplan-Meier estimate $\hat{R}(t)$ is shown.

As we can see, there is good "agreement" between the Kaplan-Meier estimates and the Nelson estimates for the survivor function in this data set. Defined in this way, $R^*(t)$ will be continuous from the right. □

Table 9.6 Nelson Estimator for a Censored Data Set Compared with the Kaplan-Meier Estimator

j	ν	Time to Failure	Nelson Estimate $\hat{Z}(t_j)$		Nelson $R^*(t_{(j)})$	Kaplan-Meier $\hat{R}(t_{(j)})$
				= 0.0000	1.000	1.000
1	1	31.7		$\frac{1}{16}$ = 0.0625	0.939	0.938
2	2	39.2		$\frac{1}{16} + \frac{1}{15}$ = 0.1292	0.879	0.875
3	3	57.5		$\frac{1}{16} + \frac{1}{15} + \frac{1}{14}$ = 0.2006	0.818	0.813
4						
5	5	65.8		$\frac{1}{16} + \frac{1}{15} + \frac{1}{14} + \frac{1}{12}$ = 0.2839	0.753	0.745
6	6	70.0		$\frac{1}{16} + \frac{1}{15} + \frac{1}{11}$ = 0.3748	0.687	0.677
7						
8						
9						
10						
11						
12						
13	13	105.8		$\frac{1}{16} + \frac{1}{11} = \frac{1}{4}$ = 0.6248	0.535	0.508
14						
15	15	110.0		$\frac{1}{16} + \frac{1}{4} + \frac{1}{2}$ = 1.1248	0.320	0.254
16						

Justification for the Nelson Estimator

We now make the same considerations as we did earlier when we justified the Kaplan-Meier estimator. The time axis is divided into small time intervals $(u_j, u_{j+1}]$ for $j = 0, 1, \ldots$, where the intervals are so short that one can disregard these possibilities:

1. Two or more units fail in the *same* interval
2. One unit fails and another is censored in the *same* interval

In addition we assume that the intervals are so short that the failure rate in the interval $(u_j, u_{j+1}]$ may be considered constant and equal to λ_j, $j = 1, 2, \ldots$

Now suppose that $t \in (u_m, u_{m+1}]$. As on page 399

$$R(t) = P(T > u_1 | T > u_0) \cdots P(T > t | T > u_m) \qquad (9.77)$$

As before, the idea is to estimate each single factor on the right-hand side of (9.77) and use the product of these estimators as an estimator of $R(t)$. What will now be a reasonable estimator of $p_j = P(T > u_{j+1} | T > u_j)$? With the same approach we used for justifying the Kaplan-Meier estimator, the only intervals for which it will be natural to estimate p_j with something other than 1 will be the intervals $(u_j, u_{j+1}]$ where a *failure* occurs. If we denote the number of components that have either failed or have been censored in the course of $(0, u_j]$ with $(r - 1)$, there will be $(n - r + 1)$ active components at the beginning of the interval. The total functioning time in such an interval will be approximately equal to $(n - r + 1)(u_{j+1} - u_j)$, and hence a natural estimator of λ_j will be

$$\hat{\lambda}_j = \frac{1}{(n - r + 1)(u_{j+1} - u_j)} \qquad (9.78)$$

A natural estimator of p_j when a failure occurs in $(u_j, u_{j+1}]$ is therefore

$$\hat{p}_j = \exp(-\hat{\lambda}_j(u_{j+1} - u_j)) = \exp\left(-\frac{1}{n - r + 1}\right) \qquad (9.79)$$

If we insert these estimators in (9.77), this leads to the estimator $R^*(t)$ given in (9.76).

Nelson Plot

From (9.66) and (9.67) it follows that

$$z(t) \text{ increasing in } t \Leftrightarrow Z(t) \text{ convex} \qquad (9.80)$$

Correspondingly

$$z(t) \text{ decreasing in } t \iff Z(t) \text{ concave} \qquad (9.81)$$

If we plot the points $(t_{(i)}, \hat{Z}(t_{(i)}))$ in a rectangular coordinate system and if the pattern of the plot is as shown in Figure 9.14a, this indicates that $Z(t)$ is convex, which again means that we are dealing with an IFR life distribution function. Similarly a plot such as the one depicted in Figure 9.14b indicates that the life distribution is DFR, while the plot in Figure 9.14c indicates a life distribution with bathtub-shaped failure rate.

Often we are interested in checking whether or not it is reasonable to assume that a specified life distribution (normal distribution, exponential distribution, or Weibull distribution) is the basis for the observed life data. Special graph paper for the different distributions has been developed for this purpose. The graph paper for distribution $F(t)$ is so designed that if we plot $(t_{(j)}, \hat{Z}(t_{(j)}))$, the pattern of the plot will be approximately a straight line provided that the life data originates from the distributiton $F(t)$. Then the parameters of the distribution may be read directly from the plot.

Another way to check whether the life distribution $F(t)$ is appropriate is by the following two steps: First, we estimate the parameters of the distribution $F(t)$, for example, by using the maximum likelihood principle. Next, we draw the estimated cumulative failure rate of the distribution in the same coordinate system as the plot of $(t_{(j)}, \hat{Z}(t_{(j)}))$. If the plotted values are close to the estimated cumulative failure rate function, this indicates that the distribution $F(t)$ is the basis for the observed data.

Nelson (1969) suggested displaying the results of the estimation by plotting $(\hat{Z}(t_{(j)}), t_{(j)})$ instead of $(t_{(j)}, \hat{Z}(t_{(j)}))$. The plot produced in this way is often called a *Nelson plot* or a *Hazard plot*. Several computer programs have adopted Nelson's somewhat illogical convention of changed axes. Among these is the SAREPTA program. A Nelson plot of the data in Example 9.12, produced by SAREPTA, is given in Figure 9.15. The parameters α and λ of the Weibull distribution is by SAREPTA estimated to be $\hat{\alpha} = 2.38$ and $\hat{\lambda} = 8.12 \cdot 10^{-3}$. The estimated cumulative failure rate function $\hat{Z}(t) = (\hat{\lambda} t)^{\hat{\alpha}}$ is by SAREPTA drawn as an overlay curve to the plot. As seen from Figure 9.15, the fit to the Weibull distribution seems to be acceptable.

Total Time on Test Plot for Censored Data

The total time on test (TTT) plot was derived for complete data sets in Section 9.2. When the data set is incomplete and the censoring is of type IV (stochastic), one may argue as follows to obtain a TTT plot: The TTT transform, as defined in (9.9), is valid for a wide range of distribution functions $F(t)$ and also for step functions. Instead of estimating the TTT transform $H_F^{-1}(t)$ by introducing the empirical distribution function $F_n(t)$ as we did in (9.15), we could estimate $F(t)$ by $1 - \hat{R}(t)$, where $\hat{R}(t)$ is the Kaplan-Meier estimator of $R(t)$.

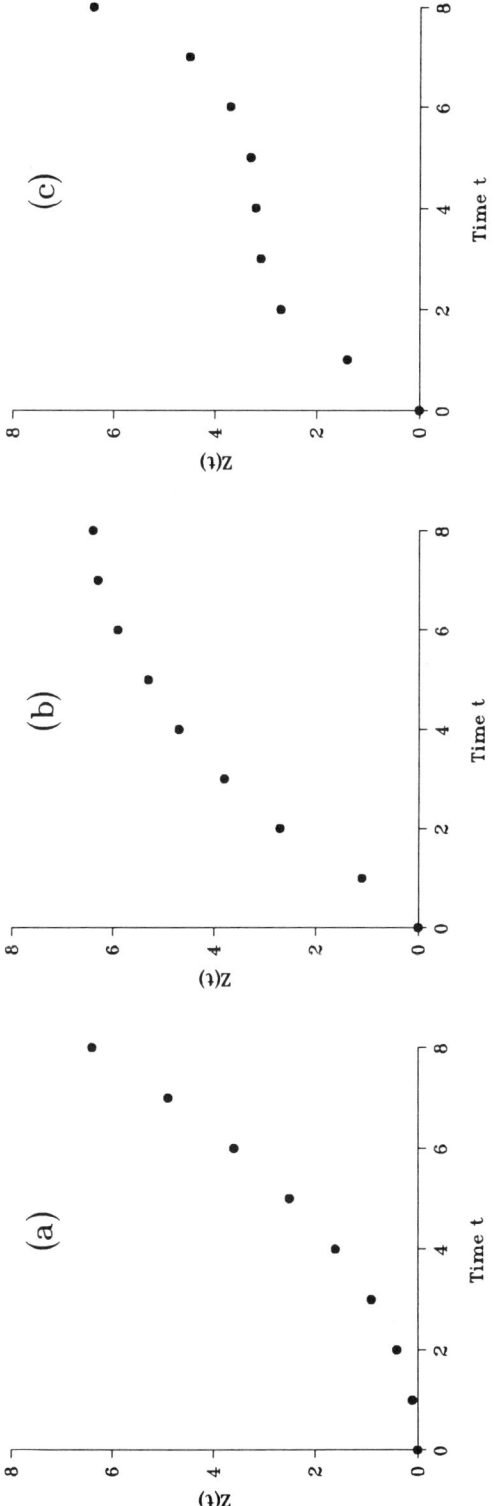

Figure 9.14 Estimated cumulative failure rate $\hat{Z}(t)$ indicating (*a*) increasing failure rate (IFR), (*b*) decreasing failure rate (DFR), and (*c*) Bathtub-shaped failure rate

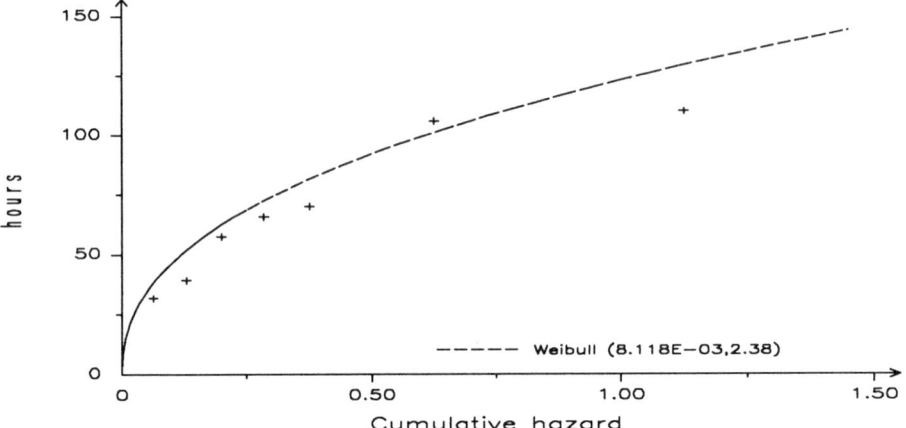

Figure 9.15 Nelson plot of the data in Example 9.12, together with an overlay curve for the Weibull distribution (the plot is made by SAREPTA)

Technically the plot is obtained as follows: Let $T_{(1)}, T_{(2)}, \ldots, T_{(k)}$ denote the k ordered failure times among $T_1, T_2, \ldots, T_n$, and let

$$v_{(i)} = 1 - \hat{R}(T_{(i)}) \quad \text{for} \quad i = 1, 2, \ldots, k \quad (9.82)$$

Define

$$\hat{H}^{-1}(v_{(i)}) = \int_0^{T_{(i)}} \hat{R}(u)\,du = \sum_{j=1}^{i-1} (T_{(j+1)} - T_{(j)})\hat{R}(T_{(j)}) \quad (9.83)$$

where $T(0) = 0$.

The TTT plot is now obtained by plotting the points

$$\left(\frac{v_{(i)}}{v_{(k)}}, \frac{\hat{H}^{-1}(v_{(i)})}{\hat{H}^{-1}(v_{(k)})} \right) \quad \text{for} \quad 1 = 1, 2, \ldots, k \quad (9.84)$$

Note that when $k = n$ (i.e. when the data set is complete),

$$v_{(i)} = \frac{i}{n}$$

$$\hat{H}^{-1}(v_{(i)}) = \mathcal{T}(T_{(i)})$$

OTHER APPLICATIONS

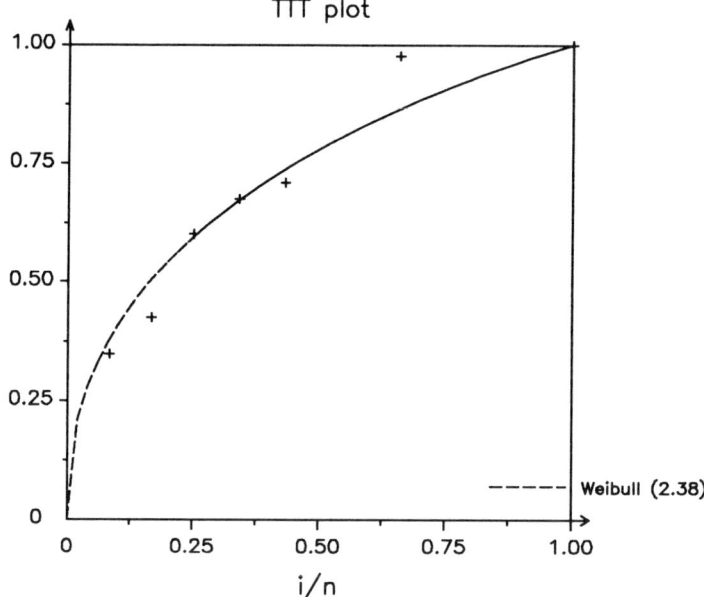

Figure 9.16 TTT plot for the censored data set in Example 9.12, together with an overlay curve for the Weibull distribution (the plot is made by SAREPTA)

and we get the same TTT plot as we got in Section 9.2 for data sets without censored observations.

A TTT plot of the censored data set in Example 9.12 is presented in Figure 9.16. The TTT plot is made by SAREPTA. The parameters α and λ of the Weibull distribution are also estimated by SAREPTA, and the TTT transform of the Weibull distribution with these parameters is drawn as an overlay curve to the TTT plot. The fit to the Weibull distribution seems to be acceptable.

9.4 OTHER APPLICATIONS

The methods described in this chapter can obviously also be applied to a general data set of independent and identically distributed nonnegative random variables. It is not necessary to restrict the data set to be lifetimes. In this section we will illustrate the application of the methods on repair times.

Repair Time Analysis

Failure time data from a compressor were discussed in Example 7.2. All compressor failures at a certain process plant in the time period from 1968 until

1989 have been recorded as part of a student thesis at the Norwegian Institute of Technology. In this period a total of 90 critical failures occurred. In this context a critical failure is defined to be a failure causing compressor downtime. The compressor is very important for the operation of the process plant, and every effort is taken to restart a failed compressor as soon as possible. The 90 repair times (in hours) are presented chronologically in Table 9.7. The repair time associated to the first failure was 1.25 hours, the second repair time was 135.00 hours, and so on.

Before applying the analytical methods described in this chapter, one should always check whether or not the repair times are independent and identically distributed. A number of graphical methods and formal statistical tests have been developed for this purpose. See, for example, Ascher and Feingold (1984), Akersten (1991), and the references cited therein. Here we will be content with a very simple graphical check of whether or not there is any trend in the data. This is done since there is reason to suspect that the repair times may increase with the age of the compressor.

A graph showing the repair times in chronological order is given in Figure 9.17. It seems obvious from this graph that there is no significant trend in the length of the repair times. Hence we shall assume in the following that the repair times are independent and identically distributed.

Since the data set is complete, the empirical distribution function of the repair times is easily determined (see p. 395). A graph of the empirical distribution function is presented in Figure 9.18. The plot is constructed by the SAREPTA program. In Section 2.11 it was claimed that the lognormal distribution is often a realistic repair time distribution. To check this assertion in our case, the parameters of the lognormal distribution have been estimated from the data in Table 9.7, and the corresponding estimated lognormal distribution function is drawn as an overlay curve to the empirical distribution function. From this graph the fit to the lognormal distribution seems to be very good.

A Kaplan-Meier, a TTT, or a Hazard plot may obviously be constructed for

Table 9.7 Repair Times (Hours) in Chronological Order

1.25	135.00	0.08	5.33	154.00	0.50	1.25	2.50	15.00
6.00	4.50	32.50	9.50	0.25	81.00	12.00	0.25	1.66
5.00	7.00	39.00	106.00	6.00	5.00	17.00	5.00	2.00
2.00	0.33	0.17	0.50	18.00	2.50	0.33	0.50	2.00
0.33	4.00	20.00	6.00	6.30	15.00	23.00	4.00	5.00
28.00	16.00	11.50	0.42	38.33	10.50	9.50	8.50	17.00
34.00	0.17	0.83	0.75	1.00	0.25	0.25	2.25	13.50
0.50	0.25	0.17	1.75	0.50	1.00	2.00	2.00	38.00
0.33	2.00	40.50	4.28	1.62	1.33	3.00	5.00	120.00
0.50	3.00	3.00	11.58	8.50	13.50	29.50	29.50	112.00

OTHER APPLICATIONS

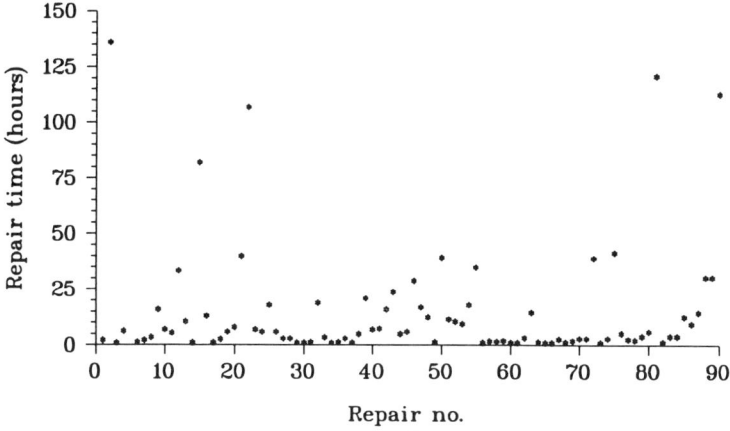

Figure 9.17 Repair times in chronological order

this data set. As an illustration, a TTT plot of the data set is presented in Figure 9.19, together with an overlay curve for the corresponding lognormal distribution. It appears from Figure 9.19 that the fit between the TTT plot and the overlay curve for the lognormal distribution is not quite good. This is due to the fact that the TTT plot is much more sensitive than the empirical distribution function with respect to detecting deviations from an assumed distribution.

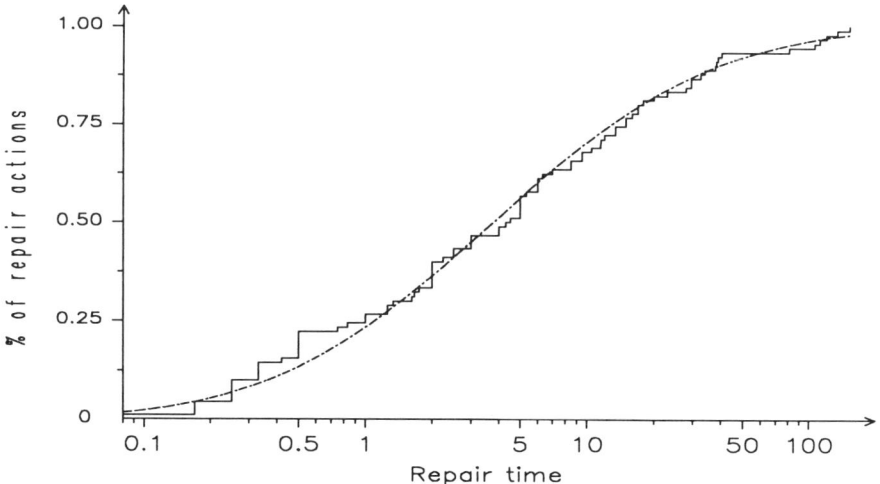

Figure 9.18 Empirical distribution function of the repair time data, with overlay curve for the lognormal distribution (printout from the SAREPTA program)

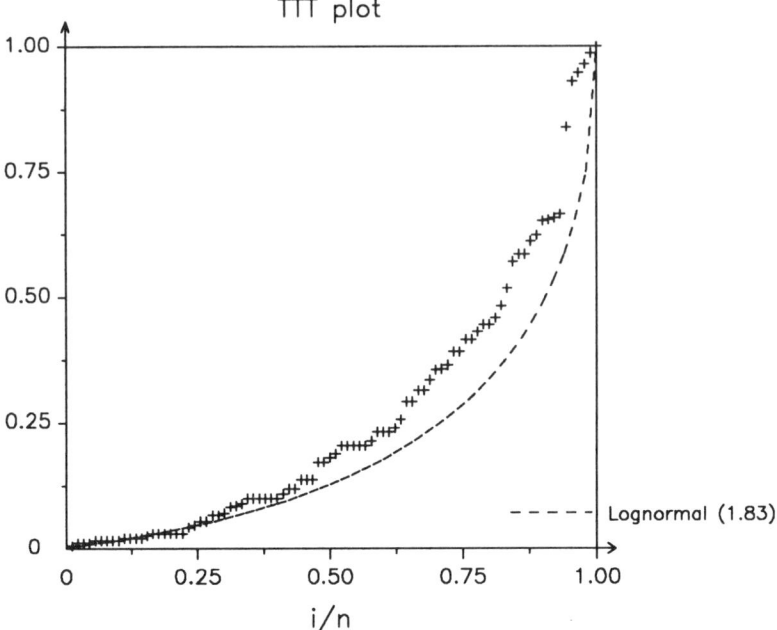

Figure 9.19 TTT plot of the repair time data, with overlay curve for the lognormal distribution (printout from the SAREPTA program)

9.5 PROBLEMS

1. Assume that you have determined the lifetimes for a total of 12 identical components and obtained the following results (given in hours):

   10.2, 89.6, 54.0, 96.0, 23.3, 30.4, 41.2, 0.8, 73.2, 3.6, 28.0, 31.6

   Construct the TTT plot for this set. What conclusion can you draw on this basis about the corresponding life distribution?

2. Prove property 2 of Corollary 9.1 by using Theorem 9.1.

3. Prove property 2 of Theorem 9.3.

4. Prove Theorem 9.5.

5. Suppose that we have experienced that the lifetime T for a certain type of component is exponentially distributed with unknown failure rate $\lambda > 0$. Furthermore suppose that we have recorded lifetimes for a total of 12 such components and interpret the result as n independent observations of T. The observed lifetimes (in hours) are

10.2, 89.6, 54.0, 96.0, 23.3, 30.4, 41.2, 0.8, 73.2, 3.6, 28.0, 31.6

Then

(a) Estimate λ.

(b) Test the hypothesis $\lambda \geq 0.025$ against $\lambda < 0.025$ (choose significance level = 0.05).

(c) Determine a 95% confidence interval for λ.

6. Consider a homogeneous Poisson process (HPP) with intensity λ. Let $N(t)$ denote the number of failures (events) in a time interval of length t. $N(t)$ is hence Poisson distributed with parameter λt. Assume that the process is observed in a time interval of length $t = 2$ years. In this time period a total of 7 failures have been observed.

    (a) Find an estimate of λ.

    (b) Determine a 90% confidence interval for λ.

7. Let X have a Poisson distribution with parameter λ. A test for the hypothesis

$$H_0 : \lambda = 3 \quad \text{against the alternative} \quad H_1 : \lambda \neq 3$$

is wanted. Use an exact test to find out whether or not you would reject H_0 when X is observed and found equal to 6 (choose significance level $\alpha = 0.05$).

8. Let X have a Poisson distribution with parameter λ.

    (a) Determine an exact 90% confidence interval for λ when X is observed and found equal to 6. For comparison also determine an approximate 90% confidence interval for λ, using the approximation of the Poisson distribution to $\mathcal{N}(\lambda, \lambda)$.

    (b) Solve the same problem as stated in (a) when X is observed and found equal to 14.

9. Denote the distribution function of the Poisson distribution with parameter λ by $\mathcal{P}_o(x; \lambda)$, and the distribution function of the chi-square distribution with ν degrees of freedom by $\Gamma_\nu(z)$.

    (a) Show that $\mathcal{P}_o(x; \lambda) = 1 - \Gamma_{2(x+1)}(2\lambda)$. (Hint: First show that $1 - \Gamma_{2(x+1)}(2\lambda) = \int_{2\lambda}^{\infty} u^x e^{-u}/x! \, du$, and next apply repeated partial integrations to the integral.)

    (b) Let $\lambda_1(X)$ and $\lambda_2(X)$ be defined by

$$\mathscr{P}_o(x; \lambda_1(x)) = \frac{\alpha}{2}$$

$$\mathscr{P}_o(x-1; \lambda_2(x)) = 1 - \frac{\alpha}{2}$$

Use the result of (a) to show that

$$\lambda_1(x) = \tfrac{1}{2} z_{\alpha/2, 2x}$$
$$\lambda_2(x) = \tfrac{1}{2} z_{1-\alpha/2, 2(x+1)}$$

where $z_{\epsilon, \nu}$ denotes the ϵ quantile of the chi-square distribution with ν degrees of freedom.

10. Let X be binomially distributed $(20, p)$. X is observed and found to be equal to 3.
    (a) Determine the corresponding exact 90% confidence interval for p.
    (b) Determine an approximate 90% confidence interval for p based on the formulas in Section 9.2.

11. Let X binomially distributed $(25, p)$. A test for the hypothesis

    $$H_0: p = 0.10 \quad \text{against the alternative} \quad H_1: p \neq 0.10$$

    is wanted. Use an exact test to find out whether or not you would reject H_0 when X is observed and found equal to 1 (choose significance level $\alpha = 0.05$).

12. Reconsider the situation in Example 9.4, but now assume that the times to failure are those that are not starred:

    $31.7, 39.2^*, 57.5, 65.5, 65.8^*, 70.0, 75.0^*, 75.2^*, 87.5^*, 88.3^*, 94.2, 101.7^*,$
    $105.8^*, 109.2, 110.0, 130.0^*$

    Calculate $\hat{R}(t)$, and display it graphically.

13. Reconsider the situation in Problem 5. Calculate $\hat{\hat{R}}(t)$, and depict it graphically.

14. Suppose that the data set in Problem 5 was obtained by simultaneously activating 20 identical units but that the test was terminated at the 12th failure.
    (a) What type of censoring is this?
    (b) Estimate λ in this situation.

(c) Calculate a 95% confidence interval for λ.

(d) Compare the results with those derived in Problem 9.4.

15. Consider the set of material strength data presented by Crowder et al. (1991, p. 46). An experiment has been carried out to gain information on the strength of a certain type of braided cord. Forty-eight pieces of cord were investigated. Seven cords were damaged during the experiment, implying right-censored strength values:

26.8*	29.6*	33.4*	35.0*	36.3	40.0*	41.7	41.9*	42.5*
43.9	49.9	50.1	50.8	51.9	52.1	52.3	52.3	52.4
52.6	52.7	53.1	53.6	53.6	53.9	53.9	54.1	54.6
54.8	54.8	55.1	55.4	55.9	56.0	56.1	56.5	56.9
57.1	57.1	57.3	57.7	57.8	58.1	58.9	59.0	59.1
59.6	60.4	60.7						

(a) Establish a Kaplan-Meier plot of the material strength data.

(b) Establish a TTT plot of the material strength data.

(c) Discuss the effect of this type of censoring.

(d) Describe the form of the failure rate function.

16. Establish a graph paper such that the Nelson plot of Weibull distributed life data is close to a straight line. Describe how the Weibull parameters α and λ may be estimated from the plot.

17. Establish a graph paper such that the Nelson plot of normally distributed, $\mathcal{N}(\mu, \sigma^2)$, life data is close to a straight line. Describe how the parameters μ and σ may be estimated from the plot.

18. Let $T_1, T_2, \ldots, T_n$ be independent and identically distributed $IG(\mu, \lambda)$. Assume λ to be known.

(a) Show that the MLE of μ, μ^* is given by

$$\mu^* = \frac{1}{n} \sum_{j=1}^{n} T_j = \overline{T}$$

(b) Determine a $(1 - \alpha)$-confidence interval for μ.

(Hint: $\lambda n(\mu^* - \mu)^2 / \mu^2 \mu^*$ is chi-square distributed with one degree of freedom).

19. Let $T_1, T_2, \ldots, T_n$ be independent and identically distributed $IG(\mu, \lambda)$. Assume μ to be known.

(a) Show that the MLE of λ, λ^*, is given by

$$\frac{1}{\lambda^*} = \frac{1}{n} \sum_{j=1}^{n} \frac{T_j - \mu}{\mu^2 T_j}$$

and that

$$E\left(\frac{1}{\lambda^*}\right) = \frac{1}{\lambda}$$

$$\text{var}\left(\frac{1}{\lambda^*}\right) = \frac{2}{n\lambda^2}$$

(b) Determine a $(1-\alpha)$-confidence interval for λ. (Hint: Use the hint given in the previous problem.)

20. Lieblein and Zelen (1956) analyzed data on the endurance of deep groove ball bearings. They registered the number X_i of million revolutions before failure for 23 such ball bearings. Their distinct ordered failure times are

17.88	28.92	33.00	41.52	42.12	45.60	48.48	51.84
51.96	54.12	55.56	67.80	68.64	68.64	68.88	84.12
93.12	98.64	105.12	105.84	127.92	128.04	173.40	

Previously these data have been analyzed assuming that they follow a Weibull distribution or a lognormal distribution, respectively. Now let us analyze this data set under the assumption that they are independent $IG(\mu, \lambda)$.

(a) Determine the MLE μ^* of μ and the MLE λ^* of λ.

(b) What would be your estimate of the variance of this distribution? Compare this estimate with the general estimate $S^2 = \sum_{j=1}^{n}(X_j - \bar{X})^2/(n-1)$.

(c) To test whether or not the inverse Gaussian distribution gives a good model for this data set, determine a percentage–percentage (P–P) plot of the data.

To get this P–P plot, first estimate $F(x_{(j)})$ by the empirical distribution function

$$\hat{F}(x_{(j)}) = \frac{j - (1/2)}{n} \quad \text{for} \quad j = 1, 2, \ldots, n$$

Next estimate $F(x_{(j)})$ by $IG(x_{(j)}, \mu^*, \lambda^*)$. Plot $\hat{F}(x_{(j)})$ against $IG(x_{(j)}, \mu^*, \lambda^*)$ for $j = 1, 2, \ldots, n$. If the model fits, the curve $(\hat{F}(x_{(j)}), IG(x_{(j)}, \mu^*, \lambda^*))$ will approximately fall along a straight line. Does it? What is your conclusion? (P–P plots are further described by Crowder et al. 1991.)

CHAPTER 10

Accelerated Life Testing

10.1 INTRODUCTION

Many of the devices produced today for complex technical systems have very high reliability under normal use conditions. Such devices may have a mean time to failure of 100,000 hours ($\approx$11.5 years) or more. The time involved in a life test such as those described in Section 9.1 would therefore be exorbitant. Furthermore the device is likely to be out of date and therefore of no interest by the time the test is completed. The questions then arise of how to make the optimal choice between several types or designs of a device and how to collect information about the corresponding life distributions under normal use conditions.

A common way of tackling these problems is to expose the device to sufficient overstress to bring the mean time to failure down to an acceptable level. Thereafter one tries to "extrapolate" from the information obtained under overstress to normal use conditions. This approach is called *accelerated life testing (ALT)* or overstress testing. Recent books describing statistical methods, test plans, and data analysis for accelerated life testing include Mann, Schafer, and Singpurwalla (1974), Kalbfleisch and Prentice (1980), Lawless (1982), Jensen and Petersen (1982), Cox and Oakes (1984), Viertl (1988), and Nelson (1990).

Depending on the kind of device in question, the accelerated testing conditions may involve a higher level of temperature, pressure, voltage, load, vibration, and so on, than the corresponding levels occurring in normal use conditions. These variables are called *stressors* (stress variables or covariates). In a specific situation there may be one or several (m) stressors $s_1, s_2, \ldots, s_m$ acting simultaneously. The vector $s = (s_1, s_2, \ldots, s_m)$ is called the *stress vector*.

In simple situations there is only one stressor s occurring on two levels $s^{(1)}$ and $s^{(2)}$, where $s^{(1)} < s^{(2)}$. Let $s^{(0)} \leq s^{(1)}$ denote normal stress. The situation becomes somewhat complicated when m stressors $s_1, s_2, \ldots, s_m$ are involved and stressor s_j occurs on n_j levels,

$$s_j^{(1)} < s_j^{(2)} < \cdots < s_j^{(n_j)} \quad \text{for} \quad j = 1, 2, \ldots, m$$

Let $s_j^{(0)}(\leq s_j^{(1)})$ denote normal stress for stressor j, for $j = 1, 2, \ldots, m$. The situation becomes more complicated when the stressors are continuously increasing with time (see Figure 10.3). The first two cases lead to *step-stress accelerated tests* (SALT); the last one leads to *progressive-stress accelerated tests* (PALT).

10.2 EXPERIMENTAL DESIGNS FOR ALT

Let us for the sake of simplicity suppose that there is only *one* stressor s. The testing experiment can be conducted according to different designs. We will discuss three such designs.

Design I

The experiment involves use of k stress levels $s^{(1)} < s^{(2)} < \cdots < s^{(k)}$ (see Figure 10.1). Let $s^{(0)} \leq s^{(1)}$ denote normal stress. A (large) number of test units are assumed to be available for the experiment, and n_j of these are to be exposed to the stress $s^{(j)}$. Censoring of type II (see Section 9.3) is applied. The experiment is then carried out as follows:

1. One stress level $s^{(i)}$ is *chosen at random* among $s^{(1)}, s^{(2)}, \ldots, s^{(k)}$, and n_i test units are chosen at random among the test units at hand. These n_i units are then exposed to stress level $s^{(i)}$. The test is terminated when $r_i (\leq n_i)$ failures have occurred. Let $T_{i1}, T_{i2}, \ldots, T_{in_i}$ denote the times to failure or censoring.

Figure 10.1 Design I for accelerated tests

2. Another stress level $s^{(j)}$ is chosen at random among the remaining levels, n_j test units are chosen at random among the remaining units and exposed to stress level $s^{(j)}$. The test is terminated when $r_j (\le n_j)$ failures have occurred. Let $T_{j1}, T_{j2}, \ldots, T_{jn_j}$ denote the times to failure or censoring. This procedure is continued until k stress levels have been selected.

If the number of test units at hand is large compared to $n = \sum_{j=1}^{k} n_j$, it seems reasonable to assume that $T_{01}, T_{02}, \ldots, T_{k r_k}$ are independent, which simplifies the analysis.

Design II

Fix k points of time $0 < t_1 < t_2 < \cdots < t_k < t$. Put n randomly chosen test units on test at time 0. In the time interval $(0, t_1]$ the units are subject to stress $s^{(1)}$; in the interval $(t_1, t_2]$ the units that have not failed by time t_1 are kept in operation under stress $s^{(2)}$. In the next interval $(t_2, t_3]$ the units that still have not failed by time t_2 are kept in operation under stress $s^{(3)}$, and so on it goes (see Figure 10.2). In the time interval $(t_k, \infty]$ the units that have not failed by time t_k are kept in operation under stress $s^{(k+1)}$ until they have all failed (hence no censoring). The lifetimes of the n test units are denoted $T_1, \ldots, T_n$.

Design III

A number n of test units are chosen at random among the test units at hand and exposed to a stress $s(t)$, which is increasing with time until the units have all failed. $s(t)$ is assumed known (Figure 10.3). The lifetimes of the n test units are observed and denoted $T_1, T_2, \ldots, T_n$.

If n is small compared to the number of units at hand and if the n units are operating independently, it seems reasonable to assume that $T_1, T_2, \ldots, T_n$ are independent in design II and design III.

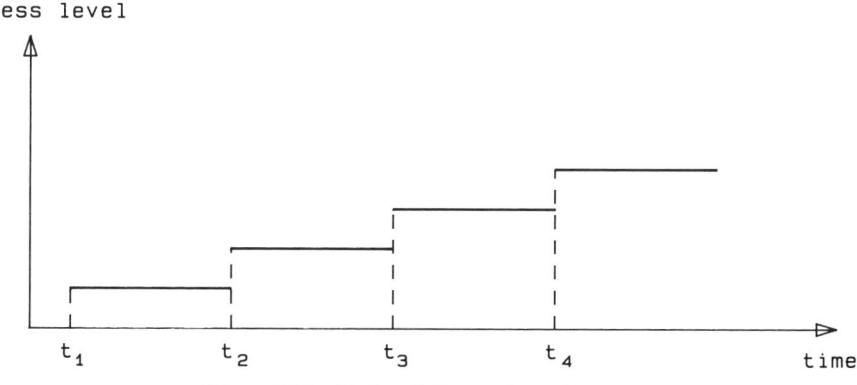

Figure 10.2 Design II for accelerated tests

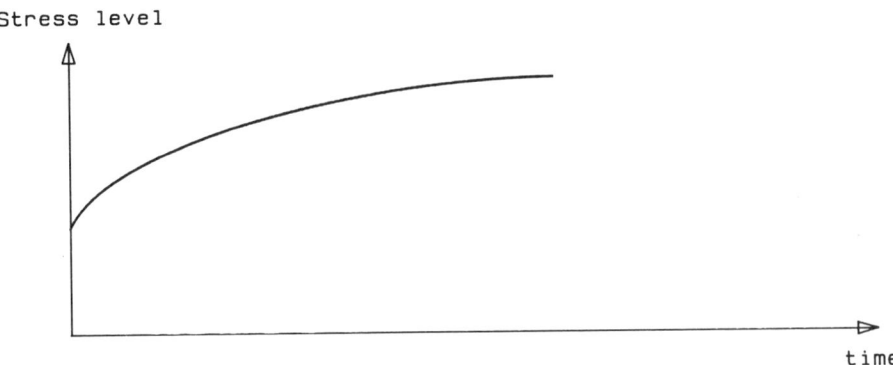

Figure 10.3 Design III for accelerated tests

10.3 PARAMETRIC MODELS USED IN SALT

The first step in any statistical analysis usually consists of formulating a stochastic model of the situation based on a priori knowledge and the experimental design used. In this case we are concerned with life distributions. Hence the data obtained through the SALT is supposed to give information about:

1. the life distribution function

$$F_T(t; \underline{s}) = P(T \leq t; \underline{s})$$

2. the survivor function

$$R_T(t; \underline{s}) = 1 - F_T(t; \underline{s})$$

3. or, the failure rate

$$z(t; \underline{s})$$

which are all more or less unknown.

The first question to be asked is: What do we know a priori about the life distribution under normal use conditions? For example, do we have reason to believe that it will belong to the exponential, the Weibull, the lognormal or the inverse Gaussian family of distributions? Are we able to derive this life distribution from the physical conditions at hand? One way of searching for a suitable model is to display the data on different kinds of probability paper and pick the family of distributions accordingly. Another way to find a suitable model is to try different models and select the one that best fits the data. The question of model discrimination has been discussed in Webster and Van Parr

(1965), Hunter and Box (1965), Hunter and Reiner (1965), Box and Hill (1967), Mann, Schafer, and Singpurwalla (1974, ch. 7), and Box, Hunter, and Hunter (1978, ch. 16), among others.

For the sake of simplicity, let us suppose that we succeed in establishing an appropriate *parametric* family of lifetime distributions under normal use conditions. The next question is: How does overstress affect this family of distributions? For example, will the life distribution under overstress belong to the same parametric family as the one obtained under normal stress? If so, the only effect of stress on the life distribution is that different stress levels lead to different parameter vectors in this family.

In an early paper on ALT, Levenbach (1957) was able to conclude that the life distributions occurring in the case he was studying belonged to the lognormal family under normal stress as well as under overstress at certain levels. Let us suppose we are as lucky as Levenbach and find that our lifetime distributions belong to a specific parametric family. Then the next question which needs to be answered is: In which way does the parameter vector of this family depend on the stress vector s? Here we only intend to give an introduction to the subject of ALT and hence will content ourselves with considering two simple examples.

Example 10.1: Design I

Suppose that the experiment is carried out as described in design I where only one stressor s has been used and that the family of life distributions is the exponential with mean $\theta(s)$, and hence failure rate $\lambda(s) = \theta(s)^{-1}$. What is required is the function $\lambda(s)$ describing the relation between the stress and the failure rate. In principle any function $\lambda(s)$ may do. However, three of the most commonly used relations are

$$\theta(s) = cs^{-a} \quad \text{(power rule model)} \tag{10.1}$$
$$\lambda(s) = ce^{-b/s} \quad \text{(Arrhenius model)} \tag{10.2}$$
$$\lambda(s) = cse^{-(b/s)} \quad \text{(simple Eyring model)} \tag{10.3}$$

where a, b, and c denote constants which are more or less unknown. These models may be derived from the physics of failure of the device in question. The power rule model has mostly been applied to dielectric breakdown of capacitors and fatigue testing of materials. The Arrhenius model has been applied to thermal aging and is also applicable to semiconductor materials. The simple Eyring model has been applied to devices exposed to constant thermal stress. A more thorough discussion of these three models and of other more general models may be found in Mann, Schafer, and Singpurwalla (1974). □

The constants $a, b, c, \ldots$ appearing in these models have to be estimated on the basis of the recorded life lengths under overstress. This may be done by inserting the appropriate expression for $\lambda(s)$ in the lifetime distribution. Given the stress level $s^{(j)}$ for $j = 1, 2, \ldots, k$, $F_T(t_j; s^{(j)})$ is now known except for the values of the constants $a, b, c, \ldots$. The occurring constants may then be estimated from the data, for example, by applying the maximum likelihood principle or the least squares principle. Let us denote the estimates by $a^*, b^*, c^*, \ldots$.

The final step then consists of inserting these estimates together with the normal stress $s^{(0)}$ in the lifetime distribution function. The result is an "estimate" of the *normal stress* life distribution, $\hat{F}_T(t; s^{(0)})$ based on the *overstress* life data.

ooo OOO ooo

Before we specify the relation between the stressor s and the failure rate $\lambda(s)$ in our simple example, let us remind ourselves of a few results about exponentially distributed random variables. The total information that is obtained through the SALT is expressed by $(s^{(j)}, n_j, r_j, T_{j1}, \ldots, T_{jn_j})$ for $j = 1, 2, \ldots, k$. From what we learned in Section 9.3, it seems natural to summarize this and say that the information obtained by the SALT is expressed by

$$(s^{(j)}, n_j, r_j, T_j) \qquad \text{for} \quad j = 1, 2, \ldots, k \qquad (10.4)$$

where

$$T_j = \mathcal{T}(T_{jn_j}), \qquad \text{for} \quad j = 1, 2, \ldots, k \qquad (10.5)$$

is the total time on test at stress level $s^{(j)}$.[1]

Let $\theta(s_j)$ denote the mean time to failure at stress level $s^{(j)}$, and let $\lambda(s^{(j)})$ denote the corresponding failure rate. Then we know from Section 9.3 that $Z_j = 2\lambda(s^{(j)})T_j$ is χ^2 distributed with $2r_j$ degrees of freedom, $j = 1, 2, \ldots, k$:

$$f_{Z_j}(z_j) = \frac{1}{2^{r_j}\Gamma(r_j)} z_j^{r_j-1} e^{-z_j/2} \qquad \text{for} \quad z_j > 0, \, j = 1, 2, \ldots, k \qquad (10.6)$$

Accordingly

$$\begin{aligned} f_{T_j}(t_j) &= \frac{1}{2^{r_j}\Gamma(r_j)} (2\lambda(s^{(j)})t_j)^{r_j-1} e^{-\lambda(s^{(j)})t_j} \cdot 2\lambda(s^{(j)}) \\ &= \frac{1}{\Gamma(r_j)} \cdot \lambda(s^{(j)})^{r_j} t_j^{r_j-1} e^{-\lambda(s^{(j)})t_j} \qquad \text{for} \quad t_j > 0, \, j = 1, 2, \ldots, k \end{aligned} \qquad (10.7)$$

---

[1] $T = (T_1, T_2, \ldots, T_k)$ is a complete, sufficient statistic in our model. $E(T_j/\lambda_j) = \lambda(s_j)$, for $j = 1, 2, \ldots, k$. Using the Lehmann-Scheffé theorem, it follows that T_j/λ_j is uniformly minimum variance unbiased estimate for $\lambda(s_j)$. See, for example, Bickel and Doksum (1977, p. 122). Hence no information is lost by summarizing this way.

Hence

$$f_{T_1,\ldots,T_k}(t_1,\ldots,t_k) = \prod_{j=1}^{k} \left(\frac{1}{\Gamma(r_j)} \cdot \lambda(s^{(j)})^{r_j} \cdot t_j^{r_j-1} e^{-\lambda(s^{(j)})t_j} \right)$$
$$\text{for } t_j > 0, j = 1, 2, \ldots, k \qquad (10.8)$$

As an example, let us consider the case where the relation between the stressor s and the mean $\theta(s)$ is described by the power rule model:

$$\theta(s^{(j)}) = c_1 (s^{(j)})^{-a} \qquad \text{for } j = 1, 2, \ldots, k \qquad (10.9)$$

Then

$$\lambda(s^{(j)}) = \frac{1}{c_1} (s^{(j)})^a \qquad (10.10)$$

If we amend the power rule slightly, without changing its basic character, to

$$\lambda(s^{(j)}) = \frac{1}{c} \left(\frac{s^{(j)}}{\dot{s}} \right)^a \qquad (10.11)$$

where $\dot{s}$ is the weighted geometric mean of the s_j's

$$\dot{s} = \prod_{j=1}^{k} (s^{(j)})^{r_j / \sum_{i=1}^{k} r_i} \qquad (10.12)$$

it will later on turn out that the MLE of a and c, a^* and c^*, become asymptotically independent. Hence this change is worthwhile.

Inserting (10.11) into (10.8) leads to

$$f_{T_1,\ldots,T_k}(t_1,\ldots,t_k,a,c) = \prod_{j=1}^{k} \frac{1}{\Gamma(r_j)} \left[\frac{1}{c} \left(\frac{s^{(j)}}{\dot{s}} \right)^a \right]^{r_j} t_j^{r_j-1} e^{-(s^{(j)}/\dot{s})^a t_j / c}$$
$$(10.13)$$

The corresponding likelihood function is

$$\ell(a,c; t_1,\ldots,t_k) = \prod_{j=1}^{k} \frac{1}{\Gamma(r_j)} \left[\frac{1}{c} \left(\frac{s^{(j)}}{\dot{s}} \right)^a \right]^{r_j} t_j^{r_j-1} e^{-(s^{(j)}/\dot{s})^a t_j / c} \qquad (10.14)$$

and the log likelihood function is

$$\ln \ell(a, c; t_1, \ldots, t_k)$$
$$= \sum_{j=1}^{k} \left[-\ln \Gamma(r_j) - r_j \ln c + a r_j \ln \left(\frac{s^{(j)}}{\dot{s}} \right) \right.$$
$$\left. + (r_j - 1) \ln t_j - \frac{1}{c} \left(\frac{s^{(j)}}{\dot{s}} \right)^a \cdot t_j \right] \qquad (10.15)$$

The MLE of a and c, a^* and c^*, are obtained by solving the two equations:

$$\frac{\partial \ln \ell}{\partial a} = \sum_{j=1}^{k} -r_j (\ln s^{(j)} - \ln \dot{s}) - \sum_{j=1}^{k} \frac{1}{c} \left(\frac{s^{(j)}}{\dot{s}} \right)^a \ln \left(\frac{s^{(j)}}{\dot{s}} \right) t_j = 0$$

$$(10.16)$$

and

$$\frac{\partial \ln \ell}{\partial c} = \sum_{j=1}^{k} -\frac{r_j}{c} + \sum_{j=1}^{k} \left(\frac{s^{(j)}}{\dot{s}} \right)^a \cdot t_j \cdot \frac{1}{c^2} = 0 \qquad (10.17)$$

with respect to a and c.

From (10.12), we realize that

$$\ln \dot{s} = \sum_{j=1}^{k} \frac{r_j}{\sum_{i=1}^{k} r_i} \ln s^{(j)} \qquad (10.18)$$

That is,

$$\sum_{j=1}^{k} r_j \ln \dot{s} = \sum_{j=1}^{k} r_j \ln s^{(j)} \qquad (10.19)$$

or

$$\sum_{j=1}^{k} r_j (\ln s^{(j)} - \ln \dot{s}) = 0 \qquad (10.20)$$

Hence (10.16) is reduced to

$$\sum_{j=1}^{k} \left(\frac{s_j}{\dot{s}}\right)^a \ln\left(\frac{s^{(j)}}{\dot{s}}\right) \cdot t_j = 0 \qquad (10.21)$$

which determines the MLE of a, a^*. c is then determined by (10.17)

$$c^* = \frac{1}{\sum_{i=1}^{k} r_i} \sum_{j=1}^{k} \left(\frac{s^{(j)}}{\dot{s}}\right)^{a^*} \cdot t_j \qquad (10.22)$$

Equations (10.21) and (10.22) do not allow us to determine a^* and c^* analytically. Iterative procedures must be used.

It can be shown (see Mann, Schafer, and Singpurwalla 1974, p. 426) that the asymptotic variances of a^* and c^* are

$$\text{as var}(a^*) = \left[\sum_{j=1}^{k} r_j \left(\ln \frac{s^{(j)}}{\dot{s}}\right)^2\right]^{-1} \qquad (10.23)$$

$$\text{as var}(c^*) = c^2 \left[\sum_{j=1}^{k} r_j\right]^{-1} \qquad (10.24)$$

and that the asymptotic covariance is

$$\text{as cov}(a^*, c^*) = 0 \qquad (10.25)$$

Furthermore it can be shown that (a^*, c^*) are asymptotically distributed as bivariate normal variables. See Mann, Schafer, and Singpurwalla (1974, p. 83). Hence a^* and c^* are asymptotically independent.

A reasonable estimate of the failure rate under normal stress $s^{(0)}$ is then

$$\lambda_0^* = \frac{1}{c^*} \left(\frac{s^{(0)}}{\dot{s}}\right)^{a^*} \qquad (10.26)$$

Therefore the density of the lifetime under normal stress may be estimated by

$$f_T(t) = \lambda_0^* e^{-\lambda_0^* t} \qquad \text{for} \quad t > 0$$

ooo OOO ooo

Table 10.1 Some References to Papers Where SALT-Parametric Models have been Studied

Life Distribution			Reference
$F_T(t,s) = 1 - e^{-\lambda(s)\cdot t}$	$\lambda(s) = cs^a$		Mann, Schafer, and
	$\lambda(s) = ce^{-b/s}$		Singpurwalla (1974);
	$\lambda(s) = cse^{-b/s}$		Singpurwalla (1973)
$F_T(t;s) = 1 - e^{-[\lambda(s)\cdot t]^\alpha}$	$\lambda(s) = cs^a$		Mann, Schafer, and
	α constant		Singpurwalla (1974);
			Nelson (1975);
			Singpurwalla and
			Al-Khayyal (1977)
$\ln T \sim \mathcal{N}(\nu(s),\tau^2)$	$\nu(s) = \alpha + \beta s$		Nelson and
	τ^2 constant		Kielpinski (1976)

Table 10.1 gives references to a few papers where other SALT-parametric models have been studied.

Example 10.2: Design III

Let us consider experiments where n identical units, operating independently, are put on test at time 0. In the time interval $(0,t]$ the units are subject to stress $s^{(0)}$, while in the interval (t,∞) the units that have not failed by time t are kept in operation under stress $s^{(1)}$ ($> s^{(0)}$) until they all have failed. Typically $s^{(0)}$ corresponds to normal stress and $s^{(1)}$ to accelerated stress.

Suppose furthermore that the accumulated fatigue in the material subject to wear is modeled as a Wiener process $\{W_0(y), y \geq 0\}$, with drift $\eta > 0$ and diffusion constant $\delta^2 > 0$. Failure occurs when the fatigue process $W_0(y)$ crosses a critical boundary ω. [The Wiener process $W_0(y)$ is defined to be an independent increment Gaussian process with $W_0(0) = 0$ and mean $E(W_0(y)) = \eta y$. Moreover each increment, $W_0(y_2) - W_0(y_1)$ for $0 < y_1 < y_2$, has variance $\delta^2(y_2 - y_1)$.]

The basic result, whose history and proof can be found in Chhikara and Folks (1989), is that if we define the fatigue failure time Y as the first time that the fatigue process $W_0(y)$ crosses the critical boundary ω, and if we set $\mu = \omega/\mu$ and $\lambda = \omega^2/\sigma^2$, then Y has the inverse Gaussian distribution $IG(y, \mu, \lambda)$ with probability density

$$f_Y(y;\mu,\lambda) = \frac{\lambda}{\sqrt{2\pi y^3}} e^{-(\lambda/\mu 2)[(y-\mu)^2/y]} \quad \text{for} \quad y > 0, \mu > 0, \lambda > 0.$$

We now make the assumption that the fatigue process changes from one Wiener process to another at the stress change point t. More precisely, in the interval $(0,t]$ we suppose that the failure occurs if the process $W_0(y)$ crosses the critical boundary $\omega > 0$, where $W_0(y)$ is a Wiener process with drift $\eta > 0$ and diffusion constant $\delta^2 > 0$. At the stress change point t, if $W_0(y)$ has not yet crossed ω in $(0, t]$, the stress is changed from $s^{(0)}$ to $s^{(1)}$, and a new Wiener process starts out at the point $(t, W_0(t))$ (see Figure 10.4). We assume that

$$W_1(y) = W_0(t + \alpha(y - t)) \quad \text{for } y > t, \alpha > 1$$

Hence our SALT fatigue process is

$$W(y) = \begin{cases} W_0(y) & \text{for } y \leq t \\ W_0(t + \alpha(y - t)) & \text{for } y > t \end{cases}$$

In Doksum and Høyland (1992) it is shown that the distribution $F_T(t)$ of the stress failure time T in this situation is

$$F_T(y) = \begin{cases} F_0(y) & \text{for } 0 \leq y \leq t \\ F_0(t + \alpha(y - t)) & \text{for } y > t \end{cases}$$

where $F_0(y) = IG(y; \mu, \lambda)$ is the inverse Gaussian distribution whose probability density is given by (2.62).

In Figure 10.5 graphs of the failure distribution $F_T(y)$ for $\mu = 43, \lambda = 485$ and $t = 0$, $\alpha = 1, 2, 3,$ and 4, are given. Suppose now that the stress s is changed from $s^{(0)}$ to $cs^{(0)}$ for some known constant $c > 1$ as time crosses the point t. In this situation it might be reasonable to model the fatigue process $W(y)$ as having drift η in $(0, t]$ and drift $c\eta$ in (t, ∞). This means that $\alpha = c$ (known).

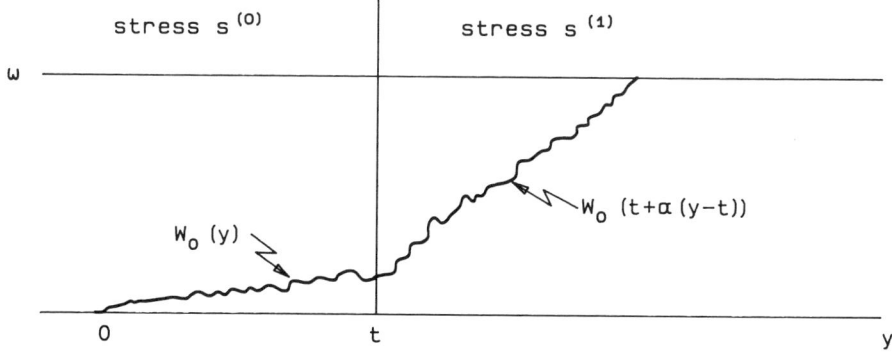

Figure 10.4 A fatigue process $W(y)$ with stress level increased from $s^{(0)}$ to $s^{(1)}$ at time t

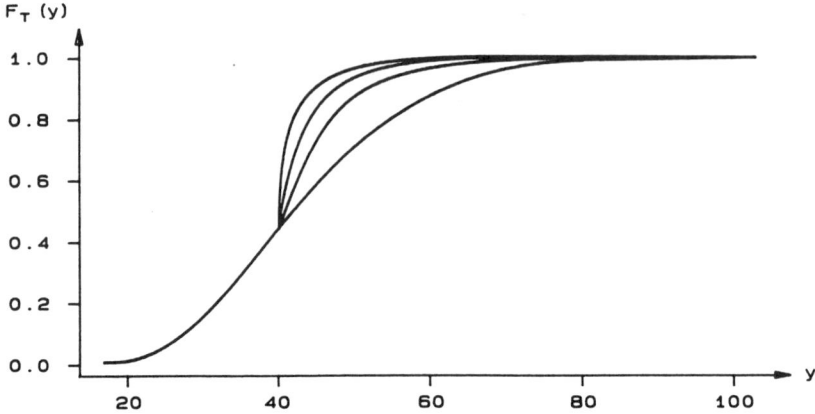

Figure 10.5 Two-stress failure distribution $F_T(y)$ with $\mu = 43$, $\lambda = 485$ and $t = 40$. *From top to bottom*: The four curves represent $F_T(y)$ for $\alpha = 4, 3, 2,$ and 1

Let $y_1, \ldots, y_n$ be the observed failure times, and introduce

$$y_j(\alpha) = y_j \quad \text{for} \quad y_j \leq t$$

and

$$y_j(\alpha) = t + \alpha(y_j - t) \quad \text{for} \quad y_j > t. \tag{10.27}$$

In this α known case the likelihood function, which we label $\ell_\alpha(\mu, \lambda)$, can be written

$$\ell_\alpha(\mu, \lambda) = \alpha^m \prod_{j=1}^{n} f_0(y_j(\alpha))$$

$$m = \sum_{s=1}^{n} I(y_j > t)$$

where m is the number of y's greater than t.

We note that the likelihood is proportional to the usual one-sample inverse Gaussian situation with the only change that y_j is replaced by $y_j(\alpha)$ for $j = 1, 2, \ldots, n$. Using (9.40), we see that the MLE for μ and $1/\lambda$ are given by

$$\mu_\alpha^* = \frac{1}{n} \sum_{j=1}^{n} y_j(\alpha)$$

$$\frac{1}{\lambda_\alpha^*} = \frac{1}{n} \sum_{j=1}^{n} \left(\frac{1}{y_j(\alpha)} - \frac{1}{\mu^*(\alpha)} \right) \tag{10.28}$$

Hence an estimate of the life distribution of the failure fatigue time T under stress $s^{(0)}$ (normal stress) is $IG(\mu_\alpha^*, \lambda_\alpha^*)$.

It also follows that μ_α^* has the $IG(\mu, n\lambda)$ distribution, that $n\lambda/\lambda_\alpha^*$ has a χ^2 distribution with $n-1$ degrees of freedom and that μ_α^* and λ_α^* generally have all the desirable properties of one-sample inverse Gaussian estimates. In particular, the estimator μ_α^* of the mean time to failure under stress $s^{(0)}$ has variance $\mu^3/n\lambda$.

A more realistic situation would be the one where α is unknown. The case is discussed in Doksum and Høyland (1992). They also discuss step-stress models with m (>2) stress levels and models with continuously increasing stress, and they include a discussion of the stochastic censoring case.

10.4 NONPARAMETRIC MODELS USED IN ALT

Two main assumptions need to be true if a successful parametric model is to be established for ALT:

1. The life distributions have to belong to the *same* known parametric family under normal stress as well as under overstress.
2. The relation between the parameter vector of this family and the stressors has to be known.

If these conditions are met, the solution of the problem is considerably simplified. Hence there is a danger that these assumptions may be made too often for mathematical convenience only and with no basis in reality. To prevent this from happening, considerable efforts have been made in recent years to develop models and methods that do not require assumptions about the functional form of the lifetime distribution in question. We will conclude this chapter with some comments on such models and methods and with references to literature that contains more detailed information.

It may be argued in certain situations—even in situations where we do not know the family of life distributions in question—that the ratio between the failure rates $z(t, s^{(i)})$ and $z(t, s^{(j)})$, corresponding to any two stress levels $s^{(i)}$ and $s^{(j)}$, is constant over time:

$$\frac{z(t; s^{(i)})}{z(t; s^{(j)})} = g(s^{(i)}, s^{(j)}) \tag{10.29}$$

where g is a function of $(s^{(i)}, s^{(j)})$ only. Such models are denoted proportional hazards (P–H) models. If we replace $z(t; s^{(0)})$ by $z_0(t)$ and $g(s^{(1)}, s^{(0)})$ by $g_1(s)$, a P–H model is characterized by the relation

$$z(t; s) = z_0(t) \cdot g_1(s) \tag{10.30}$$

g_1 being a function of s only.

Then the survivor function may be written [see (2.8)] as follows:

For stress $s^{(0)}$,

$$R(t; s^{(0)}) = e^{-\int_0^t z_0(u)du} \qquad (10.31)$$

For stress s,

$$R(t; s) = e^{-\int_0^t z_0(u)g_1(s)du} \qquad (10.32)$$

Accordingly in a P–H model the survivor functions $R(t; s)$ and $R(t; s^{(0)})$ have to satisfy the relation

$$R(t; s) = R(t; s^{(0)})^{g_1(s)} \qquad (10.33)$$

[A parametric model where (10.30) will be satisfied is the two-parameter Weibull family with constant shape parameter α (independent of s) and a scale parameter λ, which may depend on the stressor.]

If the life distribution function is only known to be continuous but is otherwise unspecified, the failure rate $\lambda_0(t)$ of the P–H model is likewise unspecified. The corresponding nonparametric P–H model is quite flexible and may be applicable to many situations occurring in practice. For a more thorough discussion of these models and their applications, the reader is referred to Viertl (1988) and Nelson (1990).

As already indicated in (10.30) the P–H models are characterized by the relation

$$z(t; s) = z_0(t) \cdot g_1(s)$$

Note that $z_0(t)$ and $g_1(s)$ may contain unknown parameters. Cox (1972) discussed a special subclass of nonparametric P–H models, now usually referred to as Cox models, which also are of particular interest in biomedical applications.

Cox considers a situation with m different stressors $s_1, s_2, \ldots, s_m$. In order to come into line with the notation used in connection with Cox models, the notation used in the introduction to this chapter must be altered slightly. Let stressor

s_1 occur at level x_1

s_2 occur at level x_2

$\vdots$

and so on. Then Cox makes the assumption that in the P–H model

$$g(x) = \exp\left(\sum_{i=1}^{m} \beta_i x_i\right) = \exp(\boldsymbol{\beta}'x) \quad (10.34)$$

where $x = (x_1, x_2, \ldots, x_m)$ represents an $(m \times 1)$ vector of stressor levels, and $\boldsymbol{\beta}' = (\beta_1, \beta_2, \ldots, \beta_m)$ a vector of regression coefficients. Since Cox introduces parameters in one *part* only of the nonparametric model, his model is sometimes denoted *semiparametric*.

Note that the x_i's may be transformed values of the real stress levels. Hence Cox models are very general and may be applicable to a large number of practical situations. With this assumption (10.32) may be rewritten as

$$R(t; x, \boldsymbol{\beta}) = R(t, s^{(0)})^{\exp(\boldsymbol{\beta}'x)} \quad (10.35)$$

Suppose that n randomly chosen units have been exposed to the m stressors and that the corresponding life lengths of the units are denoted $T_1, T_2, \ldots, T_n$. Then associated with each life time T_i is a regression vector $x_i = (x_{1i}, x_{2i}, \ldots, x_{mi})$ of data values of the m stressors, $i = 1, 2, \ldots, n$. The question is then how to estimate $F_0(t)$ and the regression coefficients.

An application of the maximum likelihood principle turns out to be complicated. Cox suggests the following approach:

First, he defines on a heuristic basis what he calls a "partial likelihood," which is not a proper likelihood function and which in a P–H model does not involve the failure rate. The partial likelihood is then used for estimating the regression coefficients. He then reads the estimated values $\hat{\beta}_1, \ldots, \hat{\beta}_m$ as true values of the regression coefficients and concludes by estimating $F_0(t)$ from the recorded life data, using Kaplan-Meiers approach (see Section 9.3). Application of this procedure is rather complicated. Readers are referred to Kalbfleisch and Prentice (1980, pp. 76–78) or Lawless (1982, pp. 345–347).

Estimates for the β's in the Cox model may be calculated by using various computer codes, for example, SYSTAT Survival and the BMDP Statistical package. These codes allow for two important generalizations of the Cox model (10.35):

1. The assumption of a simple underlying failure rate for all the life data is relaxed. The code allows for several subgroups within the recorded data.

2. The stressors x are allowed to vary with time.

10.5 PROBLEMS

1. (The sole purpose of considering the following rather unrealistic situation is to illustrate how accelerated life data may be analyzed in a very simple situation.) The device in question is supposed to have an exponential life distribution with failure rate $\lambda(s) = c \cdot s$ when exposed to the stress s, for all stresses. c is an unknown positive constant, and the purpose of the study is to estimate $\lambda(s^{(0)}) = c \cdot s^{(0)}$, where $s^{(0)}$ denotes the normal use stress. The expected life length is expected to be very long under stress s_0. Therefore an accelerated life test is carried out according to design I, where the selected stresses $s^{(1)} < s^{(2)} < \cdots < s^{(k)}$ are very much larger than $s^{(0)}$, to obtain short life lengths under these stresses.

    The experiment leads to the result in Table 10.2. T_{ij}, for $i = 1, 2, \ldots, k$, $j = 1, 2, \ldots, n_i$ are all assumed to be independent.

    (a) Show that $2cs^{(i)} \sum_{j=1}^{n_i} T_{ij}$, $i = 1, 2, \ldots, k$ are independent and χ^2 distributed with $2n_i$ degrees of freedom.

    (b) Use the result in (a) to derive an estimator $\hat{c}$ of c.

    (c) Finally, estimate the expected life length of the device in question under normal stress $s^{(0)}$.

    Table 10.2 Observed Lifetimes at the Various Stress Levels

Stress Level	Observed Lifetimes
$s^{(1)}$	$T_{11}, T_{12}, \ldots, T_{1n_1}$
$s^{(2)}$	$T_{21}, T_{22}, \ldots, T_{2n_2}$
$\vdots$	$\vdots$
$s^{(k)}$	$T_{k1}, T_{k2}, \ldots, T_{kn_k}$

CHAPTER 11

Bayesian Reliability Analysis

11.1 INTRODUCTION

The first step in almost any statistical analysis is to establish a stochastic model of the situation at hand. The observations to be collected are then considered to be realizations of random variables $X_1, X_2, \ldots, X_n$. So far in this book we have assumed it to be possible to derive the joint distribution function, $F(x_1, x_2, \ldots, x_n; \theta_1, \theta_2, \ldots, \theta_r)$ of the X_i's through basic scientific knowledge of the phenomenon to be analyzed, information obtained from exploratory data, and possibly some simplifying assumptions. Here $\boldsymbol{\theta} = (\theta_1, \theta_2, \ldots, \theta_r)$ denotes a vector of constants belonging to some subspace Ω of the r-dimensional euclidean space. In this model no vector $\boldsymbol{\theta}$ in Ω is more likely to occur than any other.

A natural question to ask is whether or not this approach always is the most appropriate one when one wants to express a priori knowledge of the phenomenon. When following this line of action, essential parts of a priori knowledge may not be taken into account. Suppose, for example, that p denotes the reliability of a certain component at time t. p will then be assumed to belong to [0, 1], but no values of p in this interval is given preference, even if one is quite certain that p is close to 1, say. This a priori knowledge easily get lost in the model.

In Bayesian inference one can introduce this kind of knowledge into the model by interpreting p as a random variable with some density $f(p)$, expressing what one thinks (believes) about the occurring value of p (see Figure 11.1). In Section 11.7 we will discuss possible interpretations of such distributions. For the time being we will only study the immediate consequences of such models.

The purpose of this chapter is mainly to illustrate the Bayesian philosophy and some of its consequences. Readers who want a comprehensive presentation of Bayesian inference are referred to DeGroot (1970), Martz and Waller (1982), or Berger (1985). Here we will restrict ourselves to considering Bayesian estimation in some simple situations.

A central tool required for the application of Bayesian methods is Bayes's theorem.

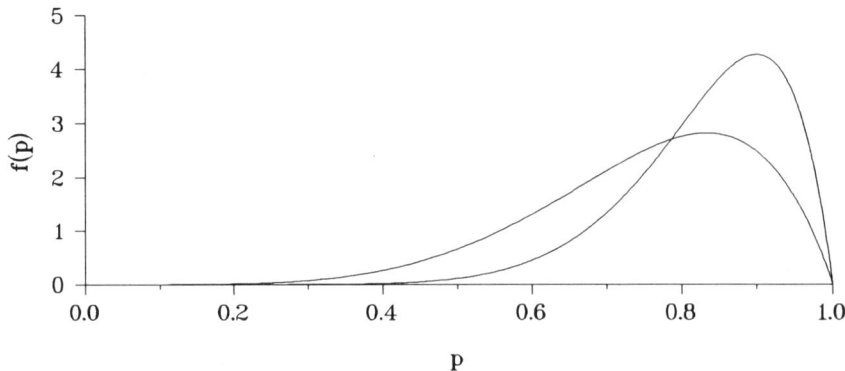

Figure 11.1 Possible densities of p

Theorem 11.1. (Bayes' Theorem) Let $B_1, B_2, \ldots$ be mutually exclusive and exhaustive events contained in a sample space S:

$$P\left(\bigcup_{i=1}^{\infty} B_i\right) = 1$$
$$B_i \cap B_j = \emptyset \quad \text{for } i \neq j$$
$$P(B_i) > 0 \quad \text{for each } i$$

and let A be an event such that $P(A) > 0$. Then for each k,

$$P(B_k|A) = \frac{P(A|B_k)P(B_k)}{\sum_{i=1}^{\infty} P(A|B_i)P(B_i)} \tag{11.1}$$

Bayes's theorem is named after the Reverend Thomas Bayes, who used the theorem in the fundamental paper, *An Essay toward Solving a Problem in the Doctrine of Chances*, which was published in 1763. A proof of Bayes's theorem is given by Dudewicz and Mishra (1988, p. 41).

11.2 BASIC CONCEPTS

In our non-Bayesian setup, let X be a random variable with probability density $f(x,\theta)$, $\theta \in \Omega$. According to the Bayesian point of view, θ is interpreted as a realization of a random variable Θ in Ω with some density $f_\Theta(\theta)$. $f_\Theta(\theta)$ expresses what one thinks (believes) about the occurring value of Θ, *before* any observation has been taken, that is, *a priori*. $f_\Theta(\theta)$ is called the *prior density*

BASIC CONCEPTS

of Θ. $f(x, \theta)$ is then read as the conditional density of X, given $\Theta = \theta$, and rewritten as $f_{X|\Theta}(x|\theta)$.

With this interpretation, the joint density of X and Θ, $f_{X,\Theta}(x, \theta)$ is given by

$$f_{X,\Theta}(x, \theta) = f_{X|\Theta}(x|\theta) \cdot f_\Theta(\theta) \tag{11.2}$$

Proceeding on this basis, the marginal density of X, $f_X(x)$ is

$$\begin{aligned} f_X(x) &= \int_\Omega f_{X,\Theta}(x, \theta) \, d\theta \\ &= \int_\Omega f_{X|\Theta}(x|\theta) f_\Theta(\theta) \, d\theta \end{aligned} \tag{11.3}$$

The conditional density of Θ, given $X = x$, becomes

$$f_{\Theta|X}(\theta|x) = \frac{f_{X,\Theta}(x, \theta)}{f_X(x)} \tag{11.4}$$

or

$$f_{\Theta|X}(\theta|x) = \frac{f_{X|\Theta}(x|\theta) f_\Theta(\theta)}{f_X(x)} \tag{11.5}$$

which is seen to be a simple form of Bayes's theorem (11.1).

By $f_{\Theta|X}(\theta|x)$ we express our *belief* concerning the distribution of Θ *after* having observed $X = x$, that is, *a posteriori*, and this is called the *posterior density* of Θ. Note that when X is observed, $f_X(x)$ occurs in (11.5) as a constant. Hence $f_{\Theta|X}(\theta|x)$ is always *proportional* to $f_{X|\Theta}(x|\theta) \cdot f_\Theta(\theta)$, which we write as:

$$f_{\Theta|X}(\theta|x) \propto f_{X|\Theta}(x|\theta) f_\Theta(\theta). \tag{11.6}$$

The Bayesian approach may be characterized as an "updating" process.

First, a probability density for Θ is assigned before any observations of X is taken. Then, as soon as the first X is observed and becomes available, the prior distribution of Θ is updated to the posteriori distribution of Θ, given $X = x$. The observed value of X has therefore changed our belief regarding the value of Θ. This process may be repeated. In the next step our posterior distribution of Θ, given $X = x$, is chosen as the new prior distribution, another X is being observed, and one is lead to a second posterior distribution, and so on.

Example 11.1

A nonrepairable valve is assumed to have constant failure rate λ. Experience (or belief) leads us to think that the failure rate is a random variable Λ which is gamma distributed with parameters $\alpha_1 = 2$, $\beta_1 = 1$ (see Figure 11.2):

$$f_\Lambda(\lambda) = \lambda e^{-\lambda} \qquad \text{for } \lambda > 0 \qquad (11.7)$$

Given $\Lambda = \lambda$, the lifetime T_1 of the valve is known to be exponentially distributed with density

$$f_{T_1|\Lambda}(t_1|\lambda) = \lambda e^{-\lambda t_1} \qquad \text{for } t_1 > 0, \lambda > 0 \qquad (11.8)$$

The joint density of T_1 and Λ becomes

$$\begin{aligned} f_{T_1,\Lambda}(t_1,\lambda) &= \lambda e^{-\lambda t_1} \cdot \lambda e^{-\lambda} \\ &= \lambda^2 e^{-\lambda(t_1+1)} \qquad \text{for } t_1 > 0, \lambda > 0 \end{aligned} \qquad (11.9)$$

The marginal density of T_1 is

$$\begin{aligned} f_{T_1}(t_1) &= \int_0^\infty \lambda^2 e^{-\lambda(t_1+1)} d\lambda \\ &= \frac{\Gamma(3)}{(t_1+1)^3} = \frac{2}{(t_1+1)^3} \qquad \text{for } t > 0 \end{aligned} \qquad (11.10)$$

The conditional density of Λ, given $T_1 = t_1$, the posterior density, is

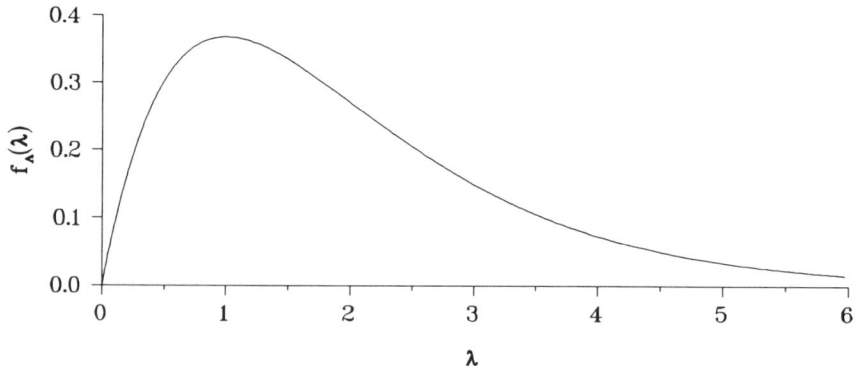

Figure 11.2 The gamma density with parameters (2, 1)

BASIC CONCEPTS 439

$$f_{\Lambda|T_1}(\lambda|t_1) = \frac{\lambda^2 e^{-\lambda(t_1+1)}}{2} (t_1 + 1)^3 \qquad (11.11)$$

$$= \frac{(t_1 + 1)^3}{\Gamma(3)} \lambda^{3-1} e^{-\lambda(t_1+1)} \qquad \text{for } \lambda > 0$$

which also represents a gamma density, now with parameters α_2, β_2 where

$$\alpha_2 = 3 = \alpha_1 + 1 \quad \text{since} \quad \alpha_1 = 2$$
$$\beta_2 = (t_1 + 1) = \beta_1 + t_1 \quad \text{since} \quad \beta_1 = 1$$

The procedure may now be repeated with (11.11) as our new prior distribution. Then we observe the lifetime T_2 of a similar valve and are lead to a new posterior distribution which is a gamma distribution with parameters

$$\alpha_3 = \alpha_2 + 1 = \alpha_1 + 2$$
$$\beta_3 = \beta_2 + t_2 = \beta_1 + (t_1 + t_2)$$

and so on.

The posterior density (11.11) could also have been derived directly using (11.6):

$$f_{\Lambda|T_1}(\lambda|t_1) \propto f_{T_1|\Lambda}(t_1|\lambda) f_\Lambda(\lambda) \qquad (11.12)$$
$$\propto \lambda e^{-\lambda t_1} \cdot \lambda e^{-\lambda}$$
$$\propto \lambda^2 e^{-\lambda(t_1+1)}$$

Hence

$$f_{\Lambda|T_1}(\lambda|t_1) = k(t_1) \cdot \lambda^2 e^{-\lambda(t_1+1)} \qquad \text{for } t_1 > 0 \qquad (11.13)$$

and since (11.13) is a density, $k(t_1)$ is easily determined to be $(1 + t_1)^3/2$. This leads to the same posterior density as we derived in (11.11).

Using the fact that the mean of a gamma distribution with parameters α and

β is α/β, we can furthermore conclude that

$$\left.\begin{aligned} E(\Lambda) &= \frac{2}{1} \\ E(\Lambda|T_1 = t_1) &= \frac{2+1}{1+t_1} \\ E(\Lambda|T_1 = t_1, T_2 = t_2) &= \frac{2+1+1}{1+t_1+t_2} \\ &\vdots \end{aligned}\right\} \quad (11.14)$$

Equation (11.14) illustrates how we update our belief about the mean of Λ, as observations of T become available. $\square$

11.3 BAYESIAN POINT ESTIMATION

Let us return to the general setup in Section 11.2. Recall that θ is a realization of a random variable $\Theta \in \Omega$ with some prior density $f_\Theta(\theta)$ and that X is a random variable with continuous density given $\Theta = \theta$, $f_{X|\Theta}(x|\theta)$. Our task is now to estimate the value θ of Θ that belongs to an observed value x of X. We shall denote this estimator by $\hat{\theta}(X)$.

As is usual, we prefer an estimator that minimizes the mean quadratic loss:

$$E[(\hat{\theta}(X) - \Theta)^2]$$

Such an estimator is denoted a *Bayesian estimator* (of θ) (with minimum expected quadratic loss). Note that in the Bayesian framework, X and Θ are both random variables. How should $\hat{\theta}(X)$ be chosen?

$$E[(\hat{\theta}(X) - \Theta)^2] = \int_{-\infty}^{+\infty} \int_\Omega (\hat{\theta}(X) - \theta)^2 f_{X,\Theta}(x,\theta) \, dx \, d\theta$$

By using (11.2), we get

$$E[(\hat{\theta}(X) - \Theta)^2] = \int_{-\infty}^{+\infty} f_X(x) \left[\int_\Omega (\theta - \hat{\theta}(X))^2 f_{\Theta|X}(\theta|x) \, d\theta \right] dx$$

Obviously $E[(\hat{\theta}(X) - \Theta)^2]$ becomes minimized if, for each x, $\hat{\theta}(x)$ is chosen

BAYESIAN POINT ESTIMATION

to minimize

$$\int_\Omega (\theta - \hat{\theta}(x))^2 f_{\Theta|X}(\theta|x)\, d\theta$$

In probability theory the following lemma is well-known:

Lemma 11.1
Let Y be a random variable with density $f_Y(y)$ and finite variance τ^2. Then

$$\psi(\eta) = \int_{-\infty}^{+\infty} (y - \eta)^2 f_Y(y)\, dy$$

is minimized when η it is chosen as $E(Y)$.

This lemma, applied to our problem, tells us that $E[(\hat{\theta}(X) - \Theta)^2]$ is minimized for

$$\hat{\theta}(X) = E(\Theta|X).$$

Hence we can conclude that *the Bayesian estimator of θ is the mean of the posterior distribution of Θ*.

Example 11.1. (cont.)
If we apply Lemma 11.1 to the iteration in Example 11.1, the Bayesian estimate of Λ, after having observed $T_1 = t_1$, is

$$\hat{\lambda}(t_1) = \frac{3}{1 + t_1} \qquad \square$$

ooo OOO ooo

Let us return to our Bayesian model where θ represents a realization of a random variable $\Theta \in \Omega$ with some prior density $f_\Theta(\theta)$. We are now considering a situation where our data $D = (x_1, \ldots, x_n)$ consist of observations of n random variables $X_1, \ldots, X_n$, assumed to be independent and identically distributed, *conditional on θ*, with density $f_{X|\Theta}(x|\theta)$. Then

$$f_{X_1,\ldots,X_n|\Theta}(x_1,\ldots,x_n|\theta) = \prod_{j=1}^n f_{X|\Theta}(x_j|\theta) \qquad (11.15)$$

The posterior distribution of Θ, given $X_1, \ldots, X_n$, may be obtained by the same procedure as we used for one single X, and we get

$$f_{\Theta|X_1,\ldots,X_n}(\theta|x_1,\ldots,x_n) \propto \left[\prod_{j=1}^{n} f_{X|\Theta}(x_j|\theta)\right] \cdot f_\Theta(\theta) \qquad (11.16)$$

Considering the right-hand side of (11.16) as a function of θ, given $x_1, \ldots, x_n$, this can also be written

$$f_{\Theta|X_1,\ldots,X_n}(\theta|x_1,\ldots,x_n) \propto \ell(\theta|x_1,\ldots,x_n) f_\Theta(\theta) \qquad (11.17)$$

where $\ell(\theta|x_1,\ldots,x_n)$ denotes the likelihood function in the usual meaning. For brevity in the following discussion we will write $\ell(\theta|D)$ instead of $\ell(\theta|x_1,\ldots,x_n)$.

Example 11.1 (cont.)
Reconsider the valve in Example 11.1, with constant (unknown) failure rate λ, where λ represents a realization of a random variable Λ with (prior) density

$$f_\Lambda(\lambda) = \lambda e^{-\lambda} \quad \text{for} \quad \lambda > 0$$

Let $T_1, \ldots, T_n$ denote the lifetimes of n such valves. Assume $T_1, \ldots, T_n$ to be independent and identically distributed, conditional on λ, with density

$$f_{T_i|\Lambda}(t_i|\lambda) = \lambda e^{-\lambda t_i} \quad \text{for} \quad \lambda > 0, i = 1, \ldots, n$$

We want to determine the Bayesian estimate of λ based on these lifetimes.
In this case

$$f_{T_1,\ldots,T_n|\Lambda}(t_1,\ldots,t_n|\lambda) = \prod_{j=1}^{n}(\lambda e^{-\lambda t_i}) = \lambda^n e^{-\lambda \sum_{i=1}^{n} t_i}$$

and

$$f_{\Lambda|T_1,\ldots,T_n}(\lambda|t_1,\ldots,t_n) \propto \lambda^n e^{-\lambda \sum_{i=1}^{n} t_i} \cdot \lambda e^{-\lambda} \qquad (11.18)$$
$$\propto \lambda^{n+1} e^{-\lambda(1+\sum_{i=1}^{n} t_i)}$$

To be a proper density, the right-hand side of (11.18) must be multiplied by the

CHOICE OF PRIOR DISTRIBUTION

constant

$$\frac{(1 + \sum_{i=1}^{n} t_i)^{n+2}}{\Gamma(n+2)}$$

Hence

$$f_{\Lambda|T_1,\ldots,T_n}(\lambda|t_1,\ldots,t_n) = \frac{(1 + \sum_{i=1}^{n} t_i)^{n+2}}{\Gamma(n+2)} \lambda^{n+1} \cdot e^{-\lambda(1+\sum_{i=1}^{n} t_i)} \quad \text{for } \lambda > 0$$

which we recognize as a gamma distribution with parameters $\alpha = (n+2)$ and $\beta = (1 + \sum_{i=1}^{n} t_i)$. Since the mean of this gamma distribution is α/β, the Bayesian estimator of λ is

$$\hat{\lambda}(T_1,\ldots,T_n) = \frac{2+n}{1 + \sum_{i=1}^{n} T_i} \quad (11.19)$$

$\square$

11.4 CREDIBILITY INTERVALS

A credibility interval is the Bayesian analogue to a confidence interval. A credibility interval for Θ, at level $1 - \varepsilon$, is an interval $(a(D), b(D))$ such that given the data D

$$P(a(D) < \Theta < b(D)|D) = \int_{a(D)}^{b(D)} f_{\Theta|D}(\theta|D)\, d\theta = 1 - \varepsilon \quad (11.20)$$

Hence the interval $(a(D), b(D))$ is an interval estimate of θ in the sense that the conditional probability of Θ belonging to the interval, given the data, is equal to $1 - \varepsilon$.

11.5 CHOICE OF PRIOR DISTRIBUTION

Conjugate Families of Distributions

First we state a useful definition:

Definition 11.1. A parametric family $\mathscr{P}$ of distributions $f_\Theta(\theta)$ is said to be closed in sampling with respect to a family $\mathscr{F}$ of distributions $f_{X|\Theta}(x|\theta)$ if

$$f_\Theta(\theta) \in \mathscr{P} \Rightarrow f_{\Theta|X}(\theta|x) \in \mathscr{P} \quad (11.21)$$

In that case $\mathscr{P}$ is also said to be a conjugate family to $\mathscr{F}$ or for short, conjugate to $\mathscr{F}$. □

Example 11.2
Let $\mathscr{F}$ be defined by the probability density

$$f_{T|\Lambda}(t|\lambda) = \lambda e^{-\lambda t} \quad \text{for } t > 0 \tag{11.22}$$

Let us show that the class $\mathscr{P}$ of distributions, defined by

$$f_\Lambda(\lambda) = \frac{\beta^\alpha}{\Gamma(\alpha)} \lambda^{\alpha-1} e^{-\lambda\beta} \quad \text{for } \lambda > 0, \alpha > 0, \beta > 0 \tag{11.23}$$

is conjugate to $\mathscr{F}$.

That is, we have to show that the corresponding $f_{\Lambda|T}(\lambda|t)$ is a gamma distribution. In this case

$$\begin{aligned} f_{T,\Lambda}(t,\lambda) &= \lambda e^{-\lambda t} \cdot \frac{\beta^\alpha}{\Gamma(\alpha)} \lambda^{\alpha-1} e^{-\lambda\beta} \\ &= \frac{\beta^\alpha}{\Gamma(\alpha)} \lambda^\alpha e^{-\lambda(\beta+t)} \quad \text{for } t > 0, \lambda > 0 \end{aligned} \tag{11.24}$$

Furthermore

$$\begin{aligned} f_T(t) &= \int_0^\infty \frac{\beta^\alpha}{\Gamma(\alpha)} \lambda^\alpha e^{-\lambda(\beta+t)} d\lambda \\ &= \frac{\alpha \cdot \beta^\alpha}{(\beta+t)^{\alpha+1}} \quad \text{for } t > 0 \end{aligned} \tag{11.25}$$

Hence

$$\begin{aligned} f_{\Lambda|T}(\lambda|t) &= \frac{\beta^\alpha(\lambda^\alpha e^{-\lambda(\beta+t)})/\Gamma(\alpha)}{\alpha \cdot \beta^\alpha/(\beta+t)^{\alpha+1}} \\ &= \frac{(\beta+t)^{\alpha+1}}{\Gamma(\alpha+1)} \lambda^\alpha e^{-\lambda(\beta+t)} \quad \text{for } t > 0 \end{aligned} \tag{11.26}$$

which is a gamma density with parameters $\alpha + 1$, and $\beta + t$. □

The result in Example 11.2 may be stated more formally as:

Theorem 11.2. The family of gamma distributions (α, β) is conjugate to the family of exponential distributions.

CHOICE OF PRIOR DISTRIBUTION

Clearly the assumption of a gamma density as a prior distribution in connection with an exponential distribution is mathematically convenient. Our intension with this prior distribution, however, is to express a priori knowledge of λ, and we raise the question whether or not this purpose is taken care of by using the gamma density as prior. The answer to this question is that the gamma distribution (α, β) is a very flexible distribution. It may take on a wide variety of shapes through varying the parameters α and β. Almost any conceivable shape of the density of Λ, can essentially be obtained by proper choice of α and β.

Example 11.3

Consider a plant which has a specified number of identical and independent valves with constant failure rate λ, where λ represents a realization of a random variable Λ with the Gamma prior density

$$f_\Lambda(\lambda) = \frac{\beta^\alpha}{\Gamma(\alpha)} \lambda^{\alpha-1} e^{-\beta\lambda} \quad \text{for } \lambda > 0, \alpha > 0, \beta > 0 \quad (11.27)$$

The parameters α and β of the prior distribution is usually "estimated" based on prior experience with the same type of valves, combined with information gained from various reliability data sources (see Chapter 12).

When a valve fails, it will be replaced with a valve of the same type. The associated downtime is considered to be negligible. Valve failures are assumed to occur according to a homogeneous Poisson process with intensity λ. The number of valve failures $N(t)$ during an accumulated time t in service thus has the Poisson distribution

$$P(N(t) = n | \Lambda = \lambda) = \frac{(\lambda t)^n}{n!} e^{-\lambda t} \quad \text{for } n = 0, 1, \ldots \quad (11.28)$$

The marginal distribution of $N(t)$ then becomes

$$\begin{aligned}
P(N(t) = n) &= \int_0^\infty P(N(t) = n | \Lambda = \lambda) \cdot f_\Lambda(\lambda) \, d\lambda \\
&= \int_0^\infty \frac{(\lambda t)^n}{n!} e^{-\lambda t} \cdot \frac{\beta^\alpha}{\Gamma(\alpha)} \lambda^{\alpha-1} e^{-\beta\lambda} \, d\lambda \\
&= \frac{\beta^\alpha t^n}{\Gamma(\alpha) n!} \int_0^\infty \lambda^{\alpha+n-1} e^{-(\beta+t)\lambda} \, d\lambda \\
&= \frac{\beta^\alpha t^n}{\Gamma(\alpha) n!} \frac{\Gamma(n+\alpha)}{(\beta+t)^{n+\alpha}}
\end{aligned} \quad (11.29)$$

By combining (11.27), (11.28), and (11.29), we get the posterior density of Λ, given $N(t) = n$,

$$f_{\Lambda|N(t)}(\lambda|n) = \frac{(\beta+t)^{\alpha+n}}{\Gamma(\alpha+n)} \lambda^{\alpha+n-1} e^{-(\beta+t)\lambda} \tag{11.30}$$

which is recognized as the gamma distribution with parameters $(\alpha + n)$ and $(\beta + t)$. □

Hence we have shown the following theorem:

Theorem 11.3. The family of gamma distributions (α, β) is conjugate to the family of Poisson distributions.

Example 11.3 (cont.)
The Bayesian estimate of λ when $N(t) = n$ is thus

$$\hat{\lambda} = E(\Lambda|N(t) = n) = \frac{\alpha+n}{\beta+t} \tag{11.31}$$

Furthermore the conditional distribution (given $N(t) = n$) of the variable $Z = 2(\beta + t)\Lambda$ is

$$f_{Z|N(t)}(z|n) = \frac{1}{2^{\alpha+n}\Gamma(\alpha+n)} z^{\alpha+n-1} e^{-z/2} \quad \text{for} \quad z > 0$$

which is recognized as the chi-square distribution with $2(\alpha+n)$ degrees of freedom (e.g., see Dudewicz and Mishra, 1988, p. 140).

A $1 - \varepsilon$ credibility interval for the failure rate is thus easily obtained as

$$P\left(\frac{z_{1-\varepsilon/2,2(\alpha+n)}}{2(\beta+t)} < \Lambda < \frac{z_{\varepsilon/2,2(\alpha+n)}}{2(\beta+t)} \,\bigg|\, N(t) = n\right) = 1 - \varepsilon \tag{11.32}$$

where $z_{\varepsilon,\nu}$ denotes the upper $100\varepsilon\%$ percentile of the chi-square distribution (χ_ν^2) with ν degrees of freedom; that is, $P(Z > z_{\varepsilon,\nu}) = \varepsilon$ when $Z \sim \chi_\nu^2$. A one-sided upper credibility interval for the failure rate is obtained in the same way:

$$P\left(\Lambda < \frac{z_{\varepsilon,2(\alpha+n)}}{2(\beta+t)} \,\bigg|\, N(t) = n\right) = 1 - \varepsilon \tag{11.33}$$

Assume that we have estimated the parameters α and β of the prior gamma distribution to be (all time units in hours)

CHOICE OF PRIOR DISTRIBUTION

$$\alpha = 3$$
$$\beta = 1 \cdot 10^4$$

The prior mean and standard deviation is thus

$$E(\Lambda) = 3 \cdot 10^{-4} \quad \text{(prior mean)}$$
$$SD(\Lambda) \approx 1.73 \cdot 10^{-4} \quad \text{(prior standard deviation)}$$

Further assume that $n = 2$ failures have been observed during a total time in service $t = 5 \cdot 10^3$ hours.

The Bayesian estimate of λ is, from (11.31),

$$\hat{\lambda} = \frac{\alpha + n}{\beta + t} = \frac{3 + 2}{10^4 + 5 \cdot 10^3} \approx 3.33 \cdot 10^{-4}$$

A 90% credibility interval for Λ with these data is

$$P\left(\frac{z_{0.95,10}}{2(\beta + t)} < \Lambda < \frac{z_{0.05,10)}}{2(\beta + t)} \,\bigg|\, N(t) = 2\right) = 0.90$$

$$P\left(\frac{3.94}{2 \cdot 1.5 \cdot 10^4} < \Lambda < \frac{18.31}{2 \cdot 1.5 \cdot 10^4} \,\bigg|\, N(t) = 2\right) = 0.90$$

$$P(1.31 \cdot 10^{-4} < \Lambda < 6.10 \cdot 10^{-4} \,|\, N(t) = 2) = 0.90$$

The prior and posterior distribution of Λ for these data are presented in Figure 11.3, with a 90% credibility interval for Λ. □

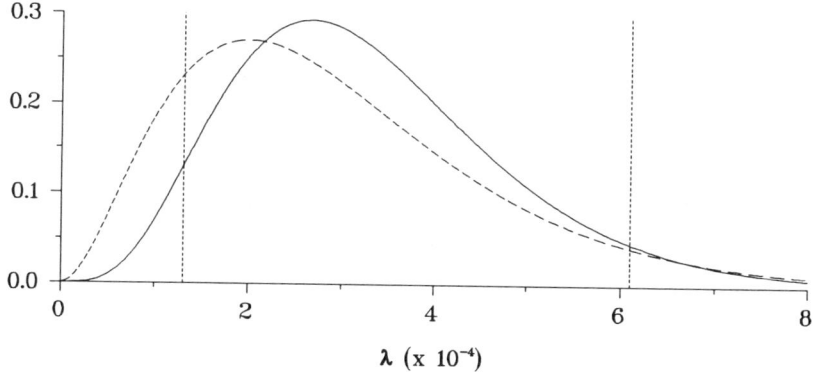

Figure 11.3 Prior (*dotted line*) and posterior density of the failure rate in Example 11.3 and a 90% credibility interval for the failure rate

Example 11.4

Let $\mathscr{F}$ be the binomial distribution defined by

$$f_{X|\Theta}(x|\theta) = \binom{n}{x} \theta^x (1-\theta)^{n-x}, \qquad x = 0, 1, \ldots, n,\ 0 \le \theta \le 1 \quad (11.34)$$

Let us show that the class $\mathscr{P}$ of beta distributions, defined by the density

$$f_\Theta(\theta) = \frac{\Gamma(r+s)}{\Gamma(r) \cdot \Gamma(s)} \theta^{r-1}(1-\theta)^{s-1}, \qquad 0 \le \theta \le 1, r > 0, s > 0 \quad (11.35)$$

is a conjugate family to $\mathscr{F}$.

To prove this statement, we have to show that the corresponding density $f_{\Theta|X}(\theta|x)$ represents a beta distribution. To do so, we make use of (11.6):

$$f_{\Theta|X}(\theta|x) \propto \binom{n}{x} \theta^x (1-\theta)^{n-x}$$
$$\cdot \frac{\Gamma(r+s)}{\Gamma(r)+\Gamma(s)} \theta^{r-1}(1-\theta)^{s-1}, \qquad 0 < \theta < 1 \quad (11.36)$$

Put differently,

$$f_{\Theta|X}(\theta|x) \propto \theta^{x+r-1}(1-\theta)^{n-x+s-1}$$

Hence

$$f_{\Theta|X}(\theta|x) = k(x) \cdot \theta^{(x+r)-1}(1-\theta)^{(n-x+s)-1}, \qquad 0 < \theta < 1 \quad (11.37)$$

Knowing (11.37) to be a probability density, the "constant" $k(x)$ has to be

$$k(x) = \frac{\Gamma(n+r+s)}{\Gamma(x+r)\Gamma(n-x+s)} \quad (11.38)$$

Equation (11.38) introduced into (11.37) gives the beta density with parameters $x+r$ and $n-x+s$. □

We have thus proved the following theorem:

Theorem 11.4. The family of beta distributions (r, s) is conjugate to the family of binomial distributions.

The assumption of a beta prior in connection with a binomial distribution is

CHOICE OF PRIOR DISTRIBUTION

mathematically convenient. Furthermore the beta distribution is a very flexible distribution. Its density can take on a wide variety of shapes by proper choice of r and s. Note that by choosing $r = 1$, $s = 1$, the beta density represents a uniform density over $[0, 1]$ which corresponds to no a priori preference for any θ in $[0, 1]$. (In this case we have a *noninformative* prior distribution for θ.)

The mean of the beta distribution (r, s) is easily found to be $r/(r + s)$ (e.g., see Dudewicz and Mishra 1988, p. 224). The prior mean is, from (11.35),

$$E(\Theta) = \frac{r}{r+s}$$

and the Bayesian estimate of the probability θ is, from (11.37),

$$\hat{\theta}(x) = \frac{r+x}{r+s+n} \quad (11.39)$$

Noninformative Prior Distribution

Example 11.5

A certain type of valves is assumed to have constant (unknown) failure rate λ, where λ represents a realization of a random variable Λ. Let $T_1, \ldots, T_n$ denote the observed lifetimes of n such valves. Assume $T_1, \ldots, T_n$ to be independent and identically distributed, conditional on λ, with density

$$f_{T_i|\Lambda}(t_i|\lambda) = \lambda e^{-\lambda t_i} \quad \text{for} \quad t_i > 0, \lambda > 0$$

When there is no prior information available about the true value of Λ, use of a noninformative prior distribution may be appropriate.

A noninformative prior distribution gives no preference to any of the possible parameter values. Hence, if the possible parameter values constitute a finite interval, the parameter is assumed to be uniformly distributed over that interval. Hence a noninformative prior distribution for the failure rate Λ may, for example, be given by the uniform density

$$f_\Lambda(\lambda) = \begin{cases} \dfrac{1}{M} & \text{for } 0 \le \lambda \le M \\ 0 & \text{otherwise} \end{cases} \quad (11.40)$$

where M is taken to be a very large number, say, 10^{10}. Then the posterior density of Λ, given the data D is, by (11.16), approximately

$$f_{\Lambda|T_1,\ldots,T_n}(\lambda|t_1,\ldots,t_n) \propto \ell(\lambda|D) \cdot f_\Lambda(\lambda)$$

$$f_{\Lambda|T_1,\ldots,T_n}(\lambda|t_1,\ldots,t_n) \propto \lambda^n e^{-\lambda \sum_{i=1}^n t_i} \cdot \frac{1}{M}$$

$$f_{\Lambda|T_1,\ldots,T_n}(\lambda|t_1,\ldots,t_n) \propto \lambda^n e^{-\lambda \sum_{i=1}^n t_i} \quad (11.41)$$

To become a proper density, the right-hand side of (11.41) must be multiplied by $(\sum_{i=1}^n t_i)^{n+1}/\Gamma(n+1)$. Hence

$$f_{\Lambda|T_1,\ldots,T_n}(\lambda|t_1,\ldots,t_n) = \frac{(\sum_{i=1}^n t_i)^{n+1}}{\Gamma(n+1)} \cdot \lambda^n e^{-\lambda \sum_{i=1}^n t_i} \quad \text{for } \lambda > 0 \quad (11.42)$$

which we recognize as the density of a gamma distribution with parameters $n+1$ and $\sum_{i=1}^n t_i$.

Hence the Bayesian estimator of the failure rate (with minimum expected quadratic loss) and a noninformative prior becomes

$$\hat{\lambda}(T_1,\ldots,T_n) = \frac{n+1}{\sum_{i=1}^n T_i} \quad (11.43)$$

The MLE of the failure rate λ in this situation is earlier determined to be

$$\lambda^*(T_1,\ldots,T_n) = \frac{n}{\sum_{i=1}^n T_i}$$

Note that the MLE of λ, $\lambda^*(T_1,\ldots,T_n)$, coincides with the mode of the posterior distribution (11.42). □

11.6 BAYESIAN LIFE TEST SAMPLING PLANS

Introduction

The density

$$f_\Theta(\theta) = \frac{b^a}{\Gamma(a)} \cdot \left(\frac{1}{\theta}\right)^{a+1} e^{-b/\theta} \quad \text{for } \theta > 0, a > 0, b > 0 \quad (11.44)$$

is called the *inverted gamma density* with parameters a and b, since the variable Θ^{-1} then has an ordinary gamma density with parameters a and $1/b$. Since $\text{var}(\Theta)$ does not exist when $a \leq 2$, we will assume that $a > 2$. Then

$$E(\Theta) = \frac{b}{a-1} = \theta_0 \quad \text{(prior mean)}$$

$$\text{var}(\Theta) = \frac{b^2}{(a-1)^2(a-2)} = \sigma_0^2 \quad \text{(prior variance)} \quad (11.45)$$

Since (11.44) can be shown to be conjugate to the family of exponential distributions, defined by the density

$$f_{T|\Theta}(t|\theta) = \frac{1}{\theta} e^{-t/\theta} \quad \text{for } t > 0, \theta > 0 \quad (11.46)$$

Equation (11.46) is often used as a prior distribution when estimating θ, which represents the MTTF of (11.46). The inverted gamma density is rather flexible and can take on a wide variety of shapes by proper choice of a and b.

Complete Data Sets

Suppose that n units are life tested until failure, and let $T_1, T_2, \ldots, T_n$ denote the observed lifetimes. Hence $D = (t_1, t_2, \ldots, t_n)$. Given that $\Theta = \theta$, $T_1, T_2, \ldots, T_n$ are assumed to be independent and identically distributed with density (11.46). We select (11.44) as prior distribution of Θ. Then, according to (11.16),

$$f_{\Theta|T_1,\ldots,T_n}(\theta|(t_1,\ldots,t_n)) \propto \left(\frac{1}{\theta}\right)^n e^{-(\sum_{i=1}^n t_i)/\theta} \frac{b^a}{\Gamma(a)} \left(\frac{1}{\theta}\right)^{a+1} e^{-b/\theta}$$

$$\text{for } \theta > 0$$

which implies that

$$f_{\Theta|T_1,\ldots,T_n}(\theta|t_1,\ldots,t_n) \propto \left(\frac{1}{\theta}\right)^{n+a+1} \cdot e^{-(b+\sum_{i=1}^n t_i)/\theta} \quad \text{for } \theta > 0$$

$$(11.47)$$

In this case $\sum_{i=1}^n t_i$ expresses the total time on test at the last failure

$$\sum_{i=1}^n t_i = \mathscr{T}(t_n).$$

Hence (11.47) may be written

$$f_{\Theta|T_1,\ldots,T_n}(\theta|(t_1,\ldots,t_n)) \propto \left(\frac{1}{\theta}\right)^{n+a+1} \cdot e^{-(b+\mathcal{T}(t_n))/\theta} \quad \text{for} \quad \theta > 0$$

(11.48)

To become a proper density, the right-hand side of (11.48) has to be multiplied by $[b + \mathcal{T}(t_n)]^{n+a}/\Gamma(n+a)$.
Therefore

$$f_{\Theta|T_1,\ldots,T_n}(\theta|t_1,\ldots,t_n) = \frac{(b+\mathcal{T}(t_n))^{n+a}}{\Gamma(n+a)} \left(\frac{1}{\theta}\right)^{n+a+1} \cdot e^{-(b+\mathcal{T}(t_n))/\theta}$$

for $\theta > 0$ (11.49)

which is recognized as the density of an inverted gamma distribution with parameters $n+a$ and $b+\mathcal{T}(t_n)$. The Bayesian estimator of MTTF = θ (minimizing the mean quadratic loss) coincides with the mean value of the distribution (11.49):

$$\hat{\theta}(T_1,\ldots,T_n) = \frac{b+\mathcal{T}(t_n)}{n+a-1} = \frac{b+n\bar{T}}{n+a-1} \quad (11.50)$$

The maximum likelihood estimator of the MTTF = θ is earlier determined to be

$$\theta^*(T_1,\ldots,T_n) = \frac{1}{n}\sum_{j=1}^{n} T_j = \bar{T}$$

We note that

$$\hat{\theta}(T_1,\ldots,T_n) = (1-w) \cdot \frac{b}{a-1} + w\bar{T}$$

or

$$\hat{\theta}(T_1,\ldots,T_n) = (1-w) \cdot \theta_0 + w\theta^* \quad (11.51)$$

where θ_0 denotes the prior mean (11.45) and

$$w = \frac{n}{n+a-1}$$

Hence the Bayesian estimator of θ is a weighted average of the prior mean θ_0 and the MLE θ^*. We also note that

$$\hat{\theta}(T_1,\ldots,T_n) \to \theta^* \quad \text{as} \quad n \to \infty$$

In words, the influence of the prior mean tends to zero as $n \to \infty$.

Type-II Censored Data Sets

Let the situation be as above, except for the fact that the life test is terminated at the rth failure, which corresponds to censoring of type II (see Section 9.3). Let $T_{(1)} \leq T_{(2)} \leq \cdots \leq T_{(r)}$ denote the recorded lifetimes.

According to (9.46) the joint probability density of $T_{(1)} \leq T_{(2)} \leq \cdots \leq T_{(r)}$, given $\Theta = \theta$, becomes

$$f_{T_{(1)},\ldots,T_{(r)}|\Theta}(t_1,\ldots,t_r|\theta) = \frac{n!}{(n-1)!} \cdot \frac{1}{\theta^r} e^{-(\sum_{i=1}^r t_i + (n-r)t_r)/\theta}$$

$$\text{for} \quad t_1 \leq \cdots \leq t_r \qquad (11.52)$$

Hence

$$f_{T_{(1)},\ldots,T_{(r)}|\Theta}(t_1,\ldots,t_r|\theta)$$

$$= \frac{n!}{(n-1)!} \cdot \frac{1}{\theta^r} e^{-(\sum_{i=1}^r t_i + (n-r)t_r)/\theta} \frac{b^a}{\Gamma(a)} \cdot \left(\frac{1}{\theta}\right)^{a+1} e^{-b/\theta}$$

$$\text{for} \quad t_1 \leq \cdots \leq t_r, \theta > 0, a > 2, b > 0 \qquad (11.53)$$

which implies that

$$f_{\Theta|T_{(1)},\ldots,T_{(r)}}(\theta|t_1,\ldots,t_r) \propto \left(\frac{1}{\theta}\right)^{r+a+1} e^{-(b+\sum_{i=1}^r t_i + (n-r)t_r)/\theta} \qquad (11.54)$$

If we introduce the total time on test concept $\mathcal{T}(t_r)$ (see Section 9.2), then (11.54) can be written

$$f_{\Theta|T_{(1)},\ldots,T_{(r)}}(\theta|t_1,\ldots,t_r) \propto \left(\frac{1}{\theta}\right)^{r+a+1} e^{-(b+\mathcal{T}(t_r))/\theta}, \quad \theta > 0 \qquad (11.55)$$

We immediately notice the similarity between (11.55) and (11.48), derived for complete data sets and are able to conclude that the Bayesian estimator of the

MTTF θ in the case of type-II censoring is

$$\hat{\theta}(T_1,\ldots,T_r) = \frac{b+\mathscr{T}(T_r)}{r+a-1} \qquad (11.56)$$

when the inverted gamma density with parameters a and b is used as prior.

11.7 INTERPRETATION OF THE PRIOR DISTRIBUTION

We will confine ourselves to discussing two essentially different interpretations of the prior distribution and illustrate them by discussing an example.

Prior Distribution Based on Empirical Data

Example 11.6

Suppose a commodity is delivered to customers in lots of size a. The relative number of defect units in a lot obviously will vary from one lot to another, and we may interpret the relative number of defect units in a lot as a random variable Θ. Every lot is controlled by taking a random sample of size n of units from the lot, and recording the number X of defect units in the sample.

Given $\Theta = \theta$, X will, if a is not too small, be approximately binomially distributed

$$p_{X|\Theta}(x|\theta) = \binom{n}{x}\theta^x(1-\theta)^{n-x}, \qquad x = 0,1,\ldots,n,\ 0 \leq \theta \leq 1 \qquad (11.57)$$

As already indicated, Θ is likely to vary from lot to lot. However, if the production process is "in control," the θ values will show a statistical regularity that may be expressed by some probability density $p_\Theta(\theta)$. Looking at the situation in this way, *two* random variables X and Θ, are attached to every lot.

In principle we can imagine that over a certain period of time we have effectively carried out 100% control of the lot and thereby observed a large number of realizations of Θ's. The histogram of these values gives us a picture of the possible density of Θ. Even in the case where we have only observed a few such X's, this data set supplies us with some information about the distribution of Θ.

The main point in this situation is that the prior distribution of Θ is empirically motivated on the basis of observed data. The procedure is then the following: First, a prior distribution of Θ is selected from the conjugate family of the binomial distribution (11.57), namely a beta distribution with unspecified parameters r and s. Next these parameters have to be estimated, one way or another, based on the observed Θ values (exploratory data). The beta distribution with these estimated parameters, is then chosen to be the prior distribution of Θ. Presumably few statisticians will have essential objections to Bayesian inference on this basis, in particular since the beta distribution is very flexible.

Subjective Bayesian Inference

Example 11.6 (cont.)

The relative number of defect units in a certain "big" lot is denoted by θ, and the number of defect units in a sample of size n from this lot, is denoted by X. Then approximately

$$P(X = x) = \binom{n}{x} \theta^x (1 - \theta)^{n-x} \quad \text{for} \quad x = 0, 1, \ldots, n, 0 \leq \theta \leq 1$$

In subjective Bayesian inference, the appearing value of θ is an unknown constant in $[0, 1]$. The statistician (the Bayesian analyst), however, considers certain θ values in $[0, 1]$ to be more likely to occur than others and expresses this subjective belief on which θ value is to be expected through a (prior) distribution of Θ, $p_\Theta(\theta)$. The successive statistical reasoning is based on a *subjective* concept of probability, based on the statistician's degree of belief and cannot be given a frequency interpretation. In principle there is nothing illogical in a situation where two (Bayesian) statisticians reach different conclusions as a consequence of different choices of priors. For mathematical convenience even the subjectivist chooses the prior from the conjugate family of distributions to the binomial. Note, however, that the subjectivist even has to choose the parameters r and s in the beta distribution on a subjective basis. This subjective approach is far more controversial among statisticians than the one where the prior is based on exploratory data. □

Situations that we meet in analysis of reliability and risk are sometimes characterized by lack of data. Despite this lack of data, decisions have to be made. In such situations it may be necessary to exploit all a priori insight in the matter, even personal judgments, expert opinions, and the like. Then the subjective approach may be the best (and only) solution to the decision problem.

11.8 THE PREDICTIVE DENSITY

Let T and Θ be two random variables where the conditional density of T, given $\Theta = \theta$ is $f_{T|\Theta}(t|\theta)$. Let $T_1, T_2, \ldots, T_n$ be n observations of T. Given $\Theta = \theta$, these are assumed to be independent.

If a prior density of Θ, $f_\Theta(\theta)$ is assumed, then the joint density of $T_1, T_2, \ldots, T_n$ and Θ is

$$f_{T_1,\ldots,T_n,\Theta}(t_1,\ldots,t_n,\theta) = \left[\prod_{i=1}^{n} f(t_i|\theta)\right] \cdot f_\Theta(\theta) \qquad (11.58)$$

Let us for short denote the data set $T_1, T_2, \ldots, T_n$ by D. After having observed D, how should we predict the next value of T?

We may argue as follows: In the Bayesian setup we determine the posterior density of Θ, given D:

$$f_{\Theta|D}(\theta|D)$$

Then we define *the predictive density* of T, given D as

$$f_{T|D}(T|D) = \int_0^\infty f(t|\theta) \cdot f_{\Theta|D}(\theta|D)\, d\theta \qquad (11.59)$$

Example 11.1 (cont.)
Reconsider the valves in Example 11.1. Given $\Lambda = \lambda$, the lifetimes $T_1, T_2, \ldots, T_n$ of the n valves are independent and exponentially distributed with density

$$f_{T|\Lambda}(t|\lambda) = \lambda e^{-\lambda t} \quad \text{for} \quad t > 0, \lambda > 0$$

where λ represents a realization of a random variable Λ with prior density

$$f_\Lambda(\lambda) = \lambda e^{-\lambda} \quad \text{for} \quad \lambda > 0 \qquad (11.60)$$

Suppose that we have observed the lifetimes $(T_1, T_2, \ldots, T_n)$ of n such valves. According to (11.18) the posterior density of Λ, given D,

$$f_{\Lambda|D}(\lambda|D) = \frac{(1+\sum_{i=1}^n t_i)^{n+2}}{\Gamma(n+2)} \lambda^{n+1} e^{-\lambda(1+\sum_{i=1}^n t_i)} \quad \text{for} \quad \lambda > 0 \qquad (11.61)$$

Our guess, based on D, is that the next observation T has density

$$f_{T|D}(t|D) = \int_0^\infty \lambda e^{-\lambda t} \cdot \frac{(1+\sum_{i=1}^n t_i)^{n+2}}{\Gamma(n+2)} \lambda^{n+1} e^{-\lambda(1+\sum_{i=1}^n t_i)} d\lambda$$

$$= \frac{(1+\sum_{i=1}^n t_i)^{n+2}}{\Gamma(n+2)} \int_0^\infty \lambda^{n+2} e^{-\lambda((1+\sum_{i=1}^n t_i)+t)} d\lambda$$

$$= \frac{(n+2)(1+\sum_{i=1}^n t_i)^{n+2}}{(1+\sum_{i=1}^n t_i + t)^{n+3}} \quad \text{for } t > 0 \quad (11.62)$$

Hence our guess is that the survival function for a given new valve of the same type is

$$P(T > t|D) = R(t|D) = \int_t^\infty \frac{(n+2)(1+\sum_{i=1}^n t_i^{n+2})}{(1+\sum_{i=1}^n t_i + t)^{n+3}} du$$

$$= \left(\frac{1+\sum_{i=1}^n t_i}{1+\sum_{i=1}^n t_i + t} \right)^{n+2}$$

$$= \left(1 + \frac{t}{1+\sum_{i=1}^n t_i} \right)^{-(n+2)}$$

$$\text{for } t > 0 \quad (11.63)$$

Comparison of (11.63) to (2.30) leads to the conclusion that

$$Y = \left(\frac{T}{\sum_{i=1}^n t_i + 1} + 1 \right), \quad Y > 1 \quad (11.64)$$

has the Pareto distribution $P(1, n+2)$. □

11.9 PROBLEMS

1. Prove Lemma 11.1.

2. Show that the Bayesian estimator of θ which minimizes the mean absolute-error loss $E(|\hat{\theta}(X) - \Theta|)$ is equal to the median of the posterior distribution of Θ (given $X = x$).

3. Determine the mean and the variance of the inverted gamma distribution (11.44)

4. Assume that X has a binomial distribution (n, p), where p represents a real-

ization of a random variable P. The prior distribution of P is $f_P(p) = 1$ for $0 \leq p \leq 1$. Determine the posterior density of P when $X = x$ is observed, and determine the Bayesian estimate for p.

5. (Kapur and Lamberson 1977, p. 402). Seven automobiles are each run over a 36,000-kilometer test schedule. The testing produced a total of 19 failures. Assuming an exponential failure distribution and a gamma prior with parameters $\alpha = 30,000$ and $\beta = 3$, answer the following:
   (a) What is the Bayesian point estimate for the MTTF?
   (b) What is the 90% lower confidence (credibility) limit on the 10,000 kilometer reliability?

6. Let $X_1, X_2, \ldots, X_n$ be independent and identically distributed $\mathcal{N}(\theta, \sigma_0^2)$, where σ_0^2 is unknown, and θ represents a realization of a random variable Θ with normal distribution $\mathcal{N}(\mu_0, \tau_0^2)$ where μ_0 and τ_0^2 are known.
   Show that the Bayesian estimate of Θ (minimizing the mean quadratic loss) is a weighted average of the prior mean and the MLE of θ

$$\hat{\theta}(X_1, \ldots, X_n) = \frac{n/\sigma_0^2}{n/\sigma_0^2 + 1/\tau_0^2} \cdot \overline{X} + \frac{1/\tau_0^2}{n/\sigma_0^2 + 1/\tau_0^2} \cdot \mu_0$$

Note that the Bayesian estimate of Θ is a weighted average of the hypothetical estimates of Θ based on the following:
- Data alone (i.e., the standard estimator $\overline{X}$)
- Prior information of Θ but no data, μ_0 (i.e., the Bayes estimator of μ before any observations are taken)

Again note that the influence of the prior mean μ_0 tends to zero as $n \to \infty$.

7. Let $X_1, X_2, \ldots, X_n$ be independent and identically distributed $\mathcal{N}(0, \sigma^2)$.
   (a) Show that the joint density of $X_1, X_2, \ldots, X_n$ can be written

$$C_T r^r e^{-\tau \sum_{i=1}^n x_i^2} \quad \text{where} \quad r = n/2, \tau = 1/(2\sigma^2).$$

   (b) Choose the gamma distribution (k, λ) with density

$$\frac{\lambda}{\Gamma(k)} (\lambda \tau)^{k-1} e^{-\lambda \tau} \quad \text{for} \quad \tau > 0$$

as prior density of τ.
Show that the posterior density of τ, given $X_1, X_2, \ldots, X_n$ then

becomes a gamma distribution $(k + r, \lambda + \sum_{i=1}^{n} x_i^2)$ with density

$$C(x_1, x_2, \ldots, x_n) \cdot \tau^{r+k-1} e^{-\tau(\lambda + \sum_{i=1}^{n} x_i^2)} \quad \text{for} \quad \tau > 0$$

(c) Use the result in (b) to show that the Bayes estimator of σ^2 (with minimum expected quadratic loss) becomes

$$\frac{\lambda + \sum_{i=1}^{n} X_i^2}{n + 2k - 2}$$

(Hint: Since $2\sigma^2 = 1/\tau$, the Bayes estimator of $2\sigma^2$ is the posterior expectation of $1/\tau$). This problem is based on an example in Lehmann (1983, p. 246).

8. Let X have a binomial distribution (n, θ) where θ represents a realization of a random variable Θ with a beta distribution (r, s). Denote the prior mean of Θ by θ_0.

Show that the Bayesian estimate of Θ (minimizing the mean quadratic loss) is a weighted average of the prior mean and the MLE of θ:

$$\hat{\theta}(X) = \frac{n}{r+s+n} \cdot \frac{X}{n} + \frac{r+s}{r+s+n} \cdot \theta_0$$

Note that the Bayesian estimate of Θ as in problem 6 is a weighted average of the hypothetical estimates of Θ based on the following:
- Data D alone (i.e., the standard estimator of θ, X/n)
- Prior information of Θ but no data, θ_0 (i.e., the Bayesian estimator of θ before any observations are taken)

Note that the influence of the prior mean θ_0 tends to zero as $n \to \infty$.

9. (Sequential binomial sampling) Consider a sequence of binomial trials with success probability θ where the number of trials may depend on the observations. The stopping rule is assumed to be *closed* (see Lehmann 1983, pp. 93, 243).

A priori θ is assumed to represent a realization of a random variable Θ with a beta distribution (r, s). Let the number of successes, the number of failures, and the total number of trials at the moment when the sampling stops, be denoted by X, Y, and N, respectively. Show that the posterior distribution of Θ, given X and Y, is a beta distribution $(r+s, s+n-x)$ and consequently that the Bayes estimator of θ, given X and Y, is the same regardless of the closed stopping rule.

CHAPTER 12

Reliability Data Sources

To assess the reliability or availability of a component, two categories of data are needed:

- Failure rates, possibly time dependent
- Repair/restoration times

For components that are tested periodically, we also need to know the time intervals between consecutive tests.

There are a number of sources for failure rate information:

1. Public databooks and databanks
2. Performance data from the actual plant
3. "Expert" judgments
4. Laboratory testing

An introduction to reliability data collection and management is given in EuReDatA (1986).

12.1 DATA BOOKS AND DATA BANKS

Some relevant data books and data banks are listed in this section. All of these sources present only constant failure rates. The only one of these data sources that also supplies information about repair times is OREDA (1992). A detailed presentation and discussion of some European reliability data banks is given by Cannon and Bendell (1991).

System Reliability Service Data Bank
The System Reliability Service (SRS) Data Bank was established by the United Kingdom Atomic Energy Authority (UKAEA) in the 1960s. It is now operated

DATA BOOKS AND DATA BANKS 461

by AEA Technology/SRD. The data bank is accessible by virtue of membership. The data are partly supplied by the members of the SRS Data Bank and cover a wide range of component types.

OREDA (1992)

The Offshore Reliability Data (OREDA) handbook contains data from a wide range of components and systems used on offshore installations, collected from installations in the North Sea and in the Adriatic Sea. OREDA (1992) is based on actual field data collected in the time period 1981 to 1991. The data are classified under the following main headings:

- Process systems
- Safety systems
- Electrical systems
- Utility systems
- Crane systems
- Drilling systems

An example of how the data are presented in the OREDA handbook is shown in Figures 12.1 and Figure 12.2. The left-hand page (Figure 12.1) in the handbook presents a brief description of the component, its application, and the internal and external environment of the component. A brief description of the testing and maintenance strategies is also supplied. An important feature of the OREDA handbook is the specification (including a drawing) of the physical boundaries of the component. The right-hand page (Figure 12.2) presents the recorded data. Whenever possible, both the calendar time and the actual time in operation are recorded, and the failure rate is estimated with respect to both times. Figure 12.2 presents data from 19 ("Population") turbine driven pumps that are installed on 6 ("Installation") different offshore platforms. The failure modes are classified in the three groups: *critical*, *degraded*, and *incipient*. The failure rate is estimated for each failure mode together with 90% confidence intervals. The failure data are mainly collected from maintenance records. This means that both component specific failures (primary failures) and common cause failures are included. It also implies that spurious failures such as false alarms are not included in full detail. The frequency of false alarms and other types of spurious operations is thus underestimated in the OREDA handbook. Repair times are recorded whenever possible. For most of the component types, only worker hours were available. To estimate the active repair time and downtime, one needs to have more details about the repair and the maintenance organization.

The data collected during the project have been stored in a computerized database. This database is only available to the ten oil companies participating in the OREDA project.

Taxonomy no. and item
1.3.2.2
Process Systems
Pumps
Turbine Driven
Oil Handling

Description
Single- and multi-stage turbine-driven centrifugal pumps. Direct and step-down gear transmission. Power rating in the range 3000-12000 kW.

Application
Crude oil export pumps. Pressure boosting for export pipelines or transfer of oil from platform storage tanks to distribution lines.

Operational mode
Continuous for periods of time, with intermittent periods in standby. Generally, redundant pumps with one in standby.

Internal environment
Crude oil.
Suction pressure: 8 -140 barg.
Discharge pressure: 23 - 181 barg.

External environment
Generally partially enclosed.

Item boundary specification
Only failures within the boundary indicated by the dashed line in the figure below are included in the reliability data source.

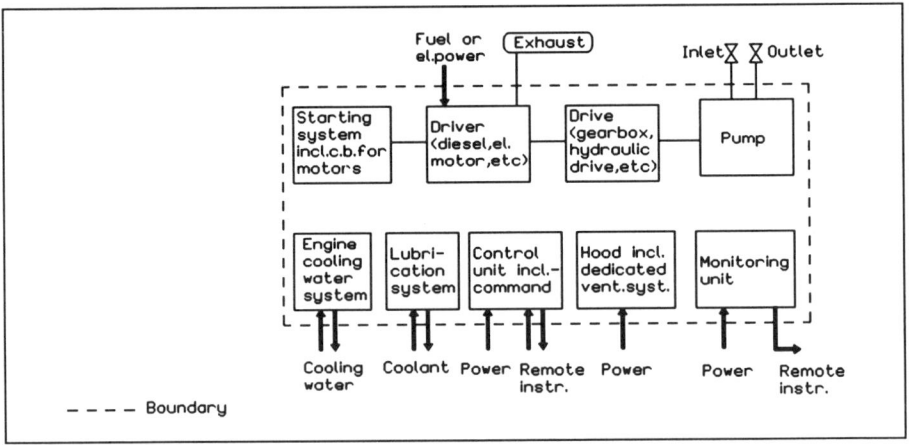

Figure 12.1 Example of data from OREDA (1992): Left-hand page (Reprinted with the kind permission of the OREDA Participants.)

DATA BOOKS AND DATA BANKS

Taxonomy no 1.3.2.2		Item Process Systems Pumps Turbine Driven Oil Handling						
Population	Installations	Aggregated time in service (10⁶ hours)				No of demands		
19	6	Calendar time * 0.3635		Operational time † 0.2214		1800		
Failure mode		No of failures	Failure rate (per 10⁶ hours)			Active repair (hours)	Repair (manhours)		
			Lower	Mean	Upper		Min	Mean	Max
Critical		187 * 187 †	867.25 1090.38	1200.48 1516.92	1541.84 1956.40	17.5	2.0	30.4	120.0
External leakage		1 * 1 †	0.66 1.07	2.99 4.77	13.40 21.40	15.0	21.0	21.0	21.0
Fail to start		3 * 3 †	1.43 1.19	9.39 16.76	25.44 45.25	4.0	4.0	4.0	4.0
Failed while running		183 * 183 †	856.67 1071.67	1188.10 1495.38	1527.61 1932.00	17.6	2.0	30.7	120.0
Degraded		13 * 13 †	14.74 27.50	39.29 61.21	71.95 107.87	9.9	4.0	10.9	28.0
Low output		13 * 13 †	14.74 27.50	39.29 61.21	71.95 107.87	9.9	4.0	10.9	28.0
Incipient		62 * 62 †	185.35 293.76	300.86 487.71	424.48 694.61	7.7	1.0	10.3	65.0
External leakage		15 * 15 †	14.13 30.07	56.60 120.79	107.73 225.64	-	-	-	-
Failed while running		22 * 22 †	47.32 67.56	99.90 136.28	161.12 219.14	8.6	1.0	10.9	65.0
Other modes		24 * 24 †	64.61 102.08	141.17 223.41	226.36 358.88	6.1	1.0	9.2	24.0
Vibration		1 * 1 †	0.48 0.00	3.20 7.24	14.58 28.70	-	-	-	-
Unknown		5 * 5 †	4.83 7.50	16.80 32.90	36.89 71.26	-	-	-	-
Other modes		4 * 4 †	2.15 3.54	13.77 28.18	33.59 65.59	-	-	-	-
Unknown		1 * 1 †	0.45 1.05	3.03 4.72	13.81 21.15	-	-	-	-
All modes		267 * 267 †	1156.27 1538.83	1557.42 2098.74	1966.03 2670.88	15.1	1.0	25.3	120.0

Comments

On demand probability (per 1000 demands). Lower ; Mean ; Upper
Critical - Fail to start: 0.028 ; 0.56 ; 2.6
The estimates were based on the subset of items for which the number of demands were recorded.

Note the sparse data for the estimates, i.e. small population or short aggregated time in service.

The estimates for the different failure modes were in some cases based on different subsets of data. This results from tests of statistical consistence among items, see sub chapter 3.4.

Figure 12.2 Example of data from OREDA (1992): Right-hand page (Reprinted with the kind permission of the OREDA Participants.)

IEEE Std. 500 (1984)

The data presented in the rather voluminous handbook *IEEE Guide to the Collection and Presentation of Electrical, Electronic, Sensing Components, and Mechanical Equipment Reliability Data for Nuclear-Power Generating Stations* are, as indicated by this title, applicable to components in nuclear power stations. The constant failure rates for the relevant failure modes are presented in a way similar to the right-hand page presentation in OREDA (1992). An example of a data page is shown in Figure 12.3. The failure modes are classified in the same three groups, *catastrophic (critical), degraded*, and *incipient*, as in the OREDA handbook. There is no description of the components, their application, and the like, as in the OREDA handbook. The data have been obtained partly from recorded field data, and partly from expert judgment and the so-called Delphi technique. The Delphi technique is described in Appendix B of the IEEE Std. 500 (1984). The failure rates in the IEEE Std. 500 cover only component specific (primary) failures. Failures due to external stresses and common cause failures are not sufficiently covered. The total failure rate for a component in practical use thus tends to be underestimated when based on IEEE Std. 500.

MIL-HDBK 217F (1991)

The U.S. Military Handbook 217 is the standard reference for reliability data for electronic components. The data are not related to specific failure modes. Most of the data come from tests in a laboratory with controlled environmental stresses. The failure rates in MIL-HDBK 217 are thus only related to component specific (primary) failures. Failures due to external stresses and common cause failures are not included. The handbook gives formulas and data to adjust the failure rate of a component to a specified environment. The general MIL-HDBK 217 failure rate model is of the form

$$\lambda_P = \lambda_B \cdot \pi_Q \cdot \pi_E \cdot \pi_A \ldots$$

where λ_B is the basic failure rate, which is estimated from reliability tests performed on components under standard environmental conditions. λ_B is thus given for standardized stresses (e.g., voltage and humidity) and temperature conditions.

$\pi_Q, \pi_E, \pi_A, \ldots$ are often called *influence* factors, and these factors take into account impact of part quality, equipment environment, application stress, and so on. The values of the basic failure rates and the various factors in the handbook are kept up to date by analysis of failure data on components and systems.

Contractors to the U.S. Department of Defense are required to use this handbook to estimate the reliability of their products. Producers of nonmilitary electronic equipment, such as instruments and avionic gear, also often elect to adhere to the handbook's guidelines because it offers a convenient and standard way of estimating reliability.

Detailed studies have shown that the MIL-HDBK 217 data often are too

CHAPTER: 11 Driven Equipment SECTION: 11.1 Pumps SUBSECTION: 11.1.2 Centrifugal

ITEM OR EQUIPMENT DESCRIPTION: 11.1.2.4.2 Alternating Service, 11.1.2.4.2.5 Service Water, Motor Driven, Includes Driver

FAILURE MODE	FAILURE RATE								(*) OUT OF SERVICE (†) REPAIR TIME OR (§) RESTORE (HOURS)			
	FAILURES/10^6 HOURS				FAILURES/10^3 CYCLES							
	LOW	REC	HIGH	REF	LOW	REC	HIGH	REF	LOW	REC	HIGH	REF
ALL MODES	173.0	314.0	558.0	17	70.0	123.0	213.0	17	1.0 †	24.0	651.0	17
CATASTROPHIC	10.0	38.0	98.0		0	1.3	17.0					
Fails While Running	10.0	38.0	98.0									
Fails to Start					0	1.3	17.0					
DEGRADED	33.0	76.0	150.0		15.0	33.0	66.0					
INCIPIENT	130.0	200.0	310.0		55.0	89.0	130.0					

(Composite of Ref 17)

Figure 12.3 Example of data from the IEEE Std. 500 (Figure reprinted from IEEE Std 500-1984, *IEEE Guide to the Collection and Presentation of Electrical, Electronic, Sensing Component and Mechanical Equipment Reliability Data for Nuclear Power Generating Stations* (Reaff 1991) copyright G1984 by the Institute of Electrical and Electronics Engineers, Inc. The IEEE takes no responsibility for and will assume no liability for damages resulting from the reader's misinterpretation of said information resulting from the placement and context in this publication. Information is reproduced with the permission of the IEEE.)

pessimistic for commercial devices, for example, see Bodsberg (1987) and O'Connor (1991). According to Watson (1992): "Opponents of MIL-HDBK 217 say that it is inaccurate, leads to costly overdesign, actually prevents higher reliability levels from being achieved, and does not address the true causes of failures. Moreover, they say, the handbook's models are out of date and do not reflect emerging technologies."

MIL-HDBK 217 also describes a special method for predicting the reliability of a system. The method is called *parts count reliability prediction*, and it assumes that system success can be achieved only if all the system components are operating, that is, if the system is a series structure. The system failure rate λ_S is obtained by adding the failure rates of the n system components:

$$\lambda_S = \sum_{i=1}^{n} \lambda_i$$

When the system is not a series system, λ_S will give an upper bound of the failure rate. The parts count method is further discussed in Anderson (1976). The parts count method has been heavily criticized, for example, by Luthra (1990) who claims: "One area that MIL-HDBK 217 really needs to be cleaned up is the Parts Count method. I think it should not be used by any engineer without lots of changes and going into the details of the models. To use canned software programs with the Parts Count method is horrifying. I have tried several programs and am forewarning others."

Several competing data handbooks have been issued. Among these are these four:

1. *HRD4 (1987)*

   Handbook of Reliability Data for Components used in Telecommunications Systems published by British Telecom. The HRD4 has some similarities with MIL-HDBK 217 in using weighting factors to adjust a basic failure rate to a specific application. The HRD4 data have approximately the same limitations as the data in MIL-HDBK 217.

2. *CNET (1983)*

   Recueil de données de fiabilité du CNET by the Centre National d'Etudes des Télécommunication. The CNET standard was originally derived from MIL-HDBK 217 but includes various modifications and enhancements.

3. *NTT (1985)*

   Standard Reliability Table for Semiconductor Devices by Nippon Telegraph and Telephone Corporation.

4. *Bellcore (1988)*

   Reliability Prediction Procedure for Electronic Equipment, TR-TSY-

000332, Bell Communications Research (Bellcore). This handbook is developed for use in nonmilitary applications.

Several computer codes have been developed to aid the analyst is using the MIL-HDBK 217 and its competitors. Some of these programs are listed below. The list is not complete. Some of the program suppliers also have a wide range of associated programs which are not listed.

1. Relex 217. MIL-HDBK 217 support, available from Innovative Software Designs, Inc. Two English Elm Court, Baltimore, MD 21228. The associated *Relex parts count* program supports the parts count method in MIL-HDBK 217.
2. Relex Bellcore. Bellcore handbook support, available from Innovative Software Design, Inc. (see above).
3. Relex CNET. CNET handbook support, available from Innovative Software Design, Inc. (see above).
4. RPP. MIL-HDBK 217 support, available from Powertronic Systems, Inc. 13700 Chef Menteur Highway, New Orleans, LA 70129.
5. RPC. Parts count method, available from Powertronic Systems, Inc. (see above).
6. RBC. Bellcore handbook support, available from Powertronic Systems, Inc. (see above).
7. MilStress. MIL-HDBK 217 support available from OMI Logistics Limited, Item Software, Wellow House, 14 Little Park Road, Fareham, Hampshire, PO15 5TD, England. The associated *PartsCount* program supports the parts count method in MIL-HDBK 217.
8. REAP. MIL-HDBK 217 support, available from Systems Effectiveness Associates, Inc. 20 Vernon Street, Norwood, MA 02062.
9. Reliability Toolbox. Includes MIL-HDBK 217 and Bellcore handbook support. Available from Innovative Timely Solutions, 6401 Lakerest Court, Raleigh, NC 27612.

NPRD-91

The *Nonelectronic Part Reliability Databook* is available from RAC, Reliability Analysis Center, 201 Mill St. Rome, NY 14440-8200. The NPRD-91 handbook contains reliability data for mechanical, electromechanical, and discrete parts and assemblies, mainly in military application. The handbook concentrates on items not included in MIL-HDBK 217. The data are grouped according to application as ground fixed, ship sheltered, airborne, and so on. The failure rates are not given for specified failure modes. Overall failure mode distributions are, however, presented in the separate 370 page handbook *Fail-*

ure Mode/Mechanism Distributions, which is also available from RAC. The NPRD-91 handbook is also available as a PC database.

T-Book *(1992)*

The *Reliability Data of Components in Nordic Nuclear Power Plants* handbook provides failure rate estimates for pumps, valves, instruments and electropower components in Nordic nuclear power plants. The data are presented as constant failure rates, with respect to the most significant failure modes. Mean active repair times are also recorded.

WASH-1400 *(1975)*

The *Reactor Safety Study*, WASH-1400, is the final report from a comprehensive study of U.S. nuclear power plants. Failure rates for some electrical and mechanical components in nuclear power plants are given in Appendixes 3 and 4 of this report.

Reliability data may also be found in some textbooks and research reports. Among these are the following:

- Smith (1988), *Reliability and Maintainability in Perspective*. This textbook has a specific chapter with failure rate data. The data are mainly compiled from the sources listed above. The data are also available as a PC computer program called FARADIP. Reference to this program is found in the textbook.

- Green and Bourne (1972), *Reliability Technology*. This textbook has an appendix with failure rate estimates for electronic, electrical, and mechanical components. Despite its date this textbook is a classic reference for reliability data.

- Molnes et al. (1986), *Reliability of Surface Controlled Subsurface Safety Valves (SCSSV)*. This research report presents SCSSV reliability data collected from offshore applications in the North Sea area, and it is one of the most detailed studies that have been performed on collection and analysis of field performance data.

A number of companies and organizations have established their own proprietary failure event and reliability data banks. Such data banks are normally not available for external companies. An association of reliability data bank holders has, however, been established in Europe with the intention to facilitate exchange of data between the members. The association is called ESREDA (European Safety and Reliability Data Association; address: Commission of the European Communities, Joint Research Centre, 21020 ISPRA, Varese, Italy.) A thorough discussion on reliability databases and problems connected to the collection and analysis of reliability data may be found in Amendola and Keller (1985) and Flamm and Luisi (1992).

12.2 DATA QUALITY

A significant effort has been devoted to the collection and processing of reliability data during the last ten years. Despite this great effort the quality of the data available is still not good enough. In this section we will only highlight some of the problems.

Detailed studies have shown that the reliability often is significantly dependent on a wide range of environmental and operational factors in addition to the materials used, design, surface treatment, and so on. As an example, see the discussion of SCSSV reliability in Molnes et al. (1986), where the failure rate of a specific valve make was found to differ over a decade from one installation to another. There is reason to suspect a similiar variation in other types of components. Environmental and operational conditions are normally not specified in the data books (except in OREDA 1992). The data supplied are thus, at best, average values, and the uncertainty of the estimates is often not specified correctly. The data in the data books are often pooled from a number of samples (plants/installations). In OREDA (1992) and several other data books, the uncertainty is given, for example, by a 90% confidence interval. This confidence interval is calculated under the assumption that all the data come from a homogeneous sample with a common constant failure rate. When the samples are nonhomogeneous, which is usually the case, these confidence intervals are without value. This problem is discussed by Lydersen and Rausand (1989).

Another problem is how to estimate the failure rate of a component that is in a standby mode part of the time. Still another problem is common cause failures. Common cause failures are sometimes included in the failure rate estimates. In practice estimating the frequency of common cause failures is very difficult to do. Likewise it is very difficult to assess the number of components affected by the common cause. In some situation all components in a subsystem will fail due to a common cause, and in other situations only a proportion of the components will fail. Common cause failures are often neglected in risk and availability studies. This is a significant limitation. Common cause failures will often have a dominating effect on the system reliability.

APPENDIX A

The Gamma and Beta Functions

A.1 THE GAMMA FUNCTION

The gamma function $\Gamma(\alpha)$ is defined for all real $\alpha > 0$ by the integral

$$\Gamma(\alpha) = \int_0^\infty t^{\alpha-1} e^{-t} \, dt \tag{A.1}$$

By partial integration it is easy to show that

$$\Gamma(\alpha + 1) = \alpha \Gamma(\alpha) \quad \text{for} \quad \alpha > 0 \tag{A.2}$$

In probabilistic and statistical applications there is a particular interest for the value of the gamma function when $\alpha = m/2$, where m is a positive integer.

Let k denote a positive integer. Then if $m = 2k$, repeated use of (A.2) leads to

$$\Gamma(k+1) = k \cdot (k-1) \cdots 2 \cdot 1 \cdot \Gamma(1) \tag{A.3}$$

However,

$$\Gamma(1) = \int_0^\infty e^{-t} \, dt = 1$$

Hence

$$\Gamma(k+1) = k! \tag{A.4}$$

Next, consider the case $m = (2k+1)/2$. By repeated use of (A.2), it leads to

$$\Gamma\left(\frac{2k+1}{2}\right) = \frac{2k-1}{2} \cdot \frac{2k-3}{2} \cdots \frac{1}{2} \cdot \Gamma\left(\frac{1}{2}\right) \quad (A.5)$$

However,

$$\Gamma\left(\frac{1}{2}\right) = \int_0^\infty t^{-1/2} e^{-t} \, dt$$

By introducing $u = \sqrt{2t}$ as a new variable of integration, we get

$$\int_0^\infty t^{-1/2} e^{-t} \, dt = \sqrt{2} \int_0^\infty e^{-u^2/2} \, du = \frac{1}{\sqrt{2}} \int_{-\infty}^\infty e^{-u^2/2} \, du = \sqrt{\pi}$$

Hence

$$\Gamma\left(\frac{2k+1}{2}\right) = \frac{2k-1}{2} \cdot \frac{2k-3}{2} \cdots \frac{1}{2} \cdot \sqrt{\pi} \quad (A.6)$$

In Table A.1 the gamma function $\Gamma(\alpha)$ is given for values of α between 1.00 and 2.00.

A.2 THE BETA FUNCTION

The beta function $B(r, s)$ is defined for all real $r > 0$, $s > 0$ by the integral

$$B(r, s) = \int_0^1 u^{r-1} (1-u)^{s-1} \, du \quad (A.7)$$

It may be shown (e.g., see Cramér 1945, p. 127) that

$$B(r, s) = \frac{\Gamma(r) \cdot \Gamma(s)}{\Gamma(r+s)}$$

Hence

$$f_U(u; r, s) = \frac{\Gamma(r) \cdot \Gamma(s)}{\Gamma(r+s)} u^{r-1}(1-u)^{s-1} \, du,$$

$$0 \leq u \leq 1, r > 0, s > 0 \quad (A.8)$$

may obviously be interpreted as a probability density since $\int_0^1 f_U(u; r, s) \, du = 1$.

APPENDIX A

Table A.1 Gamma Function $\Gamma(\alpha)$ for α between 1.00 and 2.00

α	$\Gamma(\alpha)$	α	$\Gamma(\alpha)$	α	$\Gamma(\alpha)$	α	$\Gamma(\alpha)$
1.00	1.00000	1.25	0.90640	1.50	0.88623	1.75	0.91906
1.01	0.99433	1.26	0.90440	1.51	0.88659	1.76	0.92137
1.02	0.98884	1.27	0.90250	1.52	0.88704	1.77	0.92376
1.03	0.98355	1.28	0.90072	1.53	0.88757	1.78	0.92623
1.04	0.97844	1.29	0.89904	1.54	0.88818	1.79	0.92877
1.05	0.97350	1.30	0.89747	1.55	0.88887	1.80	0.93138
1.06	0.96874	1.31	0.89600	1.56	0.88964	1.81	0.93408
1.07	0.96415	1.32	0.89464	1.57	0.89049	1.82	0.93685
1.08	0.95973	1.33	0.89338	1.58	0.89142	1.83	0.93969
1.09	0.95546	1.34	0.89222	1.59	0.89243	1.84	0.94261
1.10	0.95135	1.35	0.89115	1.60	0.89352	1.85	0.94561
1.11	0.94740	1.36	0.89018	1.61	0.89468	1.86	0.94869
1.12	0.94359	1.37	0.88931	1.62	0.89592	1.87	0.95184
1.13	0.93993	1.38	0.88854	1.63	0.89724	1.88	0.95507
1.14	0.93642	1.39	0.88785	1.64	0.89864	1.89	0.95838
1.15	0.93304	1.40	0.88725	1.65	0.90012	1.90	0.96177
1.16	0.92980	1.41	0.88676	1.66	0.90167	1.91	0.96523
1.17	0.92670	1.42	0.88636	1.67	0.90330	1.92	0.96877
1.18	0.92373	1.43	0.88604	1.68	0.90500	1.93	0.97240
1.19	0.92089	1.44	0.88581	1.69	0.90678	1.94	0.97610
1.20	0.91817	1.45	0.88566	1.70	0.90864	1.95	0.97988
1.21	0.91558	1.46	0.88560	1.71	0.91057	1.96	0.98374
1.22	0.91311	1.47	0.88563	1.72	0.91258	1.97	0.98768
1.23	0.91075	1.48	0.88575	1.73	0.91467	1.98	0.99171
1.24	0.90852	1.49	0.88595	1.74	0.91683	1.99	0.99581
						2.00	1.00000

Note: $\Gamma(\alpha)$ for other positive values of α may be calculated from formula (A.2).

A random variable with density (A.8) is said to be beta distributed (r, s). Then

$$E(U) = \frac{r}{r+s}$$
$$\text{var}(U) = \frac{rs}{(r+s)^2(r+s+1)} \quad \text{(A.9)}$$

The beta density may exhibit a large number of shapes, including the uniform distribution (choose $r = 1$ and $s = 1$), U-shaped densities, and unimodal right-skewed, and unimodal left-skewed densities.

In Chapter 11 (Theorem 11.4) the family of beta distributions $B(r, s)$ is shown to be the conjugate to the family of binomial distributions (n, p). Hence in Bayesian reliability analysis, the beta distribution is frequently used as a prior density for p in a binomial distribution.

APPENDIX B

Distribution Theorems

First we refer (without proof) two important theorems from distribution theory:

Theorem B.1. Let X be a continuous random variable with probability density $f_X(x)$ and sample space S_X. Furthermore let $a(x)$ be strictly monotonous in x and differentiable with respect to x for all x. Then $y = a(x)$ is a one-to-one transformation on x with inverse $x = b(y)$ which maps S_X into S_Y, and the density of $Y = a(X)$ is given by

$$f_Y(y) = f_X(b(y))|b'(y)| \qquad (B.1)$$

The extension of this theorem to multivariate distributions is given in the next theorem

Theorem B.2. Let $X_1,\ldots,X_n$ be continuously distributed with joint probability density $f_{X_1,\ldots,X_n}(x_1,\ldots,x_n)$ and sample space $S_{X_1,\ldots,X_n}$. If

$$y_i = a_i(x_1,\ldots,x_n) \quad \text{for} \quad i = 1,2,\ldots,n \qquad (B.2)$$

is a one-to-one transformation on the x's with inverse

$$x_i = b_i(y_1,\ldots,y_n) \quad \text{for} \quad i = 1,2,\ldots,n \qquad (B.3)$$

that maps the sample space $S_{X_1,\ldots,X_n}$ into $S_{Y_1,\ldots,Y_n}$, then the joint density of

$$Y_i = a_i(X_1,\ldots,X_n) \quad \text{for} \quad i = 1,2,\ldots,n$$

is given by

$$f_{Y_1,\ldots,Y_n}(y_1,\ldots,y_n) \\ = f_{X_1,\ldots,X_n}(b_1(y_1,\ldots,y_n),\ldots,b_n(y_1,\ldots,y_n)) \cdot |J| \qquad (B.4)$$

where J denotes the Jacobian

$$J = \begin{vmatrix} \dfrac{\partial b_1(y_1,\ldots,y_n)}{\partial y_1} & \cdots & \dfrac{\partial b_n(y_1,\ldots,y_n)}{\partial y_n} \\ \vdots & & \vdots \\ \dfrac{\partial b_n(y_1,\ldots,y_n)}{\partial y_1} & \cdots & \dfrac{\partial b_n(y_1,\ldots,y_n)}{\partial y_n} \end{vmatrix} \quad (B.5)$$

We will now use Theorems B.1 and B.2 to derive some useful theorems.

Theorem B.3. Let $X_1,\ldots,X_n$ be independently and identically distributed with common distribution function $F_X(x)$ and probability density $f_X(x)$. Denote the observations ordered according to magnitude by $X_{(1)} < \cdots < X_{(r)}$, and let r be any integer such that $r \leq n$. Then the joint density of $X_{(1)},\ldots,X_{(n)}$ is given by

$$f_{X_{(1)},\ldots,X_{(r)}}(x_1,\ldots,x_r) = \frac{n!}{(n-r)!} [1 - F_X(x_r)]^{n-r} \cdot \prod_{i=1}^{r} f_X(x_i)$$

$$\text{for } 0 < x_1 < x_2 < \cdots < x_r \quad (B.6)$$

The proof may be based on a multinomial argument.

Theorem B.4. Let $X_1,\ldots,X_n$ be independently and identically distributed with density $f_X(x) = \lambda e^{-\lambda x}$ for $x > 0$, $\lambda > 0$. Let the corresponding order statistic be denoted by $X_{(1)} < \cdots < X_{(n)}$. Now introduce the random variables

$$D_j = X_{(j)} - X_{(j-1)} \quad \text{for } j = 1, 2, \ldots, n \quad (B.7)$$

$$X_{(0)} \stackrel{\text{def}}{=} 0$$

Then

1. $D_1,\ldots,D_n$ are independent random variables
2. D_j is exponentially distributed with parameter $(n - j + 1)\lambda$ for $j = 1, 2, \ldots, n$
3. $D_j^* = (n - j + 1)D_j$ for $j = 1, 2, \ldots, n$ are exponentially distributed with parameter λ

Proof
According to Theorem B.3

$$f_{X_{(1)},\ldots,X_{(n)}}(x_1,\ldots,x_n) = n!\lambda^n e^{-\lambda \sum_{i=1}^n x_i} \quad \text{for} \quad 0 < x_1 < x_2 < \cdots < x_n$$

(B.8)

The mapping $D_j = X_{(j)} - X_{(j-1)}$ for $j = 1, 2, \ldots, n$ is one-to-one, since $X_{(j)} = \sum_{i=1}^{j} D_i$ for $j = 1, 2, \ldots, n$. The Jacobian of this mapping is easily shown to be 1.

Furthermore

$$\sum_{j=1}^n X_j = \sum_{j=1}^n X_{(j)} = nD_1 + (n-1)D_2 + \cdots + (n-j+1)D_j + \cdots + D_n$$

$$= \sum_{j=1}^n (n-j+1)D_j$$

Then according to Theorem B.2

$$f_{D_1,\ldots,D_n}(d_1,\ldots,d_n) = n!\lambda^n e^{-\lambda \sum_{j=1}^n (n-j+1)d_j}$$

$$= \prod_{j=1}^n [(n-j+1)\lambda e^{-\lambda(n-j+1)d_j}]$$

$$= \prod_{j=1}^n f_{D_j}(d_j) \quad \text{for} \quad d_j > 0, j = 1, 2, \ldots, n$$

where we have introduced

$$(n-j+1)\lambda e^{-\lambda(n-j+1)d_j} = f_{D_j}(d_j)$$

Hence properties 1 and 2 of Theorem B.4 are proved. Property 3 follows directly, using the transform

$$D_j^* = (n-j+1)D_j \quad \text{for} \quad j = 1, 2, \ldots, n$$

The D_j^* for $j = 1, 2, \ldots, n$ are sometimes called the *normed time differences* of the process. □

APPENDIX C

Maximum Likelihood Estimation

An important general method for constructing estimators is based on the *maximum likelihood* principle. Although as early as 1821 the German mathematician C. F. Gauss was first to apply the idea, the method is usually credited to the English statistician R. A. Fisher who introduced this principle in 1912 in a short paper, and later on, in a series of papers, where he investigated the properties of the estimators so obtained. Here we will content ourselves with giving a short presentation of the method and will restrict ourselves to regular parametric models.

To fix the idea, let $X_1, X_2, \ldots, X_n$ denote n independent, identically distributed random variables with density (alternatively frequency function) $f(x; \theta_1, \theta_2, \ldots, \theta_m)$, where f is of known form and $\boldsymbol{\theta} = (\theta_1, \theta_2, \ldots, \theta_m)$ belongs to a subset Θ of m-dimensional space but otherwise is unknown. $X_1, X_2, \ldots, X_n$ may, for example, represent the lifetimes of n identical units of some kind.

Let us first consider the case of *no censoring*. The joint density (frequency function) of $X_1, X_2, \ldots, X_n$ is given by $\prod_{i=1}^{n} f(x_i; \boldsymbol{\theta})$. Say this expression is a function of $\boldsymbol{\theta}$ for fixed $x_1, x_2, \ldots, x_n$. We denote this function as

$$\ell(\boldsymbol{\theta}; x_1, x_2, \ldots, x_n) = \ell(\boldsymbol{\theta}; x) \tag{C.1}$$

Then $\ell(\boldsymbol{\theta}; x)$ is called the *likelihood function*. If X has a discrete distribution, $\ell(\boldsymbol{\theta}; x)$ directly expresses the probability of observing the values $x_1, x_2, \ldots, x_n$ for a given $\boldsymbol{\theta}$ and hence indicates "how likely" it is to obtain the observations $x_1, x_2, \ldots, x_n$ for each given $\boldsymbol{\theta}$. A similar argument applies in the case where X has a continuous distribution.

The idea of the maximum likelihood principle is now as follows: Search for the value $\hat{\theta}(x_1, x_2, \ldots, x_n)$ which is most likely to have produced the observations $x_1, x_2, \ldots, x_n$;

$$\ell(\hat{\theta}(x); x) \geq L(\boldsymbol{\theta}; x) \quad \text{for} \quad \boldsymbol{\theta} \in \Theta \tag{C.2}$$

$\hat{\theta}(X_1, X_2, \ldots, X_n)$ is then called the *maximum likelihood estimator* of θ (MLE of θ).

In the case of *censored* observations, the likelihood function is somewhat modified. Suppose that some of the observations are left censored and that some are right censored. Then we can split the observation numbers, $1, 2, \ldots, n$, into three disjoint sets, U corresponding to the uncensored observations, C_R for right censored and C_L for left censored. The likelihood function is defined as

$$\ell(\theta; x) = \prod_{j \in C_L} F(x_j; \theta) \prod_{j \in U} f(x_j; \theta) \prod_{j \in C_R} R(x_j; \theta) \qquad (C.3)$$

The modification of the likelihood function is as follows: For the left censored observations replace the corresponding densities by the distribution function F; for the right-censored observations replace the densities by the survivor function $R = 1 - F$. Since $\ln \ell(\theta; x)$ attains its maximum for the same value of θ as does $\ell(\theta; x)$, $\hat{\theta}(x)$ may also be found from

$$\ln \ell(\hat{\theta}(x); x) \geq \ln \ell(\theta; x) \qquad \text{for } \theta \in \Theta \qquad (C.4)$$

Usually it is more convenient from a mathematical point of view to work with $\ln \ell(\theta; x)$ than with $\ell(\theta; x)$.

In commonly occurring situations the first step in the search for the MLE of θ, is to solve the likelihood equations

$$U_j = \frac{\partial \ln \ell(\theta; x)}{\partial \theta_j} = 0 \qquad \text{for } j = 1, 2, \ldots, m \qquad (C.5)$$

This approach will frequently involve numerical methods such as Newton and quasi-Newton algorithms for solving equations.

Suppose that we succeed in determining the MLE $\hat{\theta}_1, \hat{\theta}_2, \ldots, \hat{\theta}_m$ of $\theta_1, \theta_2, \ldots, \theta_m$ and that we are interested in estimating some function $\psi(\theta) = g(\theta_1, \theta_2, \ldots, \theta_m)$, where g is a specified one-to-one function. Then the MLE $\hat{\psi}$ of ψ will be given by $g(\hat{\theta}_1, \hat{\theta}_2, \ldots, \hat{\theta}_m)$.

Under mild regularity conditions, $\hat{\theta}$ is a consistent estimator of θ. Asymptotic properties of $\hat{\theta}$ are discussed in several textbooks in statistics, for example, by Cramér (1946), Mann, Schafer, and Singpurwalla (1974), Kalbfleish and Prentice (1980), Lawless (1982), and Nelson (1990).

Example C.1

Let $T_1, T_2, \ldots, T_k$ be independent and identically distributed with an inverse Gaussian distribution $IG(\mu, \lambda)$, where $\mu > 0$ and $\lambda > 0$ but are otherwise unknown. The ML estimators μ^* and λ^* for μ and λ may be derived the following way:

The likelihood function in this case is

$$\ell(\mu,\lambda;t_1,t_2,\ldots,t_k) = \lambda^{k/2} \prod_{j=1}^{k} \sqrt{\frac{1}{2\pi t_j^3}} \exp\left(-\frac{\lambda}{2}\frac{(t_j-\mu)^2}{\mu^2 t_j}\right)$$

and

$$\ln \ell(\mu,\lambda;t_1,t_2,\ldots,t_k) = \frac{k}{2} \ln \lambda - \frac{1}{2} \sum_{j=1}^{k} \ln 2\pi t_j^3 - \frac{\lambda}{2} \sum_{j=1}^{k} \frac{(t_j-\mu)^2}{\mu^2 t_j}$$

The ML estimates of μ and λ, μ^* and λ^* may now be determined as the solutions of the equations:

$$\frac{\partial \ln \ell}{\partial \mu} = 0 \qquad (C.6)$$

and

$$\frac{\partial \ln \ell}{\partial \lambda} = 0 \qquad (C.7)$$

Equation (C.6) becomes

$$-\frac{\lambda}{2} \sum_{j=1}^{k} \frac{-2(t_j-\mu)\mu^2 t_j - 2\mu t_j(t_j-\mu)^2}{\mu^4 t_j^2} = 0$$

$$\sum_{j=1}^{k} \frac{-2\mu^2 t_j^2 + 2\mu^3 t_j - 2\mu t_j^3 + 4\mu^2 t_j^2 - 2\mu^3 t_j}{\mu^4 t_j} = 0$$

that is,

$$\sum_{j=1}^{k} t_j = k\mu$$

Hence

$$\mu^* = \frac{1}{k} \sum_{j=1}^{k} t_j = \bar{t}$$

We next introduce $\mu = \bar{t}$ into (C.7) and differentiate with respect to λ. Then (C.7) becomes

$$\frac{k}{2} \cdot \frac{1}{\lambda} - \frac{1}{2} \sum_{j=1}^{k} \frac{(t_j - \bar{t})^2}{\bar{t}^2 t_j} = 0$$

that is,

$$\frac{1}{\lambda} = \frac{1}{k} \sum_{j=1}^{k} \frac{(t_j - \bar{t})^2}{\bar{t}^2 t_j} = \frac{1}{k} \sum_{j=1}^{k} \left(\frac{1}{t_j} - \frac{1}{\bar{t}} \right)$$

Hence λ^* is given by

$$\frac{1}{\lambda^*} = \frac{1}{k} \sum_{j=1}^{k} \left(\frac{1}{t_j} - \frac{1}{\bar{t}} \right)$$

It may be verified that the likelihood function attains its maximum for $\mu = \mu^*$ and $\lambda = \lambda^*$. Hence μ^* and λ^* are the MLE of μ and λ. □

In some cases the likelihood equations (C.5) are nonlinear and have to be iteratively solved by use of a computer. This situation is illustrated by the following example.

Example C.2
Let $T_1, T_2, \ldots, T_k$ be independent and identically Weibull distributed with probability density

$$f_{T_j}(t) = \alpha \lambda (\lambda t)^{\alpha-1} e^{-(\lambda t)^\alpha} \quad \text{for } t > 0, \alpha > 0, \lambda > 0$$

Suppose that we want to determine the ML estimates α^* and λ^* for α and λ based on $T_1, T_2, \ldots, T_k$. The likelihood function is

$$\ell(\alpha, \lambda; t_1, t_2, \ldots, t_k) = \lambda^{\alpha k} \alpha^k \prod_{j=1}^{k} (t_j^{\alpha-1} \exp(-\lambda t_j)^\alpha)$$

and the log likelihood is

$$\ln \ell(\alpha,\lambda;t_1,t_2,\ldots,t_k) = \sum_{j=1}^{k} (\alpha \ln \lambda + \ln \alpha + (\alpha - 1)\ln t_j - (\lambda t_j)^\alpha)$$

The likelihood equations now become

$$0 = \frac{\partial \ln \ell}{\partial \lambda} = \sum_{j=1}^{k} \left(\alpha \frac{1}{\lambda} - \alpha \lambda^{\alpha-1} \cdot t_j^\alpha \right)$$

$$0 = \frac{\partial \ln \ell}{\partial \alpha} = \sum_{j=1}^{k} \left(\ln \lambda + \frac{1}{\alpha} + \ln t_j - (\lambda t_j)^\alpha \cdot \ln(\lambda t_j) \right)$$

These equations are clearly nonlinear in α and λ and have to be solved iteratively on a computer. □

APPENDIX D

Laplace Transforms

Let $f(t)$ be a function that is defined on the interval $(0, \infty)$. The Laplace transform $f^*(s)$ of the function $f(t)$ is defined by

$$f^*(s) = \int_0^\infty e^{-st} f(t)\, dt$$

where s is a real number. In more advanced treatments of the Laplace transform, s is permitted to be a complex number. All functions do not have a Laplace transform. For instance, if $f(t) = e^{t^2}$, the integral diverges for all values of s.

The Laplace transform of $f(t)$ is also often denoted by $\mathscr{L}[f(t)]$:

$$\mathscr{L}[f(t)] = f^*(s) = \int_0^\infty e^{-st} f(t)\, dt \qquad (D.1)$$

to indicate the relation between the functions f and f^*. When $f(t)$ is the probability density of a nonnegative random variable T, the Laplace transform of $f(t)$ is seen to be equal to the expected value of the random variable e^{-sT}.

$$E(e^{-sT}) = \int_0^\infty e^{-st} f(t)\, dt = f^*(s)$$

The function $f(t)$ is called the inverse Laplace transform of $f^*(s)$, and is written

$$f(t) = \mathscr{L}^{-1}[f^*(s)] \qquad (D.2)$$

Theorem D.1. Let $f(t)$ be a function that is piecewise continuous on every finite interval in the range $t \geq 0$ and satisfies

APPENDIX D

$$|f(t)| \leq Me^{\alpha t} \quad \text{for all} \quad t \geq 0$$

and for some constants α and M. Then the Laplace transform of $f(t)$ exists for all $s > \alpha$.

Example D.1
Consider the function $f(t) = e^{\alpha t}$, where α is a constant. We have

$$f^*(s) = \int_0^\infty e^{-st} e^{\alpha t}\, dt = \int_0^\infty e^{-t(s-\alpha)}\, dt$$

$$= \lim_{T \to \infty} \left[\frac{-1}{s-\alpha} e^{-t(s-\alpha)} \right]_0^T$$

$$= \frac{1}{s-\alpha} \quad \text{when} \quad s > \alpha$$

Thus

$$\mathscr{L}[e^{\alpha t}] = \frac{1}{s-\alpha} \quad \text{when} \quad s > \alpha \qquad \square$$

Some elementary functions and their Laplace transforms are given in Table D.1. Some important properties of the Laplace transform are listed below. The proofs are left to the reader who may consult standard textbooks on mathematical analysis.

1. $\mathscr{L}[f_1(t) + f_2(t)] = \mathscr{L}[f_1(t)] + \mathscr{L}[f_2(t)]$
2. $\mathscr{L}[\alpha f(t)] = \alpha \mathscr{L}[f(t)]$
3. $\mathscr{L}[f(t-\alpha)] = e^{-\alpha s} \mathscr{L}[f(t)]$
4. $\mathscr{L}[e^{\alpha t} f(t)] = f^*(s - \alpha)$
5. $\mathscr{L}[f'(t)] = s \mathscr{L}[f(t)] - f(0)$
6. $\mathscr{L}[\int_0^t f(u)\, du] = \mathscr{L}[f(t)]/s$
7. $\mathscr{L}[\int_0^t f_1(t-u) f_2(u)\, du] = \mathscr{L}[f_1(t)] \cdot \mathscr{L}[f_2(t)]$
8. $\lim_{s \to \infty} s f^*(s) = \lim_{t \to 0} f(t)$
9. $\lim_{s \to 0} s f^*(s) = \lim_{t \to \infty} f(t)$

Table D.1 Some Laplace Transforms

$f(t), t \geq 0$	$f^*(s) = \mathscr{L}([f(t)])$
1	$\dfrac{1}{s}$
t	$\dfrac{1}{s^2}$
t^2	$\dfrac{2!}{s^3}$
t^n	$\dfrac{n!}{s^{n+1}}$ for $n = 0, 1, 2, \ldots$
t^α	$\dfrac{\Gamma(\alpha + 1)}{s^{\alpha+1}}$ for $\alpha > -1$
$e^{\alpha t}$	$\dfrac{1}{s - \alpha}$
$e^{\alpha t} t^n$	$\dfrac{n!}{(s - \alpha)^{n+1}}$
$\cos \omega t$	$\dfrac{s}{s^2 + \omega^2}$
$\sin \omega t$	$\dfrac{\omega}{s^2 + \omega^2}$
$\cosh \alpha t$	$\dfrac{s}{s^2 - \alpha^2}$
$\sinh \alpha t$	$\dfrac{\alpha}{s^2 - \alpha^2}$

APPENDIX E

Basic Concepts

Accelerated Test
A test in which the applied stress level is chosen to exceed that stated in the reference conditions in order to shorten the time required to observe the stress response of the item, or to magnify the responses in a given time. To be valid, an accelerated test must not alter the basic modes and mechanisms of failure, or their relative prevalence (BS 4778).

Active Redundancy
That redundancy wherein all means for performing a required function are intended to operate simultaneously (IEC 50).

Availability
The ability of an item (under combined aspects of its reliability, maintainability and maintenance support) to perform its required function at a stated instant of time or over a stated period of time (BS 4778).

Coherent System
A system whose structure function is nondecreasing as a function of all state variables. A system is coherent when

- All the components are in a failed state, the system is in a failed state
- All the components are functioning, the system is functioning
- The system is in a failed state, no additional component failures will cause the system to function
- The system is functioning, no component repair will cause the system to fail

Common Cause Failure
Multiple failures attributable to a common cause.

485

Corrective Maintenance
The actions performed, as a result of failure, to restore an item to a specified condition (MIL-STD-2173(AS)). The maintenance carried out after a failure has occurred and intended to restore an item to a state in which it can perform its required function (BS 4778).

Dependability
The collective term used to describe the availability performance and its influencing factors: reliability performance, maintainability performance, and maintenance support performance (IEC 300-1).

Design Review
A formal, documented, comprehensive, and systematic examination of a design to evaluate the design requirements and the capability of the design to meet these requirements and to identify problems and propose solutions (ISO 8402).

Distribution Function
Consider a random variable X. The distribution function of X is

$$F_X(x) = P(X \leq x)$$

Downtime
The period of time during which an item is not in a condition to perform its required function (BS 4778).

Fail Safe
A design property of an item that prevents its failures being critical failures (BS 4778).

Failure
The termination of its ability to perform a required function (BS 4778).

Failure Cause
The physical or chemical processes, design defects, quality defects, part misapplication, or other processes that are the basic reason for failure or that initiate the physical process by which deterioration proceeds to failure (MIL-STD-1629). The circumstances during design, manufacture, or use that have led to a failure (IEC 50).

Failure Effect
The consequence(s) a failure mode has on the operation, function, or status of an item (MIL-STD-1629).

Failure Mode
The effect by which a failure is observed on the failed item (EuReDatA, 1983).

Failure Mode and Effect Analysis (FMEA)
A procedure by which each potential failure mode in a system is analyzed to determine the results or effects thereof on the system and to classify each potential failure mode according to its severity (MIL-STD-1629).

Failure Rate
The rate at which failures occur as a function of time. If T denotes the time to failure of an item, the failure rate $z(t)$ is defined as

$$z(t) = \lim_{\Delta t \to \infty} \frac{P(t < T \leq t + \Delta t | T > t)}{\Delta t}$$

The failure rate is sometimes called *force of mortality* (FOM).

Failure Symptom
An identifiable physical condition by which a potential failure can be recognized (MIL-STD-2173(A)).

Fatigue
Reduction in resistance to failure of a material over time as a result of repeated or cyclic applied loads (MIL-STD-2173(AS)).

Fatigue Life
For an item subject to fatigue, the total time to functional failure of the item (MIL-STD-2173(AS)).

Force of Mortality
Same as "failure rate."

Functioning State
The state when an item is performing a required function (see also "operating state").

Gradual Failure
Failure that could be anticipated by prior examination or monitoring (BS 4778).

Hazard Rate
Same as "failure rate."

Hidden Failure
A failure not evident to the crew or operator during the performance of normal duties (MIL-STD-2173(AS)).

Infant Mortality
The relatively high conditional probability of failure during the period immediately after an item enters service. Such failures are due to defects in manufacturing not detected by quality control (MIL-STD-2173(AS)).

Inspection
Activities such as measuring, examining, testing, and gauging one or more characteristics of a product or service and comparing these with specified requirements to determine conformity (ISO 8402).

Intermittent Failure
Failure of an item for a limited period of time, after which the item restores its required function without being subjected to any external corrective action (BS 4778).

Life Cycle Cost (LCC)
The cost of acquisition and ownership of a product over a defined period of its life cycle. It may include the cost of development, acquisition, operation, support, and disposal of the product (IEC 300-3).

Maintainability
The ability of an item, under stated conditions of use, to be retained in, or restored to, a state in which it can perform its required functions, when maintenance is performed under stated conditions and using prescribed procedures and resources (BS 4778).

Maintenance
The combinations of all technical and corresponding administrative actions, including supervision actions, intended to retain an entity in, or restore it to, a state in which it can perform its required function (IEC 50).

Maintenance Support Performance
The ability of a maintenance organization, under given conditions, to provide upon demand, the resources required to maintain an entity, under a given maintenance policy (IEC 50).

Mean Time to Failure (MTTF)
Let T denote the time to failure of an item, with probability density $f(t)$ and survivor function $R(t)$. The mean time to failure MTTF is the mean (expected) value of T which is given by

$$\text{MTTF} = \int_0^\infty t \cdot f(t)\,dt = \int_0^\infty R(t)\,dt$$

Mean Time to Repair (MTTR)

Let D denote the downtime (or repair time) after a failure of an item. Let $f_D(d)$ denote the probability density of D, and let $F_D(d)$ denote the distribution function of D. The mean time to repair MTTR is the mean (expected) value of D which is given by

$$\text{MTTR} = \int_0^\infty t \cdot f_D(t)\,dt = \int_0^\infty (1 - F_D(t))\,dt$$

MTTR is also sometimes called the *mean downtime (MDT)* of the item. In some situations MTTR is used to denote the mean *active* repair time instead of the mean down time of the item.

Operating State

The state when an entity is performing a required function (IEC 50).

Percentile

Let X be a random variable with distribution function $F(x)$. The upper $100\epsilon\%$ percentile x_ϵ of the distribution $F(x)$ is defined such that

$$P(X > x_\epsilon) = \epsilon$$

Preventive Maintenance

The maintenance carried out at predetermined intervals or corresponding to prescribed criteria and intended to reduce the probability of failure or the performance degradation of an item (BS 4778).

Probability Density

Consider a random variable X. The probability density function $f_X(x)$ of X is

$$f_X(x) = \frac{dF_X(x)}{dx} = \lim_{\Delta x \to \infty} \frac{P(x < X \leq x + \Delta x)}{\Delta x}$$

where $F_X(x)$ denotes the distribution function of X and is assumed differentiable.

Quality

The totality of features and characteristics of a product or service that bear on its ability to satisfy stated or implied needs (ISO 8402).

Redundancy

In an entity the existence of more than one means for performing a required function (IEC 50).

Reliability
The ability of an item to perform a required function, under given environmental and operational conditions and for a stated period of time (ISO 8402).

Reliability Centered Maintenance (RCM)
A disciplined logic or methodology used to identify preventive maintenance tasks to realize the inherent reliability of equipment at least expenditure of resources (MIL-STD-2173(AS)).

Repair
The part of corrective maintenance in which manual actions are performed on the entity (IEC 50).

Required Function
A function or a combination of functions, of an entity, that is considered necessary to provide a given service (IEC 50).

Safety
Freedom from those conditions that can cause death, injury, occupational illness, or damage to or loss of equipment or property (MIL-STD-882).

Security
Dependability with respect to the prevention of unauthorized access and/or handling of information (Laprie 1992).

Severity
The consequences of a failure mode. Severity considers the worst potential consequence of a failure determined by the degree of injury, property damage, or system damage that could ultimately occur (MIL-STD-1629)

Single Failure Point
The failure of an item that would result in failure of the system and is not compensated for by redundancy or alternative operational procedure (MIL-STD-1629).

State Variable
A variable $X(t)$ associated with an item such that

$$X(t) = \begin{cases} 1 & \text{if item is functioning at time } t \\ 0 & \text{if item is in a failed state at time } t \end{cases}$$

State Vector
A vector $X(t) = (X_1(t), X_2(t), \ldots, X_n(t))$ of the state variables of the n components comprising the system.

Step Stress Test
A test consisting of several stress levels applied sequentially for periods of equal duration to one sample. During each period a stated stress level is applied and the stress level is increased from one period to the next (BS 4778).

Structure Function
A variable $\phi(X(t))$ associated with a system, with state vector $X(t)$, such that

$$\phi(X(t)) = \begin{cases} 1 & \text{if system is functioning at time } t \\ 0 & \text{if system is in a failed state at time } t \end{cases}$$

Sudden Failure
A failure that could not be anticipated by prior examination or monitoring (IEC 50).

Survivor Function
Let T denote the time to failure of an item. The survivor function $R(t)$ of the item is

$$R(t) = P(T > t) \quad \text{for } t \geq 0$$

$R(t)$ is sometimes called the *reliability function* or the *survival probability at time t* of the item.

Test Frequency
The number of tests of the same type per unit time interval; the inverse of the test interval (IEEE Std. 352).

Test Interval
The elapsed time between the initiation of identical tests on the same sensor, channel, and so forth (IEEE Std. 352).

Wear-out Failure
A failure whose probability of occurrence increases with the passage of time as a result of processes inherent in the entity (IEC 50).

APPENDIX F

Statistical Tables

The Cumulative Standard Normal Distribution

$$\Phi(z) = P(Z \leq z) = \int_{-\infty}^{z} \frac{1}{\sqrt{2\pi}} e^{-u^2/2} \, du$$

z	0.00	0.01	0.02	0.03	0.04	0.05	0.06	0.07	0.08	0.09
0.0	0.500	0.504	0.508	0.512	0.516	0.520	0.524	0.528	0.532	0.536
0.1	0.540	0.544	0.548	0.552	0.556	0.560	0.564	0.567	0.571	0.575
0.2	0.579	0.583	0.587	0.591	0.595	0.599	0.603	0.606	0.610	0.614
0.3	0.618	0.622	0.626	0.629	0.633	0.637	0.641	0.644	0.648	0.652
0.4	0.655	0.659	0.663	0.666	0.670	0.674	0.677	0.681	0.684	0.688
0.5	0.691	0.695	0.698	0.702	0.705	0.709	0.712	0.716	0.719	0.722
0.6	0.726	0.729	0.732	0.736	0.739	0.742	0.745	0.749	0.752	0.755
0.7	0.758	0.761	0.764	0.767	0.770	0.773	0.776	0.779	0.782	0.785
0.8	0.788	0.791	0.794	0.797	0.800	0.802	0.805	0.808	0.811	0.813
0.9	0.816	0.819	0.821	0.824	0.826	0.829	0.831	0.834	0.836	0.839
1.0	0.841	0.844	0.846	0.849	0.851	0.853	0.855	0.858	0.860	0.862
1.1	0.864	0.867	0.869	0.871	0.873	0.875	0.877	0.879	0.881	0.883
1.2	0.885	0.887	0.889	0.891	0.893	0.894	0.896	0.898	0.900	0.901
1.3	0.903	0.905	0.907	0.908	0.910	0.911	0.913	0.915	0.916	0.918
1.4	0.919	0.921	0.922	0.924	0.925	0.926	0.928	0.929	0.931	0.932
1.5	0.933	0.934	0.936	0.937	0.938	0.939	0.941	0.942	0.943	0.944
1.6	0.945	0.946	0.947	0.948	0.949	0.951	0.952	0.953	0.954	0.954
1.7	0.955	0.956	0.957	0.958	0.959	0.960	0.961	0.962	0.962	0.963
1.8	0.964	0.965	0.966	0.966	0.967	0.968	0.969	0.969	0.970	0.971
1.9	0.971	0.972	0.973	0.973	0.974	0.974	0.975	0.976	0.976	0.977
2.0	0.977	0.978	0.978	0.979	0.979	0.980	0.980	0.981	0.981	0.982
2.1	0.982	0.983	0.983	0.983	0.984	0.984	0.985	0.985	0.985	0.986
2.2	0.986	0.986	0.987	0.987	0.987	0.988	0.988	0.988	0.989	0.989
2.3	0.989	0.990	0.990	0.990	0.990	0.991	0.991	0.991	0.991	0.992
2.4	0.992	0.992	0.992	0.992	0.993	0.993	0.993	0.993	0.993	0.994
2.5	0.994	0.994	0.994	0.994	0.994	0.995	0.995	0.995	0.995	0.995
2.6	0.995	0.995	0.996	0.996	0.996	0.996	0.996	0.996	0.996	0.996
2.7	0.997	0.997	0.997	0.997	0.997	0.997	0.997	0.997	0.997	0.997
2.8	0.997	0.998	0.998	0.998	0.998	0.998	0.998	0.998	0.998	0.998
2.9	0.998	0.998	0.998	0.998	0.998	0.998	0.999	0.999	0.999	0.999
3.0	0.999	0.999	0.999	0.999	0.999	0.999	0.999	0.999	0.999	0.999

$\Phi(-z) = 1 - \Phi(z)$

Percentage Points of the Chi-square (χ^2) Distribution
$P(Z > z_{\alpha,\nu}) = \alpha$

$\nu \backslash \alpha$	0.995	0.990	0.975	0.950	0.05	0.025	0.010	0.005
1	0.00	0.00	0.00	0.00	3.84	5.02	6.63	7.88
2	0.01	0.02	0.05	0.10	5.99	7.38	9.21	10.60
3	0.07	0.11	0.22	0.35	7.81	9.35	11.34	12.84
4	0.21	0.30	0.48	0.71	9.49	11.14	13.28	14.86
5	0.41	0.55	0.83	1.15	11.07	12.38	15.09	16.75
6	0.68	0.87	1.24	1.64	12.59	14.45	16.81	18.55
7	0.99	1.24	1.69	2.17	14.07	16.01	18.48	20.28
8	1.34	1.65	2.18	2.73	15.51	17.53	20.09	21.96
9	1.73	2.09	2.70	3.33	16.92	19.02	21.67	23.59
10	2.16	2.56	3.25	3.94	18.31	20.48	23.21	25.19
11	2.60	3.05	3.82	4.57	19.68	21.92	24.72	26.76
12	3.07	3.57	4.40	5.23	21.03	23.34	26.22	28.30
13	3.57	4.11	5.01	5.89	22.36	24.74	27.69	29.82
14	4.07	4.66	5.63	6.57	23.68	26.12	29.14	31.32
15	4.60	5.23	6.27	7.26	25.00	27.49	30.58	32.80
16	5.14	5.81	6.91	7.96	26.30	28.85	32.00	34.27
17	5.70	6.41	7.56	8.67	27.59	30.19	33.41	35.72
18	6.26	7.01	8.23	9.39	28.87	31.53	34.81	37.16
19	6.84	7.63	8.91	10.12	30.14	32.85	36.19	38.58
20	7.43	8.26	9.59	10.85	31.41	34.17	37.57	40.00
25	10.52	11.52	13.12	14.61	37.65	40.65	44.31	46.93
30	13.79	14.95	16.79	18.49	43.77	46.98	50.89	53.67
40	20.71	22.16	24.43	26.51	55.76	59.34	63.69	66.77
50	27.99	29.71	32.36	34.76	67.50	71.42	76.15	79.49
60	35.53	37.48	40.48	43.19	79.08	83.30	88.38	91.95
70	43.28	45.44	48.76	51.74	90.53	95.02	100.42	104.22
80	51.17	53.54	57.15	60.39	101.88	106.63	112.33	116.32
90	59.20	61.75	65.65	69.13	113.14	118.14	124.12	128.30
100	67.33	70.06	74.22	77.93	124.34	129.56	135.81	140.17

References

Reliability engineering is a rather young scientific dicipline. New theories and methods are continually being developed. The basic theory and the most fundamental methods are presented in this book. Practicing reliability engineers should, however, keep themselves updated about the latest developments.

A number of journals and periodicals are presenting papers and information on reliability and related topics:

Reliability Engineering and System Safety (Elsevier Science Publishers)
IEEE Transactions on Reliability
Quality and Reliability International (John Wiley & Sons)
Technometrics (American Society of Quality Control, ASQC, and American Statistical Association, ASA)
Quality and Reliability Management (MCB University Press Ltd.)
IEEE Reliability Society Newsletter
Reliability Review (published quarterly by the Reliability Division of ASQC).

ooo OOO ooo

Aalen, O. O. 1975. *Statistical Inference for a family of Counting Processes.* Ph.D. dissertation. Department of Statistics, University of California, Berkeley.

Aalen, O. O. 1978. Non-parametric inference for a family of counting processes. *Ann. Statist.* **6**:701–726.

Akersten, P. A. 1991. *Repairable Systems Reliability, Studied by TTT-Plotting Techniques.* Ph.D. dissertation. Division of Quality Technology, Department of Mechanical Engineering. Linköping University, Sweden.

Amendola, A., and A. Z. Keller. 1985. *Reliability Data Bases.* Reidel, Dordrecht, The Netherlands.

American Institute of Chemical Engineers. 1985. *Guidelines for Hazard Evaluation Procedures.* American Institute of Chemical Engineers, Center for Chemical Process Safety, New York.

American Institute of Chemical Engineers. 1989. *Guidelines for Chemical Process Quantitative Risk Analysis*. American Institute of Chemical Engineers, Center for Chemical Process Safety, New York.

Andersen, P. K., and Ø. Borgan. 1985. Counting process models for life history data: A review. *Scand. J. Statist.* **12**:97–158.

Anderson, R. T. 1976. *Reliability Design Handbook*. NO RDH 376. Reliability Analysis Center (RAC), 201 Mill St. Rome, NY 14440-8200.

Anderson, R. T., and L. Neri. 1990. *Reliability-Centered Maintenance: Management and Engineering Methods*. Elsevier Applied Science, London.

Apostolakis, G., and P. Moieni. 1987. The foundation of models of dependence in probabilistic safety assessment. *Reliability Engineering*, **18**:177–195.

Ansell, J. I., and M. J. Phillips. 1989. Practical problems in the statistical analysis of reliability data (with discussion). *Applied Statistics* **38**:205–247.

Ascher, H., and H. Feingold. 1984. *Repairable Systems Reliability: Modeling, Inference, Misconceptions and Their Causes*. Marcel-Dekker, New York.

Atwood, C. L. 1986. The binomial failure rate common cause model. *Technometrics* **28**:139–148.

Atwood, C. L. 1992. Parametric estimation of time-dependent failure rates for probabilistic risk assessment. *Reliability Engineering and System Safety*, **37**:181–194.

Aven, T. 1983. Optimal replacement under a minimal repair strategy—A general failure model. *Advances in Applied Probability* **15**:198–211.

Aven, T. 1986. Reliability/availability evaluation of coherent systems based on minimal cut sets. *Reliability Engineering* **13**:93–104.

Aven, T. 1992. *Reliability and Risk Analysis*. Elsevier Science Publishers, New York.

Bain, L. J. M. Engelhardt, and F. T. Wright. 1985. Tests for an increasing trend in the intensity of a Poisson process. *Journal of the American Statistical Association* **80**: 419–422.

Barlow, R. E. and R. Campo. 1975. Total time on test processes and applications to failure data analysis. In *Reliability and Fault Tree Analysis*, ed. by R. E. Barlow, J. B. Fussell, and N. D. Singpurwalla. SIAM, Philadelphia.

Barlow, R. E., J. B. Fussell, and N. D. Singpurwalla. 1975. *Reliability and Fault Tree Analysis*. SIAM-Society for Industrial and Applied Mathematics, Philadelphia.

Barlow, R. E., and H. E. Lambert. 1975. *Introduction to Fault Tree Analysis* in *Reliability and Fault Tree Analysis*, ed. by R. E. Barlow, J. B. Fussell, and N. D. Singpurwalla. SIAM, Philadelphia.

Barlow, R. E., and F. Proschan. 1969. A note on tests for monotone failure rate. *Annals of Mathematical Statistics* **40**:595–600.

Barlow, R. E., and F. Proschan. 1975. *Statistical Theory of Reliability and Life Testing, Probability Models*. Holt, Rinehart and Winston, New York.

Bendell, A., and L. A. Walls. 1985. Exploring reliability data. *Quality and Reliability Engineering International* **1**:37–51.

Berger, J. O. 1985. *Statistical Decision Theory and Bayesian Analysis*. Springer-Verlag, New York.

REFERENCES

Bergman, B., and B. Klefsjö. 1982. A graphical method applicable to age replacement problems. *IEEE Transactions on Reliability*. **R-31**:478–481.

Bergman, B., and B. Klefsjö. 1984. The total time on test concept and its use in reliability theory. *Operations Research* **32**:596–606.

Bateman manuscript project. 1954. *Tables of Integral Transforms*. McGraw-Hill, New York.

Bhattacharayya, G. K., and A. Fries. 1982. Fatigue failure models—Birnbaum-Saunders vs. inverse Gaussian. *IEEE Trans. on Reliability* **31**:439–441.

Bickel, P. J., and K. A. Doksum. 1977. *Mathematical Statistics: Basic Ideas and Selected Topics*. Holden-Day, San Francisco.

Billinton, R., and R. N. Allan. 1983. *Reliability Evaluation of Engineering Systems: Concepts and Techniques*. Longman Scientific & Technical, Essex, England.

Birnbaum, Z. W. 1969. On the importance of different components in a multicomponent system. In *Multivariate Analysis*, ed. by P. R. Krishnaiah. Academic Press, San Diego, pp. 581–592.

Birnbaum, Z. W., and S. C. Saunders. 1969. A new family of life distributions. *Journal of Applied Probability* **6**:319–327.

Blackwell, D. 1948. A renewal theorem. *Duke Mathemathical Journal* **15**:145–150.

Bodsberg, L. 1987 *Failure Rate Prediction of Electronic Components*. SINTEF Report STF75 A87006. SINTEF, N-7034 Trondheim, Norway.

Bodsberg, L., and P. Hokstad. 1988. *Reliability of Safety Shutdown Systems. Models for Dependent and Undetected Failures*. SINTEF Report STF75 A88011, Trondheim, Norway.

Borowkow, A. A. 1972. Prozesse der Bedienungstheorie, Moscow.

Bourne, A. J., G. T. Edwards, D. M. Hunns, D. R. Poulter, and I. A. Watson. 1981. *Defences against Common-Mode Failures in Redundancy Systems*. SRD R 196. U.K. Atomic Energy Authority, Warrington.

Box, G. E. P., and W. J. Hill. 1967. Discrimination among mechanistic models. *Technometrics* **9**:57–71.

Box, G. E. P., W. G. Hunter, and J. S. Hunter. 1978. *Statistics for Experimenters*. Wiley, New York.

British Standard, BS 4778. *Glossary of Terms Used in Quality Assurance Including Reliability and Maintainability Terms*. British Standards Institution, London.

British Standard, BS 5760. *Reliability of Constructed or Manufactured Products, Systems, Equipments and Components*. British Standards Institution, London.

Brown, M., and F. Proschan. 1983. Imperfect repair. *Journal of Applied Probability* **20**:851–859.

Buttersworth, R. W., and K. T. Marshall. 1964. A survey of renewal theory with emphasis on approximations, bounds and applications. Naval Postgraduate School, Monterey, California.

Cannon, A. G., and A. Bendell. 1991. *Data Banks*. Elsevier Applied Science, London.

CARA (Computer Aided Reliability Analysis). PC-program for FMECA, fault tree analysis, cause-consequence analysis and failure rate estimation developed by SINTEF Safety and Reliability N-7034 Trondheim, Norway.

Chatterjee, P. 1975. Modularization of fault trees: A method to reduce the cost of analysis. In *Reliability and Fault Tree Analysis*, Ed. by R. E. Barlow, J. B. Fussell, and N. D. Singpurwalla. SIAM, Philadelphia.

Chhikara, R. S., and J. L. Folks. 1974. Estimation of the inverse Gaussian distribution. *JASA* **69**:250–254.

Chhikara, R. S., and J. L. Folks. 1977. The inverse Gaussian distribution as a lifetime model. *Technometrics* **19**:461–468.

Chhikara, R. S., and J. L. Folks. 1989. *The Inverse Gaussian Distribution. Theory, Methodology and Applications*. Marcel-Dekker, New York.

CNET. 1983. *Recueil de données de fiabilité du CNET*. Centre National d'Etudes des Télécommunications, CNET, LAB, Centre de Fiabilité, BP 40, 22301 Lannion Cedex, France.

Cox, D. R. 1962. *Renewal Theory*. Methuen, London.

Cox, D. R. 1972. Regression models and life tables with discussion. *Journal of the Royal Statistical Society* **B 21**:411–421.

Cox, D. R., and V. Isham. 1980. *Point Processes*. Chapman and Hall, London.

Cox, D. R., and P. A. Lewis. 1966. *The Statistical Analysis of Series of Events*, Methuen, London.

Cox, D. R., and H. D. Miller. 1965. *The Theory of Stochastic Processes*. Methuen, London.

Cox, D. R., and D. Oakes. 1984. *Analysis of Survival Data*. Chapman and Hall, London.

Cramér, H. 1946. *Mathematical Methods of Statistics*. Princeton University Press, Princeton.

Crow, L. H. 1974. Reliability Analysis of Complex Repairable Systems. In *Reliability and Biometry*, ed. by F. Proschan and R. J. Serfling. SIAM, Philadelphia, pp. 379–410.

Crowder, M. J., A. C. Kimber, R. L. Smith, and T. J. Sweeting. 1991. *Statistical Analysis of Reliability Data*. Chapman and Hall, London.

DeGroot, M. H. 1970. *Optimal Statistical Decisions*. McGraw-Hill, New York.

Dhillon, B. S., and C. Singh. 1981. *Engineering Reliability*. Wiley, New York.

Doksum, K. A., and A. Høyland. 1992. Models for variable-stress accelerated life testing experiments based on Wiener processes and the inverse Gaussian distribution. *Technometrics* **34**:74–82.

Doulliez, P., and J. Jamoulle. 1972. Transportation networks with random arc capacities. *RAIRO* **3**:45–60.

Drenick, R. F. 1960. The failure law of complex equipment. *Journal of the Society for Industrial Applied Mathematics* **8**:680–690.

Duane, J. T. 1964. Learning curve approach to reliability monitoring. IEEE Transactions on Aerospace **2**:563–566.

Dudewicz, E. J., and S. A. Mishra. 1988. *Modern Mathematical Statistics*. Wiley, New York.

Edwards, G. T., and I. A. Watson. 1979. A study of common-mode failures. *UKAEA SRD R 146*.

Endrenyi, J. 1978. *Reliability Modeling in Electric Power Systems*. Wiley, New York.

Epstein, B., and M. Sobel. 1953. Life-testing. *Journal of the American Statistical Association* **48**:486–502.

Esary, J. D., F. Proschan, and D. W. Walkup. 1967. Association of Random Variables. *Annals of Mathematical Statistics* **38**:1466–1474.

EuReDatA. 1983. *Reference Classification Concerning Components' Reliability.* EuReDatA Project Report no. 1. Commission of the European Communities, Joint Research Centre, Ispra (Varese), Italy.

EuReDatA. 1986. *Guide to Reliability Data Collection and Management.* EuReDatA Project Report no. 3. Commission of the European Communities, Joint Research Centre, Ispra Varese, Italy.

Feller, W. 1968. *An Introduction to Probability Theory and Its Applications*, vol. 1. Wiley, New York.

Flamm, J., and T. Luisi, eds. 1992. *Reliability Data Collection and Analysis.* Kluwer, Deventer, The Netherlands.

Fleming, K. N. 1974. A reliability model for common mode failures in redundant safety systems. General Atomic Report, GA-13284.

Fleming, K. N., A. Mosleh, and R. K. Deremer. 1986. A systematic procedure for the incorporation of common cause events into risk and reliability models. *Nucl. Engineering and Design* **93**:245–279.

FMD-91 1991. *Failure Mode/Mechanism Distribution.* Reliability Analysis Center RAC 201 Mill St. Rome, NY 14440-8200.

Fussell, J. B. 1975. How to hand calculate system safety and reliability characteristics. *IEEE Trans. on Reliability* **R-24**:169–174.

Gaudoin, O. 1992. Optimal properties of the Laplace trend test for software-reliability models. *IEEE Transactions on Reliability* **41**:525–532.

Gertsbakh, I. 1989. *Statistical Reliability Theory.* Marcel-Dekker, New York.

Green, A. E., and A. J. Bourne. 1972. *Reliability Technology.* Wiley, New York.

Gumbel, E. J. 1958. *Statistics of Extremes.* Colombia University Press, New York.

Guthrie, V. H., J. A. Farquharson, R. W. Bonnett, and E. E. Bjoro. 1990. Guidelines for integrating RAM considerations into an engineering project. *IEEE Transactions on Reliability* **39**:133–139.

Hammer, W. 1972. *Handbook of System and Product Safety.* Prentice-Hall, Englewood Cliffs, NJ.

Harris, B. 1986. Stochastic Models for Common Failures. In *Reliability and Quality Control*, ed. by A. P. Basu. Elsevier Science Publishers, New York, pp. 185–200.

Henley, E. J., and H. Kumamoto. 1981. *Reliability Engineering and Risk Assessment.* Prentice-Hall, Englewood Cliffs, NJ. (Reprinted and distributed by IEEE Press, 1991, with title: *Probabilistic Risk Assessment—Reliability Engineering, Design, and Analysis.*)

Hokstad, P. 1988. A shock model for common-cause failures. *Reliability Engineering and System Safety* **23**:127–145.

Holen, A. T., A. Høyland, and M. Rausand. 1988. *Reliability Analysis* (in Norwegian). Tapir, Trondheim.

HRD4. 1987. *Handbook of Reliability Data for Components used in Telecommunication Systems.* British Telecom. London Information (Rowse Muir) Ltd. Index House. ASCOT, Berks SL5 7EU, England.

Humphreys, R. A. 1987. Assigning a numerical value to the beta factor common cause evaluation. *Reliability '87.* Proceedings paper 2C.

Hunter, W. G., and G. E. Box. 1965. The experimental study of physical mechanisms. *Technometrics* **7**:23–42.

Hunter, W. G., and A. M. Reiner. 1965. Designs for discrimination between two rival models. *Technometrics* **7**:307–323.

IEC 50(191). 1990. *International Electrotechnical Vocabulary (IEV). Chapter 191: Dependability and Quality of Service.* International Electrotechnical Commission, Geneva.

IEC 271. 1974. *List of Basic Terms, Definitions and Related Mathematics for Reliability.* International Electrotechnical Commission, Geneva.

IEC 300. 1992. *Dependability Management.* International Electrotechnical Commission, Geneva.

IEC 812. 1985. *Analysis Techniques for System Reliability: Procedures for Failure Mode and Effect Analysis FMEA.* International Electrotechnical Commission, Geneva.

IEEE Std. 352. 1982. *IEEE Guide for General Principles of Reliability Analysis of Nuclear Power Generating Station Protection Systems.* The Institute of Electrical and Electronic Engineers, Inc. New York.

IEEE Std. 500. 1984. *IEEE Guide to the Collection and Presentation Of Electrical, Electronic, Sensing Component, and Mechanical Equipment Reliability Data for Nuclear Power Generating Stations.* Wiley, New York.

Ireson, W. G., and C. F. Coombs, eds. 1988. *Handbook of Reliability Engineering and Management.* McGraw-Hill, New York.

ISO 8402 International Standard. *Quality Vocabulary.* International Standards Organization.

ISO 9000 International Standard. *Quality Management and Quality Assurance Standards: Guidelines for Selection and Use.* International Standards Organization.

Jelinski, Z., and P. B. Moranda. 1972. Software Reliability Research in *Statistical Computer Performance Evaluation,* ed. by W. Freiberger. Academic Press, San Diego.

Jensen, F., and N. E. Petersen. 1982. *Burn-in: An Engineering Approach to the Design and Analysis of Burn-in Procedures.* Wiley, New York.

Johnson, N. L., and S. Kotz. 1970a. *Distributions in Statistics. Continuous Univariate Distributions,* vol. 1. Houghton Mifflin, Boston.

Johnson, N. L. and S. Kotz. 1970b. *Distributions in Statistics: Continuous Univariate Distributions,* vol. 2. Houghton Mifflin, Boston.

Kalbfleisch, J. D., and R. L. Prentice. 1980. *The Statistical Analysis of Failure Time Data.* Wiley, New York.

Kaplan, E. L., and P. Meier. 1958. Nonparametric estimation from incomplete observations. *Journal of the American Statistical Association* **53**:457–481.

Kaplan, S. 1990. Bayes is for eagles. *IEEE Transactions on Reliability* **39**:130–131.

Kapur, K. C., and L. R. Lamberson. 1977. *Reliability in Engineering Design.* Wiley, New York.

Keene, S. J. 1992. Concurrent engineering for better product development. *Reliability Review* **12**:3–5.

Klefsjö, B., and U. Kumar. 1992. Goodness-of-fit tests for the power-law process based on the TTT-plot. *IEEE Transactions on Reliability* **41**:593–598.

Knight, C. R. 1991. Four decades of reliability progress. *Proceedings Annual Reliability and Maintainability Symposium*, IEEE, Reliability Society, pp. 156–159.

Lakner, A. A., and R. T. Anderson. 1985. *Reliability Engineering for Nuclear and Other High Technology Systems*. Elsevier Applied Science, London.

Lambert, H. E. 1975. Measures of importance of events and cut sets in fault trees. In *Reliability and Fault Tree Analysis*, ed. by R. E. Barlow, J. B. Fussell, and N. D. Singpurwalla. SIAM, Philadelphia.

Laprie, J. C., ed. 1992. *Dependability: Basic Concepts and Terminology*. Springer-Verlag, New York.

Lawless, J. F. 1982. *Statistical Models and Methods for Lifetime Data*. Wiley, New York.

Lehmann, E. L. 1983. *Theory of Point Estimation*. Wiley, New York.

Levenbach, G. J. 1957. Accelerated life testing of capacitors. *IRA-Trans. on Reliability and Quality Control, PGRQC*, no. 10, pp. 9–20.

Lieblein, J., and M. Zelen. 1956. Statistical investigation of the fatigue life of deep groove ball bearings. *Journal of Research, National Bureau of Standards* **57**:273–316.

Lloyd, D. K., and M. Lipow. 1962. *Reliability: Management, Methods and Mathematics*. Prentice-Hall, Englewood Cliffs, NJ.

Lorden, G. 1970. On Excess over the Boundary. *Anuals of Mathematical Statistics* **41**:520–527.

Luthra, P. 1990. MIL-HDBK-217: What is wrong with it? *IEEE Transactions on Reliability* **39**:518.

Lydersen, S., and M. Rausand. 1989. Failure rate estimation based on data from different environments and with varying quality. In *Reliability Data Collection and Use in Risk and Availability Assessment*, ed. by V. Colombari. Springer-Verlag, New York.

Mann, N. R., R. E. Schafer, and N. D. Singpurwalla. 1974. *Methods for Statistical Analysis of Reliability and Life Data*. Wiley, New York.

Marshall, A. W., and I. Olkin. 1967. A multivariate exponential distribution. *Journal of the American Statistical Association* **62**:30–44.

Martz, H. F., and R. A. Waller. 1982. *Bayesian Reliability Analysis*. Wiley, New York.

MIL-HDBK-189. 1981. *Reliability Growth Management*. U.S. Department of Defense, Washington, DC.

MIL-HDBK-217F. 1991. *Reliability Prediction of Electronic Equipment*. U.S. Department of Defense, Washington, DC.

MIL-STD-756B. 1981. *Reliability Modeling and Prediction*. U.S. Department of Defense, Washington, DC.

MIL-STD-882B. 1984. *System Safety Program Requirements*. U.S. Department of Defense, Washington, DC.

MIL-STD-1629A 1980. *Procedures for Performing a Failure Mode, Effects and Criticality Analysis*. U.S. Department of Defense, Washington, DC.

MIL-STD-2173(AS). 1986. *Reliability-Centered Maintenance: Requirements for Naval Aircraft, Weapon Systems and Support Equipment*. U.S. Department of Defense, Washington, DC.

Miner, M. A. 1945. Cumulative damage in fatigue. *Journal of Applied Mechanics* **12**:A159–A164.

Mitrani, I. 1982. *Simulation Techniques for Discrete Event Systems*. Cambridge University Press, Cambridge.

Molnes, E., M. Rausand, and B. Lindqvist. 1986. *Reliability of Surface Controlled Subsurface Safety Valves*. SINTEF Report STF75 A86024, SINTEF, N-7034 Trondheim, Norway.

Mosleh, A., K. N. Fleming, G. W. Parry, H. M. Paula, D. H. Worledge, and D. M. Rasmuson. 1988. *Procedures for Treating Common Cause Failures in Safety and Reliability Studies: Procedural Framework and Examples*, vol. 1. NUREG/CR-4780, EPRI NP-5613.

Mosleh, A., K. N. Fleming, G. W. Parry, H. M. Paula, D. H. Worledge, and D. M. Rasmuson. 1989. *Procedures for Treating Common Cause Failures in Safety and Reliability Studies: Analytic Background and Techniques*, vol. 2. NUREG/CR-4780, EPRI NP-5613.

Mosleh, A. 1991. Common cause failures: An analysis methodology and examples. *Reliability Engineering and System Safety* **34**:249–292.

Moubray, J. 1991. *Reliability-Centered Maintenance*. Butterworth-Heinemann, Oxford.

Natvig, B. 1979. A suggestion of a new measure of importance of system components. *Stochastic Processes and Their Applications* **9**:319–330.

Natvig, B. 1985a. Multistate coherent systems. In *Encyclopedia of Statistical Sciences*, vol. 5, ed. by N. L. Johnson and S. Kotz. Wiley, New York, pp. 732–735.

Natvig, B. 1985b. Recent developments in multistate reliability theory. In *Probabilistic Models in the Mechanics of Solids and Structures*, ed. by S. Eggwertz and N. C. Lind. Springer-Verlag, New York, pp. 385–393.

Neal, A. C., and F. B. Wright. 1992. *The European Communities' Health and Safety Legislation*. Chapman and Hall, London.

Nelson, W. 1969. Hazard plotting for incomplete failure data. *Journal of Quality Technology* **1**:27–52.

Nelson, W. 1971. Analysis of accelerated life test data. Part I: The Arrhenius model and graphical methods. *IEEE Transactions on Electrical Insulation* **El-6**:165–187.

Nelson, W. 1972. Theory and application of hazard plotting for censored failure data. *Technometrics* **14**:945–966.

Nelson, W. 1975. Analysis of accelerated life test data—Least squares methods for the inverse power law model. *IEEE Transactions on Reliability*. **R-24**:103–107.

Nelson, W. 1982. *Applied Life Data Analysis*. Wiley, New York.

Nelson, W. 1990. *Accelerated Testing: Statistical Models, Test Plans, and Data Analyses*. Wiley, New York.

Nelson, W. and T. J. Kielpinski. 1976. Theory for optimum censored accelerated life tests for normal and lognormal life distributions. *Technometrics* **18**:105–114.

Neyman, J. 1945. On the problem of estimating the number of schools of fish. *Publications in Statistics*, vol. 1. University of California, Berkeley.

REFERENCES

Nowlan, F. S., and H. F. Heap. 1978. *Reliability Centered Maintenance.* DDC No. AD-A066579, Defense Documentation Center, Defense Logistics Agency, Alexandria, VA.

NPRD-91. 1991. *Nonelectronic Part Reliability Databook.* Reliability Analysis Center RAC, 201 Mill St. Rome, NY 14440-8200.

O'Connor, P. D. T. 1991. *Practical Reliability Engineering.* Wiley, New York.

OREDA-1992 (Offshore Reliability Data). 1992. DNV Technica, P.O. Box 300, N-1322 Høvik, Norway.

Parry, G. W. 1991. Common cause failure analysis: A critique and some suggestions. *Reliability Engineering and System Safety* **34**:309–326.

Peterson, J. L. 1981. *Petri Net Theory and the Modelling of Systems.* Prentice-Hall, Englewood Cliffs, NJ.

Prabhu, N. U. 1965. *Stochastic Processes.* Macmillan, New York.

Ravichandran, N. 1990. *Stochastic Methods in Reliability Theory.* Wiley, New York.

Ripley, B. D. 1987. *Stochastic Simulation.* Wiley, New York.

Rippon, S. 1975. Browns Ferry fire. *Nuclear Engineering International* (May): 461.

ROCOF. PC-program for analysis of life data for repairable systems developed by SINTEF Safety and Reliability, N-7034 Trondheim, Norway (β-version available).

Ross, S. M. 1970. *Applied Probability Models with Optimization Applications.* Holden Day, San Francisco.

Ross, S. M. 1983. *Stochastic Processes.* Wiley, New York.

Samset, O. 1988. *Reliability Estimation Based on Operating History of Repairable Systems.* Diploma thesis. Division of Mathematical Sciences, Norwegian Institute of Technology, Trondheim.

Sandtorv, H., and M. Rausand. 1991. RCM—Closing the loop between design reliability and operational reliability. *Maintenance* **6**:13–21.

SAREPTA (Survival and Repair Time Analysis). PC-program developed by SINTEF Safety and Reliability, N-7034 Trondheim, Norway.

Satyanarayana, A., and A. Prabhakar. 1978. New topological formula and rapid algorithm for reliability analysis. *IEEE Transactions on Reliability* **R-27**:82–100.

Schrödinger, E. 1915. Zur Theorie der Fall- und Steigversuche an Teilchen mit Brownscher Bewegung. *Physikalische Zeitschrift* **16**:289–295.

Shooman, M. 1968. *Probabilistic Reliability: An Engineering Approach.* McGraw-Hill, New York.

Siegel, G., and S. Wünsche. 1979. Abschätzungen der Erneuerungsfunction. *Mathematische Operationsforschung und Statistik, Series Optimization* **10**:265–275.

Singpurwalla, N. D. 1973. Inference from accelerated life tests using Arrhenius type reparametrization. *Technometrics* **15**:289–299.

Singpurwalla, N. D., and F. A. Al-Khayyal. 1977. Accelerated life tests using the power law model for the Weibull distribution. In *The Theory and Applications of Reliability*, vol. 2, ed. by C. P. Tsokos and I. N. Shimi. pp. 381–399.

Smith, A. M., J. E. Mott, and G. L. Crellin. 1988. *Defensive Strategies for Reducing Susceptibility to Common-Cause Failures: Defensive Strategies*, vol. 1. EPRI NP-5777.

Smith, D. J. 1988. *Reliability and Maintainability in Perspective: Practical, Contractual, Commercial and Software Aspects*. Macmillan Education, London.

Smith, W. L. 1958. Renewal theory and its ramifications. *Journal of Royal Statistical Society* **B 20**:243–302.

Smith, W. L., and M. R. Leadbetter. 1963. On the renewal function for the Weibull distribution. *Technometrics* **5**:393–396.

SRS Data Bank. *AEA Technology/SRD*. National Centre of Systems Reliability, Safety and Reliability Directorate, Wigshaw Lane, Culcheth, Warrington WA3 4NE, England.

Strandberg, K. 1992. Elements of a Dependability Programme. Paper presented to NSDCS'92, The Nordic Seminar on Dependable Computing Systems, 19–21 August, Trondheim, Norway.

Sverdrup, E. 1967. *Law and Chance Variations*, vol. 1. North Holland, Amsterdam.

Sweet, A. L. 1990. On the hazard rate of the lognormal distribution. *IEEE Transactions on Reliability* **39**:325–328.

T-Book. 1992. *Reliability Data of Components in Nordic Nuclear Power Plants*. 3d ed. ATV Office, Vattenfall AB, S-16287 Vällingby, Sweden.

Takács, L. 1956. On a probability problem arising in the theory of counters. *Cambridge Philosophical Society* **32**:488–489.

Taylor, H. M., and S. Karlin. 1984. *An Introduction to Stochastic Modeling*. Academic Press, San Diego.

Thompson, W. A., Jr. 1981. On the foundations of reliability. *Technometrics* **23**:1–13.

Thompson, W. A., Jr. 1988. *Point Process Models with Applications to Safety and Reliability*. Chapman and Hall, London.

Tobias, P. A., and D. C. Trindade. 1986. *Applied Probability*. Van Nostrand Reinhold, New York.

Trivedi, K. S. 1982. *Probability and Statistics with Reliability, Queuing, and Computer Science Applications*. Prentice-Hall, Englewood Cliffs, NJ.

Tweedie, M. C. K. 1946. Inverse statistical variates. *Nature* **155**:453.

Tweedie, M. C. K. 1957. Statistical properties of inverse Gaussian distributions. Parts I and II. *Annals of Mathematical Statistics* **28**:362–377, 696–705.

Vatn, J. 1992. Finding minimal cut sets in fault trees. *Reliability Engineering and System Safety* **36**:59–62.

Vesely, W. E. 1977. Estimating common cause failure probabilities in reliability and risk analysis: Marshall-Olkin specializations. In *Nuclear Systems Reliability Engineering and Risk Assessment*, ed. by J. B. Fussell and G. R. Burdick. SIAM, Philadelphia, pp. 314–341.

Vesely, W. E. 1991. Incorporating aging effects into probabilistic risk analysis using a Taylor expansion approach. *Reliability Engineering and System Safety* **32**:315–337.

Vesely, W. E., F. F. Goldberg, N. H. Roberts, and D. F. Haasl. 1981. *Fault Tree Handbook*. U.S. Nuclear Regulatory Commission. Available from NTIS.

Vesely, W. E., and R. E. Narum 1970. *PREP and KITT, Computer Codes for the Automatic Evaluation of a Fault Tree*. Idaho Nuclear Corp., IN-1349.

Viertl, R. 1988. *Statistical Methods in Accelerated Life Testing*. Vandenhoeck & Ruprecht, Göttingen.

Villemeur, A. 1992. *Reliability, Availability, Maintainability and Safety Assessment.* Vol. 1: *Methods and Techniques.* Vol. 2: *Assessment, Hardware, Software and Human Factors.* Wiley, New York.

Wald, A. 1947. *Sequential Analysis.* Wiley, New York.

WASH-1400. 1975. Reactor Safety Study. U.S. Nuclear Regulatory Commission, NUREG-75/014.

Watson, G. F. 1992. MIL reliability: A new approach. *IEEE Spectrum* (August): 46–49.

Webster, J. T., and V. B. Parr. 1965. A method for discrimination between failure density functions used in reliability predictions. *Technometrics* **7**:1–10.

Weibull, W. 1951. A statistical distribution function of wide applicability. *Journal of Applied Mechanics* **18**:293–297.

Whitmore, G. A., and V. Seshadre. 1987. A heuristic derivation of the inverse Gaussian distribution. *American Statistician* **41**:280–281.

Author Index

Aalen, O.O., 317
Akersten, P.A., 271, 356, 412
Al–Khayyal, F.A., 428
Allan, R.N., 159, 244
Amendola, A., 468
Andersen, P.K., 318
Anderson, R.T., 3, 6, 16, 79, 466
Apostolakis, G., 339
Ascher, H., 22, 264, 268, 271, 273, 277, 293, 298, 302, 314, 319, 320, 356, 412
Atwood, C.L., 319, 320, 339
Aven, T., 151, 184, 271

Bain, L.J., 320
Barlow, R.E., 61, 64, 88, 98, 236, 273, 274, 275, 288, 301, 342, 358, 359, 361, 390, 391, 392, 404
Bendell, A., 268, 460
Berger, J.O., 435
Bergman, B., 372
Bhattacharayya, G.K., 50
Bickel, P.J., 381, 424
Billinton, R., 159, 244
Birnbaum, Z.W., 46, 111, 195
Bjoro, E.E., 6
Blackwell, D., 290
Bodsberg, L., 328, 466
Bonnett, R.W., 6
Borgan, Ø., 318
Borowkow, A.A., 288
Bourne, A.J., 328, 468
Box, G.E.P., 13, 423
Brown, M., 271
Buttersworth, R.W., 288

Campo, R., 358, 359, 361
Cannon, A.G., 460

Chatterjee, P., 117
Chhikara, R.S., 49, 52, 382, 428
Coombs, C.F., 6
Cox, D.R., 27, 29, 50, 216, 220, 222, 225, 264, 277, 289, 298, 302, 320, 356, 373, 393, 399, 401, 419, 432
Cramér, H., 53, 54, 478
Crellin, G.L., 328
Crow, L.H., 319
Crowder, M.J., 319, 320, 324, 393, 417, 418

DeGroot, M.H., 435
Deremer, R.K., 328
Dodge, H.F., 1
Doksum, K.A., 381, 424, 431
Drenick, R.F., 301
Duane, J.T., 319
Dudewicz, E.J., 45, 47, 51, 148, 155, 162, 176, 177, 217, 280, 375, 436, 446, 449

Edwards, G.T., 326, 327, 328, 332, 352
Endrenyi, J., 159, 226
Engelhardt, M., 320
Esary, J.D., 340

Farquharson, J.A., 6
Feingold, H., 22, 264, 268, 271, 273, 277, 293, 298, 302, 314, 319, 320, 356, 412
Feller, W., 151, 290
Flamm, J., 468
Fleming, K.N., 328, 329, 333
Folks, J.L., 49, 52, 382, 428
Fries, A., 50
Fussell, J.B., 164, 170

Gaudoin, O., 321
Gertsbakh, I., 64

507

Goldberg, F.F., 81
Green, A.E., 468
Gumbel, E.J., 55, 57
Guthrie, V.H., 6

Haasl, D.F., 81
Hammer, W., 78
Harris, B., 332, 352
Heap, H.F., 79
Henley, E.J., 12, 13, 81, 170, 185, 195
Hill, W.J., 423
Hokstad, P., 328, 339
Humphreys, R.A., 336, 337
Hunns, D.M., 328
Hunter, J.S., 423
Hunter, W.G., 423
Høyland, A., 431

Ireson, W.G., 6
Isham, V., 264, 302

Jelinski, Z., 324
Jensen, F., 419
Johnson, N.L., 55

Kalbfleisch, J.D., 356, 398, 399, 401, 433, 478
Kaplan, E.L., 396, 399, 401
Kaplan, S., 3
Kapur, K.C., 59, 458
Karlin, S., 29, 216, 273
Keene, S.J., 9
Keller, A.Z., 468
Kielpinski, T.J., 428
Kimber, A.C., 319, 320, 324, 393, 417, 418
Klefsjö, B., 319, 372
Knight, C.R., 2
Kotz, S., 55
Kumamoto, H., 12, 13, 81, 170, 185, 195
Kumar, U., 319

Lakner, A.A., 3, 6, 16
Lamberson, L.R., 59, 458
Lambert, H.E., 88, 195
Laprie, J.C., 5, 6, 490
Lawless, J.F., 55, 356, 393, 399, 401, 419, 433
Leadbetter, M.R., 285
Lehmann, E.L., 459
Levenbach, G.J., 423
Lewis, P.A., 320
Lieblein, J., 368, 418

Lindqvist, B., 468, 469
Lipow, M., 59
Lloyd, D.K., 59
Lorden, G., 288
Luisi, T., 468
Luthra, P., 466
Lydersen, S., 469

Mann, N.R., 40, 53, 55, 59, 390, 393, 419, 423, 427, 428, 478
Marshall, A.W., 338
Marshall, K.T., 288
Martz, H.F., 435
Meier, P., 396, 399, 401
Miller, H.D., 27, 29, 50, 216, 220, 222, 225
Miner, M.A., 46
Mishra, S.A., 45, 47, 51, 148, 155, 162, 176, 177, 217, 280, 375, 436, 446, 449
Mitrani, I., 183
Moieni, P., 339
Molnes, E.M., 468, 469
Moranda, P.B., 324
Mosleh, A., 328, 329
Mott, J.E., 328

Narum, R.E., 153
Natvig, B., 15, 195
Neal, A.C., 17
Nelson, W., 55, 317, 382, 395, 404, 419, 428, 432, 478
Neri, L., 79
Neyman, J., 12
Nowlan, F.S., 79

Oakes, D., 356, 373, 393, 399, 401, 419
O'Connor, P.D.T., 6, 466
Olkin, I., 338

Parr. B. Van, 422
Parry, G.W., 328, 329
Paula, H.M., 328
Petersen, N.E., 419
Peterson, J.L., 184
Poulter, D.R., 328, 329
Prabhakar, A., 151
Prabhu, N.U., 289, 297
Prentice, R.L., 356, 398, 399, 401, 419, 433, 478
Proschan, F., 61, 64, 98, 236, 271, 273, 274, 275, 288, 301, 340, 342, 390, 391, 392, 404

AUTHOR INDEX

Rasmuson, D.M., 328, 329
Rausand, M., 79, 468, 469
Ravichandran, N., 159
Reiner, A.M., 423
Ripley, B.D., 182, 183
Rippon, S., 326
Roberts, N.H., 81
Romig, H.G., 1
Ross, S.M., 29, 216, 263, 264, 271, 272, 273, 289, 293, 295, 297, 300, 393, 306, 315

Samset, O., 318
Sandtorv, H., 79
Satyanarayana, A., 151
Saunders, S.C., 46
Schafer, R.E., 40, 53, 55, 59, 390, 393, 419, 423, 427, 428, 478
Schrödinger, E., 51, 380
Seshadre, V., 51
Shewhart, W., 1
Shooman, M., 105
Siegel, G., 288
Singpurwalla, N.D., 40, 53, 55, 59, 390, 393, 419, 423, 427, 428, 478
Smith, D.J., 3, 4, 6, 468
Smith, R.L., 319, 320, 324, 393, 417, 418
Smith, W.L., 285, 289, 290, 291
Strandberg, K., 6, 9
Sverdrup, E., 373, 383

Sweet, A.L., 43, 49
Sweeting, T.J., 319, 320, 324, 393, 417, 418

Takáks, L., 282
Taylor, H.M., 29, 216, 273
Thompson, W.A., 22, 29, 273, 302, 316
Tweedie, M.C.K., 51, 72

Vatn, J., 90
Vesely, W.E., 81, 153, 319, 338
Viertl, R., 419, 432
Villemeur A., 2, 6, 12

Wald, A., 51
Walkup, D.W., 340
Waller, R.A., 435
Walls, L.A., 268
Watson, G.F., 466
Watson, I.A., 326, 327, 328, 332, 352
Webster, J.T., 422
Weibull, W., 37
Whitmore, G.A., 51
Worledge, D.H., 328, 329
Wright, F.B., 17
Wright, F.T., 320
Wünsche, S., 288

Zelen, M., 368, 418

Subject Index

Absorbing state, 239, 243, 244, 246
Accelerated life tests (ALT), 419–485
  experimental design for ALT, 420, 421
  nonparametric models, 431–433
  progressive stress (PALT), 420, 428–430
  proportional hazards (P-H) models, 431–433
  step-stress (SALT), 420, 422, 424, 428, 491
Active redundancy, 154, 485
Age, 293, 297
Age replacement, 291–293, 369–372
AIChE, 82, 93, 144
ALT, 419–434, 485
Alternating renewal process, 303–313
AMSTAT News, 393
AND gate, 82, 140
A posteriori distribution, *see* Posterior distribution
Applications:
  ball bearing data (Lieblein and Zelen), 368–369, 418
  compressor data, 264–268, 411–414
  downhole safety valves (DHSV), 193–194
  emergency shutdown (ESD) system, 190–193
  fatigue analysis, 45–46
  fire detector system, 84–87, 96, 98
  material strength data, 417
  motorette data (Nelson), 382, 383
  offshore separator, 90–92, 118–119
  pitting corrosion, 57–59
  water chlorination system, 260–261
A priori distribution, *see* Prior distribution
Arrhenius model, 423
Arrival frequency, 227
As bad as old, 271

As good as new, 32, 171, 215, 275
Associated variables, 340–352, 354
Asynchronous sampling, 298
Availability, 4, 6, 160, 305
  average, 5, 161–163, 305
  limiting, 161–163, 224, 304–305

Barlow-Proschan's test, 390–393
Basic event (fault tree), 82
Basic event probability, 139
Bathtub curve, 23, 24
Bayes' theorem, 436, 437
Bayesian estimator, 440–441, 443, 447, 449, 450, 452, 454, 458
Bayesian life test sampling plan, 450–454
Bellcore, 466
Beta distribution, 339, 448, 449, 459, 473
β-factor model, 330, 333–336, 352
Beta function, 472, 473
Binary representation, 15
Binary variable, 15, 98
Binomial distribution, 30, 131, 322, 338, 383–384, 448, 454
  parameter estimation, 383–384
Binomial failure rate model, 330, 338–339, 353
Birnbaum's measure:
  of reliability importance of components, 195–200, 210–211, 213
  of structural importance, 109–114, 200
Birnbaum–Saunders distribution, 46–48, 50, 71
Blackwell's theorem, 290
BMDP, 393, 433
Bottom-up approach (FMECA), 74–75
Boundary conditions (fault tree analysis), 83
Bridge structure, 106

SUBJECT INDEX

Bridge structure (*Continued*)
  pivotal decomposition, 113–114
  structure function, 108–109
Brownian motion, 51
Browns Ferry, 326, 328
BS 4778, 3, 4, 5, 10, 485, 486, 487, 488, 489, 491
BS 5760, 6, 7, 8, 16
Burn-in period, 24

CAFTA, 145
CAFTAN, 86, 92, 143, 145, 151, 199
Calendar time, 263, 461, 463
CARA, 80, 86, 87, 90, 92, 143, 145, 151, 199
Cascading failures, 325, 328, 330
Catastrophic failure, 10, 78, 464
Censored data, 385–386
  stochastic, *see* type IV
  type I, 385
  type II, 385
  type III, 385, 386
  type IV, 386
Central limit theorem, 47, 155, 280
Chance failure period, 24
Chapman-Kolmogorov equation, 217
Characteristic life (of Weibull distribution), 38
Chi-square (χ^2) distribution, 53, 321, 373, 375, 378, 388, 446, 493
Closed stopping rule, 459
CNET, 466
Coherent module, 116, 117
Coherent structure/system, 101–103, 485
Command fault, 11, 13, 84
Common cause candidates, 329
Common cause failures, 80, 247–249, 259, 325–339, 469, 485
Completeness, 381
Compound Poisson process, 273–275, 302
Concurrent engineering, 9
Confidence interval, 373, 374, 377, 383
Congruential generator, 182
Conjugate families of distributions, 443–449
Convolution, 277–278, 294
Corrective maintenance, 159, 486
Corrosion, 53
Counting process, 263, 271
Covariates, 419
Cox model, *see* Proportional hazards model
Cox-Lewis model, 319
Credibility interval, 443, 446, 447
Critical failure, 10, 78, 461, 463
Critical path vector, 110, 199

Critical path set, 110
Criticality importance, 200–203, 210–211
Criticality matrix (FMECA), 79
Criticality ranking (of cut sets), 93
Cumulant-generating function, 51, 52, 71
Cumulative damage model, 273
Cumulative distribution function, *see* Distribution function
Cumulative failure rate, 63, 402
Cumulative intensity (of Poisson process), 314
Cut set, 88, 106

Decreasing failure rate, *see* DFR
Decreasing failure rate average, *see* DFRA
Degraded failure, 10, 461, 463, 464
Departure rate, 220, 226–227
Dependability, 5
Dependability management, 6
Dependence:
  positive, 325, 340
  negative, 324, 326
Design review, 486
DFR, 60–64, 70, 358, 359, 366
DFRA, 63–64, 393
Distribution function, 19, 486
Distributions:
  binomial, 30, 131, 322, 338, 383–384, 448, 454
  Birnbaum-Saunders, 46–48, 50, 71
  Chi-square (χ^2), 53, 321, 373, 375, 378, 388, 446, 493
  Erlangian, 34, 62
  exponential, 27, 31–33, 61, 273, 309–313, 422, 438, 444
  extreme value, 53–59
  F-distribution, 383
  gamma, 29, 33–35, 36, 62, 273–275, 279, 298, 438, 439, 443, 444, 445, 446, 458
  geometric, 177, 180
  Gompertz, 320
  Gumbel (of the smallest extreme), 55–56, 320
  truncated, 56, 59
  Gumbel (of the largest extreme), 57
  inverse Gaussian, 48–53, 71, 418, 422, 428, 429, 430, 431, 478–480
  inverted gamma, 450–452
  lognormal, 42–46, 49, 68, 422
  normal (Gaussian), 40–42, 47, 51, 155, 280–282, 317–318, 321, 384, 391, 458, 492
  left truncated, 41

SUBJECT INDEX 513

Pareto, 35–37, 457
Poisson, 28, 67, 272, 280, 314
Rayleigh, 38
uniform, 30, 60, 68, 70, 403
weakest link, 39
Weibull, 37–40, 59, 61, 67, 285, 286, 422, 480–481
    three parameter Weibull, 39–40, 67
Diversity, 329
Domino effect, 325
Double expectation, 176
Downtime, 159, 486
Drenick's theorem, 301
Dual structure, 123

Elementary renewal theorem, 289, 302
Empirical distribution function, 359, 394, 395, 413
Empirical bathtub curve, 23
Empirical survivor function, 395, 396
Environmental protection, 7
Equilibrium renewal process, 298, 301
ERAC, 151
Erlangian distribution, 34, 62
Error factor, 44
ESREDA, 468
Euler's constant, 56
EuReDatA, 10, 460, 487
European Union, 9, 17
Evident failure, 10, 77
Eyring model, 423
Exact system reliability, 145–153
Excess life, 270, 293
Exponential distribution, 27, 31–33, 61, 273, 309–313, 422, 438, 444
    as good as new property, 32
    right truncated, 58
    parameter estimation, 372–379
Exponential family, 381
Extreme value, 53
Extreme value distribution, 53–59

F-distribution, 383
Fail safe, 91, 118, 486
Failed state, 15
FailMode, 80
Failure, 10, 486
    catastrophic failure, 10, 78, 464
    critical failure, 10, 78, 461, 463
    degraded failure, 10, 461, 463, 464
    evident failure, 10, 77
    gradual failure, 10, 487
    hidden failure, 10, 77, 487

incipient failure, 10, 461, 463, 464
intermittent failure, 488
major failure, 78
marginal failure, 10
minor failure, 78
negligible failure, 10
primary failure, 11, 13, 84
secondary failure, 11, 13, 84
sudden failure, 10, 491
Failure cause, 486
Failure effect, 74, 77, 486
Failure mechanism, 12, 77
Failure mode, 10, 74–80, 461, 463, 486
Failure Mode and Effect Analysis, see FMEA
Failure Mode, Effect and Criticality Analysis, see FMECA
Failure rate, 6, 20–24, 487
    failure rate classification, 78
Failure symptom, 487
False alarm, 171, 191–193
FARADIP, 468
Fatigue, 44–47, 487
Fatigue failure, 44, 429
Fatigue life, 487
Fatigue limit, 46
Fault tree analysis, 81–93
    AND gate, 82, 140
    OR gate, 82, 140, 141
    basic event, 82, 139
    boundary conditions, 83
    construction, 81–82
    logic gates, 81–82
    relation to reliability block diagram, 95–97
    TOP event, 81, 83–84, 86, 139
    top structure, 84
    transfer symbol, 82
    undeveloped event, 82, 84
FaultrEASE, 145
FaultTree, 145
Fixed time increments, 182
FME, 80
FMEA, 7, 10, 12, 73–80, 487
FMECA, 74–80
    bottom-up approach, 74–75
    criticality matrix, 79
    FMECA worksheet, 75
    functional approach, 76
    hardware approach, 76
    top-down approach, 74–75
FMECA (computer program—SINTEF), 80
FMECA (computer program—System Effectiveness Associates), 80
Force of mortality (FOM), 21, 316, 487

Forward recurrence time, 270, 293
FRANTIC ABC, 145
Frequency of system failures, 170, 229, 232, 233, 234, 238, 239
Functioning state, 15, 228, 487
Functional approach (FMECA), 76
Fundamental renewal theorem, 282–283

Gamma distribution, 29, 33–35, 36, 62, 273–275, 279, 298, 438, 439, 443, 444, 445, 446, 458
Gamma function, 471–473
Gaussian distribution, see Normal distribution
General addition theorem, 148
Generalized β-factor method, 330, 337, 338
Geometric distribution, 177, 180
Gompertz distribution, 320
Goodness of fit, 48, 49
Gradual failure, 10, 487
Greenwood's formula, 401
Guarantee parameter, 39
Gumbel distribution of the smallest extreme, 55–56, 320
 truncated, 56, 59
Gumbel distribution of the largest extreme, 57

Happy system, 264, 314, 321
Hardware approach (FMECA), 76
Hazard plot, 408–409
Hazard rate, 22, 487
Hidden failure, 10, 77, 487
Homogeneous Poisson process, see HPP
HPP, 26–31, 33, 271–275, 279–280, 284–285, 301, 316–317, 322, 445, 446
HRD4, 466
Human errors, 14, 80, 93
Human reliability, 2
Hypothesis testing, 320–321, 377

IEC 50, 485, 486, 488, 489, 490, 491
IEC 300, 5, 6, 9, 16, 486, 488
IEC 812, 74
IEEE Std. 352, 16, 74, 491
IEEE Std. 500, 464, 465
IEEE Reliability Society Newsletter, 495
IEEE Spectrum, 393
IEEE Transactions on Reliability, 2, 495
IFR, 60–64, 358, 359, 366
IFRA, 63–64, 275, 392
Imperfect repair, 271
Improvement potential, 208–211
Incipient failure, 10, 461, 463, 464
Inclusion-exclusion principle, 148–153

Increasing failure rate, see IFR
Increasing failure rate average, see IFRA
Increment, 268
Independent increments, 268, 272
Indicator variable, 396
Infant mortality, 23, 488
Inspection, 488
Intensity of Poisson process, 26–27, 314, 322
Interarrival time, 263, 276
Intermittent failure, 488
Inverse Gaussian distribution, 48–53, 71, 418, 422, 428, 429, 430, 431, 478–480
 parameter estimation, 379–383
Inverted Gamma distribution, 450–452
"Ip" (LI), 100
IRRAS, 145
Irreducible process, 225
Irrelevant component, 101
ISO 8402, 3, 486, 488, 489, 490
ISO 9000, 9, 16

k-out-of-n structure, 100
Kaplan-Meier estimator, 397–401, 402, 406
Kaplan-Meier plot, 397, 398
Key renewal theorem, 290, 296
Kinetic tree theory, 153
KITT, 153

Laplace test, 320–321
Laplace transform, 25, 71, 220–221, 223, 241–242, 278–279, 482–484
Lattice distribution, 290, 295, 300, 305, 307
Law of large numbers, 162
Lethal shocks, 339, 353
Life Cycle Cost (LCC), 9, 488
Life Cycle Profit (LCP), 9
Likelihood function, 375, 379, 387, 425, 430, 442, 477–481
Limit distribution, 54–55
Limiting availability, 224
Lindeberg–Levy's central limit theorem, see Central limit theorem
Linear model, 318–319, 323
Local time, 263
Logic gates (fault tree), 81–82
Logistic support, 9
Log-likelihood function, 426
Log-linear model, 318–320, 323
Lognormal distribution, 42–46, 49, 68, 422

Machine Directive, 17
Maintainability, 5, 488
Maintenance, 488

SUBJECT INDEX

Maintenance support, 4, 488
Major failure, 78
Major Hazards Directive, 17
Marginal distribution, 30, 437, 438
Marginal failure, 10
Markov diagram, 214
Markov model, 181
Markov process, 216
Markov property, 216
MAROS, 185
Marshall-Olkin model, 338
Maximum likelihood, 433, 477–481
Maximum likelihood estimator, see MLE
Mean downtime (MDT), 489
Mean duration of a system failure, 229, 232, 234, 235, 238, 239
Mean duration of a visit, 227, 228
Mean fractional dead-time, see MFDT
Mean time between failures, see MTBF
Mean time between replacements, see MTBR
Mean time between system failures, 229
Mean time to failure, see MTTF
Mean time to repair, see MTTR
Median, 43, 44, 289
Memoryless property, 32, 217
MFDT, 174–177, 180–181, 352, 353
MIL-HDBK test, 320–321, 323
MIL-HDBK 189, 321
MIL-HDBK 217, 337, 464, 466, 467
MIL-STD 756, 16
MIL-STD 785, 6, 16
MIL-STD 882, 5, 16, 78, 490
MIL-STD 1629, 74, 486, 487, 490,
MIL-STD 2173(AS), 486, 487, 488, 490
MilStress, 467
Miner's rule, 46
Minimal cut representation, 108–109
Minimal cut parallel structure, 108
Minimal cut set, 88, 106
Minimal path representation, 107–108
Minimal path series structure, 107
Minimal path set, 90, 107
Minimal repair, 271, 314–315
Minor failure, 78
MIRIAM, 184
Mixture of distributions, 69, 70
MLE, 379, 380, 425, 426, 427, 430, 450, 453, 478, 480
MOCUS, 88–90, 122, 125
Modified renewal process, 276, 297–298, 304
Modular decomposition, 114–117
Module, 116
Moment-generating function, 51, 52

Moments (of a random variable), 68
Monte Carlo simulation, 181–185
MTBF, 24, 270
MTBR, 291–292, 370
MTTF, 5, 6, 24, 239–243, 488
MTTR, 5, 24, 489
Multiplicity of failures, 325, 338
Multistate theory, 15

NBU, 63–64, 300–301
NBUE, 64, 288–289
Negative dependence, 324, 326
Negligible failure, 10
Nelson's estimator of the cumulative failure rate, 402, 403, 405, 406–408
Nelson plot, 408, 410
Nelson–Aalen estimator, 317–318
Nelson–Aalen plot, 264, 324
New better than used, see NBU
New better than used in expectation, see NBUE
New worse than used, see NWU
New worse than used in expectation, see NWUE
Next event simulation, 182
NHPP, 272, 314–323
Nonhomogeneous Poisson process, see NHPP
Non-informative prior distribution, 449–450
Nonparamctric model, 394
Nonrepairable system, 127, 132–138
  series, 133
  parallel, 133–135
  2-out-of-3, 135–136
  k-out-of-n, 136–138
Nonstationary process, 268
Normal distribution, 40–42, 47, 51, 155, 280–282, 317–318, 321, 384, 391, 458, 492
  left truncated, 41
  standard normal, 40
NPRD, 467
NTT, 466
NWU, 63–64, 300
NWUE, 64, 288–289

Operating state, 489
Operating time, 266
Operational mode, 77
OR gate, 82, 140–141
Order of cut set, 88
Order of path set, 90
Order of system, 98
Order statistics, 355, 403–404

Orderly, 269
Ordinary renewal process, 276, 283–284, 287–288, 304
Organizing structure, 132
OREDA, 337, 460–463, 464, 469
Overlay curve, 369, 408, 413

Parallel structure, 94, 99
Pareto distribution, 35–37, 457
Parts count method, 466, 467
Partial likelihood, 433
Path set, 90, 106
pcFOSP, 185
Percentile, 489
Peril rate, 314
Periodic testing, 171–181
Petri net, 184
P–H model, 431–433
Pitting corrosion, 57–59
Pivotal decomposition, 113–114, 125, 146, 198
Plotting techniques, see Hazard plot; Kaplan-Meier plot; TTT-plot
Poisson distribution, 28, 67, 272, 280, 314
Poisson process, 26–29, 217
Positive dependence, 325, 340
Posterior distribution, 437. See also A posteriori distribution
Potential lifetime, 437
Power function, 378
Power law model, 318–319, 323
Power rule model, 59, 423
Predictive density, 455–457
Preventive maintenance, 329, 489
Primary failure, 11, 13, 84
Prime module, 117
Prior distribution, 436. See also A priori distribution, 436
Probability density, 19, 489
Probability distribution, 19
Product Liability Directive, 17
Product limit estimator, see Kaplan–Meier estimator
Product Safety Directive, 17
Production regularity, 5, 184
Propagating failures, 326
Progressive-stress accelerated tests (PALT), 420, 428–430
Proper module, 116
Proportional hazards model, see P–H model
Pseudo-random numbers, 182

Qualitative evaluation (of fault tree), 93

Quality, 4, 489
Quality and Reliability International, 495
Quality and Reliability Management, 495
Quantitative fault tree analysis, 138–145
  approximation formulas, 141–143

RAM, 5
RAMS, 6
Random shocks, 33
Rayleigh distribution, 38
Rate of occurrence of failures, see Rocof
RBC, 467
RCM, 9, 79, 490
Reachable state, 225
Reactor Safety Study, see WASH-1400
REAP, 467
Redundancy, 104–105, 153–159, 328, 489
  active, 154
  passive, 154–158
  partly loaded, 154, 158–159
Regular process, 269, 272
Relevant component, 101
RELEX, 80, 467
Reliability, 3, 128, 490
Reliability block diagram, 93–97
  relation to fault tree, 95–97
Reliability Centered Maintenance, see RCM
Reliability Engineering and System Safety, 326, 495
Reliability function, 20
Reliability management, 6
Reliability Review, 495
Reliability Toolbox, 80, 145, 467
Remaining lifetime, 270, 293, 295, 300
  mean, 64
Renewal density, 276, 283
Renewal function, 276, 282, 288
Renewal period, 276
Renewal process, 275–313
  alternating, 303–313
  equilibrium, 298, 301
  modified, 276, 297–298, 304
  ordinary, 276, 283–284, 287–288, 304
  stationary, 298, 300
  superimposed, 301
Renewal reward process, 302–303
Repair, 490
Repair rate, 43, 161, 166
Repair time, 43, 159, 187
Repair time distribution, 43–44, 160, 412
Repairable system, 159–185
Replacement, 159
Required function, 490

SUBJECT INDEX 517

Residual life, 293
Risk analysis, 7, 16
Risk Spectrum Fault Tree, 145
ROCOF, 21, 167, 269, 272, 314, 316, 318, 319, 323, 324
Root cause, 325
RPC, 467
RPP, 467

S–N diagram, 45–46
Sad system, 264, 314, 321
SAE-ARP 926, 74, 76
Safety, 5, 490
SALP-PC, 145
SAREPTA, 364, 369, 393, 408, 410, 411, 413, 414
SAS, 394
Scale parameter, 37
Scaled TTT-transform, 362
  empirical, 367
  of exponential distribution, 362
  of Weibull distribution, 364
Secondary failure, 11, 13, 84
Security, 5, 490
Seed, 183
Sequential probability ratio test, 51
Series structure, 94, 99
Severity, 490
Severity ranking, 77, 78
Seveso Directive, 17
Shape parameter, 37
Shock model, 33
Single failure point, 490
Singular matrix, 220
Software reliability, 2
S-PLUS, 394
SPSS, 394
Square-root method, 330–333, 352
Staggered testing, 192
Standby, 154–159, 251–259
  cold, 154–158, 251–257
  partly loaded, 154, 158–159, 257–259
  warm (loaded), 154
Standby system, 153–154
STATA, 394
State equations, 218–219
State frequency, *see* Visit frequency
State space, 214
State space diagram, *see* Markov diagram
State variable, 18, 98, 490
State vector, 98, 127, 138–139, 490
Stationary increments, 268
Stationary process, 268

Steady state, 15
Steady state availability, 225
Step-stress accelerated tests (SALT), 420, 422, 424, 428, 491
Stressor, 59, 419, 420, 425, 432
Stressor dependent model, 59
Structural importance, 109–113
Structural reliability, 2
Structure, 99
Structure function, 98, 127, 491
Subjective Bayesian inference, 455
Sudden failure, 10, 491
Sufficiency, 381
Superimposed process, 301
Survival probability, 6
Survivor function, 20, 128, 243, 491
  conditional, 63–64
Synchronous sampling, 298
SYSTAT, 394, 433
System availability, 164–171, 228
  approximation formulas, 165–171
System reliability, 127–138
  series system, 128–129
  parallel system, 129–130
  k-out-of-n system, 130–131
System Reliability Service (SRS), 460
System structure, 101

T-book, 468
Test frequency, 491
Test interval, 491
Ties, 401
Time to failure, 18
Threshold parameter, 39
Top-down approach (FMECA), 74–75
TOP event, 81, 83–84, 86, 139
TOP event frequency, 143–144
TOP event probability, 139–143
  approximation formulas, 141–143, 166
Top structure (fault tree), 84
Total time on test, 356, 388, 389, 424, 451, 452, 453
Total time on test plot, *see* TTT-plot
Total time on test transform, *see* TTT-transform
Transfer symbol (fault tree), 82
Transition, 215–220
Transition probability, 216–220
Transition rate, 217
Transition rate matrix, 219
  steady state, 225–226
TRAP, 151
Truncated normal distribution, 41

TTT-plot, 357, 358, 414
 of censored data, 408, 410, 411
TTT-transform, 361–369

UMVU, 381, 382
Unavailability, 160
Undeveloped event (fault tree), 82, 84
Uniform distribution, 30, 60, 68, 70, 403
Uniformly minimum variance unbiased, *see* UMVU
Unimodal distribution, 67
Up time, 159
Upper bound approximation, 143, 152, 165, 170
Useful life period, 24

Variable time increments, 182
Vesely–Fussell's measure, 203–208, 210–211

Visit frequency, 226–227, 232
Voting system, *see* k-out-of-n system.
Voting unit, 86, 92, 131

Wald's equation, 273–274, 288, 302
WASH-1400, 2, 44, 81, 330, 468
Weakest link distribution, 39
Wearout failure, 491
Wearout period, 24
Weibull distribution, 37–40, 59, 61, 67, 285, 286, 422, 480–481
 rth moment, 67
 three parameter, 39–40, 67
Weibull distribution of the smallest extreme, 57
Wiener process, 50, 428
Wöhler curve, *see* S–N diagram

WILEY SERIES IN PROBABILITY
AND MATHEMATICAL STATISTICS

ESTABLISHED BY WALTER A. SHEWHART AND SAMUEL S. WILKS
Editors
*Vic Barnett, Ralph A. Bradley, Nicholas I. Fisher, J. Stuart Hunter,
J. B. Kadane, David G. Kendall, David W. Scott, Adrian F. M. Smith,
Stephen M. Stigler, Jozef L. Teugels, Geoffrey S. Watson*

Probability and Mathematical Statistics
   ANDERSON · An Introduction to Multivariate Statistical Analysis, *Second Edition*
 *ANDERSON · The Statistical Analysis of Time Series
   ARNOLD, BALAKRISHNAN, and NAGARAJA · A First Course in Order
     Statistics
   BACCELLI, COHEN, OLSDER, and QUADRAT · Synchronization and Linearity:
     An Algebra for Discrete Event Systems
   BARNETT · Comparative Statistical Inference, *Second Edition*
   BERNARDO and SMITH · Bayesian Statistical Concepts and Theory
   BHATTACHARYYA and JOHNSON · Statistical Concepts and Methods
   BILLINGSLEY · Convergence of Probability Measures
   BILLINGSLEY · Probability and Measure, *Second Edition*
   BOROVKOV · Asymptotic Methods in Queuing Theory
   BRANDT, FRANKEN, and LISEK · Stationary Stochastic Models
   CAINES · Linear Stochastic Systems
   CHEN · Recursive Estimation and Control for Stochastic Systems
   CONSTANTINE · Combinatorial Theory and Statistical Design
   COOK and WEISBERG · An Introduction to Regression Graphics
   COVER and THOMAS · Elements of Information Theory
 *DOOB · Stochastic Processes
   DUDEWICZ and MISHRA · Modern Mathematical Statistics
   ETHIER and KURTZ · Markov Processes: Characterization and Convergence
   FELLER · An Introduction to Probability Theory and Its Applications, Volume I,
     Third Edition, Revised; Volume II, *Second Edition*
   FULLER · Introduction to Statistical Time Series
   FULLER · Measurement Error Models
   GIFI · Nonlinear Multivariate Analysis
   GUTTORP · Statistical Inference for Branching Processes
   HALD · A History of Probability and Statistics and Their Applications before 1750
   HALL · Introduction to the Theory of Coverage Processes
   HANNAN and DEISTLER · The Statistical Theory of Linear Systems
   HEDAYAT and SINHA · Design and Inference in Finite Population Sampling
   HOEL · Introduction to Mathematical Statistics, *Fifth Edition*
   HUBER · Robust Statistics
   IMAN and CONOVER · A Modern Approach to Statistics
   JUREK and MASON · Operator-Limit Distributions in Probability Theory
   KAUFMAN and ROUSSEEUW · Finding Groups in Data: An Introduction to Cluster
     Analysis
   LARSON · Introduction to Probability Theory and Statistical Inference, *Third Edition*
   LESSLER and KALSBEEK · Nonsampling Error in Surveys
   LINDVALL · Lectures on the Coupling Method
   MANTON, WOODBURY, and TOLLEY · Statistical Applications Using Fuzzy Sets
   MORGENTHALER and TUKEY · Configural Polysampling: A Route to Practical
     Robustness
   MUIRHEAD · Aspects of Multivariate Statistical Theory
   OLIVER and SMITH · Influence Diagrams, Belief Nets and Decision Analysis
 *PARZEN · Modern Probability Theory and Its Applications

*Now available in a lower priced paperback edition in the Wiley Classics Library.

Probability and Mathematical Statistics (Continued)
    PILZ · Bayesian Estimation and Experimental Design in Linear Regression Models
    PRESS · Bayesian Statistics: Principles, Models, and Applications
    PUKELSHEIM · Optimal Experimental Design
    PURI and SEN · Nonparametric Methods in General Linear Models
    PURI, VILAPLANA, and WERTZ · New Perspectives in Theoretical and Applied Statistics
    RENCHER · Methods of Multivariate Analysis
    RAO · Asymptotic Theory of Statistical Inference
    RAO · Linear Statistical Inference and Its Applications, *Second Edition*
    ROBERTSON, WRIGHT, and DYKSTRA · Order Restricted Statistical Inference
    ROGERS and WILLIAMS · Diffusions, Markov Processes, and Martingales, Volume II: Ito Calculus
    ROHATGI · A Introduction to Probability Theory and Mathematical Statistics
    ROSS · Stochastic Processes
    RUBINSTEIN · Simulation and the Monte Carlo Method
    RUZSA and SZEKELY · Algebraic Probability Theory
    SCHEFFE · The Analysis of Variance
    SEBER · Linear Regression Analysis
    SEBER · Multivariate Observations
    SEBER and WILD · Nonlinear Regression
    SERFLING · Approximation Theorems of Mathematical Statistics
    SHORACK and WELLNER · Empirical Processes with Applications to Statistics
    SMALL and McLEISH · Hilbert Space Methods in Probability and Statistical Inference
    STAUDTE and SHEATHER · Robust Estimation and Testing
    STOYANOV · Counterexamples in Probability
    STYAN · The Collected Papers of T. W. Anderson: 1943–1985
    WHITTAKER · Graphical Models in Applied Multivariate Statistics
    YANG · The Construction Theory of Denumerable Markov Processes

Applied Probability and Statistics
    ABRAHAM and LEDOLTER · Statistical Methods for Forecasting
    AGRESTI · Analysis of Ordinal Categorical Data
    AGRESTI · Categorical Data Analysis
    ANDERSON and LOYNES · The Teaching of Practical Statistics
    ANDERSON, AUQUIER, HAUCK, OAKES, VANDAELE, and WEISBERG · Statistical Methods for Comparative Studies
    *ARTHANARI and DODGE · Mathematical Programming in Statistics
    ASMUSSEN · Applied Probability and Queues
    *BAILEY · The Elements of Stochastic Processes with Applications to the Natural Sciences
    BARNETT · Interpreting Multivariate Data
    BARNETT and LEWIS · Outliers in Statistical Data, *Second Edition*
    BARTHOLOMEW, FORBES, and McLEAN · Statistical Techniques for Manpower Planning, *Second Edition*
    BATES and WATTS · Nonlinear Regression Analysis and Its Applications
    BELSLEY · Conditioning Diagnostics: Collinearity and Weak Data in Regression
    BELSLEY, KUH, and WELSCH · Regression Diagnostics: Identifying Influential Data and Sources of Collinearity
    BHAT · Elements of Applied Stochastic Processes, *Second Edition*
    BHATTACHARYA and WAYMIRE · Stochastic Processes with Applications
    BIEMER, GROVES, LYBERG, MATHIOWETZ, and SUDMAN · Measurement Errors in Surveys
    BIRKES and DODGE · Alternative Methods of Regression
    BLOOMFIELD · Fourier Analysis of Time Series: An Introduction
    BOLLEN · Structural Equations with Latent Variables

*Now available in a lower priced paperback edition in the Wiley Classics Library.

Applied Probability and Statistics (Continued)

BOULEAU · Numerical Methods for Stochastic Processes
BOX · R. A. Fisher, the Life of a Scientist
BOX and DRAPER · Empirical Model-Building and Response Surfaces
BOX and DRAPER · Evolutionary Operation: A Statistical Method for Process Improvement
BOX, HUNTER, and HUNTER · Statistics for Experimenters: An Introduction to Design, Data Analysis, and Model Building
BROWN and HOLLANDER · Statistics: A Biomedical Introduction
BUCKLEW · Large Deviation Techniques in Decision, Simulation, and Estimation
BUNKE and BUNKE · Nonlinear Regression, Functional Relations and Robust Methods: Statistical Methods of Model Building
CHATTERJEE and HADI · Sensitivity Analysis in Linear Regression
CHATTERJEE and PRICE · Regression Analysis by Example, *Second Edition*
CLARKE and DISNEY · Probability and Random Processes: A First Course with Applications, *Second Edition*
COCHRAN · Sampling Techniques, *Third Edition*
*COCHRAN and COX · Experimental Designs, *Second Edition*
CONOVER · Practical Nonparametric Statistics, *Second Edition*
CONOVER and IMAN · Introduction to Modern Business Statistics
CORNELL · Experiments with Mixtures, Designs, Models, and the Analysis of Mixture Data, *Second Edition*
COX · A Handbook of Introductory Statistical Methods
*COX · Planning of Experiments
CRESSIE · Statistics for Spatial Data, *Revised Edition*
DANIEL · Applications of Statistics to Industrial Experimentation
DANIEL · Biostatistics: A Foundation for Analysis in the Health Sciences, *Sixth Edition*
DAVID · Order Statistics, *Second Edition*
DEGROOT, FIENBERG, and KADANE · Statistics and the Law
*DEMING · Sample Design in Business Research
DILLON and GOLDSTEIN · Multivariate Analysis: Methods and Applications
DODGE and ROMIG · Sampling Inspection Tables, *Second Edition*
DOWDY and WEARDEN · Statistics for Research, *Second Edition*
DRAPER and SMITH · Applied Regression Analysis, *Second Edition*
DUNN · Basic Statistics: A Primer for the Biomedical Sciences, *Second Edition*
DUNN and CLARK · Applied Statistics: Analysis of Variance and Regression, *Second Edition*
ELANDT-JOHNSON and JOHNSON · Survival Models and Data Analysis
EVANS, PEACOCK, and HASTINGS · Statistical Distributions, *Second Edition*
FISHER and VAN BELLE · Biostatistics: A Methodology for the Health Sciences
FLEISS · The Design and Analysis of Clinical Experiments
FLEISS · Statistical Methods for Rates and Proportions, *Second Edition*
FLEMING and HARRINGTON · Counting Processes and Survival Analysis
FLURY · Common Principal Components and Related Multivariate Models
GALLANT · Nonlinear Statistical Models
GLASSERMAN and YAO · Monotone Structure in Discrete-Event Systems
GROSS and HARRIS · Fundamentals of Queueing Theory, *Second Edition*
GROVES · Survey Errors and Survey Costs
GROVES, BIEMER, LYBERG, MASSEY, NICHOLLS, and WAKSBERG · Telephone Survey Methodology
HAHN and MEEKER · Statistical Intervals: A Guide for Practitioners
HAND · Discrimination and Classification
*HANSEN, HURWITZ, and MADOW · Sample Survey Methods and Theory, Volume I: Methods and Applications
*HANSEN, HURWITZ, and MADOW · Sample Survey Methods and Theory, Volume II: Theory

*Now available in a lower priced paperback edition in the Wiley Classics Library.

Applied Probability and Statistics (Continued)

HEIBERGER · Computation for the Analysis of Designed Experiments
HELLER · MACSYMA for Statisticians
HINKELMAN and KEMPTHORNE: Design and Analysis of Experiments, Volume I: Introduction to Experimental Design
HOAGLIN, MOSTELLER, and TUKEY · Exploratory Approach to Analysis of Variance
HOAGLIN, MOSTELLER, and TUKEY · Exploring Data Tables, Trends and Shapes
HOAGLIN, MOSTELLER, and TUKEY · Understanding Robust and Exploratory Data Analysis
HOCHBERG and TAMHANE · Multiple Comparison Procedures
HOEL · Elementary Statistics, *Fifth Edition*
HOGG and KLUGMAN · Loss Distributions
HOLLANDER and WOLFE · Nonparametric Statistical Methods
HOSMER and LEMESHOW · Applied Logistic Regression
HØYLAND and RAUSAND · System Reliability Theory: Models and Statistical Methods
HUBERTY · Applied Discriminant Analysis
IMAN and CONOVER · Modern Business Statistics
JACKSON · A User's Guide to Principle Components
JOHN · Statistical Methods in Engineering and Quality Assurance
JOHNSON · Multivariate Statistical Simulation
JOHNSON and KOTZ · Distributions in Statistics
   Continuous Univariate Distributions—2
   Continuous Multivariate Distributions
JOHNSON, KOTZ, and BALAKRISHNAN · Continuous Univariate Distributions, Volume 1, *Second Edition*
JOHNSON, KOTZ, and KEMP · Univariate Discrete Distributions, *Second Edition*
JUDGE, GRIFFITHS, HILL, LÜTKEPOHL, and LEE · The Theory and Practice of Econometrics, *Second Edition*
JUDGE, HILL, GRIFFITHS, LÜTKEPOHL, and LEE · Introduction to the Theory and Practice of Econometrics, *Second Edition*
KALBFLEISCH and PRENTICE · The Statistical Analysis of Failure Time Data
KASPRZYK, DUNCAN, KALTON, and SINGH · Panel Surveys
KISH · Statistical Design for Research
KISH · Survey Sampling
LANGE, RYAN, BILLARD, BRILLINGER, CONQUEST, and GREENHOUSE · Case Studies in Biometry
LAWLESS · Statistical Models and Methods for Lifetime Data
LEBART, MORINEAU, and WARWICK · Multivariate Descriptive Statistical Analysis: Correspondence Analysis and Related Techniques for Large Matrices
LEE · Statistical Methods for Survival Data Analysis, *Second Edition*
LePAGE and BILLARD · Exploring the Limits of Bootstrap
LEVY and LEMESHOW · Sampling of Populations: Methods and Applications
LINHART and ZUCCHINI · Model Selection
LITTLE and RUBIN · Statistical Analysis with Missing Data
MAGNUS and NEUDECKER · Matrix Differential Calculus with Applications in Statistics and Econometrics
MAINDONALD · Statistical Computation
MALLOWS · Design, Data, and Analysis by Some Friends of Cuthbert Daniel
MANN, SCHAFER, and SINGPURWALLA · Methods for Statistical Analysis of Reliability and Life Data
MASON, GUNST, and HESS · Statistical Design and Analysis of Experiments with Applications to Engineering and Science
McLACHLAN · Discriminant Analysis and Statistical Pattern Recognition
MILLER · Survival Analysis
MONTGOMERY and PECK · Introduction to Linear Regression Analysis, *Second Edition*
NELSON · Accelerated Testing, Statistical Models, Test Plans, and Data Analyses
NELSON · Applied Life Data Analysis

Applied Probability and Statistics (Continued)
  OCHI · Applied Probability and Stochastic Processes in Engineering and Physical Sciences
  OKABE, BOOTS, and SUGIHARA · Spatial Tesselations: Concepts and Applications of Voronoi Diagrams
  OSBORNE · Finite Algorithms in Optimization and Data Analysis
  PANKRATZ · Forecasting with Dynamic Regression Models
  PANKRATZ · Forecasting with Univariate Box-Jenkins Models: Concepts and Cases
  PORT · Theoretical Probability for Applications
  PUTERMAN · Markov Decision Processes: Discrete Stochastic Dynamic Programming
  RACHEV · Probability Metrics and the Stability of Stochastic Models
  RÉNYI · A Diary on Information Theory
  RIPLEY · Spatial Statistics
  RIPLEY · Stochastic Simulation
  ROSS · Introduction to Probability and Statistics for Engineers and Scientists
  ROUSSEEUW and LEROY · Robust Regression and Outlier Detection
  RUBIN · Multiple Imputation for Nonresponse in Surveys
  RYAN · Statistical Methods for Quality Improvement
  SCHUSS · Theory and Applications of Stochastic Differential Equations
  SCOTT · Multivariate Density Estimation: Theory, Practice, and Visualization
  SEARLE · Linear Models
  SEARLE · Linear Models for Unbalanced Data
  SEARLE · Matrix Algebra Useful for Statistics
  SEARLE, CASELLA, and McCULLOCH · Variance Components
  SKINNER, HOLT, and SMITH · Analysis of Complex Surveys
  STOYAN · Comparison Methods for Queues and Other Stochastic Models
  STOYAN, KENDALL, and MECKE · Stochastic Geometry and Its Applications
  STOYAN and STOYAN · Fractals, Random Shapes and Point Fields
  THOMPSON · Empirical Model Building
  THOMPSON · Sampling
  TIERNEY · LISP-STAT: An Object-Oriented Environment for Statistical Computing and Dynamic Graphics
  TIJMS · Stochastic Modeling and Analysis: A Computational Approach
  TITTERINGTON, SMITH, and MAKOV · Statistical Analysis of Finite Mixture Distributions
  UPTON and FINGLETON · Spatial Data Analysis by Example, Volume I: Point Pattern and Quantitative Data
  UPTON and FINGLETON · Spatial Data Analysis by Example, Volume II: Categorical and Directional Data
  VAN RIJCKEVORSEL and DE LEEUW · Component and Correspondence Analysis
  WEISBERG · Applied Linear Regression, *Second Edition*
  WESTFALL and YOUNG · Resampling-Based Multiple Testing: Examples and Methods for *p*-Value Adjustment
  WHITTLE · Optimization Over Time: Dynamic Programming and Stochastic Control, Volume I and Volume II
  WHITTLE · Systems in Stochastic Equilibrium
  WONNACOTT and WONNACOTT · Econometrics, *Second Edition*
  WONNACOTT and WONNACOTT · Introductory Statistics, *Fifth Edition*
  WONNACOTT and WONNACOTT · Introductory Statistics for Business and Economics, *Fourth Edition*
  WOODING · Planning Pharmaceutical Clinical Trials: Basic Statistical Principles
  WOOLSON · Statistical Methods for the Analysis of Biomedical Data

Tracts on Probability and Statistics
  BILLINGSLEY · Convergence of Probability Measures
  TOUTENBURG · Prior Information in Linear Models